Technical Communication

A Practical Approach

7th EDITION

William Sanborn Pfeiffer

Warren Wilson College

Kaye E. Adkins

Missouri Western State University

Prentice Hall
Upper Saddle River, New Jersey
Columbus, Ohio

Library of Congress Cataloging-in-Publication Data

Pfeiffer, William S.
  Technical communication : a practical approach / William Sanborn
Pfeiffer, Kaye Adkins. — 7th ed.
    p. cm.
  ISBN-10: 0-13-500050-5
  ISBN-13: 978-0-13-500050-2
  1. English language—Technical English—Problems, exercises, etc.   2. Communication of
technical information—Problems, exercises, etc.   3. English language—Rhetoric—Problems,
exercises, etc.   4. Technical writing—Problems, exercises, etc.   I. Adkins, Kaye E.   II. Title.
  PE1475.P47 2010
  808'.0666—dc22

                                                                                    2008045410

Editor in Chief: Vernon Anthony
Acquisitions Editor: Gary Bauer
Development Editor: Erin Curtis/
  Ohlinger and Associates
Editorial Assistant: Megan Heintz
Production Coordination: Peggy Kellar/
  Aptara, Inc.
Project Manager: Rex Davidson
Senior Operations Supervisor: Pat Tonneman
Art Director: Diane Ernsberger
Interior Design: Wanda Espana/
  Wee Design Group

Cover Designer: Wanda Espana/
  Wee Design Group
Cover art: iStockPhoto
IRC Manager, Rights and Permissions:
  Zina Arabia
Manager, Visual Research: Beth Brenzel
Image Permission Coordinator:
  Angelique Sharps
Director of Marketing: David Gesell
Marketing Manager: Leigh Ann Sims
Marketing Assistant: Alicia Wozniak
Photo credits: Appear on pages 661 to 663.

This book was set in Perpetua by Aptara® Inc., and was printed and bound by C. J. Krehbiel. The cover was printed by Phoenix Color Corp.

Pearson Education Ltd., London
Pearson Education Singapore Pte. Ltd.
Pearson Education Canada, Inc.
Pearson Education—Japan

Pearson Education Australia Pty. Limited
Pearson Education North Asia Ltd., Hong Kong
Pearson Educación de Mexico, S.A. de C.V.
Pearson Education Malaysia Pte. Ltd.

Prentice Hall
is an imprint of

www.pearsonhighered.com

10 9 8 7 6 5 4 3 2 1
ISBN–13: 978-0-13-500050-2
ISBN–10:   0-13-500050-5

Dedication

Deepest thanks go to my family—Evelyn, Zachary, and Katie—for their love and support throughout this and every writing project I take on.

—Sandy

To my family—Perry, Ian, and Evan—for their support and patience during this project.

—Kaye

Preface

Good writing is always a breaking of the soil, clearing away prejudices, pulling up of sour weeds of crooked thinking, stripping the turf so as to get at what is fertile beneath.

—Henry Seidel Canby (1878–1961), "Cultivate Your Garden"

Most writers agree with Henry Seidel Canby that writing is hard work, but well-crafted writing makes the effort worthwhile. Clear writing, the kind we call *technical writing* or *technical communication,* helps businesses run more smoothly, helps government run more effectively, and helps all of us accomplish our goals.

To help you become an effective technical communicator, all editions of this book have stressed one simple principle: You learn to write well by doing as much writing as possible. This seventh edition adds new features that make it even more usable, without changing what has made the book work in all editions—updated models and references, clear explanations of the writing process, advice for using technology, and a new chapter on collaborative writing.

In the seventh edition, big changes are happening at McDuff, Inc., the fictional company that serves as the basis for many examples and assignments. Rob McDuff, founder of the company, has retired and turned the company over to his son Jim. Rob's granddaughter Jeannie has become vice president of domestic operations and is being groomed to take over the company. Most visible to McDuff's customers and clients is its name change—McDuff has become M-Global, Inc., a change that reflects its international scope and global focus. These changes correspond to the changes in the authorship of *Technical Communication: A Practical Approach.* This new edition is distinguished by the addition of a co-author, Kaye Adkins, a professor of technical communication with expertise in documentation writing and experience in the oil and gas, software, and banking industries.

At the start of our classes, we sometimes ask students to describe their professional goals for the next 10 years. As you might expect, they hope to rise to important positions in the workplace and make genuine contributions to their professions. Such long-term thinking is crucial, keeping you on course in your life.

Yet, ultimately, the way you handle the small details of daily life most influences the contribution you make in the long run. If you do good work, believe in what you do, and communicate well with others—both interpersonally and in writing—success will come your way. The author Robert Pirsig put it this way in his 1974 classic, *Zen and the Art of Motorcycle Maintenance:* "The place to improve the world is first in one's own heart and head and hands, and then work outward from there."

We believe—and this book tries to show—that clear, concise, and honest writing is one of the most powerful tools of your heart, head, and hands.

Kaye Adkins, Associate Professor of English/Technical Communication
Missouri Western State University

William S. Pfeiffer, President
Warren Wilson College

>>> New Features of *Technical Communication: A Practical Approach, Seventh Edition*

Every student who plans to work in business and industry must master the art of technical writing. Indeed, in this new century, effective writing remains a major criterion for success in all professions. What follows is a summary of the main features of this book that help you become a better technical writer. Some were brought forward from the previous edition, and others are new features developed for this edition.

NEW! Collaboration Chapter

With collaborative writing being essential in the workplace, the new edition provides students with guidelines for working in teams (pp. 447–468).

NEW! "Write About It" Assignments in Each Chapter

Each "Communication Challenge" now includes a writing assignment that asks students to analyze and respond to the challenge and the discussion questions (examples: pp. 120, 274, 442).

NEW! M-Global

The new name of McDuff, Inc. reflects a broadening of the scope and scale of the McDuff company to allow for the introduction of different types of writing situations, document types, and designs.

Reorganized Chapters on the Writing Process

The discussion of the writing process in chapters 1 through 4 has been completely revised. Information about patterns of organization is now in one chapter instead of two. The Planning Form has also been revised to help students plan the organization and style of their documents (page 10).

An Increased Emphasis on Audience Analysis and Contextual Writing

Throughout the textbook, greater emphasis is placed on analyzing the audience and writing for context, reinforcing the need for students to analyze writing situations properly.

Updated Coverage of Research

With constant changes in print and electronic databases, it is important that students have current information on the availability and use of major sources for class papers. In particular, guidelines for using technology in the research process have been updated (pp. 495–550).

Guidelines and Examples of CSE, MLA, and APA Documentation

Now students will have information available on three of the most commonly used forms of documentation in the classroom. Furthermore, the main example in the research chapter is now produced in APA style (pp. 532–537).

>>> Five Core Features of *Technical Communication: A Practical Approach*

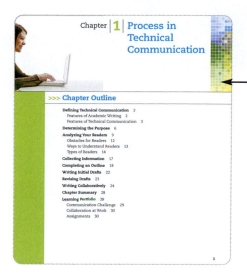

Chapter | **1** | Process in Technical Communication

>>> **Chapter Outline**

Defining Technical Communication 2
  Features of Academic Writing 2
  Features of Technical Communication 3
Determining the Purpose 6
Analyzing Your Readers 9
  Obstacles for Readers 12
  Ways to Understand Readers 13
  Types of Readers 14
Collecting Information 17
Completing an Outline 18
Writing Initial Drafts 22
Revising Drafts 23
Writing Collaboratively 24
Chapter Summary 28
Learning Portfolio 29
  Communication Challenge 29
  Collaboration at Work 30
  Assignments 30

1

Focus on Process and Product

This book has students practicing writing early (chapter 1). The text immerses them in the process of technical writing while teaching practical formats for getting the job done.

A Simple ABC Pattern for All Documents

The "ABC format"—Abstract, Body, and Conclusion—guides students' work in this course and throughout their careers. This underlying three-part structure provides a convenient handle for designing almost every technical document.

144 Chapter 5 Letters, Memos, and Electronic Communication

ABC Format for Email

Simply understanding that email *should* have a format puts you ahead of many writers, who consider email a license to ramble on without structure. Yes, email is casual and quick, but that does not make it formless. The following three-part ABC format resembles that used for letters:

ABC Format: Email

- **ABSTRACT:** Casual, friendly greeting if justified by relationship
  - Short, clear statement of purpose for writing
  - List of main topics to be covered
- **BODY:** Supporting information for points mentioned in abstract
  - Use of short paragraphs that start with main ideas
  - Use of headings and lists
  - Use of abbreviations and jargon only when understood by all readers
- **CONCLUSION:** Summary of main point
  - Clarity about action that comes next

Remember—your reader is confronted with many emails during the day. Furthermore, the configurations of some computers make reading a screen harder on the eyes than reading print memos. So give each email a structure that makes it simple for your reader to find important information.

Appropriate Use and Style for Email

Email is an appropriate reflection of the speed at which we conduct business today. Indeed, it mirrors the pace of popular culture as well. Following are some of the obvious advantages that using email provides:

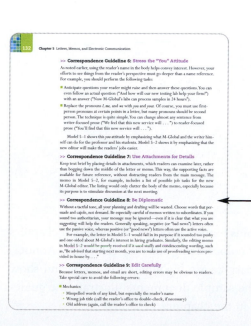

132 Chapter 5 Letters, Memos, and Electronic Communication

>> Correspondence Guideline 6: Stress the "You" Attitude

As noted earlier, using the reader's name in the body helps convey interest. However, your efforts to see things from the reader's perspective must go deeper than a name reference. For example, you should perform the following tasks:

- Anticipate questions your reader might raise and then answer these questions. You can even follow an actual question ("And how will our new testing lab help your firm?") with an answer ("Now M-Global's labs can process samples in 24 hours").
- Replace the pronouns *I, me,* and *we* with *you* and *your.* Of course, you must use first-person pronouns at certain points in a letter, but many pronouns should be second person. The technique is quite simple. You can change almost any sentence from writer-focused prose ("We feel that this new service will . . .") to reader-focused prose ("You'll find that this new service will . . .").

  Model 5–1 shows this you attitude by emphasizing what M-Global and the writer himself can do for the professor and his students. Model 5–2 shows it by emphasizing that the new editor will make the readers' jobs easier.

>> Correspondence Guideline 7: Use Attachments for Details

Keep text brief by placing details in attachments, which readers can examine later, rather than bogging down the middle of the letter or memo. This way, the supporting facts are available for future reference, without distracting readers from the main message. The memo in Model 5–2, for example, includes a list of possible job tasks for the new M-Global editor. The listing would only clutter the body of the memo, especially because its purpose is to stimulate discussion at the next meeting.

>> Correspondence Guideline 8: Be Diplomatic

Without a tactful tone, all your planning and drafting will be wasted. Choose words that persuade and cajole, not demand. Be especially careful of memos written to subordinates. If you sound too authoritarian, your message may be ignored—even if it is clear that what you are suggesting will help the readers. Generally speaking, negative (or "bad news") letters often use the passive voice, whereas positive (or "good news") letters often use the active voice.

For example, the letter in Model 5–1 would fail in its purpose if it sounded too pushy and one-sided about M-Global's interest in hiring graduates. Similarly, the editing memo in Model 5–2 would be poorly received if it used stuffy and condescending wording, such as, "Be advised that starting next month, you are to make use of proofreading services provided in-house by . . . ."

>> Correspondence Guideline 9: Edit Carefully

Because letters, memos, and email are short, editing errors may be obvious to readers. Take special care to avoid the following errors:

- Mechanics
  - Misspelled words of any kind, but especially the reader's name
  - Wrong job title (call the reader's office to double-check, if necessary)
  - Odd address (again, call the reader's office to check)

Numbered Guidelines

Many sets of short, numbered guidelines make this book easy to use to complete class projects. Each set of guidelines takes students through the process of finishing assignments, such as writing a proposal, doing research on the Internet, constructing a bar chart, and preparing an oral presentation.

M-Global, Inc.—A Fictional Company

M-Global, Inc., creates a fictional company for the classroom. Not all students have experience working in a professional or technical organization, so M-Global supplies a realistic backdrop for many of the book's examples and assignments.

■ Model 6–3 ■ Technical description (with definition included): Bunsen burner

Annotated Models

The text contains models grouped at the end of chapters on pages with color edging for easy reference. Annotations in the margins are highlighted in color and show exactly how the sample documents illustrate the guidelines set forth in the chapters.

>>> Additional Features Define the Book's Mission and Demonstrate Its Utility in the Classroom

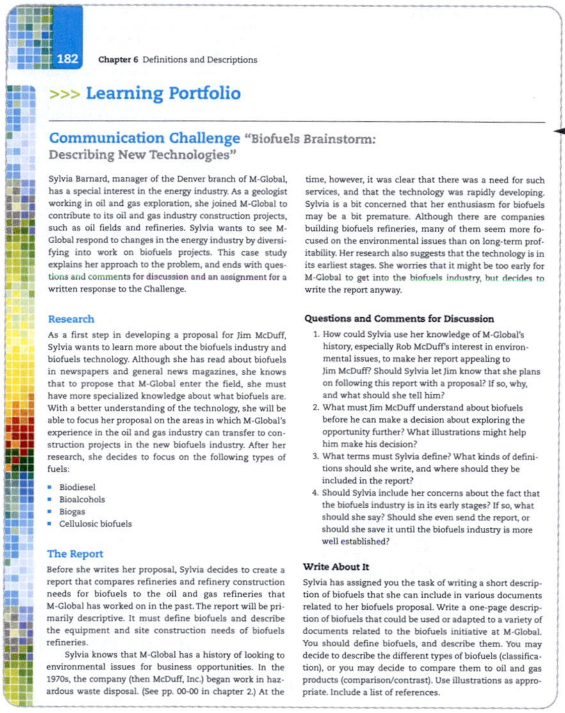

182 Chapter 6 Definitions and Descriptions

>>> Learning Portfolio

Communication Challenge "Biofuels Brainstorm: Describing New Technologies"

Sylvia Barnard, manager of the Denver branch of M-Global, has a special interest in the energy industry. As a geologist working in oil and gas exploration, she joined M-Global to contribute to its oil and gas industry construction projects, such as oil fields and refineries. Sylvia wants to see M-Global respond to changes in the energy industry by diversifying into work on biofuels projects. This case study explains her approach to the problem, and ends with questions and comments for discussion and an assignment for a written response to the Challenge.

Research

As a first step in developing a proposal for Jim McDuff, Sylvia wants to learn more about the biofuels industry and biofuels technology. Although she has read about biofuels in newspapers and general news magazines, she knows that to propose that M-Global enter the field, she must have more specialized knowledge about what biofuels are. With a better understanding of the technology, she will be able to focus her proposal on the areas in which M-Global's experience in the oil and gas industry can transfer to construction projects in the new biofuels industry. After her research, she decides to focus on the following types of fuels:

- Biodiesel
- Bioalcohols
- Biogas
- Cellulosic biofuels

The Report

Before she writes her proposal, Sylvia decides to create a report that compares refineries and refinery construction needs for biofuels to the oil and gas refineries that M-Global has worked on in the past. The report will be primarily descriptive. It must define biofuels and describe the equipment and site construction needs of biofuels refineries.

Sylvia knows that M-Global has a history of looking to environmental issues for business opportunities. In the 1970s, the company (then McDuff, Inc.) began work in hazardous waste disposal. (See pp. 00-00 in chapter 2.) At the time, however, it was clear that there was a need for such services, and that the technology was rapidly developing. Sylvia is a bit concerned that her enthusiasm for biofuels may be a bit premature. Although there are companies building biofuels refineries, many of them seem more focused on the environmental issues than on long-term profitability. Her research also suggests that the technology is in its earliest stages. She worries that it might be too early for M-Global to get into the biofuels industry, but decides to write the report anyway.

Questions and Comments for Discussion

1. How could Sylvia use her knowledge of M-Global's history, especially Rob McDuff's interest in environmental issues, to make her report appealing to Jim McDuff? Should Sylvia let Jim know that she plans on following this report with a proposal? If so, why, and what should she tell him?
2. What must Jim McDuff understand about biofuels before he can make a decision about exploring the opportunity further? What illustrations might help him make his decision?
3. What terms must Sylvia define? What kinds of definitions should she write, and where should they be included in the report?
4. Should Sylvia include her concerns about the fact that the biofuels industry is in its early stages? If so, what should she say? Should she even send the report, or should she save it until the biofuels industry is more well established?

Write About It

Sylvia has assigned you the task of writing a short description of biofuels that she can include in various documents related to her biofuels proposal. Write a one-page description of biofuels that could be used or adapted to a variety of documents related to the biofuels initiative at M-Global. You should define biofuels, and describe them. You may decide to describe the different types of biofuels (classification), or you may decide to compare them to oil and gas products (comparison/contrast). Use illustrations as appropriate. Include a list of references.

Communication Challenges

Every chapter includes an M-Global case study, with related questions and a short writing assignment. Called a "Communication Challenge," each case describes a communication problem that relates to the material in its respective chapter. These case studies can be used as a springboard for class discussion or for project assignments.

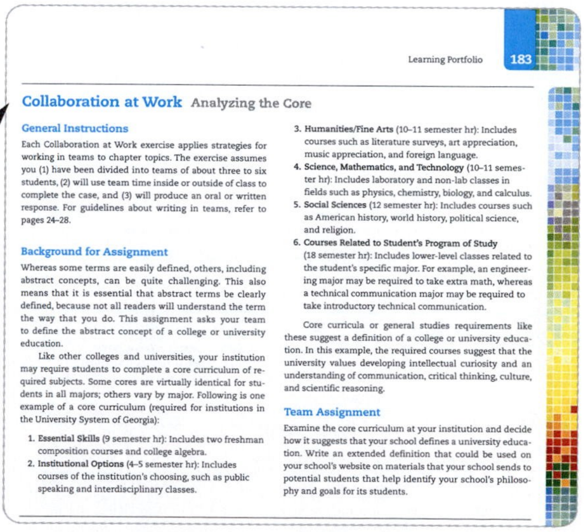

Learning Portfolio 183

Collaboration at Work Analyzing the Core

General Instructions

Each Collaboration at Work exercise applies strategies for working in teams to chapter topics. The exercise assumes you (1) have been divided into teams of about three to six students, (2) will use team time inside or outside of class to complete the case, and (3) will produce an oral or written response. For guidelines about writing in teams, refer to pages 24–28.

Background for Assignment

Whereas some terms are easily defined, others, including abstract concepts, can be quite challenging. This also means that it is essential that abstract terms be clearly defined, because not all readers will understand the term the way that you do. This assignment asks your team to define the abstract concept of a college or university education.

Like other colleges and universities, your institution may require students to complete a core curriculum of required subjects. Some cores are virtually identical for students in all majors; others vary by major. Following is one example of a core curriculum (required for institutions in the University System of Georgia):

1. **Essential Skills** (9 semester hr): Includes two freshman composition courses and college algebra.
2. **Institutional Options** (4–5 semester hr): Includes courses of the institution's choosing, such as public speaking and interdisciplinary classes.
3. **Humanities/Fine Arts** (10–11 semester hr): Includes courses such as literature surveys, art appreciation, music appreciation, and foreign language.
4. **Science, Mathematics, and Technology** (10–11 semester hr): Includes laboratory and non-lab classes in fields such as physics, chemistry, biology, and calculus.
5. **Social Sciences** (12 semester hr): Includes courses such as American history, world history, political science, and religion.
6. **Courses Related to Student's Program of Study** (18 semester hr): Includes lower-level classes related to the student's specific major. For example, an engineering major may be required to take extra math, whereas a technical communication major may be required to take introductory technical communication.

Core curricula or general studies requirements like these suggest a definition of a college or university education. In this example, the required courses suggest that the university values developing intellectual curiosity and an understanding of communication, critical thinking, culture, and scientific reasoning.

Team Assignment

Examine the core curriculum at your institution and decide how it suggests that your school defines a university education. Write an extended definition that could be used on your school's website on materials that your school sends to potential students that help identify your school's philosophy and goals for its students.

Collaboration at Work

Each chapter also includes a "Collaboration at Work" exercise that engages the student's interest in the chapter content by getting teams to complete a simple project.

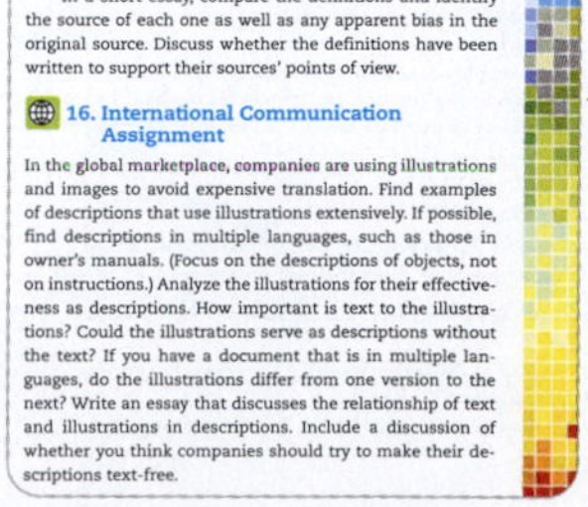

Learning Portfolio 185

an argument on a controversial issue. Examine the following definitions of *global warming* from various sources on the Internet, and find and read each organization's home page. Can you see implied biases in the definition, or does the definition appear neutral? Does this bias or neutrality support the general goals of the organization that published the definition?

In a short essay, compare the definitions and identify the source of each one as well as any apparent bias in the original source. Discuss whether the definitions have been written to support their sources' points of view.

16. International Communication Assignment

In the global marketplace, companies are using illustrations and images to avoid expensive translation. Find examples of descriptions that use illustrations extensively. If possible, find descriptions in multiple languages, such as those in owner's manuals. (Focus on the descriptions of objects, not on instructions.) Analyze the illustrations for their effectiveness as descriptions. How important is text to the illustrations? Could the illustrations serve as descriptions without the text? If you have a document that is in multiple languages, do the illustrations differ from one version to the next? Write an essay that discusses the relationship of text and illustrations in descriptions. Include a discussion of whether you think companies should try to make their descriptions text-free.

Coverage of International Communication

Because globalism continues to transform the business world, this book includes suggestions for understanding other cultures and for writing in an international context. In addition, each chapter's set of exercises ends with an "International Communication Assignment".

Assignments on Ethics

To reinforce the ethical guidelines described in chapter 2, each chapter includes an ethics assignment. No one can escape the continuous stream of ethical decisions required of every professional almost every day—such as deciding what tone to adopt in a proposal—which is why the text addresses ethical issues in these assignments.

Appendix

>>> Handbook

This handbook includes entries on the basics of writing. It contains three main types of information:

1. **Grammar:** the rules by which we edit sentence elements. Examples include rules for the placement of punctuation, the agreement of subjects and verbs, and the placement of modifiers.
2. **Mechanics:** the rules by which we make final proofreading changes. Examples include the rules for abbreviations and the use of numbers. A list of commonly misspelled words is also included.
3. **Usage:** information on the correct use of particular words, especially pairs of words that are often confused. Examples include problem words like *affect/effect, complement/compliment,* and *who/whom.*

Another editing concern, technical style, is the topic of chapter 17, including guidelines for sentence structure, conciseness, accuracy of wording, active and passive voice, and nonsexist language. Together, chapter 17 and this handbook will help you turn unedited drafts into final revised documents.

This handbook is presented in alphabetized fashion for easy reference during the editing process. Grammar and mechanics entries are in all uppercase; usage entries are in lowercase. Several exercises follow the entries.

A/An

A and *an* are different forms of the same article. *A* occurs before words that start with consonants or consonant sounds. EXAMPLES:

- a three-pronged plug
- a once-in-a-lifetime job (*once* begins with the consonant sound of *w*)
- a historic moment (many speakers and some writers mistakenly use *an* before *historic*)

An occurs before words that begin with vowels or vowel sounds. EXAMPLES:

- an eager new employee
- an hour before closing

A lot/Alot

The correct form is the two-word phrase *a lot*. Although acceptable in informal discourse, *a lot* usually should be replaced by more formal diction in technical writing. EXAMPLE: "They retrieved many [not *a lot of*] soil samples from the construction site."

613

Writing Handbook

This book provides a well-indexed, alphabetized writing handbook on grammar, mechanics, and usage that gives quick access to rules for eliminating editing errors during the revision process (pp. 613–660).

Information on English as a Second Language

A growing number of technical communication students are from other countries or cultures where English is not the first language. The English as a Second Language (ESL) section of the writing handbook focuses on three main problem areas: articles, prepositions, and verb use. It also applies ESL analysis to an excerpt from a technical report (pp. 629–637).

English as a Second Language (ESL)

Technical writing challenges native English speakers and nonnative speakers alike. The purpose of this section is to present a basic description of three grammatical forms: articles, verbs, and prepositions. These forms may require more intense consideration from international students when they complete technical writing assignments. Each issue is described using the ease-of-operation section from a memo about a fax machine. The passage, descriptions, and charts work together to show how these grammar issues function collectively to create meaning.

Ease of Operation—Article Usage

The AIM 500 is so easy to operate that **a** novice can learn to transmit **a** document to another location in about two minutes. Here's **the** basic procedure:

1. Press **the** button marked TEL on **the** face of **the** fax machine. You then hear **a** dial tone.
2. Press **the** telephone number of **the** person receiving **the** fax on **the** number pad on **the** face of **the** machine.
3. Lay **the** document face down on **the** tray at **the** back of **the** machine.

>>> Your One-Stop Source for *Technical Communication* Resources

MyTechCommLab for *Technical Communication: A Practical Approach, Seventh Edition*

PEARSON **mytechcommlab**

This comprehensive resource can be packaged at no additional cost with purchase of a new text. MyTechCommLab comes in two versions: a generic version that requires no instructor involvement and an e-book version that includes instructor grade book and classroom management tools. Both versions provide a wide array of multimedia tools, all in one place, and all designed specifically for technical communicators.

■ **More than 80 Model Documents,** most with interactive activities and annotations selected from a variety of professions and purposes (letters, memos, career correspondence, proposals, reports, instructions and procedures, descriptions and definitions, website, and presentation). MyTechCommLab also contains **50 Interactive Documents** that include rollover annotations highlighting purpose, audience, design, and other critical topics.

■ **Grammar, Mechanics, and Writing Help:** If students need more practice in basic grammar and usage, MyTechCommLab's grammar diagnostics generate a study plan linked to the thousands of test items in ExerciseZone, with results tracked by Pearson's exclusive GradeTracker.

■ **Document Design Resources:** A **Writing Process Tutorial** leads students through each stage of the writing process from prewriting to final formatting. A new **Tutorial on Writing Formal Reports** offers step-by-step guidance for creating one of the most common document types in technical communication and working with sources. **Activities and Case Studies** provide more than 65 exercises, all rooted in technical communication and many document-based, including 3 new case studies on usability. An **online reference library of e-books** includes pdf files for books on Visual Communication and Workplace Literacy.

■ **Research Help: Research Navigator**™ helps students research quickly and efficiently. Our program is complete with extensive help on the research process and includes four exclusive databases of credible and reliable source material—EBSCO's ContentSelect Academic Journal Database, The *New York Times* Search-by-Subject Archive, the FT.com archives, and a "Best of the Web" Link Library.

■ **E-book with Online Reference Sources:** Students can access the textbook within the book-specific MyTechCommLab. Embedded in the pages are links to online resources, including URLs referenced in the text and the model documents.

To order *Technical Communication: A Practical Approach,* Seventh Edition, with the generic MyTechCommLab access code, order ISBN: 032133857X.

To order *Technical Communication: A Practical Approach*, Seventh Edition, with MyTechCommLab with e-book in CourseCompass access code, order ISBN: 0135049911.

A standalone access code can be purchased online at http://www.prenhall.com. To preview MyTechCommLab, go to www.mytechcommlab.com.

>>> Companion Website: A Wealth of Open Access Online Materials

The *Technical Communication: A Practical Approach,* Seventh Edition, Companion Website contains a wealth of cases, exercises, activities, and documents for each chapter at http://www.prenhall.com/pfeiffer. Online materials for each chapter in the text include the following:

- *Chapter Outlines* provide an overview of major chapter concepts.
- *Interactive Editing and Revision Exercises:* Interactive documents allow students to see poorly done and corrected versions of documents with additional assignable document revision exercises.
- *Communication Cases:* Students encounter workplace situations with assignments in a wide range of career-oriented applications.
- *Portfolio Activities:* A variety of writing activities specific to technical and career fields allows students to practice producing communication relevant to their interests.
- *M-Global Activity:* A set of activities placed in the context of M-Global's new Alternate Energy Initiative.
- *Collaboration Exercises:* Assignments designed to provide practice writing and communicating in teams.
- *Sample Forms and Documents:* Example documents and M-Global document models in downloadable Microsoft Word format.
- *Online Resource Links:* Links to online resources, including guides to document preparation, job search information, and library research tools.
- *Chapter Quizzes:* Self-grading multiple-choice quizzes help students master chapter concepts and prepare for tests.
- *Distance Learning Solutions:* Ready-made Blackboard, WebCT, and CourseCompass online courses are available. If you adopt the text with an online course, student access cards can be packaged with the textbook at no additional charge to the student.

>>> Instructor's Resources

All instructor's resources are available for download at the Instructor's Resource Center. To access supplementary materials online, instructors need to request an instructor access code. Go to **www.pearsonhighered.com/irc**, where you can register for an instructor access code. Within 48 hours of registering you will receive a confirming e-mail, including an instructor access code. Once you have received your code, locate your text in the online catalog and click on the Instructor Resources button on the left side of the catalog product page. Select a supplement and a login page will appear. Once you have logged in, you can access instructor material for all Prentice Hall textbooks.

Instructor's Manual

An expanded Instructor's Manual loaded with helpful teaching notes for your classroom, including answers to the chapter quiz questions, a test bank, and instructor notes for assignments and activities located on the Companion Website (http://www.prenhall.com/pfeiffer).

- **Test Generator**
- **PowerPoint Lecture Presentation Package**

>>> Acknowledgments

Our thanks to the following reviewers of the seventh edition for helping with the revision of the textbook:

- Heidi Hatfield Edwards, Florida Institute of Technology
- Liz Kleinfeld, Red Rocks Community College
- Brian Van Horne, Metropolitan State College of Denver

In addition, the following reviewers have helped throughout the multiple editions of this book:

- Brian Ballentine, Case Western Reserve University
- Jay Goldberg, Marquette University
- Linda Grace, Southern Illinois University
- Darlene Hollon, Northern Kentucky University
- John Puckett, Oregon Institute of Technology
- Kirk Swortzel, Mississippi State University
- Catharine Schauer, Visiting Professor, Embry Riddle University

A special thanks goes to Craig Baehr, Texas Tech University, for contributing Chapter 11: Web Pages and Writing for the Web to the sixth edition of the text.

We also thank the following people who contributed activities, exercises, and documents to the Companion Website and PowerPoint presentation package:

- Connie Cerniglia, Guilford Technical Community College—for contributing a variety of career-related assignments and documents
- Linda Gray, Oral Roberts University—for creating the innovative Interactive Editing and Revision Exercises
- Melanie Rosen Brown, St. Johns River Community College—for contributing interesting activities and case studies
- Catharine Schauer, Visiting Professor, Embry-Riddle Aeronautical University—for contributing activities and exercises
- May Beth Van Ness, University of Toledo—for helping check and revise our existing online material
- Lesley Wadsworth, Terra Community College—for help creating the PowerPoint lecture presentation package

Friends and colleagues who contributed to this edition or other editions include Shawn Tonner, Mark Stevens, Saul Carliner, George Ferguson, Alan Gabrielli, Bob Harbort, Mike Hughes, Dory Ingram, Becky Kelly, Chuck Keller, Jo Lundy, Minoru Moriguchi, Randy Nipp, Jeff Orr, Ken Rainey, Lisa A. Rossbacher, Betty Oliver Seabolt, Hattie Schumaker, John Sloan, Herb Smith, Lavern Smith, James Stephens, John Ulrich, Steven Vincent, and Tom Wiseman.

Four companies allowed us to use written material gathered during Sandy's consulting work: Fugro-McClelland, Law Engineering and Environmental Services, McBride-Ratcliff and Associates, and Westinghouse Environmental and Geotechnical Services. Although this book's fictional firm, M-Global, Inc., has features of the world we observed as consultants, we want to emphasize that M-Global is truly an invention.

Sandy thanks the following students for allowing us to adapt their written work for use in this book: Michael Alban, Becky Austin, Corey Baird, Natalie Birnbaum, Cedric Bowden, Gregory Braxton, Ishmael Chigumira, Bill Darden, Jeffrey Daxon, Rob Duggan, William English, Joseph Fritz, Jon Guffey, Sam Harkness, Gary Harvey, Lee Harvey, Hammond Hill, Sudhir Kapoor, Steven Knapp, Wes Matthews, Kim Meyer, James Moore, Chris Owen, Scott Lewis, James Porter, James Roberts, Mort Rolleston, Chris Ruda, Barbara Serkedakis, Tom Skywark, Tom Smith, DaTonja Stanley, James Stephens, Chris Swift, and Jeff Woodward. Kaye thanks her research assistants Rachel Stancliff and Ted Koehler, who identified outdated examples and references and provided updated references, examples, and models.

For all seven editions, it has been our good fortune to have the same extraordinary developmental editor, our friend and colleague, Monica Ohlinger. The seventh edition also benefitted from the suggestions and guidance of Erin Curtis. In addition, we want to give special thanks to our Prentice Hall editor, Gary Bauer, for his continuing faith in the book.

Brief Contents

Chapter 1   Process in Technical Communication   *1*

Chapter 2   M-Global, Inc.: Ethics and Globalism in the Workplace   *34*

Chapter 3   Organizing Information   *74*

Chapter 4   Page Design   *101*

Chapter 5   Letters, Memos, and Electronic Communication   *126*

Chapter 6   Definitions and Descriptions   *170*

Chapter 7   Process Explanations and Instructions   *193*

Chapter 8   Informal Reports   *224*

Chapter 9   Formal Reports   *258*

Chapter 10   Proposals and Feasibility Studies   *302*

Chapter 11   Web Pages and Writing for the Web   *366*

Chapter 12   Graphics   *400*

Chapter 13   Collaboration and Writing   *447*

Chapter 14   Oral Communication   *469*

Chapter 15   Technical Research   *495*

Chapter 16   The Job Search   *551*

Chapter 17   Style in Technical Writing   *588*

Contents

Chapter | **1** | **Process in Technical Communication**   1

Defining Technical Communication   2

    Features of Academic Writing   2

    Features of Technical Communication   3

Determining the Purpose   6

Analyzing Your Readers   9

    Obstacles for Readers   12

    Ways to Understand Readers   13

    Types of Readers   14

Collecting Information   17

Completing an Outline   18

Writing Initial Drafts   22

Revising Drafts   23

Writing Collaboratively   24

Chapter Summary   28

>>>**Learning Portfolio   29**

    >**Communication Challenge— "Bad Chairs, Bad Backs"   29**

    >**Collaboration at Work   30**

    >**Assignments   30**

Chapter | **2** | **M-Global, Inc.: Ethics and Globalism in the Workplace**   34

Culture in Organizations   35

    Elements of an Organization's Culture   35

    Business Climate   36

The Global Workplace   37

    Understanding Cultures   37

    Communicating Internationally   42

Ethics on the Job   43

    Ethical Guidelines for Work   43

    Ethics and Legal Issues in Writing   45

Introducing M-Global   47

  History of M-Global, Inc.   48

  Projects   49

Corporate and Branch Offices   51

  Headquarters   51

  Branches   54

Writing at M-Global   57

  Examples of Internal Writing   58

  Examples of External Writing   58

Chapter Summary   59

>>>**Learning Portfolio   60**

  >**Communication Challenge—"Employee Orientation Manual: Global Dilemmas"   60**

  >**Collaboration at Work   61**

  >**Assignments   61**

M-Global, Inc. Project Sheets

Worldwide Locations of M-Global Inc., Office   65

Project 1: Sentry Dam   66

Project 2: Completed Ocean Exploration Program   67

Project 3: Monitored Construction of General Hospital   68

Project 4: Managed Construction of Nevada Gold Dome   69

Project 5: Examined Big Bluff Salt Marsh   70

Project 6: Designed and Installed Control Panel for Nuclear Plant   71

Project 7: Designed and Taught Seminar in Technical Writing   72

Project 8: Designed and Created Documentation of Data Security Procedures   73

Chapter | 3 | Organizing Information   74

Importance of Organization   75

Three Principles of Organization   76

ABC Format for Documents   80

  Document Abstract: The "Big Picture" for Decision Makers   81

  Document Body: Details for All Readers   82

  Document Conclusion: Wrap-Up Leading to Next Step   83

Tips for Organizing Sections and Paragraphs   83

  Common Patterns of Organization   84

  Document Sections   87

  Paragraphs   87

Modular Writing   88

Chapter Summary   89

>>>**Learning Portfolio   90**

>**Communication Challenge—"Telecommuting: The Last Frontier?"   90**

>**Collaboration at Work   92**

>**Assignments   90**

Models for Good Writing

Model 3–1: ABC format in whole document   97

Model 3–2: ABC format in document section   99

Model 3–3: ABC format in paragraphs   100

Chapter 4 | Page Design   101

Elements of Page Design   102

Grids   103

White Space   105

Headings   107

Lists   111

In-Text Emphasis   112

Fonts and Color   112

Type Size   113

Font Types   114

Color   114

Computers in the Page Design Process   115

Headers and Footers   115

Templates   115

Style Sheets   115

Using the Elements of Page Design   117

Chapter Summary   117

>>>**Learning Portfolio   119**

>**Communication Challenge—"The St. Paul Style Guide:
Trouble in the River City"   119**

>**Collaboration at Work   120**

>**Assignments   121**

Model for Good Writing

Model 4–1: Page design in memorandum   124

Chapter | **5** | **Letters, Memos, and Electronic Communication** 126

General Guidelines for Correspondence   127

Letters   133

Positive Letters   *134*

Negative Letters   *135*

Neutral Letters   *136*

Sales Letters   *137*

Memoranda   138

Email   140

Guidelines for Email   *140*

Appropriate Use and Style for Email   *144*

Memoranda versus Email   145

Chapter Summary   146

>>>**Learning Portfolio   147**

>**Communication Challenge—"Ethics and Sales Letters"   147**

>**Collaboration at Work   148**

>**Assignments   148**

Models for Good Writing

Model 5–1: M-Global sample letter   154

Model 5–2: M-Global sample memo   155

Model 5–3: M-Global sample email   157

Model 5–4: Block style for letters   158

Model 5–5: Modified block style (with indented paragraphs) for letters   159

Model 5–6: Simplified style for letters   160

Model 5–7: Memo style   161

Model 5–8: Positive letter in block style   162

Model 5–9: Negative letter in modified block style (with indented paragraphs)   163

Model 5–10: Neutral letter (invitation) in block style   164

Model 5–11: Neutral letter (placing order) in simplified style   165

Model 5–12: Sales letter in simplified style   166

Model 5–13: Memorandum: changes in services   167

Model 5–14: Memorandum: changes in benefits   168

Model 5–15: Email: changes in procedure   169

Chapter | **6** | # Definitions and Descriptions 170

Definitions versus Descriptions   171

Technical Definitions at M-Global   171

Descriptions at M-Global   172

Guidelines for Writing Definitions   173

Example of a Description   177

Guidelines for Writing Descriptions   177

Example of a Description   180

Chapter Summary   180

>>>**Learning Portfolio   182**

>**Communication Challenge—"Biofuels Brainstorm: Describing New Technologies"   182**

>**Collaboration at Work   183**

>**Assignments   183**

Models for Good Writing

Model 6–1: Brief description (with formal definition included)   187

Model 6–2: Description from a user's manual   188

Model 6–3: Technical description (with definition included): Bunsen burner   190

Chapter | **7** | # Process Explanations and Instructions 193

Process Explanations versus Instructions   194

Process Explanations at M-Global   194

Instructions at M-Global   196

Guidelines for Process Explanations   197

Guidelines for Instructions   201

Chapter Summary   207

>>>**Learning Portfolio   209**

>**Communication Challenge—"M-Global's Home of Hope: The Good, the Bad, and the Ugly?"   209**

>**Collaboration at Work   210**

>**Assignments   211**

Models For Good Writing

Model 7–1: M-Global process explanation: electronic mail   215

Model 7–2: M-Global instructions: electronic mail   216

Model 7–3: Process explanation   217

Model 7–4: An M-Global process explanation with a flowchart   218

Model 7–5: Instructions for making travel arrangements   219

Model 7–6: M-Global memo containing how-to instructions
for a scanner   221

Chapter | 8 | Informal Reports   224

When to Use Informal Reports   225

Letter Reports at M-Global   225

Memo Reports at M-Global   226

General Guidelines for Informal Reports   227

Specific Guidelines for Five Informal Reports   231

Problem Analyses   232

Recommendation Reports   233

Equipment Evaluations   234

Progress/Periodic Reports   235

Lab Reports   236

Chapter Summary   237

>>>**Learning Portfolio   238**

>**Communication Challenge— "A Nonprofit Job: Good Deed
or Questionable Ethics?"   238**

>**Collaboration at Work   239**

>**Assignments   239**

Models for Good Writing

Model 8–1: Recommendation report (letter format)   246

Model 8–2: Equipment evaluation (memo format)   248

Model 8–3: Problem analysis (memo format)   250

Model 8–4: Progress report (memo format)   252

Model 8–5: Periodic report (memo format)   254

Model 8–6: Lab report (letter format)   256

Chapter | **9** | **Formal Reports**   258

When to Use Formal Reports   259

Strategy for Organizing Formal Reports   261

Guidelines for the Nine Parts of Formal Reports   262

Cover / Title Page   263

Letter / Memo of Transmittal   264

Table of Contents   265

List of Illustrations   266

Executive Summary   267

Introduction   268

Discussion Sections   269

Conclusions and Recommendations   270

End Material   271

Formal Report Example   271

Chapter Summary   272

>>>**Learning Portfolio   273**

>**Communication Challenge—"The Ethics of Clients Reviewing
   Report Drafts"   273**

>**Collaboration at Work   274**

>**Assignments   274**

Models for Good Writing

Model 9–1: Title page with illustration (see explanation on page 263)   278

Model 9–2: Letter of transmittal   279

Model 9–3: Memo of transmittal   280

Model 9–4: Table of contents (all subheadings included)   281

Model 9–5: Table of contents (third-level subheadings omitted)   282

Model 9–6: List of illustrations—formal report   283

Model 9–7: Executive summary—formal report   284

Model 9–8: Introduction—formal report   285

Model 9–9: Formal report   286

Chapter | **10** | **Proposals and Feasibility Studies**   302

Proposals and Feasibility Studies at M-Global   303

M-Global Proposals   304

M-Global Feasibility Studies   305

Guidelines for Informal Proposals   306

Guidelines for Formal Proposals   311

    Cover / Title Page   311

    Letter / Memo of Transmittal   312

    Table of Contents   313

    List of Illustrations   313

    Executive Summary   313

    Introduction   314

    Discussion Sections   315

    Conclusion   315

    Appendices   316

Guidelines for Feasibility Studies   316

Chapter Summary   319

>>>**Learning Portfolio   321**

    >**Communication Challenge—"The Black Forest Proposal:**
       Good Marketing or Bad Business?"   321

    >**Collaboration at Work   322**

    >**Assignments   323**

Models for Good Writing

Model 10–1: Letter proposal   329

Model 10–2: Memo proposal   332

Model 10–3: Formal proposal   334

Model 10–4: Formal proposal   343

Model 10–5: Feasibility study (one alternative)   364

Chapter | **11** | **Web Pages and Writing for the Web**   *366*

Your Role in Developing Websites and Content   367

Planning   368

Content Development   371

    Content Chunking   371

    Guidelines for Writing Web Content   372

    Adapting Content for the Web   372

    Scripting Languages and Software Authoring Tools   373

    Document Conversion Issues and Common File Formats   374

Structure   374

    Site Structures and Types   375

Process of Developing a Structure *376*

Navigation Design *379*

Guidelines for Labeling *379*

Grouping and Arrangement Strategies *381*

Design 381

Design Conventions and Principles *382*

Finding a Theme and Developing Graphic Content *384*

File Formats and Graphics *384*

Interface Layouts *385*

Usability and Publication 388

Testing Your Site for Your User Base *389*

Performing Usability Reviews *389*

Quick Usability Checks and System Settings *389*

Accessibility Guidelines *391*

Publishing Your Site *393*

Chapter Summary 393

>>>**Learning Portfolio** **395**

>**Communication Challenge—"What Does Your Company Do Anyway?"** **395**

>**Collaboration at Work** **395**

>**Assignments** **396**

Models for Good Writing

Model 11–1: Original M-Global Website 398

Model 11–2: Proposed M-Global Website 399

Chapter | **12** | **Graphics** 400

Terms in Graphics 401

Reasons for Using Special Fonts, Color, and Graphics 402

Using Fonts 405

Font Types *405*

General Guidelines *405*

Using Color 408

The Cost and Time of Using Color *408*

Developing a Color Style Sheet *409*

Color Terms *411*

Guidelines for Using Color *411*

General Guidelines for Graphics   413

Specific Guidelines for Seven Graphics   416

   Tables   416

   Pie Charts   420

   Bar Charts   423

   Line Charts   426

   Flowcharts   428

   Organization Charts   430

   Technical Drawings   431

Misuse of Graphics   435

   Description of the Problem   435

   Examples of Distorted Graphics   435

Chapter Summary   440

>>>**Learning Portfolio   441**

   >Communication Challenge—"Massaging M-Global's Annual Report"   441

   >Collaboration at Work   443

   >Assignments   443

Chapter | 13 | Collaboration and Writing   447

Approaches to Collaboration   449

Collaboration and the Writing Process   449

   The Writing Team   450

   Planning   451

   Budgeting Time and Money   451

   Communication   454

   Modular Writing   455

Teamwork   456

   Running Effective Meetings   456

   Writers and Subject Matter Experts   459

Chapter Summary   461

>>>**Learning Portfolio   462**

   >Communication Challenge—"A Field Guide: Planning a User's Manual"   462

   >Collaboration at Work   463

   >Assignments   463

Models for Good Writing

Model 13–1: Example of M-Global modular writing   466

Model 13–2: Meeting agenda   467

Model 13–3: Meeting minutes   468

Chapter | **14** | **Oral Communication**   469

Presentations and Your Career   470

Guidelines for Preparation and Delivery   471

Guidelines for Presentation Graphics   478

Overcoming Nervousness   481

Why Do We Fear Presentations?   *482*

A Strategy for Staying Calm   *482*

An Example of an M-Global Oral Presentation   485

Chapter Summary   485

>>>Learning Portfolio   487

>Communication Challenge—"Ethics and the Technical Presentation"   487

>Collaboration at Work   488

>Assignments   488

Model for Good Writing

Model 14–1: Text and graphics of sample M-Global presentation   490

Chapter | **15** | **Technical Research**   495

Getting Started   496

Searching Online Catalogs   498

Author or Title Search   *499*

Subject Search   *499*

Keyword Search   *499*

Advanced Search Techniques   *500*

Searching in the Library   505

Library Services   *505*

Library Resources   *506*

Searching the Web   514

Fundamentals of Web Searching   *514*

Web Search Options   *516*

Using Questionnaires and Interviews   520

    Questionnaires   520

    Interviews   525

Using Borrowed Information Correctly   527

    Avoiding Plagiarism   527

    Following the Research Process   528

Selecting and Following a Documentation System   532

Writing Research Abstracts   538

    Types of Abstracts   538

    Guidelines for Writing Research Abstracts   539

Chapter Summary   541

>>>**Learning Portfolio   542**

    >**Communication Challenge—"To Cite or Not to Cite"   542**

    >**Collaboration at Work   543**

    >**Assignments   543**

Model for Good Writing

Model 15–1: Memo report citing research—APA Style   546

Chapter | 16 | The Job Search   551

Researching Occupations and Companies   552

Job Correspondence   555

    Job Letters   556

    Resumes   558

Job Interviews   562

    Preparation   562

    Performance   565

    Follow-Up-Letters   566

Negotiating   567

Chapter Summary   571

>>>**Learning Portfolio   572**

    >**Communication Challenge—"20-Something—Have Degree, Won't Travel"   572**

    >**Collaboration at Work   573**

    >**Assignments   574**

Models for Good Writing

Model 16–1: Job letter (modified block) and chronological resume   577

Model 16–2: Job letter (block style) and chronological resume   579

Model 16–3: Job letter (modified block) and functional resume   581

Model 16–4: Job letter (modified block) and functional resume   583

Model 16–5: Combined resume   585

Model 16–6: Combined resume formatted for submission online   586

Model 16–7: Resume with graphics—not effective for computer scanning   587

Chapter | 17 | Style in Technical Writing   588

Overview of Style   589
   Definition of Style   589
   Importance of Tone   589
Writing Clear Sentences   590
   Sentence Terms   591
   Guidelines for Sentence Style   591
Being Concise   592
Being Accurate in Wording   596
Using the Active Voice   598
   What Do Active and Passive Mean?   598
   When Should Actives and Passives Be Used?   599
Using Nonsexist Language   600
   Sexism and Language   600
   Techniques for Nonsexist Language   600
Plain English and Simplified English   603
   Plain English   603
   Simplified English   603
Chapter Summary   604

>>>**Learning Portfolio   605**
   >Communication Challenge—"An Editorial Adjustment"   605
   >Collaboration at Work   606
   >Assignments   607

Appendix: *Handbook*   613
Photo Credits   661
Index   665

Chapter | 1 | Process in Technical Communication

>>> Chapter Outline

Defining Technical Communication 2
  Features of Academic Writing 2
  Features of Technical Communication 3

Determining the Purpose 6

Analyzing Your Readers 9
  Obstacles for Readers 12
  Ways to Understand Readers 13
  Types of Readers 14

Collecting Information 17

Completing an Outline 18

Writing Initial Drafts 22

Revising Drafts 23

Writing Collaboratively 24

Chapter Summary 28

Learning Portfolio 29
  Communication Challenge 29
  Collaboration at Work 30
  Assignments 30

Good communication skills are essential in any career you choose. Jobs, promotions, raises, and professional prestige result from your ability to present both written and visual information effectively. With so much at stake, you need a simple road map to direct you toward writing excellence. *Technical Communication: A Practical Approach* is such a map. Chapters 1–4 of *Technical Communication: A Practical Approach* gives you an overview of technical writing and prepares you to complete the assignments in this book. Chapters 5–10 introduce common genres, or types, of technical communication documents. The rest of the book, chapters 11–17, discusses specific elements of technical documents, as well as special communication situations.

>>> Defining Technical Communication

You may have learned how to write short essays in previous writing courses. This book helps you transfer that basic knowledge to the kind of writing done on the job. Career writing is so practical, so well grounded in common sense, that it will seem to proceed smoothly from your previous work. This section highlights features of traditional academic writing on the one hand and technical communication on the other.

Features of Academic Writing

Writing you have done in school probably has had the following characteristics:

- **Purpose:** Communicating what you know about the topic, in a way that justifies a high grade
- **Your knowledge of topic:** Less than the teacher who evaluates the writing
- **Audience:** The teacher who requests the assignment and reads it from beginning to end
- **Criteria for evaluation:** Depth, logic, clarity, unity, supporting evidence, and grammar
- **Graphic elements:** Sometimes used to explain and persuade

Academic writing requires that you use words to display your learning to someone who knows more about the subject than you do. Because this person's job is to evaluate your work, you have what might be called a *captive audience*. In an academic setting, the purpose is to demonstrate knowledge, and the audience is someone already familiar with the subject or approach. In this sense, academic writing shows your command of information to someone more knowledgeable about the subject than you are. The next section examines a different kind of writing—the kind you will be doing in this course and in your career. Note similarities to the kind of writing you have been doing in other classes. Planning, drafting, and revising are important, even for short correspondence. Clear organization is essential. Finally, your purpose should be clear, and you should understand your audience, even though the purpose and audience differ considerably from those of academic writing.

Features of Technical Communication

The rules for writing shift somewhat when you begin your career. Employees unprepared for this change often flounder for years, never quite understanding the new rules. *Technical communication* is a generic term for all written and oral communications done on the job—whether in business, industry, or other professions. It is particularly identified with documents in technology, engineering, science, the health professions, and other fields with specialized vocabularies. The terms *technical writing, professional writing, business writing*, and *occupational writing* also refer to writing done in your career.

Besides projects that involve writing, your career will also bring you speaking responsibilities, such as formal speeches at conferences and informal presentations at meetings. Thus the term *technical communication* can encompass the full range of the writing and speaking responsibilities required to communicate your ideas on the job. The following discusses the main characteristics of technical communication:

- **Purpose:** Getting something done within an organization (completing a project, persuading a customer, pleasing your boss, etc.) or helping someone else (a customer, client, or colleague) get something done
- **Your knowledge of topic:** Usually greater than that of the reader
- **Audience:** Often several people with differing technical backgrounds
- **Criteria for evaluation:** Clear and simple organization of ideas and supporting detail appropriate to the needs of busy readers
- **Graphic elements:** Frequently used to explain existing conditions and to present alternative courses of action

Contrast these features with those of academic writing, listed earlier. In particular, note the following main differences:

1. Technical communication aims to help people make decisions and perform tasks, whereas academic writing aims only to display your knowledge.
2. Technical communication usually responds to the needs of the workplace, whereas academic writing usually responds to an assignment created by a teacher.
3. Technical communication is created by an informed writer conveying needed information both verbally and visually to an uninformed reader, whereas academic writing is created by a student as the learner for a teacher as the source of knowledge.
4. Technical communication often is read by many readers, whereas academic writing aims to satisfy only one person, the teacher.

Finally, technical communication places greater emphasis on techniques of organization and visual cues that help readers find important information as quickly as possible.

Figure 1–1 lists some typical on-the-job writing assignments. Although not exhaustive, the list does include many of the writing projects you will encounter. Also, see

Correspondence: In-House or External
- Memos to your boss and to your subordinates
- Routine letters to customers, vendors, etc.
- "Good news" letters to customers
- "Bad news" letters to customers
- Sales letters to potential customers
- Electronic mail (email) messages to coworkers or customers over a computer network

Short Reports: In-House or External
- Analysis of a problem
- Recommendation
- Equipment evaluation
- Progress report on project or routine periodic report
- Report on the results of laboratory or field work
- Description of the results of a company trip

Long Reports: In-House or External
- Complex problem analysis, recommendation, or equipment evaluation
- Project report on field or laboratory work
- Feasibility study

Other Examples
- Proposal to boss for new product line
- Proposal to boss for change in procedures
- Proposal to customer to sell a product, a service, or an idea
- Proposal to funding agency for support of research project
- Abstract or summary of technical article
- Technical article or presentation
- Operation manual or other manual
- Web site

Figure 1–2 for an example of a short technical document. Note that it has the five features of technical communication listed previously.

1. It is written to get something done—i.e., to evaluate a printer
2. It is sent from someone more knowledgeable about the printer to someone who needs information about it
3. Although the memo is directed to one person, the reader probably will share it with others before making a decision concerning the writer's recommendation
4. It is organized clearly, moving from data to recommendations and including headings
5. It provides limited data to describe the features of the printer

Although technical communication plays a key role in the success of all technical professionals and managers, the amount of time you devote to it will depend on your job. A survey of technical managers gives some idea of the time involved. Conducted by the National Aeronautics and Space Administration (NASA), the survey canvassed managers in profit-making and nonprofit organizations in the field of aeronautics. As Figure 1–3 shows, 100% of the profit managers and 98% of the nonprofit managers in the study consider technical communication a "somewhat important" or "very important" part of their jobs.

■ **Figure 1–2** ■ Short report

MEMORANDUM

DATE: December 6, 2008
TO: Holly Newsome
FROM: Michael Allen *MA*
SUBJECT: Printer Recommendation

Introductory Summary

Recently you asked for my evaluation of the Hemphill 5000 printer/fax/scanner/copier currently used in my department. Having analyzed the machine's features, print quality, and cost, I am quite satisfied with its performance.

Features

Among the Hemphill 5000's features, I have found these five to be the most useful:
1. Easy to use control panel
2. Print and copy speed of up to 34 pages per minute for color and black-and-white
3. Ability to print high quality documents like brochures & report covers
4. Built-in networking capability
5. Ability to scan documents to or from a USB port

In addition, the Hemphill 5000 offers high quality copies, color copies and faxes, and it uses high capacity ink cartridges to reduce costs.

Print Quality

The Hemphill 5000 produces excellent prints that rival professional typeset quality. The print resolution is 1200 x 1200 dots per inch, among the highest attainable in printer/fax/scanner/copier combinations. This memo was printed on the 5000, and, as you can see, the quality speaks for itself.

Cost

Considering the features and quality, the 5000 is an excellent network combination printer for workgroups within the firm. At a retail price of $239, it is also one of the lowest-priced combination printers, yet it comes with a two-year warranty and excellent customer support.

Conclusion

On the basis of my observation, I strongly recommend that our firm continue to use and purchase the Hemphill 5000. Please call me at ext. 204 if you want further information about this excellent machine.

M-Global Inc | 127 Rainbow Lane | Baltimore MD 21202 | 410.555.8175

Even more telling is the amount of time the NASA survey respondents spend communicating. Figure 1–3 indicates that they use (1) more than one third of their work time conveying information *to* others and (2) another one third working with technical information sent to them *by* others. Based on a 40-hour week, therefore, both groups spend roughly *two thirds* of their working week on job duties associated with technical communication.

■ **Figure 1–3** ■ Data from NASA aeronautics survey

Source: Thomas E. Pinelli et al., Technical Communications in Aeronautics: Results of an Exploratory Study—An Analysis of Profit Managers' and Nonprofit Managers' Responses (Washington, D.C.: National Aeronautics and Space Administration, NASA TM-101626, October 1989), 71. (Available from NTIS, Springfield, VA.)

TABLE 1. Importance of Technical Communications

How Important	Profit Managers		Nonprofit Managers	
	No.	%	No.	%
Very	86	92.5	43	84.3
Somewhat	7	7.5	7	13.7
Not at all	0	0.0	1	2.0
Total	93	100.0	51	100.0

TABLE 2. Time Spent Communicating Technical Information to Others

Time Spent per Week, Hours	Profit Managers		Nonprofit Managers	
	No.	%	No.	%
5 or less	13	14.3	9	18.0
6 to 10	33	36.2	16	30.0
11 to 20	37	40.7	21	42.0
21 or more	8	8.8	5	10.0
Total	91	100.0	51	100.0
Mean	13.5		13.9	

TABLE 3. Time Spent Working with Technical Information Received from Others

Time Spent per Week, Hours	Profit Managers		Nonprofit Managers	
	No.	%	No.	%
5 or less	8	8.7	6	12.0
6 to 10	42	46.2	23	46.0
11 to 20	36	39.6	18	36.0
21 or more	5	5.5	3	6.0
Total	91	100.0	50	100.0
Mean	13.0		13.0	

Now that you know the nature and importance of technical communication, the next section examines the first part of the planning stage: determining a document's purpose.

>>> Determining the Purpose

If you have already taken a basic composition course, you will see similarities between rhetorical aims studied in that course and those in technical communication. Indeed, technical communication uses the same building blocks as all other good writing. Writing assignments you have had in school have probably asked you to *inform* your reader about an event or object, to *analyze* a process or idea, or to *argue* the strength or weakness of an interpretation or theory.

Information: When readers pick up a technical communication document, they may want to know how to perform an operation or follow an established procedure. They may want to make an informed decision. Clear, reliable information is the basis of analysis and argument.

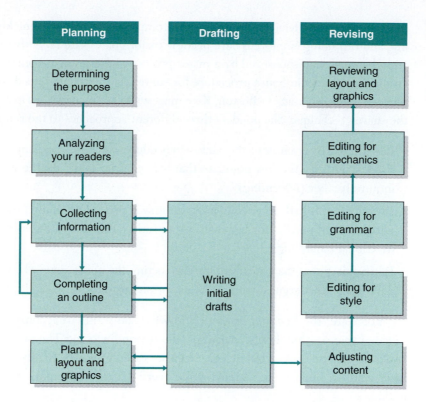

Planning	Drafting	Revising
Determining the purpose		Reviewing layout and graphics
Analyzing your readers		Editing for mechanics
Collecting information	Writing initial drafts	Editing for grammar
Completing an outline		Editing for style
Planning layout and graphics		Adjusting content

■ **Figure 1–4** ■
Flowchart for the technical communication process

Analysis: At first, it may not seem like analysis is an important purpose of workplace writing, but it is essential to problem solving and decision making. You may be asked to analyze options for a supervisor who will make a recommendation to a client, or you may be asked to use analysis to make your own recommendation.

Argument: Good argument forms the basis for all technical communication. Some people have the mistaken impression that only recommendation reports and proposals argue their case to the reader, and that all other writing should be objective rather than argumentative. The fact is, every time you commit words to paper, you are arguing your point.

Technical communication is composed of three main steps: planning, drafting, and revising. As shown on the Figure 1–4 flowchart, these main steps are further divided into eleven substeps that you follow in completing most technical communication. This section of the chapter introduces you to an essential step that must be completed before you even begin to write— determining the purpose.

Kate Paulsen works as a training supervisor for the Boston office of M-Global, Inc., a firm

described in more detail in chapter 2. The company is growing so quickly that hiring, training, and retraining employees have become major goals. Kate recently flew to Cleveland to attend a workshop sponsored by a major professional training organization. The workshop emphasized a new in-house procedure for surveying the training needs of a company's employees. After returning to Boston, Kate must write her manager a trip report that describes the survey technique. She ponders three different approaches to the report:

- **Giving an overview** of the survey procedure she studied during the three-day workshop, stressing a few key points so that her manager could decide whether to inquire further (informing)
- **Providing details** of exactly how the survey procedure could be applied to her firm, with enough specifics for her manager to see exactly how the survey could be used at M-Global (analyzing)
- **Proposing** that the procedure for conducting the needs survey be used at M-Global, in language that argues strongly for adoption (arguing)

For Kate, the first step is to decide what she wants to accomplish. Likewise, every piece of *your* writing should have a specific reason for being. The purpose may be dictated by someone else or selected by you. In either case, it must be firmly understood *before* you start writing. Purpose statements guide every decision you make while you plan, draft, and revise.

Kate Paulsen's three choices indicate some of your options, but there are others. Your choice of purpose will fall somewhere within this continuum:

Neutral, objective statement ⟵⟶ **Persuasive, subjective statement**

For example, when reporting to your boss on the feasibility of adding a new wing to your office building, you should be quite objective. You must provide facts that can lead to an informed decision by someone else. If you are an outside contractor proposing to construct such a wing, however, your purpose is more persuasive. You will be trying to convince readers that your firm should receive the construction contract.

When preparing to write, therefore, you must ask yourself two related questions about your purpose.

>> Question 1: Why Am I Writing This Document?

This question should be answered in just one or two sentences, even in complicated projects. Often the resulting purpose statement can be moved as is to the beginning of your outline and later to the first draft.

For example, Kate Paulsen finally decides on the following purpose statement, which becomes the first passage in her trip report:

> This memo will highlight main features of the training needs survey introduced at the workshop I attended in Cleveland. I will focus on several possible applications you might want to consider for our office training.

Note that Kate's purpose rests about halfway across the persuasive continuum shown earlier. Although she will not be strongly advocating M-Global's use of the survey, she will be giving information that suggests the company might benefit by using it.

>> Question 2: What Response Do I Want from Readers?

The first question about purpose leads inevitably to the second about results. Again, your response should be only one or two sentences long. Although brief, it should pinpoint exactly what you want to happen as a result of your document. Are you just giving data for the file? Will information you provide help others do their jobs? Will your document recommend a major change?

In Kate Paulsen's case, she decides on this results statement:

> Although I'm not yet sure if this training survey is worth purchasing for M-Global,
> I want my boss to consider it.

Unlike the purpose statement, the results statement may not go directly into your document. Kate's statement hints at a hidden agenda that may be implicit in her trip report but will not be stated explicitly. This statement, written for her own use, becomes an essential part of her planning. It is a concrete goal for her to keep in mind as she writes.

The answers to these two questions about purpose and results are included on the Planning Form your instructor may ask you to use for assignments. Figure 1–5 on pages 10 and 11 includes a copy of the form, along with instructions for using it. The last page of this book contains another copy you can duplicate for use with assignments.

Having established your purpose, you are now ready to consider the next part of the writing process: audience analysis.

>>> Analyzing Your Readers

One cardinal rule governs all on-the-job writing:

Write for your reader, not for yourself.

This rule especially applies to science and technology because many readers may know little about your field. In fact, experts on writing agree that most technical communication assumes too much knowledge on the part of the reader. The key to avoiding this problem is to examine the main obstacles readers face and adopt a strategy for overcoming them.

This section (1) highlights problems that readers have understanding technical communication, (2) suggests techniques to prevent these problems, and (3) describes some main classifications of technical readers. At first, analyzing your audience might seem awkward and even unproductive. You are forced out of your own world to consider that of your reader before you even put pen to paper. The payoff, however, is a document that has clear direction and gives the audience what it wants.

PLANNING FORM

Name: _____ Assignment _____

I. Purpose: Answer each question in one or two sentences.

    A. Why are you writing this document?

    B. What response do you want from readers?

II. Audience
    A. Reader Matrix: Fill in names and positions of people who may read the document

	Decision Makers	Advisers	Receivers
Managers			
Experts			
Operators			
General Readers			

    B. Information on individual readers: Answer these questions about the primary audience for this document. If the primary audience includes more than one reader (or type of reader) and there are significant differences between the readers, answer the questions for each (type of) reader. Attach additional sheets as necessary.

Primary audience:

1. What is this reader's technical or educational background?

2. What main question does this person need answered?

3. What main action do you want this person to take?

4. What features of this person's personality might affect his or her reading?

III. Document
    A. What information do I need to include in the

        1. Abstract?

        2. Body?

        3. Conclusion?

    B. What organizational patterns are appropriate to the subject and purpose?

    C. What style choices will present a professional image for me and the organization I represent?

■ **Figure 1–5** ■ Planning Form for all technical documents

Instructions for Completing the Planning Form

The Planning Form is for your use in preparing assignments in your technical communication course. It focuses only on the planning stage of writing. Complete it before you begin your first draft.

1. Use the Planning Form to help plan your strategy for all writing assignments. Your instructor may or may not require that it be submitted with assignments.

2. Photocopy the form on the back page of this book or write the answers to questions on separate sheets of paper, whatever option your instructor prefers. (Your instructor may ask you to use an electronic version or enlarged, letter-sized copies of the form that are included in the Instructor's Resource Manual.)

3. Answer the two purpose questions in one or two sentences each. Be as specific as possible about the purpose of the documents and the response you want—especially from the decision makers.

4. Note that the reader matrix classifies each reader by two criteria: (a) technical levels (shown on the vertical axis) and (b) relationship to the decision-making process (shown on the horizontal axis). Some of the boxes will be filled with one or more names whereas others may be blank. How you fill out the form depends on the complexity of your audience and, of course, on the directions of your instructor.

5. If your document is based on a simulated case from M-Global, Inc., refer to Chapter 2 for any M-Global positions and titles you may want to use in the reader matrix.

6. Note that the "Information on Individual Readers" section can be filled out for one or more readers, depending on what your instructor requires.

7. Answer the document questions in one or two sentences each. Refer to Chapter 3 for information about the ABC format and organizing patterns that can be used in documents. Refer to Chapter 17 for information about style.

■ **Figure 1–5** ■ Continued

Obstacles for Readers

As purchasing agent for M-Global, Inc., Charles Blair must recommend one automobile sedan for fleet purchase by the firm's sales force and executives. First, he will conduct research—interviewing car firm representatives, reading car evaluations in consumer magazines, and inquiring about the needs of his firm's salespeople. Then he will submit a recommendation report to the selection committee consisting of the company president, the accounting manager, several salespeople, and the supervisor of company maintenance. As Charles will discover, readers of all backgrounds often have these four problems when reading any technical document:

1. Constant interruptions
2. Impatience finding information they need
3. A different technical background from the writer
4. Shared decision-making authority with others

If you think about these obstacles every time you write, you will be better able to understand and respond to your readers.

>> Obstacle 1: Readers Are Always Interrupted

As a professional, how often will you have the chance to read a report or other document without interruption? Such times are rare. Your reading time will be interrupted by meetings and phone calls, so a report often gets read in several sittings. Aggravating this problem is the fact that readers may have forgotten details of the project.

>> Obstacle 2: Readers Are Impatient

Many readers lose patience with vague or unorganized writing. They think, "What's the point?" or "So what?" as they plod through memos, letters, reports, and proposals. They want to know the significance of the document right away.

>> Obstacle 3: Readers Lack Your Technical Knowledge

In college courses, the readers of your writing are professors who usually have knowledge of the subject on which you are writing. In your career, however, you will write to readers who lack the information and background you have. They expect a technically sophisticated response, but in language they can understand. If you write over their heads, you will not accomplish your purpose. Think of yourself as an educator; if readers do not learn from your reports, you have failed in your objective.

>> Obstacle 4: Most Documents Have More Than One Reader

If you always wrote to only one person, technical communication would be much easier than it is. Each document could be tailored to the background, interests, and technical education of just that individual. However, this is not the case in the actual world of business and industry. Readers usually share decision-making authority with others who may read all or just part of the text. Thus you must respond to the needs of many individuals—most of whom have a hectic schedule, are impatient, and have a technical background different from yours.

Ways to Understand Readers

Obstacles to communication can be frustrating, yet there are techniques for overcoming them. First, you must try to find out exactly what information each reader needs. Think of the problem this way—would you give a speech without learning about the background of your audience? Writing depends just as much, if not more, on such analysis. Follow these four steps to determine your readers' needs:

>> Audience Analysis Step 1: Write Down What You Know about Your Reader

To build a framework for analyzing your audience, you need to write down—not just casually think about—the answers to these questions for each reader:

1. What is this reader's technical or educational background?
2. What main question does this person need answered?
3. What main action do you want this person to take?
4. What features of this person's personality might affect his or her reading?

The Planning Form in Figure 1–5 includes these four questions.

>> Audience Analysis Step 2: Talk with Colleagues Who Have Written to the Same Readers

Often your best source of information about your readers is a colleague where you work. Ask around the office or check company files to discover who else may have written to the same audience. Useful information could be as close as the next office.

>> Audience Analysis Step 3: Find Out Who Makes Decisions

Almost every document requires action of some kind. Identify decision makers ahead of time so that you can design the document with them in mind. Know the needs of your *most important* reader.

>> Audience Analysis Step 4: Remember That All Readers Prefer Simplicity

Occasionally, you could be in the unenviable position of knowing little or nothing about your readers. Despite your best efforts, you either cannot find information about them or may be prohibited from doing so. For example, a proposal writer sometimes is not permitted to contact the intended reader of the proposal, for legal reasons. Even if you uncover little specific information about your readers, however, you can always rely on one basic fact: Readers of all technical backgrounds prefer concise and simple writing. The popular KISS principle (Keep It Short and Simple) is a worthy goal.

Technical Level	Decision-Making Level		
	Decision Makers	Advisers	Receivers
Managers			
Experts			
Operators			
General Readers			

Types of Readers

You have learned some typical problems readers face and some general solutions to these problems. To complete the audience-analysis stage, this section shows you how to classify readers by two main criteria: knowledge and influence. Specifically, you must answer two questions about every potential reader:

1. How much does this reader already know about the subject?
2. What part will this reader play in making decisions?

Then use the answers to these questions to plan your document. Figure 1–6 (adapted from the Planning Form in Figure 1–5) provides a reader matrix by which you can quickly view the technical levels and decision-making roles of all your readers. For complex documents, your audience may include many of the 12 categories shown on the matrix. Also, you may have more than one person in each box—that is, there may be more than one reader with the same background and decision-making role.

Technical Levels

On-the-job writing requires that you translate technical ideas into language that nontechnical people can understand. This task can be very complicated because you often have several readers, each with different levels of knowledge about the topic. If you are to "write for your reader, not for yourself," you must identify the technical background of each reader. Four categories help you classify each reader's knowledge of the topic.

>> Reader Group 1: Managers

Many technical professionals aspire to become managers. Once into management, they may be removed from hands-on technical details of their profession. Instead, they manage people, set budgets, and make decisions of all kinds. Thus you should assume that management readers are not familiar with fine technical points, have forgotten details of your project, or both. These managers often need

- Background information
- Definitions of technical terms
- Lists and other format devices that highlight points
- Clear statements about what is supposed to happen next

In chapter 3, we discuss an all-purpose ABC format for organization that responds to the needs of managers.

>> Reader Group 2: Experts

Experts include anyone with a good understanding of your topic. They may be well educated—as with engineers and scientists—but that is not necessarily the case. In the example mentioned earlier, the maintenance supervisor with no college training could be considered an expert about selecting a new automobile for fleet purchase. That supervisor understands the technical information about car models and features. Whatever their educational levels, most experts in your audience need

- Thorough explanations of technical details
- Data placed in tables and figures
- References to outside sources used in writing the report
- Clearly labeled appendices for supporting information

>> Reader Group 3: Operators

Because decision makers are often managers or technical experts, these two groups tend to get most of the attention. However, many documents also have readers who are operators. They may be technicians in a field crew, workers on an assembly line, salespeople in a department store, or drivers for a trucking firm—anyone who puts the ideas in your document into practice. These readers expect

- A clear table of contents for locating sections that relate to them
- Easy-to-read listings for procedures or instructions
- Definitions of technical terms
- A clear statement of exactly how the document affects their jobs

>> Reader Group 4: General Readers

General readers often have the least amount of information about your topic or field. For example, a report on the environmental impact of a toxic waste dump might be read by general readers who are homeowners in the surrounding area. Most will have little technical understanding of toxic wastes and associated environmental hazards. Do not assume that general readers are not well educated. They may be engineers or research chemists who are unfamiliar with the topic about which you are writing. These general readers often need

- Definitions of technical terms
- Frequent use of graphics, such as charts and photographs
- A clear distinction between facts and opinions

As with managers, general readers must be assured that (1) all implications of the document have been put down on paper and (2) important information has not been buried in overly technical language.

Decision-Making Levels

Figure 1–6 shows that your readers, whatever their technical level, can also be classified by the degree to which they make decisions based on your document. Pay special attention to those most likely to use your report to create change. Use the following three levels to classify your audience during the planning process.

>> First-Level Audience: Decision Makers

The first-level audience, the *decision makers,* must act on the information. If you are proposing a new fax machine for your office, first-level readers decide whether to accept or reject the idea. If you are comparing two computer systems for storing records at a hospital, the first-level audience decides which unit to purchase. If you are describing electrical work your firm completed in a new office building, the first-level audience decides whether the project has fulfilled agreed-on guidelines.

In other words, decision makers translate information into action. They are usually, but not always, managers within the organization. One exception occurs in highly technical companies, wherein decision makers may be technical experts with advanced degrees in science or engineering. Another exception occurs when decision-making committees consist of a combined audience. For example, the deacons' committee of a church may be charged with the task of choosing a firm to build a new addition to the sanctuary.

>> Second-Level Audience: Advisers

This second group could be called *influencers.* Although they don't make decisions themselves, they read the document and give advice to those who make the decisions. Often, the second-level audience is composed of experts, such as engineers and accountants, who are asked to comment on technical matters. After reading the summary, a decision-making manager may refer the rest of the document to advisers for their comments.

>> Third-Level Audience: Receivers

Some readers do not take part in the decision-making process, but only receive information contained in the document. For example, a report recommending changes in the hiring of fast-food workers may go to the store managers after it has been approved, just so they can put the changes into effect. This third-level audience usually includes readers defined as *operators* in the previous section—that is, those who may be asked to follow guidelines or instructions contained in a report.

Using all this information about technical and decision-making levels, you can analyze each reader's (1) technical background with respect to your document and (2) potential for making decisions after reading what you present. Then you can move on to the research and outline stages of writing.

>>> Collecting Information

Having established a clear sense of purpose and your readers' needs, you're ready to collect information for writing. Although you may want to use a scratch outline to guide the research process, a detailed outline normally gets written after you have collected research necessary to support the document.

This section lays out a general strategy for research. Details about research are included in chapter 15 ("Technical Research").

>> Research Step 1: Decide What Kind of Information You Need

There are two types of research—primary and secondary. *Primary research* is that which you collect on your own, whereas *secondary information* is generated by others and found in books, periodicals, or other sources. Figure 1–7 gives examples of both types. Use the kind of research that will be most helpful in supporting the goals of your project. Following are two examples:

- **Report context for using primary research:** A recommendation report to purchase new CAD (Computer-Assisted Design) software for the design department is supported by your survey of the designers, all of whom recommend the software you selected.

Primary	Secondary
1. **Interviews** 2. **Surveys** 3. **Laboratory Work** 4. **Field Work** 5. **Personal Observation**	1. **Bibliographies** (lists of possible sources— in print or on computer data bases) 2. **Periodical Indexes** (lists of journal and magazine articles, by subject) 3. **Newspaper Indexes** 4. **Books** 5. **Journals** 6. **Newspapers** 7. **Reference Books** (encyclopedias, dictionaries, directories, etc.) 8. **Government Reports** 9. **Company Reports**

■ **Figure 1–7** ■
Research sources

■ **Report context for using secondary research:** Your report on CAD software depends on data found in several written sources, such as an article in a mechanical engineering journal that contrasts features of three programs. On the basis of this article, you recommend a particular software package.

>> Research Step 2: Devise a Research Strategy

Before you start searching through libraries or surfing the Internet, you need a plan. In its simplest form, this plan may list the questions that you expect to answer in your quest for information. For example, a research strategy for a report on office chairs might pose these questions:

■ What kind of chair design do experts in the field of workplace environment recommend?

■ Are there any data that connect the design of chairs with the efficiency of office workers?

■ Have any specific chair brands been recommended by experts?

■ Is there information that suggests a connection between poor chair design and specific health problems?

>> Research Step 3: Record Notes Carefully

See chapter 15 for the variety of resources available to you at a well-stocked library. Once you have located the information you need in these sources, you must be very careful incorporating it into your own document. As chapter 15 denotes, you must clearly distinguish direct quotations, paraphrasing, and summaries in your notes. Then, when you are ready to translate these notes into a first draft, you know exactly how much borrowed information you used and in what form.

>> Research Step 4: Acknowledge Your Sources

The care that you took in Step 3 must be accompanied by thorough acknowledgment of the specific sources. Chapter 15 explains how to use several citation systems.

>> Research Step 5: Keep a Bibliography for Future Use

Consider any research you do for a writing project to be an investment in later efforts. Even after your research for a project is complete and you have submitted the report, keep active files on any subjects that relate to your work. Update these files every time you complete a research-related project, such as the two mentioned previously on chair design and CAD software. If you or a colleague wants to examine the subject later, you have developed your own database from which to start.

>>> Completing an Outline

After determining purpose and audience and completing your research, you are ready to write an outline. Outlines are one method for planning a piece of writing, especially long documents. They do not have to be pretty; they just have to guide your writing of the draft. If you conscientiously use outlines now, you will find it easier to organize and write documents of all kinds throughout your career. Figures 1–8 and 1–9 show the outline process in action. Refer to the following steps in preparing functional outlines:

>> Outline Step 1: Record Your Random Ideas Quickly

At first, ideas need not be placed in a pattern. Just jot down as many major and minor points as possible. For this exercise, try to use only one piece of paper, even if it is oversized. Putting points on one page helps prepare the way for the next step, in which you begin to make connections among points.

>> Outline Step 2: Show Relationships

Next, connect related ideas. Using your brainstorming sheet, follow these three steps:

1. Circle or otherwise mark the points that will become main sections.
2. Connect each main point with its supporting ideas, using lines or arrows.
3. Delete material that seems irrelevant to your purpose.

Figure 1–8 shows the results of applying steps 1 and 2 to a writing project at M-Global, Inc., the company used throughout this book. Diane Simmons, office services manager at

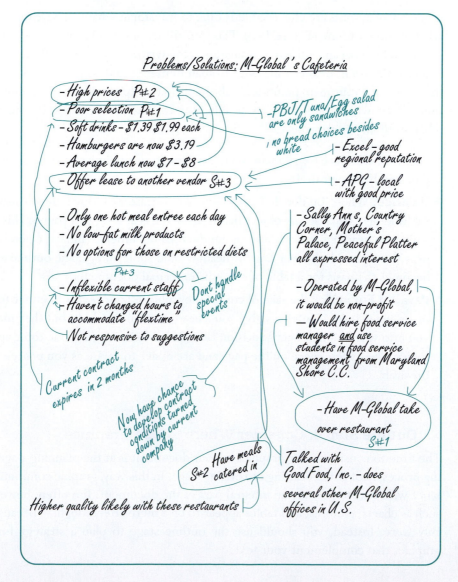

■ Figure 1–8 ■ The outlining process: Early stage

the Baltimore branch, plans to recommend a change in food service. She uses the brainstorming technique to record her major and minor points. First, she circles the six main ideas. As it happens, these ideas include three main problems and three possible solutions, so she labels them P#1 through P#3 (problems) and S#1 through S#3 (solutions). Second, she draws arrows between each main point and its related minor points. In this case, there is no material to be deleted. Although the result is messy, it prepares her for the next step of writing the formal outline.

Like Diane Simmons, you face one main question as you plan your outline: What pattern of organization best serves the material? Chapter 3 presents an *ABC format* that applies to overall structure. Each document should start with an **A**bstract (summary), move to the **B**ody (discussion), and end with a **C**onclusion.

>> Outline Step 3: Draft a Final Outline

Once related points are clustered, it is time to transform what you have done into a somewhat ordered outline (Figure 1–9). This step allows you to (1) refine the wording of your points and (2) organize them in preparation for writing the draft. Although you need not produce the traditional outline with Roman numerals and so on, some structure is definitely needed. Your main points and sub-points may help you identify sections of your document that will be identified by headings and subheadings. Abide by these basic rules when outlining your project:

- **Depth:** Make sure every main point has enough sub-points so that it can be developed thoroughly in your draft.

- **Balance:** When you decide to subdivide a point, break it down into at least two subheadings (because any object that is divided has at least two parts). This same rule applies to headings and subheadings in the final document. (In fact, a good outline provides you with the wording for headings and subheadings. The outline even becomes the basis for a table of contents in formal documents.)

- **Parallel Form:** For the sake of consistency, phrase your points in either topic or sentence form. Sentences give you a head start on the draft, but they may lock you into wording that needs revision later. Most writers prefer the topic approach; topics take up less space on the page and are easier to revise as you proceed through the draft.

>> Outline Step 4: Consider Where to Use Graphics

The time to consider using graphics in your document is at the planning stage of the writing process, not at the drafting or revising stage. In this way, graphic communication combines with text to become an integral part of the document. Too often, however, graphics such as charts, pictures, and tables appear to be a mere afterthought—because they probably were. Instead, you should use the outline stage to plan a strategy for developing graphics that complement your text.

<u>PROBLEMS AND SOLUTIONS: CURRENT CAFETERIA IN BUILDING</u>

I.   Problem #1: Poor selection
   A. Only one hot meal entree each day
   B. Only three sandwiches—PBJ, egg salad, and tuna
   C. Only one bread—white
   D. No low-fat milk products (milk, yogurt, LF cheeses, etc.)
   E. No options for those with restricted diets
II.   Problem #2: High prices
   A. Soft drinks from $1.39 to $1.99 each
   B. Hamburgers now $3.19
   C. Average lunch now $7-$8
III.   Problem #3: Inflexible staff
   A. Unwilling to change hours to meet M-Global's flexible work schedule
   B. Have not acted on suggestions
   C. Not willing to cater special events in building
IV.   Solution #1: End lease and make food service an M-Global department
   A. Hire food service manager
   B. Use students enrolled in food service management program at
      Maryland Shore Community College
   C. Operate as nonprofit operation—just cover expenses
V.   Solution #2: Hire outside restaurant to cater meals in to building
   A. Higher quality likely
   B. Initial interest by four nearby restaurants
      1. Sally Ann's
      2. Country Corner
      3. Mother's Palace
      4. Peaceful Platter
VI.   Solution #3: Continue leasing space but change companies
   A. Initial interest by three vendors
      1. Excel—good regional reputation for quality
      2. APG—close by and local, with best price
      3. Good Food, Inc.—used by two other M-Global offices with good
         results
   B. Current contract over in two months
   C. Chance to develop contract not acceptable to current company

■ **Figure 1–9** ■ The outlining process: Later stage

In the final outline in Figure 1–9, for example, the writer might discover several opportunities for reinforcing textual information with visual language. Following are a few possibilities:

■ Chart showing the increase in cafeteria prices over the last three years

■ Table contrasting prices for a few lunch items at the current cafeteria and at the four restaurants mentioned in section V. B of the outline

■ Map showing the location of the four nearby restaurants

■ Chart showing the relative costs for the contract with the vendors listed in section VI. A of the outline

As a side benefit, the exercise of planning graphics at the outline stage may uncover weaknesses in your argument—that is, places where you need to develop further statistical support. Chapter 12 covers the use of graphics in more detail.

>>> Writing Initial Drafts

With your research and outline completed, you are ready to begin the draft. This stage in the writing process should go quickly if you have planned well. Yet many writers have trouble getting started. The problem is so widespread that it has its own name—"writer's block." If you suffer from it, you are in good company; some of the best and most productive writers often face the "block."

In business and industry, the worst result of writer's block is a tendency to delay the start of writing projects, especially proposals; these delays can lead to rushed final drafts and editing errors. Outlining and other planning steps are wasted if you fail to complete drafting on time. The suggestions that follow can help you start writing and then keep the words flowing.

>> Drafting Step 1: Schedule at Least a One-Hour Block of Drafting Time

Most writers can keep the creative juices flowing for at least an hour if distractions are removed. Rather than writing for three or four hours with your door open and thus with constant interruptions, schedule an hour or two of uninterrupted writing time. Most other business can wait an hour, especially considering the importance of good writing to your success. Colleagues and staff members will adjust to your new strategy for drafting reports. They may even adopt it themselves.

>> Drafting Step 2: Do Not Stop to Edit

Later, you will have time to revise your writing; that time is not now. Instead, force yourself to get ideas from the outline to paper or computer screen as quickly as possible. Most writers have trouble getting back into their writing pace once they have switched gears from drafting to revising.

>> Drafting Step 3: Begin with the Easiest Section

In writing the body of the document, it isn't necessary that you move chronologically from beginning to end. Because the goal is to write the first draft quickly, you may want to start with the section that flows best for you. Later, you can piece together sections and adjust content.

>> Drafting Step 4: Write Summaries Last

As already noted, the outline used for drafting covers just the body of the document. Only after you have drafted the body should you write overview sections, such as summaries. You cannot summarize a report until you have actually completed it. Because most writers have trouble with the summary—a section that is geared mainly to decision makers in the audience—they may get bogged down if they begin writing it prematurely.

>>> Revising Drafts

You may have heard the old saw, "There is no writing, only rewriting." In technical communication, as in other types of writing, careful revision breeds success. The term *revision* encompasses five tasks that transform early drafts into final copy:

1. Adjusting content
2. Editing for style
3. Editing for grammar
4. Editing for mechanics
5. Reviewing layout and graphics

Following are some broad-based suggestions for revising your technical prose. For more details, consult chapter 17, "Style in Technical Writing," or the Handbook at the end of the book. Also, chapter 4 examines the use of computers during the revision process.

>> Revision Step 1: Adjust Content

In this step, go back through your draft to (1) expand sections that deserve more attention; (2) shorten sections that deserve less attention; and (3) change the location of sentences, paragraphs, or entire sections. The use of word processing has made this step considerably easier than it used to be.

>> Revision Step 2: Edit for Style

The term *style* refers to changes that make writing more engaging, more interesting, and more readable. Such changes are usually matters of choice, not correctness. For example, you might want to

- Shorten paragraphs
- Rearrange a paragraph to place the main point first
- Change passive voice sentences to active voice
- Shorten sentences
- Define technical terms
- Add headings, lists, or graphics

One stylistic error deserves special mention because of its frequency: long, convoluted sentences. As a rule, simplify a sentence if its meaning cannot be understood easily in one reading. Also, be wary of sentences that are so long you must take a breath before you complete them.

>> Revision Step 3: Edit for Grammar

You probably know your main grammatical weaknesses. Perhaps comma placement or subject–verb agreement gives you problems. Or maybe you have trouble distinguishing couplets like imply/infer, effect/affect, or complementary/complimentary. In editing the document for grammar, focus on the particular errors that have given you problems in the past.

>> Revision Step 4: Edit for Mechanics

Your last revision of text should be for mechanical errors, such as misspelled words, misplaced pages, incorrect page numbers, missing illustrations, and errors in numbers (especially cost figures). Word processing software can help prevent some of these errors, such as most misspellings, but computer technology has not eliminated the need for at least one final proofing check.

>> Revision Step 5: Review Layout and Graphics

Finally, you should review the visual elements of your document. Check that all illustrations are referred to in the text and that they are placed appropriately. You should also check for consistency in layout and design elements such as headings, list formats, fonts and use of white space. For more about using layout and graphics correctly and consistently, see chapter 12.

This five-stage revision process produces final drafts that reflect well on you, the writer.

Next are two final suggestions that apply to all stages of the process:

1. **Depend on another set of eyes besides your own.** One strategy is to form a partnership with another colleague, whether in a technical writing class or on the job. In this arrangement, you both agree that you will carefully review each other's writing. This buddy system works better than simply asking favors of friends and colleagues. Choose a colleague in whom you have some confidence and from whom you can expect consistent editing quality. However, never make changes suggested by another person unless you fully understand the reason for doing so. After all, it is your writing.

2. **Remember the importance of completing each step separately.** Revising in stages yields the best results.

>>> Writing Collaboratively

Writing can seem like a lonely act at times. Your own experience in school may reinforce the image of the solitary writer—with sweat on brow—toiling away on research, outlines, and drafts. In fact, this description does not typify much writing in the working world outside college. In many professions and organizations, writing in teams is the rule rather than the exception. Chapter 13 discusses writing in teams, or *collaborative writing,* in greater detail.

Guidelines for Team Writing

This section offers six pointers for team writing to be used in this course and throughout your career. The suggestions concern the writing process as well as interpersonal communication.

>> Team Guideline 1: Get to Know Your Team

Most people are sensitive about strangers evaluating their writing. Before collaborating on a writing project, therefore, learn as much as you can about those with whom you will be working. Drop by their offices before your first meeting, or talk informally as a group before the writing process begins. In other words, establish a personal relationship first. This familiarity helps set the stage for the spirited dialogue, group criticism, and collaborative writing to follow.

>> Team Guideline 2: Set Clear Goals and Ground Rules

Every writing team needs a common understanding of its objectives and procedures for doing business. Either before or during the first meeting, the following questions should be answered:

1. What is the team's main objective?
2. Who will serve as team leader?
3. What exactly will be the leader's role in the group?
4. How will the team's activities be recorded?
5. How will responsibilities be distributed?
6. How will conflicts be resolved?
7. What will the schedule be?
8. What procedures will be followed for planning, drafting, and revising?

The guidelines that follow offer suggestions for answering the preceding questions.

>> Team Guideline 3: Use Brainstorming Techniques for Planning

The term *brainstorming* means to pool ideas in a nonjudgmental fashion. In this early stage, participants should feel free to suggest ideas without criticism by colleagues in the group. This nonjudgmental approach does not come naturally to most people; thus, the leader may have to establish ground rules for brainstorming before the team proceeds.

Following is one sample approach to brainstorming:

Step 1: The team recorder takes down ideas as quickly as possible.

Step 2: Ideas are written on large pieces of paper affixed to walls around the meeting room so all participants can see how major ideas fit together.

Step 3: Members use ideas as springboards for suggesting other ideas.

Step 4: The team takes some time to digest ideas generated during the first session, before meeting again.

Results of a brainstorming session might look much like a nonlinear outline produced during a solo writing project (see Figure 1–8). The goal of both is to generate as many ideas as possible; these ideas can be culled and organized later.

>> Team Guideline 4: Use Storyboarding Techniques for Drafting

Storyboarding helps propel participants from the brainstorming stage toward completion of a first draft. It also makes visuals an integral part of the document. Originating in the screenwriting trade in Hollywood, the storyboard process can take many forms, depending on the profession and individual organization. In its simplest form, a *storyboard* can be a sheet of paper or an electronic template that contains (1) one draft-quality illustration and (2) a series of sentences about one topic (Figure 1–10). As applied to technical writing, the technique involves six main steps:

Step 1: The team or its leader assembles a topic outline from ideas brought forth during the brainstorming session.

Step 2: All team members are given one or more topics to develop on storyboard forms.

Step 3: Each member works independently on the boards, creating an illustration and a series of subtopics for each main topic (see Figure 1–10).

Step 4: Members meet again to review all completed storyboards, modifying them where necessary and agreeing on key sentences.

Step 5: Individual members develop draft text and related graphics from their own storyboards.

Step 6: The team leader or the entire group assembles the draft from the various storyboards.

>> Team Guideline 5: Agree on a Thorough Revision Process

As with drafting, all members usually help with revision. Team editing can be difficult, however, as members strive to reach consensus on matters of style. Following are some suggestions for keeping the editing process on track:

- Avoid making changes simply for the sake of individual preference.
- Search for areas of agreement among team members, rather than areas of disagreement.
- Make only those changes that can be supported by accepted rules of style, grammar, and use.
- Ask the team's best all-round stylist to do a final edit.

This review will help produce a uniform document, no matter how many people work on the draft.

DOCUMENT TITLE: M-Global's Training Needs
STORYBOARD TOPIC: Results of employee survey
STORYBOARD WRITER: Susan Hernandez

1. In one sentence, summarize this section of the document.

The recent survey of employees showed a strong preference for nontechnical over technical training.

2. In sentence form, include the key points to be developed in this document section. Use same order that points will appear in document.

A. The greatest interest was in the area of sales and marketing training—engineers, in particular, feel deficient here.

B. Many employees also wanted further training in project management—with emphasis on scheduling, accounting practices, and basic management.

C. The third most called-for training area was communication skills—that is, report writing, grammar, and oral presentations.

D. Many employees want training in stress management to reduce or manage on-the-job pressures and make work more enjoyable.

E. The fifth area of interest was technical training in the respondents' own area of expertise.

3. Include an illustration that supports the text in this document section.

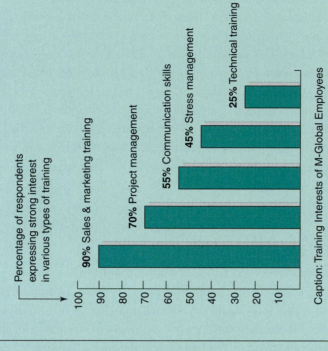

Percentage of respondents expressing strong interest in various types of training

- 90% Sales & marketing training
- 70% Project management
- 55% Communication skills
- 45% Stress management
- 25% Technical training

Caption: Training Interests of M-Global Employees

(The five most popular training topics, according to company-wide employee survey.)

■ **Figure 1–10** ■ Completed storyboard

27

>> **Team Guideline 6: Use Computers to Communicate**

When team members are at different locations, computer technology can be used to complete some or even the entire project. Team members must have personal computers and the software to connect their machines to a network, allowing members to send and receive information online. This section describes three specific computer applications that can improve communication among members of a team-writing project: email, computer conference, and groupware. Chapter 4 ("Page Design") discusses how the individual writer can use computers to plan, draft, and revise copy. Chapter 5 ("Letters, Memos, and Electronic Communication") discusses stylistic features of electronic mail.

Team writing may play an important part in your career. If you use the preceding techniques, you and your team members will build on each other's strengths to produce top-quality writing.

>>> Chapter Summary

Technical communication refers to the many kinds of writing and speaking you will do in your career. In contrast to most academic writing, technical communication aims to get something done (not just to demonstrate knowledge), relays information from someone more knowledgeable about the topic (you) to someone less knowledgeable about it (the reader), is read by people from mixed technical and decision-making levels, presents ideas clearly and simply, and often uses data and graphics to provide support.

For each writing project, you should complete a three-stage process of planning, drafting, and revising. *Planning* involves determining your purpose, knowing the readers' needs, collecting information, and outlining major and minor points. In the *drafting* stage, you use the outline to write a first draft as quickly as possible—without stopping to make changes. Finally, the *revision* process requires that you adjust content; then edit for style, grammar, and mechanics; and review layout and graphics.

Technical communication can be completed by you alone or by you as a member of a writing team. The latter approach is common in organizations, for it exploits the strengths of the varied professionals within an organization. In group writing you work closely with your team during the planning, drafting, and revising processes.

>>> Learning Portfolio

Communication Challenge "Bad Chairs, Bad Backs"

The engineers, programmers, scientists, and other employees in M-Global's Boston branch spend a lot of time in their chairs. Although all the office furniture is new and expensive, employees have experienced regular back pain since the new chairs arrived. Unfortunately, the furniture was ordered through the corporate office, so complaints cannot be handled in a routine and informal way in the Boston office. The branch manager, Richard DeLorio, mentioned the problem to his boss, Jeannie McDuff, Vice President for Domestic Operations. Predictably, Jeannie asked Richard to "put it in writing."

This case study explains their approach, presents the analysis of the memo that is part of the proposed strategy, and ends with questions and comments for discussion and an assignment for a written response to the Challenge.

Gathering Evidence

Richard knows that he must thoroughly and objectively document the problems associated with the arrival of the new chairs. He asks one of the engineers to gather information that supports his memo. The engineer creates an elaborate spreadsheet with accompanying charts that identifies the employees who have received the new chairs, as well as the frequency, types, and timing of specific complaints about back pain.

Richard knows that Jeannie does not have the time or interest to work her way through all the data. What is more, as he looks at it, he realizes that he can't draw a direct correlation between the data and the back pain that he and his fellow employees have been experiencing. He asks the engineer to revise the information and to write a discussion of the data so that Jeannie can quickly and easily see the point being made. At the same time, Richard researches information about office chairs and back pain.

Writing the Memo

Richard creates the following analysis to help him plan his memo:

Purpose: To convince my boss, Jeannie McDuff, that the new chairs at the Boston branch should be replaced.

Readers: Jeannie McDuff is the primary reader and decision maker. Other readers may include the Purchasing Officer, who could use this information to select chairs (assuming Jeannie agrees to replace them).

Information about Jeannie McDuff: Jeannie joined M-Global as a civil engineer. Even though her grandfather founded the company, she has been expected to work her way up through the ranks. However, as VP for Domestic Operations, she doesn't have much to do with the day-to-day branch operations and probably didn't have much to do with actually selecting the chairs that we have been given. The most important point to make to her is that these chairs are hurting our productivity (through absences and distraction as a result of the discomfort) and could cost M-Global money (through increased use of medical insurance—which could lead to increased insurance premiums).

Jeannie prefers clean, clear documents—only as elaborate as necessary, but that follow the guidelines for Plain English style.

Questions and Comments for Discussion

1. How could Richard have helped the engineer create usable information when he first assigned the writing task?
2. Based on Richard's planning sheet, what strategies do you think will help him convince Jeannie to replace the chairs? What information should he emphasize? How should he organize his points?
3. What other information could have been useful to help Richard make his case?
4. Even though the document written by the engineer and the document that Richard writes include some of the same information, the documents are different because they have different purposes and audiences. How are they different? How does purpose and audience contribute to those differences?

Write About It

You have been asked by Richard to create a document of no more than one page that identifies the most important concerns for safe, comfortable, and ergonomically correct computer stations. Research the recommendations for workplace health and computer use and write your recommendation in a memo of one page. Include a list of your sources.

Collaboration at Work Outline for a Consulting Report

General Instructions

Each Collaboration at Work exercise applies strategies for working in teams to chapter topics. The exercise assumes you (1) have been divided into teams of about three to six students, (2) use team time inside or outside of class to complete the case, and (3) produce an oral or written response. For guidelines about writing in teams, refer to pages 24–28.

Background for Assignment

Assume you and your team members comprise one of several teams from a private consulting firm. The firm has been hired to help plan a hotel/conference center to be built on your campus. Although the center will cater to some private clients, most customers will be associated with your institution—for example, parents of students, candidates for teaching or administrative positions, and participants in academic conferences.

Obviously, a project of this sort requires careful planning. One step in the process is to assess the needs of various people and groups that will occupy the center.

Following are listed just a few of the many groups or departments whose needs should be considered:

Accounting	Landscaping	Registration
Catering	Maintenance	Sales and marketing
Computing	Procurement	Security
Housekeeping	Recreation	Training

The topics range broadly because the facility will have multiple purposes for a diverse audience.

Team Assignment

The consulting firm—of which your team is a part—will issue a joint report that describes the needs of all groups who will work in the new hotel/conference center. Assume that your team's task is to produce just a portion of the outline—not the text—of the report. Your outline will address one or more of the needs reflected in the previous list, or other needs of your choosing that have not been listed. Your instructor will indicate whether you are to refer to the section in this chapter entitled "Completing an Outline."

Assignments

Your instructor will indicate whether Assignments 1 through 7 should serve as the basis for class discussion, for a written exercise, or for both. Assignments 8, 9, and 11 through 13 require a written response. Assignment 10 is an ongoing assignment.

1. Features of Academic Writing

Option A Select an example of writing that you wrote for a high school or college course other than this one. Then prepare a brief analysis in which you explain (1) the purpose of the writing sample, (2) the audience for which it was intended, and (3) the ways in which it differs from technical writing, as defined in this chapter.

Option B As an alternative to using your own example, complete the assignment by using the following example. Assume that the passage was written as homework or as an in-class essay in an environmental science class in college.

There are many different responses that are possible in the event toxic waste contamination is suspected or discovered at a site. First, you can simply monitor the site by periodically taking soil and/or water samples to check for contamination. This approach doesn't solve the problem and may not prove politically acceptable when contamination is obvious to the community, but it does help determine the extent of the problem. A second approach—useful when contamination is likely or proved—is to contain the toxic waste by sealing off the site in some fashion, such as by building barriers between it and the surrounding area or by "capping" it in some manner (as in the case of a toxic waste pit). Basically, this alternative depends on the ability to isolate the toxic substances effectively. A third strategy, useful when the contamination is liquefied (as with toxic groundwater), is to pump the water from under the ground or from surface ponds and then transport it to treatment systems.

A fourth method is appropriate when toxic substances need to be treated on-site, in which case they can be incinerated or they can be solidified at the site in some way. Then they can be placed in a landfill at the site. Fifth, waste can be hauled to another location where it can be incinerated or placed in some kind of secure landfill—when an off-site disposal approach is needed.

2. Features of Technical Communication

Option A Locate an example of technical communication (such as a user's guide, owner's manual, or a document borrowed from a family member or an acquaintance who works in a technical profession) and prepare a brief analysis in which you explain (1) the purpose for which the piece was written, (2) the apparent readers and their needs, (3) the way in which the example differs from typical academic writing, and (4) the relative success with which the piece satisfies this chapter's guidelines.

Option B Using the following brief example of technical writing, prepare the analysis requested in Option A.

DATE: June 15, 2008
TO: Pat Jones, Office Coordinator
FROM: Sean Parker
SUBJECT: New Productivity Software

Introductory Summary

As you requested, I have examined the FreeWork open source productivity suite software we are considering. On the basis of my observations, I recommend we secure one copy of FreeWork and test it in our office for two months. Then after comparing it to the other two packages we have tested, we can choose one of the three productivity packages to use throughout the office.

Features of FreeWork

As we agreed, my quick survey of FreeWork involved reading the user's manual, completing the orientation disk, and reviewing installation options. Here are the five features of the package that seemed most relevant to our needs:

1. Formatting Flexibility: FreeWork includes diverse "style sheets" to meet our needs in producing reports, proposals, letters, memos, articles, and even brochures. By engaging just one command on the keyboard, the user can change style sheets—whereby the program will automatically place text in a specified format.

2. Mailers: For large mailings, we can take advantage of FreeWork's "Mail Out" feature that automatically places names from mailing lists on form letters.

3. Documentation: To accommodate our staff's research needs, FreeWork has the capacity to renumber and rearrange footnotes as text is being edited.

4. Page Review: This package's "PagePeek" feature permits the user to view an entire written page on the screen. Without having to print the document, he or she can then see how every page of text will actually look on the page.

5. Tables of Contents: FreeWork can create and insert page numbers on tables of contents, created from the headings and subheadings in the text.

6. Spreadsheets: FreeWork includes a powerful spreadsheet that can be integrated into documents.

7. Database: FreeWork's database component can create forms and reports that can be integrated into documents.

8. Graphics: FreeWork includes a basic drawing program that will probably meet our needs.

Conclusion

Though I gave FreeWork only a brief look, my survey suggests that it may be a strong contender for use in our office. If you wish to move to the next step of starting a two-month office test, just let me know.

---

3. Purpose and Audience

The following examples deal with the same topic in four different ways. Using this chapter's guidelines on purpose and audience, determine the main reason for which each excerpt was written and the technical level of the intended readers.

A. You can determine the magnitude of current flowing through a resistor by use of this process:

- Connect the circuit (power supply, resistor, ammeter, voltmeter).
- Set the resistor knob to a setting of "1."

- Turn the voltage adjusting knob to the left until it stops rotating.
- Switch the voltmeter to "On" and make sure it reads "0.00 volts."
- Switch the power supply to "On."
- Slowly increase the voltage on the voltmeter from 0 to 10 volts.
- Take the reading from the ammeter to determine the amount of current flowing through the resistor.

B. After careful evaluation of several testers, I strongly recommend that Langston Electronics Institute purchase 100 Mantra Multitesters for use in our laboratories in Buffalo, Albany, and Syracuse.

C. Selected specifications for the Ames Multitester are as follows:

- Rangers. 43
- DC Voltage 0–125–250mV 1.25–2.5–10–5–125–500–1000V
- AC Voltage 0–5–25–125–250–500–1000V
- DC Current. 0–25–50µA–2.5–5–25–50–250–500mA–10amperes
- Resistance . . . . . . 0–2K–20K–200K–20 Mega ohms
- Decibels . . . . –20 to +62 in db 8 ranges
- Accuracy. . . . . ±3% on DC measurements
  ±4% on AC measurements
  ±3% on scale length on resistance
- Batteries . . . . one type AA penlight cell
- Fuse. . . . . 0.75A at 250V

Note that the accuracy rate for the Ames is within our requirements of ±6% and is considerably lower than the three other types of testers currently used by our staff.

D. Having used the Ames Multitester in my own home laboratory for the last few months, I found it extremely reliable during every experiment. In addition, it is quite simple to operate and includes clear instructions. As a demonstration of this operational ease, my 10-year-old son was able to follow the instructions that came with the device to set up a functioning circuit.

4. Audience Analysis

Find a commercial website (a website from a manufacturer or retailer) designed for children. Sites that promote cereal, toys, or snack foods are good choices.

- Is the site designed to inform, provide analysis, or to persuade? How do you know?
- What have the designers of the website done to appeal to their audience? What do their choices tell you about the results of their audience analysis?

- Is there a section on the website specifically targeted to parents? How does it differ from the web pages for children? How is it similar to the pages for children?

 ### 5. Interview

Interview a friend, relative, or recent college graduate who works as a technical professional or manager. Prepare to share your results with the class. Gather specific information on the following topics:

- The percentage of the workweek spent writing
- Types of documents that are written and their purpose
- Specific types of readers of these documents

6. Contrasting Styles

Find two articles on the same topic in a professional field that interests you. One article should be taken from a newspaper or magazine of general interest, such as one you would find on a newsstand; the other should be from a magazine or journal written mainly for professionals in the field you have chosen. Now contrast the two articles according to purpose, intended audience, and level of technicality.

7. Contrasting Audiences

Photocopy three articles from the same Sunday issue of a local or national newspaper. Choose each article from a different section of the paper—for example, you could use the sections on automobiles, business, travel, personal computers, national political events, local events, arts, editorials, or employment. Describe the intended audience of each article. Then explain why you think the author has been successful or unsuccessful in reaching the particular audience for each article.

8. Rewrite for Different Audience

Locate an excerpt from a technical article or textbook, preferably on a topic that interests you because of your background or college major. Rewrite all or part of the selection so that it can be understood by readers who have no previous knowledge of the topic.

9. Collecting and Organizing Information

Most word-processing programs include a feature that allows users to track the changes that are made to documents, as well as insert questions, comments and advice for revision. This feature is especially useful for collaborative projects because it allows team members to see who recommended various changes, as well as allowing several people to comment on drafts.

A. Identify the reviewing features that are available in your word-processing program and how they are accessed. Create a list of the features that you believe would be most useful for students working on team projects.

B. Organize the information you have collected into a single page reference for students who want to use the reviewing features in your word-processing program.

10. Revision

As noted in the chapter, it is helpful to become aware of the problems that occur most frequently in your writing. One way to do this is to keep a record, or log, of the problems that appear most often. Create a log by listing the most common broad categories of errors:

- Sentence boundary errors (fragments, run-ons, comma splices)
- Agreement (subject/verb, pronoun/reference, changes in tense)
- Word choice
- Punctuation
- Spelling

To begin collecting information about your most common problems, look at papers that you have received back from teachers. In your log, record problems that have been marked on your papers. After you have recorded information from four or five papers, you may be able to see patterns developing. For example, are most of your entries "Sentence boundary errors"? If so, are they all similar—maybe run-on sentences? Once you have identified your common problems, develop strategies for proofreading for them. To help you find run-on sentences, for example, try looking for sentences with two verbs. If you often have problems with punctuation—commas, for example—make sure you look them up in the handbook in the back of *Technical Communication: A Practical Approach*, learn the rules, and check them when you proofread your papers.

11. Ethics Assignment

Write a short essay in which you describe one or more ethical problems that might arise from collaborative writing. First review the "Collaborative Writing" section of this chapter, and then interview at least one person who has written collaboratively on the job. Each problem you cite should be supported either by a simulated case from the workplace or by a real example gained from your interview(s).

12. International Communication Assignment

World cultures differ in the way the organize information and in the visual cues they use for readers. Using a resource like newdirectory.com, find websites for newspapers from three different countries and analyze each website for the way information is presented. Do you notice any differences in how information is arranged on the pages of the site? For example, are the website's topics arranged vertically on the left of the page, as they are in most English-language websites? How are graphics treated? What other differences do you see? Write a brief essay about what you have learned about how these websites present information, and what issues companies that are creating websites for global audiences need to be aware of.

13. A.C.T.N.O.W. Assignment (Applying Communication To Nurture Our World)

As part of the annual Earth Day celebration (April 22), you and several classmates have been asked to propose an event to the Student Affairs Office on your campus. It is supposed to be an "environmentally friendly" activity conducted on campus by campus staff, students, faculty, or individuals from outside the campus community. Following the guidelines in this chapter, develop an outline of the project you plan to propose.

Chapter | 2 | M-Global, Inc.: Ethics and Globalism in the Workplace

>>> Chapter Outline

Culture in Organizations   35

Elements of an Organization's Culture   35

Business Climate   36

The Global Workplace   37

Understanding Cultures   37

Communicating Internationally   42

Ethics on the Job   43

Ethical Guidelines for Work   43

Ethics and Legal Issues in Writing   45

Introducing M-Global   47

History of M-Global, Inc.   48

Projects   49

Corporate and Branch Offices   51

Headquarters   51

Branches   54

Writing at M-Global   57

Examples of Internal Writing   58

Examples of External Writing   58

Chapter Summary   59

Learning Portfolio   60

Communication Challenge   60

Collaboration at Work   61

Assignments   61

M-Global, Inc. Project Sheets   65

Worldwide Locations of M-Global Inc., Offices   65

Project 1: Sentry Dam   66

Project 2: Completed Ocean Exploration Program   67

Project 3: Monitored Construction of General Hospital   68

Project 4: Managed Construction of Nevada Gold Dome   69

Project 5: Examined Big Bluff Salt Marsh   70

Project 6: Designed and Installed Control Panel for Nuclear Plant   71

Project 7: Designed and Taught Seminar in Technical Writing   72

Project 8: Designed and Created Documentation of Data Security Procedures   73

Chapter 1 defines technical communication and outlines the writing process. This chapter brings ethics and globalism into the context of effective communication and introduces you to the fictional company M-Global, Inc., used in examples and assignments throughout this book. Then chapters 3 and 4 cover organization and page design, respectively. Together, these four chapters provide the foundation for your work in the rest of the book.

The use of M-Global, Inc., in this textbook is intended to yield two main benefits for you as a student:

- **Real-world context:** M-Global provides you with an extended case study in modern technical communication. By placing you in actual working roles, the text prepares you for writing and speaking tasks ahead in your career.

- **Continuity:** The use of M-Global material lends continuity to class assignments and discussions throughout the term. Your use of this international organization in assignments and class emphasizes the connections among all on-the-job assignments.

Thus M-Global gives you a window into an international organization similar to one where you may work soon. This chapter addresses six main topics:

- Corporate culture
- Globalism related to the workplace
- Ethics
- Background information on M-Global
- Activities and positions at the corporate and branch offices
- Typical writing tasks at M-Global

>>> Culture in Organizations

The first part of this section presents three features common to the culture of any organization that may employ you. Then the second part concentrates on the larger context for corporate culture—the business climate.

Elements of an Organization's Culture

We use the term *organization* to remind you that in addition to commercial firms, there are many career opportunities in government and even in nonprofit organizations. As noted in chapter 1, the writing you do in an organization differs greatly from the writing you do in college. The stakes on the job are much higher than a grade on your college transcript. Writing directly influences the following:

- Your performance evaluations
- Your professional reputation
- Your organization's productivity and success in the marketplace

Given these high stakes, let's look at typical features of the organizations where you may spend your career.

Starting a job is both exciting and, sometimes, a bit intimidating. Although you look forward to practicing skills learned in college, you also wonder just how you will fare in new surroundings. Soon you discover that any organization you join has its own personality. This personality, or *culture,* can be defined as follows:

Corporate culture: The main features of life at a particular organization. An organization's culture is influenced by the firm's history, type of business, management style, values, attitude toward customers, and attitude toward its own employees. Taken together, all features of a particular organization's culture create a definable quality of life within the working world of that organization.

Let's look more closely at three features mentioned in the preceding definition: a firm's history, its type of business, and its management style.

>> Feature 1: Organization History

A firm's origin often is central to its culture. On the one hand, the culture of a 100-year-old steel firm depends on accumulated traditions to which most employees are accustomed; on the other hand, the culture of a recently established software firm may depend more on the entrepreneurial spirit of its founders. Thus the facts, and even the mythology, of an organization's origin may be central to its culture, especially if the person starting the firm remained at the helm for a long time.

>> Feature 2: Type of Business

Culture is greatly influenced by an organization's type of business. Many computer software firms, for example, are known for their flexible, nontraditional, innovative, and sometimes chaotic culture. Some of the large computer hardware firms, however, have a culture focused more on tradition, formality, and custom.

>> Feature 3: Management Style

A major component of an organization's culture is its style of leadership. Some organizations run according to a rigid hierarchy, with all decisions coming from the top. Other organizations involve a wide range of employees in the decision-making process. As you might expect, most organizations have a decision-making culture somewhere between these two extremes.

An organization's culture influences who is hired and promoted at the firm, how decisions are made, and even how company documents are written and reviewed. Now let's examine the larger context for an organization's culture—the business climate.

Business Climate

An organization's culture is not isolated from the cultures of other organizations, or from the wider culture or cultures in which it is located. Organizations, especially businesses and corporations such as M-Global, must respond to the business climate.

Business Climate: The economic and political factors that influence an organization's priorities, plans and activities. These factors include competition, investor interests, regulations, and the overall health of the economy.

To compete in today's global business climate, companies are focusing on quality and efficiency. To improve quality, companies seek to respond quickly to customer needs and to encourage employee interest in the success of the organization through an emphasis on team building and employee input. To improve efficiency, companies work to improve productivity while reducing costs. This has resulted in such strategies as just-in-time delivery and improved use of communication technology.

Two practices that are being used more often in the global business climate are outsourcing and offshoring. *Outsourcing* is the practice of purchasing goods or subcontracting services from an outside company. Both the client company and the company that is providing the goods or services may be in the same country, or they may be in different countries. *Offshoring* always includes more than one country, but in offshoring, a company moves some of its operations to another country. This practice is often done to reduce labor costs, but it may also help a company work more efficiently by creating offices closer to suppliers or clients. Although both practices are changing the workplace, they also offer opportunities for companies and employees who are prepared for the global marketplace.

>>> The Global Workplace

Very possibly, you will end up working for an organization that does some of its business beyond the borders of its home country. It may even have many international offices, as does M-Global. Such organizations face opportunities and challenges of diversity among employees or customers. They seek out employees who are able to view issues from a perspective outside their own cultural bias, which we all have. This section examines work in the global workplace, with emphasis on suggestions for writing for readers in different cultures.

Communication has entered what might be seen as its newest frontier—intercultural and international communication. More than ever before, industries that depend on good communication have moved beyond their national borders into the global community. Some people criticize internationalism and the so-called shrinking of the planet. They worry about the possible fusion of cultures and loss of national identities and uniqueness; others welcome the move toward globalism. Whatever your personal views, the phenomenon is with us for the foreseeable future. Following are some practical suggestions for dealing with it.

Understanding Cultures

In studying other cultures, we must avoid extremes of focusing exclusively on either the differences or similarities among cultures. On the one hand, emphasizing differences can lead to inaccurate stereotypes; large generalizations about people can be misinformed and thus can impede, rather than help, communication. On the other hand, emphasizing similarities can tend to mask important differences by assuming we are all alike—one big global family. The truth is somewhere in between. All cultures have both common features and distinctive differences that must be studied. Such study helps set the stage for establishing productive ties outside one's national borders, in fields such as technical communication.

Exactly how do we go about studying features of other cultures? Traditionally, there are two ways. One only touches the surface of cultural differences by offering simplistic do's and don'ts, such as the following:

1. In Japan, always bow as you greet people.
2. In Mexico, be sure to exchange pleasantries with your client before you begin to discuss business.
3. In Germany, do not be a minute late for an appointment.
4. In China, always bring gifts that are nicely wrapped.

These and hundreds of other such suggestions may be useful in daily interactions, but they do not create cultural understanding and often present inaccurate stereotypes of the way people operate.

The other, more desirable, approach goes below the surface to the deeper structure of culture. It requires that we understand not only what people do, but also why they do it. Although learning another language certainly enhances one's ability to learn about another culture, linguistic fluency does not in and of itself produce cultural fluency. One must go beyond language to grasp one essential point:

> People in different cultures have different ways of thinking, Different ways of acting, and different expectations in communication.

To be sure, there are a few basic ethical guidelines evident in most cultures with which you will do business, but other than these core values, differences abound that should be studied by employees of multinational firms. Then these differences must be reflected in communication with colleagues, vendors, and customers.

One of the ways that differences between cultures can be understood is through the concepts of low-context cultures and high-context cultures. *Low-context cultures* consist of diverse religions, ethnic backgrounds, and educational levels; as a result, communication must be explicit, because members of a group cannot assume that they share knowledge or attitudes. The United States is an example of a low-context culture. Important characteristics that affect communication in low-context cultures include

- Openness to outsiders
- A focus on actions and solving problems, with a willingness to disagree openly
- A dependence on formally established rules to govern behavior

High-context cultures are more homogenous than low-context cultures. Because members of the culture share characteristics such as religion, ethnic background, and education, their communications may be less explicit. Think about the way that you communicate with members of your family. With a few words, you can tell a whole story; for example: "It's just like when the squirrels moved into the attic." To outsiders, this means nothing, but members of your family immediately understand the situation. Important characteristics of high-context cultures include

- Clear distinctions between insiders and outsiders
- A focus on maintaining relationships, on saving face, and on helping others save face
- A dependence on internalized cultural norms to govern behavior

Although these concepts provide a starting point for learning about other cultures, interactions between cultures in the global marketplace can be very complex, as suggested by Nancy Settle-Murphy in the summary in Figure 2–1.

The concept of low-context and high-context cultures offers a general way of thinking about how to relate to clients and colleagues in other cultures and countries, but if you find yourself working in a global, intercultural setting, you should understand the specific cultural practices of those you are working with. Companies in the United States can get information about the cultures and business practices of other countries from the U.S. Commercial Service of the Department of Commerce, as well as from organizations like the Society for Intercultural Education, Training, and Research (SIETAR). However, there are some general questions you can ask to prepare you to communicate with people outside your own culture.[1] Consider these questions to be a starting point for your journey toward understanding communication in the global workplace.[2]

Question 1—*Work:* What are their views about work and work rules?

Question 2—*Time:* What is their approach to time, especially with regard to starting and ending times for meetings, being on time for appointments, expected response time for action requests, hours of the regular workday, etc.?

Question 3—*Beliefs:* What are the dominant religious and philosophical belief systems in the culture, and how do they affect the workplace?

Question 4—*Gender:* What are their views of equality of men and women in the workplace, and how do these views affect their actions?

---

[1]A good overview of this subject can be found in Emily A. Thrush (2001). High-Context and Low-Context Cultures: How Much Communication is Too Much? In Deborah S. Bosley (ed.), *Global Contexts: Case Studies in International Communication* (pp. 27–41). Boston: Allyn & Bacon.

[2]The questions in this section are drawn from information in two excellent sources for the student of international communication: Iris Varner and Linda Beamer (1995). *Intercultural Communication in the Global Workplace* (pp. 306–308). Chicago: Irvin; and David P. Victor (1992). *International Business Communication.* New York: Harper Collins.

Category	Cultural Differences
Big picture vs. details	People from "high-context" cultures tend to derive their most valuable information from the context that surrounds words rather than the actual words. Precise details may be less important than the broader context.
	People from "low-context" cultures pay more attention to the words and details than to the overall context. They see the trees, but may not always see the forest.
Order vs. chaos	"Monochronic" cultures are more comfortable taking one thing at a time. Following the correct order or using the right process can seem almost as important as achieving the desired outcome. Unstructured conversations and interruptions can be unsettling.
	"Polychronic" cultures cope well with simultaneous activities and see interruptions as a necessary and natural way of doing business.
Formal vs. informal	Some cultures have a more compartmentalized communications flow, where information is parceled out on a need-to-know basis, usually top-down.
	In other cultures, people share information more freely among all levels, back and forth and up and down, and maintain multiple channels of communications, both formal and informal.
Motivations and rewards	In some cultures, achieving personal recognition or widespread popularity may be the chief motivators.
	People from other cultures may be more motivated by their contributions toward building a stronger company or a more harmonious organization. Financial rewards are less important to some than to others.
Quality vs. quantity of decisions	People from certain cultures like to make decisions only after they have carefully solicited input and gained buy-in from multiple perspectives. Such a methodical process may take more time up front, but once decisions are made, results are usually achieved quickly.
	For others, speed trumps quality, even if it means that hurried decisions are eventually revisited and work must be redone.
Giving and receiving feedback	People from some cultures seek constant validation for the quality of their work, and may assume that the absence of feedback signals at least mild disappointment. These same people tend to provide frequent unsolicited feedback.
	Others assume that unless they hear otherwise, the quality of their work is just fine. Some feel a need to lead with the positive before delving into the negative when giving feedback, while others regard "sugarcoating" as confusing and unnecessary.
Expressing opinions	In some cultures, people tend to break in frequently to ask questions, pose challenges, or openly disagree, while others prefer to maintain group harmony by never openly disagreeing, especially in front of a group.
	Some tend to allow others to speak before voicing their own opinions, while others speak over others' voices if that's what it takes to get heard.
	Some need silence to think (and to translate into their native language and back again), and others are uncomfortable with silence, rushing in to fill a pause.
Role of managers	In cultures where egalitarianism is prized, team members tend to have equal say when making decisions and setting priorities, regardless of seniority. Managers are seen as organizers and enablers, helping to set strategy, remove roadblocks, and otherwise grease the skids for moving in the right direction.
	In cultures where hierarchy is important, managers typically make decisions and pass them down to team members, who implement the decisions and report back to management.
Willingness to sacrifice personal time	Some cultures abhor the notion of giving up personal time for work. Weeknights, weekends, holidays, and vacations are sacrosanct.
	People from other cultures quite frequently, though not necessarily happily, forgo personal time if needed.

■ **Figure 2–1** ■ Cultural differences

Source: Pejovic, J. (2006, May). Trans-Atlantic Roundtable. *Intercom, 53.* 12.

Question 5—*Personal Relationships:* What degree of value is placed on close personal relationships among people doing business with each other?

Question 6—*Teams:* What part does teamwork have in their business, and, accordingly, how is individual initiative viewed?

Question 7—*Communication Preferences:* What types of business communication are valued most—formal writing, informal writing, formal presentations, casual meetings, email, phone conversations?

Question 8—*Negotiating:* What are their expectations for the negotiation process, and, more specifically, how do they convey negative information?

Question 9—*Body Language:* What types of body language are most common in the culture, and how do they differ from your own?

Question 10—*Writing Options:* What writing conventions are most important to them, especially in prose style and the organization of information? How important is the design of the document in relationship to content and organization?

To be sure, asking these questions does not mean we bow to attitudes that conflict with our own ethical values, as in the equal treatment of women in the workplace. It only means that we first seek to comprehend cultures with which we are dealing before we operate within them. Intercultural knowledge translates into power in the international workplace. If we are aware of diversity, then we are best prepared to act.

It might help to see how some of these issues were addressed by Sarah Logan, a marketing specialist who transferred to M-Global's Tokyo office three years ago. In her effort to find new clients for M-Global's services, she discovered much about the Japanese culture that helped her and her colleagues do business in Japan. For example, she learned that Japanese workers at all levels depend more on their identification with a group than on their individual identity. Thus Sarah's marketing prospects in Japan felt most comfortable discussing their work as a corporate department or team, rather than their individual interests or accomplishments—at least until a personal relationship was established.

Sarah learned that an essential goal of Japanese employees is what they call "wa"—harmony among members of a group and, for that matter, between the firm and those doing business with it. Accordingly, her negotiations with the Japanese often took an indirect path. Personal relationships usually were established and social customs usually observed before any sign of business occurred. A notable exception, she discovered, occurred among the smaller, more entrepreneurial Japanese firms, where employees often displayed a more Western predisposition toward getting right down to business.

She also discovered that Japanese business is dominated by men more than in her own culture, and that there tends to be more separation of men and women in social contexts. Although this cultural feature occasionally frustrated her, she tried to focus on understanding behavior rather than judging it from her own perspective. Moreover, she knew Japan is making changes in the role of women. Indeed, her own considerable success in getting business for M-Global suggested that Japanese value ability and hard work most of all.

Like Sarah Logan, you should enter every intercultural experience with a mind open to learning about those with whom you will work. Adjust your communication strategies so that you have the best chance of succeeding in the international marketplace. Intercultural awareness does *not* require that you jettison your own ethics, customs, or standards; instead, it provides you with a wonderful opportunity to learn about, empathize with, and show respect for the views of others.

Communicating Internationally

This section includes guidelines for writing and designing English-language documents so that multinational readers can understand and translate them more easily.

When writing documents for other cultures, remember that your work will not be read in the cultural context in which it was written. For that matter, you may lose control of the document altogether if it is translated into a language that you do not know. In order to help solve this problem, organizations such as Intecom and the AeroSpace and Defence Industries Association of Europe have worked to develop and promote Simplified English, also known as Controlled English. The goal of Simplified English is to eliminate ambiguity, improve translation, and make reading English easier for non-native English speakers. Following are some basic guidelines to reduce the risk of misunderstanding:

1. **Simplify grammar and style rules.** It is best to write in clear language—with relatively simple syntax and short sentences—so that ideas cannot be misunderstood.

2. **Use simple verb tenses and verb constructions.** For example, constructions like gerunds and the progressive can have multiple meanings, and some languages don't have an equivalent to the passive voice.

3. **Limit vocabulary to words with clear meanings.** Avoid compound words or confusing phrases. The European Association of Aerospace Industries (AECMA) identifies a list of approved words. AECMA's specifications can be found at www.simplifiedenglish-aecma.org/Simplified_English.htm.

4. **Use language and terminology consistently.** Always use the same word to describe the same thing.

5. **Define technical terms.** All good technical writing includes well-defined terminology, but this feature is especially important in international writing. A glossary remains an effective tool for helping international readers.

6. **Avoid slang terms and idioms.** A non-native speaker or someone from outside the United States may be unfamiliar with phrases you use every day. The ever-popular sports metaphors such as "ballpark estimate," "hitting a home run," and "let's punt on this" present obvious obstacles for some readers. Use phrasing that requires little cultural context.

7. **Include visuals.** Graphics are a universal language that allows readers entry into the meaning of your document, even if they have difficulty with the text.

>>> Ethics on the Job

This section outlines the ethical context in which all workers do their jobs. The goal is (1) to present one main ethical principle and four related guidelines for the workplace, and (2) to show how ethical guidelines can be applied to a specific activity—writing. At the end of this chapter and throughout the book are assignments in which your own ethical decisions play an important role.

Ethical Guidelines for Work

As with your personal life, your professional life holds many opportunities for demonstrating your views of what is right or wrong. There is no way to escape these ethical challenges. Most occur daily and without much fanfare, but cumulatively they compose our personal approach to morality. Thus our belief systems, or lack thereof, are revealed by the manner in which we respond to this continuous barrage of ethical dilemmas.

Obviously, not everyone in the same organization—let alone the same industry or profession—has the same ethical beliefs, nor should they. After all, each person's understanding of right and wrong flows from individual experiences, upbringing, religious beliefs, and cultural values. Some *ethical relativists* even argue that ethics only makes sense as a descriptive study of what people do believe, not a prescriptive study of what they should believe. Yet there are some basic ethical guidelines that, in our view, should be part of the decision-making process in every organization. These guidelines apply to small employers, just as they apply to large multinational organizations. Although they may be displayed in different ways in different cultures, they should transcend national identity, cultural background, and family beliefs. In other words, these guidelines represent what, ideally, should be the *core* values for employees at international companies.

The guidelines in this section flow from one main tenet that Peter Singer calls the principle of Equal Consideration of Interests (ECI)[3]:

ECI: Make judgments and act in ways that treat the interests and well-being of others as no less important than your own.

Note that the ECI principle resembles similar principles espoused by religions and philosophies worldwide, a fact that makes ECI especially useful as a bedrock principle

---

[3]This definition is a slightly condensed form of the one in Singer (1979). Practical Ethics. Cambridge, England: Cambridge Univ. Press, as included in Raymond S. Pfieffer and Ralph P. Forsberg (1993). *Ethics on the Job* (pp. 12–17). Belmont, CA: Wadsworth.

for multinational organizations. Now let's examine four guidelines that flow from this principle.

>> Ethics Guideline 1: Be Honest

First, you should relate information accurately and on time—to your colleagues, to customers, and to outside parties, such as government regulators. This guideline also means you should not mislead listeners or readers by leaving out important information that relates to a situation, product, or service. In other words, give those with whom you communicate the same information that you would want presented to you.

This guideline does not prescribe the manner or form in which information will be delivered. Indeed, issues such as format, organization, and presentation change from culture to culture; the need for accurate and timely information, however, does not.

>> Ethics Guideline 2: Do No Harm

The most healthy, productive, and enjoyable workplaces are those with a positive and constructive atmosphere. One way to achieve such a working environment is to avoid words or actions calculated to harm others. For example, avoid negative, rumor-laden conversations that hurt feelings, spread unsupported information, or waste time.

Of course, different cultures and countries differ in the degree to which personal and familiar chatting takes place at work, but this cultural difference does not change the fact that you should consider the impact words and deeds have on colleagues, clients, and competitors. Our ideal goal should be to make the working world a better place at the end of each day; however, a minimum goal should be to leave the world at least as good as we found it.

>> Ethics Guideline 3: Keep Your Commitments

People expect that you will keep your word. Be careful about the commitments you make to superiors, subordinates, customers, and others. When you do make a legitimate commitment, follow through on it. *Legitimate* means a commitment wherein you do not violate other ethical guidelines, such as those of being honest and doing no harm.

Cultural guidelines differ on exactly what makes up a commitment, and you should be sensitive to such variations. In one country, a comment in a meeting may seal an agreement from the perspective of some participants. In another, multiple legal contracts are required. Make sure you know what constitutes a commitment with your audience, and then make sure you abide by it.

>> Ethics Guideline 4: Be Independent

Make no mistake; teamwork is crucial to the success of organizations for which you work. Group efforts, however, do not relieve you from the personal responsibility for making decisions that come with your job and then accepting the resulting blame or credit. We cannot simply "go along with" group decisions if we have ethical reservations. This phenomenon of excessive conformity within teams has its own name—*groupthink*—and it can be dangerous. Effective teamwork is more important than ever, but it is only as valid as the strength of individual contributors.

If you do business in non-Western countries, such as Japan, you will learn that some cultures emphasize teamwork to the point where individuals seem to be absorbed into the fabric of the group. If you are part of such teams, it may be necessary to alter your style to adapt to a pattern of decision making that is highly consensual. Yet such cultural adjustments don't change the importance of being assertive, when appropriate, and taking individual responsibility. Although you should be cooperative in teamwork, you cannot sacrifice your own values on the altar of group consensus. Be clear and direct, while still listening and adjusting to others' views.

Now let's examine the manner in which ethical considerations play a part in the *writing* responsibilities at companies such as M-Global, Inc.

Ethics and Legal Issues in Writing

In your career, you should develop and apply your own code of ethics, making certain it follows the four guidelines already noted. Writing—whether on paper, audiotape, videotape, or computer screen—presents a special ethical challenge for demonstrating your personal code of ethics. Along with speaking, there may be no more important way you display your beliefs during your career. The following section (1) lists some ethical questions related to specific documents and (2) provides responses based on the ethical guidelines noted earlier.

Being honest, doing no harm, keeping commitments, and showing independence of thought and action—all four of these ethical guidelines apply to written communication. Following are some typical examples from the working world of M-Global, followed by some discussion of your legal obligations in writing.

>> Ethics Questions in M-Global Writing

Each of the six pairs that follow presents an ethical dilemma regarding a specific document, followed by an answer to each problem.

Lab report: Should you mention a small, possibly insignificant percentage of the data that was collected but that doesn't support your conclusions?

    Answer: Yes. Readers deserve to see all the data, even (and perhaps especially) any information that doesn't support your conclusion. They need a true picture of the lab study so that they can draw their own conclusions.

Trip report: Should you mention the fact that one client you visited expressed dissatisfaction with the service he received from your team?

    Answer: Yes. Assuming that your report was supposed to present an accurate reflection of your activities, your reader deserves to hear about all your client contacts—good news and bad news. You can counter any critical comments by indicating how your team plans to remedy the problem.

Proposal: Should you include cost information, even though cost is not a strong point in your proposal?

> **Answer:** Yes. Most clients expect complete and clear cost data in a proposal. It is best to be forthright about costs, even if they are not your selling point. Then you can highlight features that are exemplary about your firm so that the customer is encouraged to look beyond costs to matters of quality, qualifications, scheduling, experience, and so on.

Feasibility study: Should you list all the criteria you used in comparing three products, even though one criterion could not be applied adequately in your study?

> **Answer:** Yes. It is unethical to adjust criteria after the fact to accommodate your inability to apply them consistently. Besides, information about a project dead end may be useful to the reader.

Technical article: Should you acknowledge ideas you derived from another article, even though you quoted no information from the piece?

> **Answer:** Yes. Your reliance on all borrowed ideas should be noted, whether the ideas are quoted, paraphrased, or summarized. The exception is common knowledge, which is general information that is found in many sources. Such common knowledge need not be footnoted.

Statement of qualifications (SOQ): Should you feel obligated to mention technical areas in which your firm does not have extensive experience?

> **Answer:** Probably not, as long as you believe the customer is not expecting such information in the statement of qualifications. Ethical guidelines do not require you to tell everything about your firm, especially in a marketing document like an SOQ (Statement of Qualifications). They require only that you provide the information that the client requests or expects.

Of course, many other types of technical writing require careful ethical evaluation. You might even consider performing an ethical review during the final process of drafting a document. Other parts of this book cover topics that apply to specific stages of such an ethical review, as well as to ethics in spoken communication. For ethics in definition and description, see chapter 6; for ethics in instructions and process explanations, see chapter 7; for ethics in the use of graphics, see chapter 12; for ethics in the research process, see chapter 15; and for ethics in negotiation, see chapter 16.

>> Legal Issues in Writing

Some countries, such as the United States, have a fairly well-developed legal context for writing, which means you must pay great attention to detail as you apply ethical principles to the writing process. This section highlights some common guidelines.

■ Acknowledge Sources for Information Other Than Common Knowledge

As noted in the technical article example in the previous section, you are obligated to provide sources for any information other than common knowledge. *Common knowledge* is usually considered to be factual and nonjudgmental information that could be found in general sources about a subject. The sources for any other types of information beyond this definition must be cited in your document. Chapter 15 offers more detail about plagiarism and the format for citing sources.

- **Seek Written Permission Before Borrowing Extensive Text** Generally, it is best to seek written permission for borrowing more than a few hundred words from a source, especially if the purpose of your document is profit. This so-called fair use is, unfortunately, not clearly defined and subject to varying interpretations. It is best to (1) consult a reference librarian or other expert for an up-to-date interpretation of the application of fair use to your situation; and (2) err on the side of conservatism by asking permission to use information, if you have any doubt. This probably hasn't been an issue in papers you have written for school, because they were for educational use and were not going to be published. However, this issue should be addressed in any writing you do outside of school.

- **Seek Written Permission Before Borrowing Graphics** Again, you probably haven't been concerned about this issue in projects you have created in school, but you must seek permission for any graphics you borrow for projects created outside of school. This guideline applies to any nontextual element, whether it is borrowed directly from the original or adapted by you from the source. Even if the graphic is not copyrighted, such as in an annual report from a city or county, you should seek permission for its use.

- **Seek Legal Advice When You Cannot Resolve Complex Questions** Some questions, such as the use of trademarks and copyright, fall far outside the expertise of most of us. In such cases it is best to consult an attorney who specializes in such law. Remember that the phrase *Ignorance is bliss* has led many a writer into problems that could have been prevented by seeking advice when it was relatively cheap—at the beginning. Concerning U.S. copyrights in particular, you might first want to consult free information provided by the U.S. Copyright Office at its website (www.copyright.gov).

In the final analysis, acting ethically on the job means thinking constantly about the way in which people are influenced by what you do, say, and write. Peter Singer's ECI principle embodies this approach perfectly: "Make judgments and act in ways that treat the interests and well-being of others as no less important than your own." These "others" can include your colleagues, your customers, your employers, or the general public. Also, remember that what you write could have a very long shelf life, perhaps to be used later as a reference for legal proceedings. Always write as if your professional reputation could depend on it, because it just might.

>>> Introducing M-Global

Thus far, this chapter has examined three topics related to most organizations: corporate culture, globalism, and ethics. Now let's look at a fictional firm—M-Global, Inc.—that deals with these topics on a daily basis. Although the rest of this chapter is mainly intended for courses in which the instructor chooses to use the fictional M-Global context for class assignments, it may be useful to all students who want a window into organizations where they may soon work.

Today, M-Global, Inc., is trying hard to develop a company culture based on a concern for quality, intercultural awareness, and ethics. The firm's management believes such an effort is crucial to the success of the firm. This section takes a detailed look at M-Global, first with a brief overview of its history and then with a description of its major project types.

History of M-Global, Inc.

Just out of Georgia Tech in 1959, Rob McDuff spent several years as a civil engineer in the U.S. Army Corps of Engineers. Then in 1963, he started a small engineering consulting firm in Baltimore, Maryland, where he grew up. This firm's specialty was doing consulting work for construction firms and real-estate developers. Specifically, McDuff, Inc., tested soils and then recommended foundation designs for structures that were being proposed.

In the early days, Rob McDuff did much of the fieldwork himself. He also analyzed the data, wrote reports, and did the marketing for new business. Work progressed well, and his new company earned a reputation for high-quality service. Now, Rob McDuff nostalgically looks back on those days as some of the most satisfying of his career. He seized the opportunity to fulfill a dream that many people still have today: starting a business and then using skill, hard work, and imagination to make it grow.

From its founding in 1963 until about 1967, the company worked mostly for construction firms in the Baltimore area. Most projects included tasks such as drilling soil borings at a site, testing soil samples back in the lab, analyzing the data, and writing a report that gave construction recommendations. The work was not glamorous, but it provided an important service. By the late 1960s, McDuff, Inc., enjoyed a first-rate reputation, with offices in Baltimore and Boston and about 80 employees.

McDuff, Inc., kept growing steadily, with a large spurt in the mid-1970s and another in the 1980s. The first was tied to increased oil exploration in all parts of the world. Oil firms needed experts to test soils, especially in offshore areas. The results of these projects were used to position oil rigs at locations where they could withstand rough seas. The second growth period was tied to environmental work required by the federal government, state agencies, and private firms. McDuff became a major player in the waste-management business, consulting with clients about ways to store or clean up hazardous waste. The third growth period has moved the firm into diverse service industries, such as security systems, hotel management, and landscaping.

In 2008, Rob McDuff announced that he was retiring and turning over the company to his son, Jim, who would become president. With the change in management, Rob McDuff announced a name change to reflect their more diversified, global scope, and McDuff, Inc. became M-Global, Inc. Although engineering and environmental services remain important to the company, it plans to expand its other operations and emphasize the company's role in the global marketplace.

Today, after 50 years of business, M-Global, Inc., has about 2,500 employees. There are nine offices in the United States and seven offices overseas, as well as a corporate headquarters in Baltimore that is separate from the Baltimore branch office. M-Global performs a wide variety of work. What started as a technical consulting engineering firm has expanded into a firm that does both technical and nontechnical work for a variety of customers.

Projects

Every company must improve its products and services to stay in business, and M-Global is no exception. If it had stayed just with soils testing work, the company would be stagnant today. Periodic slowdowns in the construction and oil industries would have taken their toll. Fortunately, the company diversified. Following are its nine main project areas today:

1. **Soils work on land:** These projects involve making design recommendations for foundations and other parts of office buildings, dams, factories, subdivisions, reservoirs, and mass-transit systems. When done well, such work helps prevent later problems, like cracks in building walls.

2. **Soils work at sea:** This geological and engineering work used to be done exclusively for oil and gas companies. It helped them place offshore platforms at safe locations or select drilling locations with the best chance of hitting oil. Now, however, M-Global is also hired by countries and states that want to preserve the ecologically sensitive offshore environment. By collecting and analyzing data from its ship, the *Dolphin*, M-Global helps clients decide whether an offshore area should be preserved or developed.

3. **Construction monitoring:** Besides designing parts of structures, M-Global also helps observe construction. Following are some services it offers during the construction process:

■ Checking the quality of concrete being poured into such structures as cooling towers for nuclear power plants

■ Watching construction workers to make sure they follow proper procedures

■ Testing the strength of concrete and other foundation materials once they are put in place

4. **Construction management:** About 15 years ago, M-Global got into the business of actually supervising projects other than its own jobs. Large construction companies hire M-Global to orchestrate all parts of a project so that it is completed on time. The work involves the following activities:

■ Establishing a schedule

■ Observing the work of subcontractors

■ Regularly informing contractors about job progress

5. **Environmental management:** In the late 1970s, Rob McDuff began to realize that garbage—all kinds of it—could mean big business for his firm. Suddenly, the United States and other countries faced major problems caused by the volume of current

wastes and by improper disposal of wastes. As M-Global's fastest growing market, environmental management work can involve one or more of these tasks:

- Testing surface soil and water for toxic wastes
- Drilling borings to see if surface pollution has filtered into the groundwater
- Designing a cleanup plan
- Supervising the cleanup
- Predicting the impact of proposed projects on the environment
- Analyzing the current environmental health of wetlands, beaches, national forests, lakes, and other areas

As Rob McDuff had hoped, managing wastes and determining the environmental impact of proposed projects proved to be excellent markets. Although there is growing competition, the company got into the business early enough to establish a good reputation for reliable and affordable work.

6. **Equipment development:** Here the firm departs from its traditional emphasis on services and instead produces products. The ED team, as it is known, designs and builds specialized equipment, both for M-Global's own project needs and for its clients. For example, it is building mechanisms as diverse as a prototype for a new device to test water pollution levels on the one hand, and a new instrument gauge to install in tractors on the other.

7. **Document development:** To support its equipment development, M-Global has put together a Publications Development team (known in-house as The Pub). This team was originally established to create documentation for the equipment designed by the ED team, but it grew to house the writing of proposals and RFPs (Requests for Proposals). Members of The Pub are assigned to project teams throughout the company, often participating in the earliest stages of development of project design. Recently, the Publications Development team began offering documentation services to clients, creating on-line and print documentation and helping clients set up content management systems.

8. **Training:** M-Global entered the training business about five years ago, when it realized that there was a good market for technical training in skills represented by the firm. Recently, the Training Department also started offering nontechnical training in areas such as report writing, because the company employs several writers who are excellent trainers.

9. **Miscellaneous service industries:** Once it achieved growth in fields clearly related to its original mission, M-Global began seeing opportunities for starting or buying out companies that provide services related only indirectly to civil engineering. The three most prominent examples are corporate and residential landscaping, security (both systems and staff), and hotel management. These businesses have grown rapidly and created a more diverse group of employees at M-Global; however, the firm's management is careful about such acquisitions, for it does not want to enter markets it does not understand.

These nine project areas reflect M-Global's diversity. Although starting as a traditional engineering firm, M-Global sought out new markets and became a scrappy competitor in many areas. As he looks back on his company's first 40 years, Rob McDuff likes to think that the entrepreneurial spirit thrives in this company he started out of his basement in the 1960s.

>>> Corporate and Branch Offices

Headquarters

M-Global, Inc., has 16 branch offices and a corporate headquarters. Although not a large company by international standards, it has become well known within its own fields. The company operates as a kind of loose confederation. Each office enjoys a good measure of independence. Yet some corporate structure is required for these purposes:

1. To coordinate projects that involve employees from several offices
2. To prevent duplication of the same work at different offices
3. To ensure fairness, consistency, and quality in the handling of human resource issues throughout the firm (salaries, benefits, workload, etc.)

The corporate office gives special attention to problems related to international communications. Among its non-U.S. clients and employees, it must respond to differences in cultures and ways of doing business. This effort can mean the difference between success and failure in negotiating deals, completing projects, hiring employees, and so forth.

The corporate office in Baltimore is housed in a building across the street from the Baltimore branch office. This separation is important because it keeps the mostly "overhead" functions of the corporate office distinct from the mostly profit-generating functions of the branches. Also, the physical separation is symbolic to offices outside of Baltimore, which already suspect that the large Baltimore branch office receives special treatment from corporate headquarters. Figure 2–2 shows an organization chart for this office. What follows is a brief description of the responsibilities of each service group in the corporate office.

>> Service 1: Human Resources

The Human Resources Department performs mostly personnel-related tasks. Its main work covers these four fields:

1. **Employment:** The office handles job advertisements, ensures that branch offices follow government guidelines that apply to hiring, gives legal advice on workers' compensation, and visits college campuses to recruit prospective graduates.

2. **Benefits:** Two staff members are responsible for handling company benefits. They send information to employees, check the accuracy of benefit deductions from paychecks, stay current about the newest benefits available for M-Global employees, and make recommendations to the corporate staff about benefit changes.

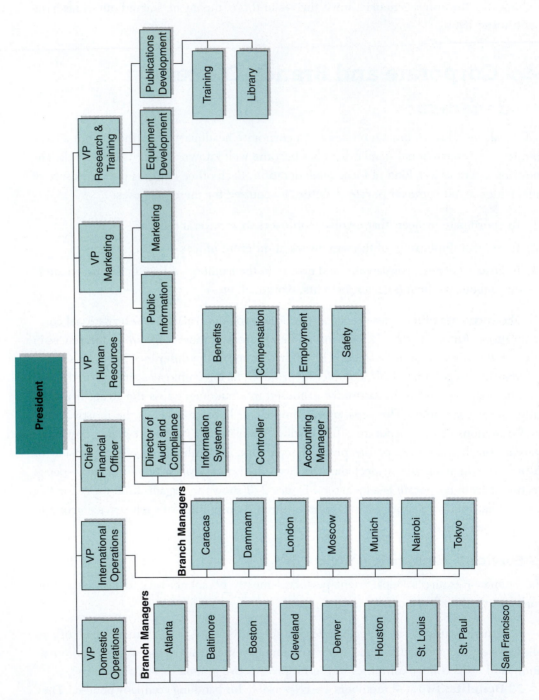

■ **Figure 2–2** ■ M-Global's corporate office organization

3. **Safety:** Given the company's interest in waste management, safety is crucial. The company hired a manager of safety in 1980 to complete the following tasks:

- Educate employees about the importance of safe work practices
- Provide proper equipment and training
- Visit job sites to make safety checks
- Respond to questions by government agencies

4. **Compensation:** The compensation expert has three main duties:

- Monitor salaries at all offices to ensure some degree of uniformity within the job classifications
- Research salary guidelines in all professions represented in the firm to make sure the company's salaries are competitive
- Monitor branch offices to make sure performance evaluation interviews are done each year for each employee, before salary decisions are made

>> Service 2: Financial Services

The Financial Services Division is responsible for

- General accounting
- Budgeting
- Internal and external financial reporting to management, investors, and regulators

>> Service 3: Project Coordination

With 16 offices spread over a wide geographic area, the company sometimes falls into the trap of the left hand not knowing what the right hand is doing. Specifically, individual offices may not know what project resources exist in another office. If a large project requires experts from several offices, the corporate office can assemble the team. To make this system work, the office keeps an accurate record of current projects at each office.

>> Service 4: Marketing

By working closely with clients, the engineers and scientists at every branch help to secure repeat work from current clients. Yet their technical responsibilities keep them from spending much time on marketing for new clients. The corporate office, however, has a marketing and proposal-writing staff that works extensively on seeking new business. Also, it helps branch offices write proposals requested by current clients.

>> Service 5: Information Services and Technology

The corporate office houses the company's mainframe computers. Each branch office has terminals tied into the mainframes, giving them direct access to corporate databases in Baltimore through M-Global's company intranet. All employees can reach each other through email. Of course, each branch also uses its own stand-alone computer systems for word processing and some other functions.

>> Service 6: Research

At M-Global, the term *research* covers the services of the Equipment Development (ED) Team and the Publications Development Team (The Pub). The ED lab and Publications Development office are housed in the corporate office; they are the only teams in the building that could be considered profit generating. The ED Team also retains the important overhead function of designing and building tools and other mechanisms that M-Global employees use on their projects. Publications Development creates documentation for those tools and mechanisms, but it also documents M-Global's in-house procedures and its team members participate in proposal writing teams. The Publications Development office also includes the M-Global library, where two full-time librarians maintain a modest corporate collection that includes the following items:

- Copies of all company reports and proposals
- More than 100 technical, business, and general periodicals
- Many reference works in technology and the sciences

Many of these company documents are available through M-Global's intranet; the rest can be sent the same day they are requested—by email, phone, or fax. In addition, each branch office usually keeps a small reference and periodical collection for its own use.

>> Service 7: Training

As noted earlier, M-Global now performs two types of training: in-house courses for its own employees, and external training for clients who need training in a number of technical and nontechnical subjects in which M-Global is proficient. The Training Department directs these efforts and helps M-Global employees find useful outside training or college courses.

Branches

Each M-Global branch is unique in its particular combination of technical and nontechnical positions, but all 16 branches include a common management structure, as shown in Figure 2–3. A branch manager, who reports to one of two corporate vice presidents, supervises a team of four or more department managers. These managers, in turn, supervise the technical and nontechnical employees at the branch.

Branch positions beneath the manager can be grouped into four categories:

1. Technical professional
2. Nontechnical professional
3. Technical staff
4. Nontechnical staff

Figure 2–4 through Figure 2–7 list some of the positions in these four groups. Although all these employees are under the supervision of their respective branch managers, some interact closely with employees at the corporate level. For example, each human resources manager reports to his or her respective branch manager and works closely with the corporate vice president of human resources.

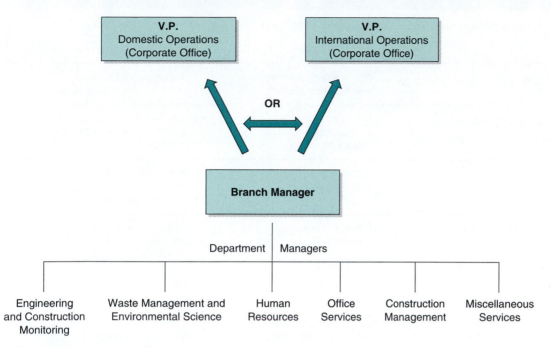

■ **Figure 2–3** ■ Branch office: Typical management structure

Position	Typical Education (Minimum)	Main Duties
1. Department Manager	• B.S. in engineering or science • M.S. in engineering or science or M.B.A.	• Oversees entire technical department in engineering or science
2. Project Manager	• B.S. in engineering or science or engineering technology	• Oversees entire projects in engineering, waste management, construction, etc.
3. Research Engineer (Equipment Development)	• B.S. in engineering or design • M.S. or Ph.D. in engineering or design	• Designs new tools, mechanisms, or other equipment at ED lab at corporate office
4. Field Engineer	• B.S. in engineering or engineering technology	• Completes site work for projects and then completes remaining work at office
5. Field Scientist	• B.S. in biology, chemistry, environmental science, etc.	• Completes site work for hazardous waste projects and then completes remaining work back at office
6. Landscape Architect	• B.A. or B.S. in landscape architecture	• Designs and oversees construction of landscape plans

■ **Figure 2–4** ■ Sample positions: Technical professionals

Position	Typical Education (Minimum)	Main Duties
1. Office Services Manager	• B.S. or B.A. in business	• Oversees accounting, word processing, purchasing, physical plant, etc.
2. Human Resources Manager	• B.S./B.A. in liberal arts or in human resources	• Oversees benefits, safety, employment, compensation
3. Technical Writer	• B.S./B.A. in technical communication or in liberal arts	• Helps write and edit reports, proposals, and other branch documents
4. Marketing Specialist	• B.S./B.A. in business or in liberal arts	• Writes to and visits potential clients • Helps with proposals
5. Training Specialist	• B.S./B.A. in education or in liberal arts	• Works with corporate office to plan in-house training and external training for clients
6. Security Manager	• B.S. in criminal justice	• Oversees design and implementation of a security plan
7. Hotel Manager	• B.S. in business or hotel management	• Manages hotel property

■ **Figure 2–5** ■ Sample positions: Nontechnical professionals

Position	Typical Education (Minimum)	Main Duties
1. Field/Lab Technician	• Vo-tech or associate's degree (in technical field)	• Recovers samples from site • Completes lab tests
2. Computer Operator	• Vo-tech or associate's degree (in technical field)	• Inputs data
3. Field Hand	• High school diploma	• Operates and maintains equipment • Orders and picks up supplies
4. Research Technician	• Vo-tech or associate's degree (in technical field)	• Assists research engineers in work at the Equipment Development Lab
5. Warehouse Supervisor	• Vo-tech or associate's degree (in technical field)	• Keeps track of and maintains equipment in a branch office

■ **Figure 2–6** ■ Sample positions: Technical staff

Position	Typical Education (Minimum)	Main Duties
1. Secretary	• High school diploma or associate's degree	• Handles paperwork for professional workers • Does some typing • Has some client contact
2. Receptionist	• High school diploma	• Oversees all of switchboard operation
3. Library Assistant	• High school diploma	• Helps librarian with cataloging, ordering books, etc.
4. Training Assistant	• High school diploma	• Helps orchestrate training activities of all kinds

■ **Figure 2–7** ■ Sample positions: Nontechnical staff

Most M-Global positions designated *professional* (Figure 2–4 and Figure 2–5) require bachelor's degrees or higher in an appropriate field. Positions designated *staff* require at least a high-school education and sometimes more (such as a certificate program, vocational–technical training, or 2-year college degree). Although these categories apply just to our fictional firm of M-Global, some may resemble jobs that exist in a real organization where you now work or will work in the future.

Now that you have viewed M-Global's structure, you must be made aware of a continuing management problem at the company—one common to many firms with widely spread offices and a corporate office. Branch employees often question whether the corporate people really understand and respond to their needs. Indeed, it is easy to feel misunderstood and even neglected when you work hundreds or thousands of miles from the company's hub. Effective written and spoken communication can go a long way toward bridging the gaps between M-Global's local and central offices.

>>> Writing at M-Global

Good writing is crucial to M-Global's existence. First, one of its main products is a written report. After the company completes a technical project, the project report stands as a permanent statement about, and reflection of, the quality of M-Global's work. Second, many company projects result from written proposals. Third, most routine activity within the firm is preceded or followed by memos, reports, in-house proposals, and manuals. As an employee at M-Global, you would be writing to readers in the following groups:

■ Superiors at your own M-Global branch

■ Subordinates at your M-Global branch

■ Employees at other branches or at the corporate office

■ Clients

■ Subcontractors and vendors

As pointed out in chapter 1, you often write to a mixed group of readers, all with different needs and backgrounds. Similarly, at M-Global, readers of the same document could come from more than one of the groups listed. For example, assume M-Global's corporate training manager must send a memo to 20 employees officially confirming their attendance at an upcoming training seminar at the corporate office. Coming to Baltimore from all domestic offices, these employees need a seminar schedule as well as information about the course. Copies of the memo must be sent to (1) the participants' managers, who must be reminded that they will be minus an employee for three days; (2) the vice president for research and training, who likes to be made aware of any company-wide training; and (3) a training assistant, who must make room arrangements for the seminar.

This memo is not unique. Most documents at M-Global and other companies are read by persons from different levels. What follows are more examples of M-Global writing directed to diverse readers. Some of these projects resemble the examples and assignments in later chapters.

Examples of Internal Writing

1. Memo about changes in benefits—from a manager of human resources at a branch to all employees at that branch

2. Memo about changes in procedures for removing asbestos from buildings—from a project manager to field engineers and technicians

3. Orientation booklet on M-Global—from the manager of employment to all new employees at the firm

4. Internal proposal for funds to develop a new piece of equipment—from a technician to the Equipment Development lab manager

5. Draft of a project report—written by a project manager for review by a department manager (before being submitted to the client)

6. Long report on future markets for M-Global—from the vice president of business and marketing to all 2,500 M-Global employees

7. Memo on a new procedure for compensating domestic employees who work on overseas projects—from the corporate manager of compensation to all branch managers

8. Manual on new accounting procedures—from the accounting manager to all branch managers

9. Article on an interesting environmental project at a national park—from a project manager to all employees who read the company's monthly newsletter

10. Trip report on a professional conference—from a biologist in the environmental science area to the corporate manager of training

Examples of External Writing

1. Sales letter—from a marketing specialist to a potential client

2. M-Global brochure describing technical services—from a marketing specialist to a potential client

3. Proposal—from a branch manager to a potential client

4. Progress report—from a project director to a client

5. Final project report—from a project director to a client

6. Refresher letter—from a marketing specialist to a previous client

7. Complaint letter—from a warehouse supervisor to a supplier

8. Article on technical subject—from a civil engineer for a technical periodical

9. Training manual—from a document writing team for client

10. Report on job site safety—from a compliance specialist to a government agency

>>> Chapter Summary

This book uses the fictional firm of M-Global, Inc., to lend realism to your study of technical writing. The many M-Global examples and assignments give you a purpose, an audience, and an organizational context that simulate what you will face in your career.

As with other organizations where you might work, M-Global has developed its own personality, or *culture,* which can be influenced by many features, including its history, type of business, and management style. Because organizations must be responsive to the business climate, many organizations today are paying attention to the global environment.

Another major concern at M-Global—and at all organizations—is ethical behavior in the workplace. Companies and their employees should follow some basic ethical guidelines in all their work, including communication with colleagues and customers.

This chapter looks specifically at the culture of M-Global, Inc. Although started as an engineering consulting firm with a narrow focus, M-Global is now an international company with 2,500 employees, 16 branches, and a corporate office. The firm works in nine main project areas: soils engineering on land, soils engineering at sea, construction monitoring, construction management, waste management (environmental science), equipment development, document development, training, and service industries. The project sheets at the end of the chapter give summary information about specific M-Global projects in all nine areas.

M-Global's corporate office is in Baltimore, Maryland. It helps the branch offices in the areas of human resources, financial services, project coordination, marketing, information technology, research, and training. Each of the 16 offices is run by a branch manager. At each branch, employees are grouped into four categories: technical professionals, nontechnical professionals, technical staff, and nontechnical staff.

M-Global employees at all levels do a good deal of writing, both to superiors and subordinates within the organization and to clients and other outside readers. Documents often have multiple readers with different backgrounds, making writing even more challenging. M-Global employees who meet this challenge have the best chance of doing valuable work for the company and succeeding in their careers.

>>> Learning Portfolio

Communication Challenge "Employee Orientation Manual: Global Dilemmas"

When McDuff became M-Global, management decided to emphasize the global nature of the organization by creating a global company culture. To help achieve this goal, Human Resources has been asked to create an employee orientation manual that will be given to employees at all 16 branches. M-Global has a Human Resources Manual that is quite specific about issues such as work hours (whether regular or flextime), vacation time, office dress, required training, and safety, and the goal is to introduce new employees to these policies.

Assume the role of a new employee in the Human Resources office who has been assigned the task of gathering and analyzing all of the current materials used for employee orientation. Although some branches in the United States share orientation materials, others, especially those in more isolated offices, such as Tokyo and Nairobi, have their own materials. Some smaller offices have no formal materials, relying instead on informal introductions to the organization's culture.

This case study explains M-Global's approach to global issues, and ends with questions and comments for discussion and an assignment for a written response to the Challenge.

Global Issues in Human Resource Policies

Because some countries have specific laws governing vacations and holidays, some orientation materials are much more specific than others. However, some policies and practices that you might take for granted can, in fact, be problematic; for example, it is important to remember that not only is the Dammam, Saudi Arabia, office in a different time zone, but the Saudi workweek is Saturday through Wednesday.

One problem you do not have to worry about is reading the orientation materials—McDuff has always had a policy that all internal documents would be written in English, and it plans to keep this policy as M-Global. However, some branch managers have taken the opportunity presented by this new project to complain about the English-only policy. They see no reason why they cannot write internal memos, reports, procedures, and other documents in the language of the country in which the office is located. Although most M-Global employees have a fair reading and writing knowledge of English, there is the issue of pride at work, and some employees at lower levels have weak English skills. Moreover, these branch managers argue, if M-Global is going embrace multiple cultures, why shouldn't it embrace multiple languages?

Questions and Comments for Discussion

Answer the following questions from your own point of view. Before doing so, however, make sure you have carefully considered the perspective of the home office and the branch managers.

1. Is M-Global's English-only policy justified? Is there any compromise that would satisfy the branch managers and the executive management?
2. Elaborate on some of the general language problems multinational firms can face.
3. The use of English does not by itself break down communication barriers with colleagues and customers at global firms—that is, English is spoken around the world by people from many different cultures. Its use does not mean that people necessarily think, write, or speak by the same conventions. Examine this view. Putting aside obvious dialect and vocabulary differences, how can one's culture and national background affect the use of English in writing and speaking?
4. Some of the communication problems at M-Global are the result of having branches in both low-context and high-context cultures. What differences in work rules might you expect in each type of culture? How can the conflicting and confusing work rules be addressed? Give your opinion on the degree to which common work rules and practices are important at M-Global's domestic offices, as well as at its international branches.
5. Identify the most important issues that should be addressed in the orientation manual. If you were a new employee, what would you want to know about the organization that you had just joined?

Write About It

As the new employee in Human Resources, write a memo to the Vice President of Human Resources, Karrie Camp, that identifies the key sections that you think should be included in the *Welcome to M-Global* handbook. (Karrie has been with M-Global for 30 years and is serious about the "resources" part of Human Resources. She believes that ensuring that employees work efficiently and effectively for the good of M-Global is an important part of her job.) Using the Internet, see what information you can find about paid leave and holidays, management styles, and general business practices in the countries where M-Global has branches. Then identify the problem areas that must be addressed before the handbook can be completed.

Collaboration at Work Defining Culture

General Instructions

Each "Collaboration at Work" exercise applies strategies for working in teams to chapter topics. The exercise assumes you (1) have been divided into teams of about three to six students, (2) will use time inside or outside of class to complete the case, and (3) will produce an oral or written response. For guidelines about writing in teams, refer to pages 24–28.

Background for Assignment

Assume you and team members work for a communication consulting firm hired by your school. Your task is to improve communications at your institution—both external communication (e.g., to prospective students, prospective employees, and the community) and internal communication (e.g., between and among current students, faculty, and administrators).

Before your team can begin to develop an action plan, you would like to describe the current culture of the school. (See the definition of *corporate culture* on page 36 of this chapter.) When applied to a college or university, the term *culture* might include some of the following features:

- History of the school
- Type of institution and variety of academic programs
- Typical background of students
- Academic structure
- Types of interaction among faculty and staff
- Enrollment patterns
- Extracurricular life on campus
- Relationship with community outside the school

Team Assignment

Your team will choose one or more features of the school's culture to describe. (Alternatively, your instructor may assign specific features to each team, such that the combined descriptions of all teams presents a composite of the school's culture.) The members of your team should (1) discuss a plan for developing the description, (2) collect information in ways that seem most appropriate (e.g., through interviews, from written documents, from discussion among your team members), and (3) assemble information into a cohesive response. Your main goal is to produce an objective observation, not to argue a point.

Assignments

1. Company Profile

Having read the information in this chapter about M-Global, conduct your own profile of a multinational company in your region. Collect information from such sources as corporate annual reports, newspaper or magazine articles, or personal contacts. Consider some or all of the following subtopics: company history, types of projects, corporate structure, common types of writing produced, and special features of the company (such as an international market or workforce). Your instructor will indicate whether your report should be presented orally or in writing.

Assignments 2, 3, 4, and 5 can be completed either as individual exercises or as team projects, depending on the directions of your instructor. Prepare a response that can be delivered as an oral presentation for discussion in class.

Analyze the context of each case by considering what you learned in chapter 1 about the context of technical writing *and* what you learned in this chapter about M-Global. In particular, answer the following questions:

- What is the purpose of the document to be written?
- What result do you hope to achieve by writing it?
- Who are your readers and what do they want from your document?
- What method of organization is most useful?
- What tone and choice of language are most effective?

2. Analysis: Memo Changing Supplies Policy

As the office services manager at the St. Louis office, you have a problem. In the past fiscal year, the office has used significantly more bond paper, computer paper, pens, mechanical pencils, eraser fluid, and file folders than in previous fiscal years. After going back through the year's projects, you can find no business-related reason why the office bought $12,000 more of these items. Given that everyone has easy access to the supplies, you have concluded that some employees are taking them home. Putting the best face on it, you assume they may be "borrowing" supplies to complete company business they take home with them, and then just keeping items at home. Putting the worst face on it, you wonder whether some employees are stealing from the company.

After consulting with the branch manager and some other managers, you decide to restrict access to office supplies. Starting next month, these supplies must be signed out through secretaries in the various departments. First, you plan to meet with the secretaries to explain how to make the system work. Then on the following day, you will send a memo explaining the change to all employees.

Would you change your approach in this memo if it were to be sent to a specific audience in the office? Why or why not? Answer this question with regard to the four employee groups shown in Figure 2–4 ("Technical Professionals"), Figure 2–5 ("Nontechnical Professionals"), Figure 2–6 ("Technical Staff"), and Figure 2–7 ("Nontechnical Staff").

3. Analysis: Letter Requesting Testimonials

As a writer in the corporate marketing department, you spend a good deal of your time preparing materials to be used in sales letters, brochures, and company proposals. Yesterday, you were assigned the task of asking 20 customers if they will write *testimonial letters* about their satisfaction with M-Global's work. In all cases, these clients used M-Global for many projects and expressed satisfaction informally with the work. Now you are going to ask them to express their satisfaction in the form of a letter, which M-Global could use as a testimonial to secure other business.

Your strategy is to write a personalized form letter to the 20 clients, and then follow it up with phone calls.

4. Analysis: Memo on Inventory Control

For five years, you supervised the supply warehouse at the Houston office. Your main job is to maintain equipment and see that it is returned after jobs are completed. When checking out equipment, each project manager is supposed to fill out part of a project equipment form that lists all equipment used on the job and the date of checkout. When returning the equipment, the project manager should complete the form by listing the date of return and any damage, no matter how small, that must be repaired before the equipment is used again. This equipment ranges from front-end loaders and pickup trucks to simple tools like hammers, wrenches, and power drills.

Lately, you have noticed that many forms you receive are incomplete. In particular, project managers are failing to record fully any equipment damage that occurred on the job. For example, if someone fails to report that a truck's alignment is out, the truck will not be in acceptable shape for the next project for which it could be used.

Your oral comments to project managers have not done much good. Apparently, the project managers do not take the warehouse problem seriously, so you believe it is time to put your concerns in writing. The goal is to inform all technical professionals who manage projects that from now on the form must be filled out correctly. You have no authority, as such, over the managers; however, you know that their

bosses would be very concerned about this problem if you chose to bring it to his or her attention.

At this point, you have decided to ask nicely one more time—this time in writing. You want your memo to emphasize issues of safety and profitability, as well as the need to follow a procedure that has helped you maintain a first-rate warehouse.

5. Analysis: Memo Report on Flextime

As branch manager of the Atlanta office, you have always tried to give employees as much flexibility as possible in their jobs—as long as the jobs got done. Recently, you have had many requests to adopt flextime. In this arrangement, the office would end its standard 8:00 A.M. to 4:30 P.M. workday (with a half-hour lunch break). Instead, each employee would fit her or his eight-hour day within the following framework: 7:00 A.M. to 8:30 A.M. arrival, a half hour or full hour for lunch, and 3:30 P.M. to 5:30 P.M. departure.

Two conditions will prevail if flextime is adopted. First, each employee's supervisor must agree on the hours chosen, because the supervisor must ensure that departmental responsibilities are covered. Second, each employee must "lock in" a specific flextime schedule until another is negotiated with the supervisor. In other words, an employee's hours will not change from day to day.

Before you spend any more time considering this change, you want the views of the employees. You decide to write a short memo report that (1) explains the changes being considered and the conditions (see previous paragraph); (2) solicits their views in writing, by a certain date; and (3) asks what particular work hours they prefer, if given the choice. Also, you want your short report to indicate that later there may be department meetings and finally a general office meeting on the subject, depending on the degree of interest expressed by employees in their memos to you.

⑦ 6. Ethics Assignment

For this assignment your instructor will place you into a team, with the goal of presenting an oral or written report.

Option A   The Society for Technical Communicators (STC) is the main U.S. professional association for technical communicators. Its ethical guidelines that follow are intended both for those who are permanent employees of organizations and also for communicators who work as consultants and contractors. Evaluate the quality, usefulness, and appropriateness of these guidelines by answering the following questions:

a. What do the guidelines suggest about the role of technical communicators in the workplace?

STC Ethical Guidelines for Technical Communicators

Introduction

As technical communicators, we observe the following ethical guidelines in our professional activities. Their purpose is to help us maintain ethical practices.

Legality

We observe the laws and regulations governing our professional activities in the workplace. We meet the terms and obligations of projects we undertake. We ensure that all terms of our contractual agreements are consistent with the STC Ethical Guidelines.

Honesty

We seek to promote the public good in our activities. To the best of our ability, we provide truthful and accurate communications. We dedicate ourselves to conciseness, clarity, coherence, and creativity, striving to address the needs of those who use our products. We alert our clients and employers when we believe material is ambiguous. Before using another person's work, we obtain permission. In cases where individuals are credited, we attribute authorship only to those who have made an original, substantive contribution. We do not perform work outside our job scope during hours compensated by clients or employers, except with their permission; nor do we use their facilities, equipment, or supplies for personal gain. When we advertise our services, we do so truthfully.

Confidentiality

Respecting the confidentiality of our clients, employers, and professional organizations, we disclose business-sensitive information only with their consent or when legally required. We acquire releases from clients and employers before including their business-sensitive information in our portfolios or before using such material for a different client or employer or for demonstration purposes.

Quality

With the goal of producing high quality work, we negotiate realistic, candid agreements on the schedule, budget, and deliverables with clients and employers in the initial project planning stage. When working on the project, we fulfill our negotiated roles in a timely, responsible manner and meet the stated expectations.

Fairness

We respect cultural variety and other aspects or diversity in our clients, employers, development teams, and audiences. We serve the business interests of our clients and employers, as long as such loyalty does not require us to violate the public good. We avoid conflicts of interest in the fulfillment of our professional responsibilities and activities. If we are aware of a conflict of interest, we disclose it to those concerned and obtain their approval before proceeding.

Professionalism

We seek candid evaluations of our performance from clients and employers. We also provide candid evaluations of communication products and services. We advance the technical communication profession through our integrity, standards, and performance.

(Approved by the STC Board of Directors, April 1995)

b. How would you adjust the depth, breadth, or balance of the items presented, if at all?

c. How does the document satisfy, or fail to satisfy, the ethical guidelines discussed in this chapter, such as the ECI principle?

d. Are all guidelines and terms clear to the reader?

e. How might the role of the U.S. technical communication professional, as described in the guidelines, differ from the role of technical communicators in several other cultures outside the United States?

Option B   Your team is to investigate the ethical climate in one or more organizations that are in the same type of business. You may decide to (a) collect organization codes of ethics, (b) do research on ethical guidelines issued by professional associations to which the organizations belong, (c) interview employees about ethical decisions they face on the job, or (d) read any available information on ethics related to the companies or profession.

Option C   For this option, your team selects (or is assigned) one of the eight project sheets at the end of this chapter. Perform a brainstorming session in your team by which you arrive at numerous potential ethical dilemmas related to your project. For example, you might consider some of these concerns: (a) decisions to be made by and about employees on the job, (b) technical questions related to the project, (c) interaction with clients, and (d) communication with any parties or agencies that are not directly connected with the project but that may be influenced by it.

 ### 7. International Communication Assignment

Refer to the 10 questions in "The Global Workplace" section of this chapter. Using them as the basis for your investigation, conduct your own research project on the cultural features of employees of a specific country. Consider using some or all of the following sources: campus library, travel agencies, consulate offices, international students' office on your campus, or individuals who have worked in or visited the country. Your instructor will indicate whether your report should be presented orally or in writing.

 ### 8. A.C.T. N.O.W Assignment (Applying Communication To Nurture Our World)

Select a topic of importance to the local or regional community where you live or you attend college. It should be one that aims to improve the culture, environment, or general livability of the area. In addition, the topic must be one about which you will be able to gather facts or opinions with relative ease from a newspaper or local library. After some preliminary research, interview two individuals to solicit their views on the topic, and then write an essay in which you (a) objectively describe the two points of view of the individuals you interviewed, (b) analyze the degree to which you believe the two opinions satisfy the Equal Consideration of Interests (ECI) Principle, and (c) give your own opinion on the topic.

THE FOLLOWING PAGES PROVIDE A GLIMPSE into the world of M-Global, Inc. In addition to the map opposite, the pages contain information about seven specific projects completed by M-Global.

Each page includes an illustration and project overview. Assignments throughout the book ask you to make use of information from the project sheets.

Worldwide Locations of M-Global, Inc., Offices

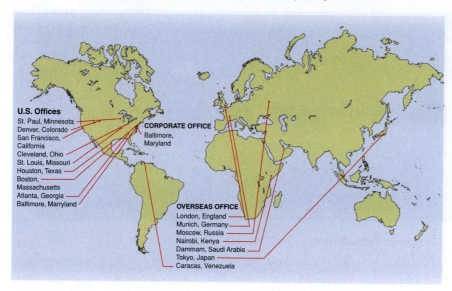

U.S. Offices
St. Paul, Minnesota
Denver, Colorado
San Francisco, California
Cleveland, Ohio
St. Louis, Missouri
Houston, Texas
Boston, Massachusetts
Atlanta, Georgia
Baltimore, Marryland

CORPORATE OFFICE
Baltimore, Maryland

OVERSEAS OFFICE
London, England
Munich, Germany
Moscow, Russia
Nairobi, Kenya
Dammam, Saudi Arabia
Tokyo, Japan
Caracas, Venezuela

U.S. Locations

1. Corporate headquarters—Baltimore, Maryland
2. Baltimore, Maryland
3. Boston, Massachusetts
4. Atlanta, Georgia
5. Houston, Texas
6. Cleveland, Ohio
7. St. Paul, Minnesota
8. St. Louis, Missouri
9. Denver, Colorado
10. San Francisco, California

Non-U.S. Locations

1. Caracas, Venezuela
2. London, England
3. Moscow, Russia
4. Munich, Germany
5. Nairobi, Kenya
6. Dammam, Saudi Arabia
7. Tokyo, Japan

M-Global Inc | 127 Rainbow Lane | Baltimore MD 21202 | 410.555.8175

PROJECT 1: Sentry Dam
CLIENT: Sanborne County Water Authority

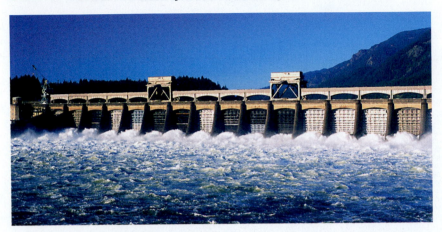

Brief Project Description

M-Global worked from January through April of 2005 on field and laboratory work preceding construction of Sentry Dam. After submitting its geologic and engineering report, the M-Global team worked with the water district on final dam design. It also gave some help during construction. The dam was completed in July 2007. Because Sentry is a high-hazard dam—meaning that its failure would cause loss of life—safety was crucial.

Main Technical Tasks

- Drilled 20 soil borings at the site to sample soil and rock
- Drilled 15 test wells to find the depth to groundwater and to check on water seepage in the dam area
- Lab-tested the soil, shale, and other samples from borings to evaluate the strength of material on which the dam would rest
- Designed an overflow spillway that would be anchored in strong bedrock, not weak shale
- Monitored water seepage in the foundation and dam during construction
- Certified the dam's safety after construction

Main Findings or Benefits

- Completed field and lab work on schedule and at budget
- Designed innovative concrete dam spillway that bypassed weak shale and connected with strong bedrock

M-Global Inc | 127 Rainbow Lane | Baltimore MD 21202 | 410.555.8175

PROJECT 2: Completed Ocean Exploration Program
CLIENT: Republic of Cameroon

Brief Project Description

During the spring of 2004, M-Global used its drillship Dolphin to examine the ocean floor over a 10-mile stretch off the Cameroon coast. M-Global collected data on-site and then tested and analyzed samples at its labs. After sending the report on the study, M-Global met with the client concerning conclusions and recommendations for further offshore use of the coastline.

Main Technical Tasks

- Kept Dolphin on-site for two months to map the seafloor, to drill borings, and to observe ocean habitats
- Used sonar to develop a profile on the surface of the ocean floor and its near-surface geology (return time of sound waves helped gauge the depth to the floor and to sediments below the floor)
- Drilled successfully for samples from Dolphin's drilling platform, often in difficult weather
- Analyzed samples from the borings to estimate geologic age and stability of the ocean floor
- Viewed ocean life and geology firsthand at some locations, using a small submersible craft with a one-person crew

Main Findings or Benefits

- Concluded that most of the zone was too environmentally and geologically sensitive to be used for offshore drilling of oil and gas
- Found two locations where a pipeline might be safely placed, with minimum damage to ocean life and minimum risk of geologic disturbance (such as earthquake or ocean avalanche)

M-Global Inc | 127 Rainbow Lane | Baltimore MD 21202 | 410.555.8175

PROJECT 3: **Monitored Construction of General Hospital**
CLIENT: **Floor County, Florida**

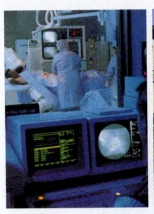

Brief Project Description

M-Global was hired to observe construction of General Hospital to make sure that all construction work was done according to agreed-on standards and legal specifications. M-Global had one or more employees on site continuously during the entire construction process.

Main Technical Tasks

- Checked quality of masonry, steel, and other materials used in the building's foundation and main structure
- Ascertained that heating, plumbing, and other systems were installed according to legal codes and according to contract
- Served as liaison between medical staff and construction personnel so that interior construction was done correctly
- Completed final "sign-off" for entire facility before it opened

Main Findings or Benefits

- Guaranteed client that all materials and procedures used in construction were up to contract standards
- Spent over 50 extra hours (without charge) consulting with doctors, nurses, and other technical staff about interior construction and placement of major pieces of equipment
- Billed client at 10% below proposed fee because of other work recently contracted with the county for the same calendar year

PROJECT 4: **Managed Construction of Nevada Gold Dome**

CLIENT: **City of Rondo, Nevada**

Brief Project Description

Rondo, Nevada, recently received permission to start a new football team, the Nevada Gamblers. The city is building a domed stadium, called the Gold Dome, for the new franchise. Because of concerns about scheduling the various construction firms and subcontractors involved in the project, the client hired M-Global to manage the entire project—from groundbreaking to occupancy.

Main Technical Tasks

- Established construction schedule that would allow for maximum overlapping work by different subcontractors
- Held daily coordination meetings with a group of principals that represented every contractor on site that day
- Handled schedule delays by dealing immediately with contractors and subcontractors
- Used M-Global's patented "TimeTrack" scheduling software to stay aware of the entire flow of work at site

Main Findings or Benefits

- Had entire project completed two weeks before deadline
- Saved client $100,000 by eliminating proposed overtime work that was not needed once scheduling was fine-tuned
- Arranged schedule so that team owners and staff members could complete a walkthrough of the facility at key points during the construction process

M-Global Inc | 127 Rainbow Lane | Baltimore MD 21202 | 410.555.8175

PROJECT 5: Examined Big Bluff Salt Marsh
CLIENT: State of Georgia

Brief Project Description

Georgia hired M-Global to explore the environmental quality of the Big Bluff Salt Marsh, located on Paradise Island off the Atlantic coast. The island is owned by the state. Developers have approached the state about buying island land for building condominiums and other tourist-related structures. M-Global's field-work, lab work, and research resulted in a report about the level of development that the marsh can tolerate.

Main Technical Tasks

- Took inventory of wildlife and grasses throughout the marsh
- Consulted with wetlands experts around the U.S. about notable features of Big Bluff Salt Marsh
- Tested soil and water for current levels of pollution
- Researched two other Atlantic coast salt marshes where development has occurred to determine compatibility of development and marshes

Main Findings or Benefits

- Concluded that Big Bluff Salt Marsh serves as a nursery and feeding ground for fish caught commercially along the coast
- Learned from Environmental Protection Agency (EPA) that any major devel-opment of Big Bluff Salt Marsh could be a violation of federal wetlands policy and thus could be challenged by the EPA
- Recommended that the state sell land next to the marsh only if it would be used for low-impact activities—such as day trips by visitors—and not for construction of homes and businesses

M-Global Inc | 127 Rainbow Lane | Baltimore MD 21202 | 410.555.8175

PROJECT 6: Designed and Installed Control Panel for Nuclear Plant
CLIENT: Russian Government

Brief Project Description

In 2002, M-Global's safety experts and mechanical engineers were hired to design a new control panel and to retrofit it into an existing plant. The panel was designed, manufactured, installed, and tested by M-Global—with the help of several subcontractors.

Main Technical Tasks

- Spent one week at site observing operators using old panel
- Hired ergonometric and nuclear power experts to help evaluate old panel design and to suggest features of new design
- Designed and manufactured panel
- Installed panel at Russian plant and observed one full week of testing, when panel was used at plant under simulated conditions
- Remained on-site for three days after full power was resumed in order to continue training operators on use of new panel

Main Findings or Benefits

- Designed panel that international experts considered to be as safe as any currently in use
- Stayed on schedule, keeping the plant out of use only two weeks

M-Global Inc | 127 Rainbow Lane | Baltimore MD 21202 | 410.555.8175

PROJECT 7: Designed and Taught Seminar in Technical Writing
CLIENT: Government of Germany

Brief Project Description

The government of Germany has greatly increased the number of technical experts in departments in the City of Bonn. M-Global's Munich office was selected to design and teach an in-house technical writing seminar for 20 mid-level government employees in Bonn. They work in agriculture, health, and engineering. Although in different fields, the individuals write the same types of reports.

Main Technical Tasks

- Met with members of the seminar and their managers to determine needs of the group
- Examined sample reports from all participants
- Studied the government's style guidelines
- Designed and taught a three-day seminar, using a manual of guidelines and samples tailored for the group
- Evaluated actual on-the-job reports written by participants after the seminar

Main Findings or Benefits

- Received "very good" to "excellent" ratings from all 20 participants on the written critiques completed on the last day of the course
- Wrote a final report to client that documented improvement shown in participants' reports written after the seminar, as compared with those written before the seminar

M-Global Inc | 127 Rainbow Lane | Baltimore MD 21202 | 410.555.8175

PROJECT 8 : **Designed and Created Documentation of Data Security Procedures**
CLIENT: **Kansas Department of Social and Health Services**

Brief Project Description

In response to public concerns about the security of private data, the Kansas Department of Social and Health Services undertook a systematic documentation of all security protocols for personal information. Using the recommendations of an Information Systems Audit, M-Global created on-line and print documentation of computer security procedures.

Main Technical Tasks

- Identified procedures to be documented
- Designed information architecture for procedural documentation
- Created on-line help files to be used by computer operators
- Created print-format guide to data security procedures

Main Findings or Benefits

- Assisted in meeting public expectations of privacy of confidential information
- New documentation contributed to improved security rating in follow-up audit
- Recognized by Kansans for Security and Privacy for contributions to security of state records.

M-Global Inc | 127 Rainbow Lane | Baltimore MD 21202 | 410.555.8175

Chapter | 3 | Organizing Information

>>> **Chapter Outline**

Importance of Organization   75

Three Principles of Organization   76

ABC Format for Documents   80

    Document Abstract: The "Big Picture"
      for Decision Makers   81

    Document Body: Details for All Readers   82

    Document Conclusion: Wrap-Up Leading to Next Step   83

Tips for Organizing Sections and Paragraphs   83

    Common Patterns of Organization   84

    Document Sections   87

    Paragraphs   87

Modular Writing   88

Chapter Summary   89

Learning Portfolio   90

    Communication Challenge   90

    Collaboration at Work   92

    Assignments   92

Models for Good Writing   97

    Model 3–1: ABC format in whole document   97

    Model 3–2: ABC format in document section   99

    Model 3–3: ABC format in paragraphs   100

Tom Kent asks the department secretary to hold his calls. Closing his door, he reaches for the report draft written by one of his staff members and sits down to read it. As an M-Global manager for 10 years, he has reviewed and signed off on every major report written by members of his department. Of all the problems that plague the drafts he reads, poor organization bothers him the most.

This problem is especially annoying at the beginning of a document and the beginning of individual sections. Sometimes he has no idea where the writer is going. His people don't seem to understand that they are supposed to be "telling a story," even in a technical report. Grammar and style errors are annoying to him, but organization problems are much more troublesome. They require extensive rewriting and time-consuming meetings with the report writer. Reaching for his red pen, Tom hopes for the best as he begins to read yet another report.

You, too, will face internal reviewers like Tom Kent when you write on the job. To help you avoid organization problems, this chapter offers strategies for organizing information as you plan, draft, and revise your writing. It builds on the discussion of the three stages of writing covered in chapter 1. Then chapter 4 completes your introduction to technical communication by showing you how to use effective page design to keep readers' attention.

>>> Importance of Organization

Poorly organized documents cost time and money at all stages of the writing process. A writer who hasn't planned the structure of a document wastes valuable time trying to decide what to include, how to divide information into meaningful sections, and how the document will focus on the reader's needs. A supervisor who receives a poorly organized document knows that he won't be able to send it on, but must ask the writer to spend more time revising it. Finally, a poorly organized document wastes the time of readers because they must wade through pages of information, trying to make sense of what they are reading. If they can't, they will probably set it to one side, unread, as a waste of their time.

As you learned in chapter 1, your documents will be read by varied readers with diverse technical backgrounds. Chapter 2 displays this technical range within M-Global and refers to an even broader technical spectrum among M-Global's clients. Given this reader diversity, this chapter aims to answer one essential question: How can you best organize information to satisfy so many different people?

Figure 3–1 shows you three possible options for organizing information for the technical expertise of a mixed technical audience, but only one is recommended in this book.

Experts	Operators	Managers	General Readers
Option A Organize information for technical readers			
		Option B Organize information for less-technical readers	
Option C Organize information for all readers			

■ **Figure 3–1** ■
Options for organizing information

Some writers, usually those with technical backgrounds themselves, choose option A. They direct their writing to the most technical people. Other writers choose option B. They respond to the dilemma of a mixed technical audience by finding the lowest common denominator—that is, they write to the level of the least technical person. Each option satisfies one segment of readers at the expense of the others.

Option C is preferred in technical writing for mixed readers. It encourages you to organize documents so that all readers—both technical and nontechnical—get what they need. The rest of this chapter provides strategies for developing this option. It describes general principles of organization and guidelines for organizing entire documents, individual document sections, and paragraphs.

>>> Three Principles of Organization

Good organization starts with careful analysis of your audience. Most readers are busy, and they skip around as they read. Think about how you examine a news organization's website or a weekly newsmagazine. You are likely to take a quick look at articles of special interest to you; then you might read them more thoroughly, if there is time. That approach also resembles how your audience treats technical reports and other work-related documents. If important points are buried in long paragraphs or sections, busy readers may miss them. Three principles respond realistically to the needs of your readers:

>> Principle 1: Write Different Parts for Different Readers

The longer the document, the less likely it is that anyone will read it from beginning to end. As shown in Figure 3–2, they use a *speed-read approach* that includes these steps:

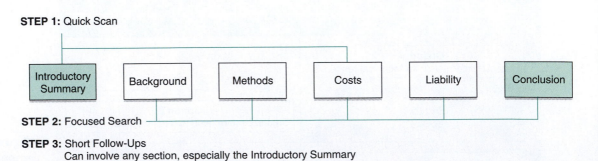

STEP 1: Quick Scan

| Introductory Summary | Background | Methods | Costs | Liability | Conclusion |

STEP 2: Focused Search

STEP 3: Short Follow-Ups
Can involve any section, especially the Introductory Summary

■ **Figure 3–2** ■ Sample speed-read approach to short proposal

Step 1: **Quick scan.** Readers scan easy-to-read sections like executive summaries, introductory summaries, introductions, tables of contents, conclusions, and recommendations. They pay special attention to beginning and ending sections, especially in documents longer than a page or two, and to illustrations.

Step 2: **Focused search.** Readers go directly to parts of the document body that give them what they need at the moment. To find information quickly, they search for navigation devices like subheadings, listings, and white space in margins to guide their reading. (See chapter 4 for a discussion of page design.)

Step 3: **Short follow-ups.** Readers return to the document, when time permits, to read or reread important sections.

Your job is to write in a way that responds to this nonlinear and episodic reading process of your audience. Most important, you should direct each section to those in the audience most likely to read that particular section. Shift the level of technicality as you move from section to section within the document to meet the needs of each section's specific readers. On the one hand, managers and general readers favor less technical language and depend most heavily on overviews at the beginning of documents; on the other hand, experts and operators expect more technical jargon and pay more attention than others to the body sections of documents.

Of course, you walk a thin line in designing different parts of the document for different readers. Although technical language and other stylistic features may change from section to section, your document must hang together as one piece of work. Common threads of organization, theme, and tone must keep it from appearing fragmented or pieced together.

>> Principle 2: Emphasize Beginnings and Endings

Suspense fiction relies on the interest and patience of readers to piece together important information. The writer usually drops hints throughout the narrative before finally revealing who did what to whom. Technical writing operates differently. Busy readers expect to find information in predictable locations without having to search for it. Their first-choice locations for important information are as follows:

- The beginning of the entire document
- The beginnings of report sections
- The beginnings of paragraphs

The reader interest curve in Figure 3–3 reflects this focus on beginnings, but the curve also shows that the readers' second choice for reading is the ends of documents, sections, and paragraphs—that is, most readers tend to remember best the first and last things they read. The ending is a slightly less-desirable location than the beginning because it is less accessible, especially in long sections or documents. Of course, some readers inevitably read the last part of a document first, for they may have the habit of fanning pages when first seeing a document. Thus, although there is no guarantee that the first document section will be read first, you can be fairly sure that either the beginning or the ending gets first attention.

■ **Figure 3–3** ■
Reader interest
curve

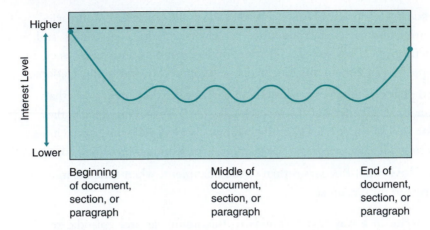

■ **Figure 3–3** ■
Reader interest
curve

Emphasizing beginnings and endings responds to the reading habits and psychological needs of readers. At the beginning, they want to know where you're heading. They need a simple road map for the rest of the passage. In fact, if you don't provide something important at the beginnings of paragraphs, sections, and documents, readers will start guessing the main point themselves. It is in your best interest to direct the reader to what *you* consider most important in what they are about to read, rather than to encourage them to guess at the importance of the passage. At the ending, readers expect some sort of wrap-up or transition; your writing shouldn't simply drop off. The following paragraph begins and ends with such information (italics added):

> *Already depleted sea turtle, marine mammal, seabird, and noncommercial fish populations are endangered by incidental capture in fishing gear.* Worldwide, about 25 percent of the catch is discarded, either because it is not commercially valuable or because of regulatory requirements that prohibit keeping undersized or nontargeted marine life. Destructive fishing practices, such as bottom trawling and dredging, are damaging vital habitat upon which fish and other living resources depend. *Taken together, overfishing, bycatch, and habitat destruction are changing relationships among species in food webs and altering the function of marine ecosystems.* [Pew Oceans Commission. (2003). *America's Living Oceans: Charting a Course for Sea Change.* Arlington, VA: Pew Oceans Commission, 5, 9.]

The first sentence gives readers an immediate impression of the two topics to be covered in the paragraph. The paragraph body explores details of both topics. Then the last sentence flows smoothly from the paragraph body by reinforcing the main point about over fishing.

Why does this top-down pattern, which seems so logical from the reader's perspective, frequently get ignored in technical writing? The answer comes from the difference between the way you complete your research or fieldwork and the way busy readers expect results of your work to be conveyed in a report, as shown in Figure 3–4. Having moved logically from data to conclusions and recommendations in technical work, many writers assume they should take this same approach in their report. They reason that the reader wants and needs all the supporting details before being confronted with conclusions and recommendations that result from these data.

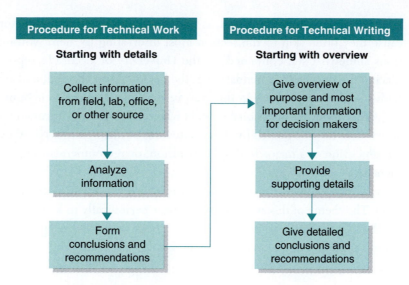

Such reasoning is wrong. Readers want the results placed first, followed by details that support your main points. Of course, you must be careful not to give detailed conclusions and recommendations at the beginning; most readers want and expect only a brief summary. This overview provides a framework within which readers can place the details presented later. In other words, readers of technical documents want the "whodunit" answer at the beginning. Recall the motto in chapter 1: *Write for your reader, not for yourself.* Now you can see that this rule governs the manner in which you organize information in everything you write.

>> Principle 3: Repeat Key Points

You have learned that different people focus on different sections of a document. Sometimes no one carefully reads the entire report. For example, managers may have time to read only the summary, whereas technical experts may skip the leadoff sections and go directly to "meaty" technical sections with supporting information. These varied reading patterns require a *redundant* approach to organization—you must repeat important information in different sections for different readers.

For example, assume you are an M-Global employee in Denver and are writing a report to the University of Colorado on choosing sites for several athletic fields. Having examined five alternatives, your report recommends one site for final consideration. Your 25-page report compares and contrasts all five alternatives according to criteria of land cost, nearness to other athletic locations, and relative difficulty of grading the site and building the required facilities. Given this context, where should your recommendation appear in the report? Following are five likely spots:

1. Executive summary

2. Cost section in the body

3. Location section in the body

4. Grading/construction section of the body

5. Concluding section

Our assumption, you recall, is that few readers move straight through a report. Because they often skip to the section most interesting to them, you must make main sections somewhat self-contained. In the University of Colorado report, that would mean placing the main recommendation at the beginning, at the end, and at one or more points within each main section. In this way, readers of all sections encounter your main point.

What about the occasional readers who read all the way through your report, word for word? Will they be put off by the restatement of main points? No, they won't. Your strategic repetition of a major finding, conclusion, or recommendation gives helpful reinforcement to readers always searching for an answer to the "So what?" question as they read. Fiction and nonfiction may be alike in this respect—writers of both genres are telling a story. The theme of this story must reappear periodically to keep readers on track.

Now we're ready to be more specific about how the three general principles of organization apply to documents, document sections, and paragraphs.

>>> ABC Format for Documents

You have learned the three principles of organization: (1) write different parts of the document for different readers, (2) emphasize beginnings and endings, and (3) repeat key points. Now let's move from principles to practice. We next develop an all-purpose pattern of organization for writing entire documents. (The next major section covers document sections and paragraphs.)

Technical documents should assume a three-part structure that consists of a beginning, a middle, and an end. This book labels this structure the *ABC Format* (for *Abstract*, *Body*, and *Conclusion*). Visually, think of this pattern as a three-part diamond structure, as shown in Figure 3–5:

- **Abstract:** A brief beginning component is represented by the narrow top of the diamond, which leads into the body.

- **Body:** The longer middle component is represented by the broad, expansive portion of the diamond figure.

- **Conclusion:** A brief ending component is represented by the narrow bottom of the diamond, which leads away from the body.

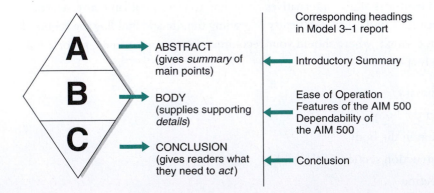

■ **Figure 3–5** ■ ABC format for all documents

Model 3–1 (pp. 97–98) includes a memo report that conforms to this structure. The following sections discuss the three ABC components in detail.

Document Abstract: The "Big Picture" for Decision Makers

Every document should begin with an overview. As used in this text, *abstract* is defined as follows:

> **Abstract:** brief summary of a document's main points. Although its makeup varies with the type and length of the document, an abstract usually includes (1) a clear purpose statement for the document, (2) the most important points for decision makers, and (3) a list or description of the main sections that follow the abstract. As a capsule version of the entire document, the abstract should answer readers' typical mental questions, such as: "How does this document concern me?" "What's the bottom line?" "So what?"

Abstract information is provided under different headings, depending on the document's length and degree of formality. Some common headings are "Summary," "Executive Summary," "Introductory Summary," "Overview," and "Introduction." The abstract may vary in length from a short paragraph to a page or so. Its purpose, however, is always the same: to provide decision makers with highlights of the document.

For example, assume you are an engineer who has evaluated environmental hazards for the potential purchaser of a shopping mall site. Here is how the summary might read:

> As you requested, we have examined the possibility of environmental contamination at the site being considered for the new Klinesburg Mall. Our field exploration revealed two locations with deposits of household trash, which can be easily cleaned up. Another spot has a more serious deposit problem of 10 barrels of industrial waste. However, our inspection of the containers and soil tests revealed no leaks.
>
> Given these limited observations and tests, we conclude that the site poses no major environmental risks and recommend development of the mall. The rest of this report details our field activities, test analyses, conclusions, and recommendations.

You have provided the reader with a purpose for the report, an overview of important information for decision-makers, and a reference to the four sections that follow. In so doing, you have answered the following questions, among others, for the readers:

- What are the major risks at the site?
- Are these risks great enough to warrant not buying the land?
- What major sections does the rest of the report contain?

This general abstract, or overview, is mainly for decision-makers. Highlights must be brief, yet free of any possible misunderstanding. On some occasions, you may need to state that further clarification is included in the text, even though that point may seem obvious. For example, if your report concerns matters of safety, the overview may not be detailed enough to prevent or eliminate risks. In this case, state this point clearly so that the reader will not misunderstand or exaggerate the purpose of the abstract.

Later chapters in this book contain guidelines for writing the following specific types of abstracts:

- Introductory summaries for short reports (chapter 8)
- Executive summaries for formal reports (chapter 9)
- Introductory summaries for short proposals (chapter 10)
- Executive summaries for formal proposals (chapter 10)
- Abstracts of technical articles (chapter 15)

Document Body: Details for All Readers

The longest part of any document is the body. As used in this book, the *body* is defined as follows:

> **Body:** the middle section(s) of the document providing supporting information to readers, especially those with a technical background. Unlike the abstract and conclusion, the body component allows you to write expansively about items such as (1) the background of the project, (2) field, lab, office, or any other work on which the document is based, and (3) details of any conclusions, recommendations, or proposals that might be highlighted at the beginning or end of the document. The body answers this main reader question: "What support is there for points put forth in the abstract at the beginning of the document?"

Managers may read much of the body, especially if they have a technical background and if the document is short. Yet the more likely readers are technical specialists who (1) verify technical information for the decision makers or (2) use your document to do their jobs. In writing the body, use the following guidelines:

- **Separate fact from opinion.** Never leave the reader confused about where opinions begin and end. Body sections usually move from facts to opinions that are based on facts. To make the distinction clear, preface opinions with phrases such as, "We believe that," "I feel that," "It is our opinion that," and the like. Such wording gives a clear signal to readers that you are presenting judgments, conclusions, and other nonfactual statements. Also, you can reinforce the facts by including data in graphics.

- **Adopt a format that reveals much structure.** Use frequent headings and subheadings to help busy readers locate important information immediately. (Chapter 4 covers these and other elements of page design.)

- **Use graphics whenever possible.** Use graphics to draw attention to important points. Today more than ever, readers expect visual reinforcement of your text, particularly in more persuasive documents like proposals. (Chapter 12 deals with graphical elements in technical documents.)

By following these guidelines, which apply to any document, you will make detailed body sections as readable as possible. They keep ideas from becoming buried in text and show readers what to do with the information they find.

Document Conclusion: Wrap-Up Leading to Next Step

Your conclusion deserves special attention, for readers often recall first what they have read last. We define the *conclusion* component as follows:

Conclusion: the final section(s) of the document bringing readers—especially decision makers—back to one or more central points already mentioned in the body. Occasionally, it may include one or more points not previously mentioned. In any case, the conclusion provides closure to the document and often leads to the next step in the writer's relationship with the reader.

The conclusion component may have any one of several headings, depending on the type and length of the document. Possibilities include "Conclusion," "Closing," "Closing Remarks," and "Conclusions and Recommendations." Chapters 8, 9, and 10 in Part 2 of this text describe the options for short and long documents of many kinds. In general, however, a conclusion component answers the following types of questions:

- What major points have you made?
- What problem have you tried to solve?
- What should the reader do next?
- What will you do next?
- What single idea do you want to leave with the reader?

Because readers focus on beginnings and endings of documents, you want to exploit the opportunity to drive home your message—just as you did in the abstract. Format can greatly affect the impact you make on decision makers. Although specific formats vary, most conclusions take one of these two forms:

- **Listings:** The listing format is especially useful when pulling together points mentioned throughout the document. Whereas the abstract often gives readers the big picture in narrative format, the conclusion may instead depend on listings of findings, conclusions, or recommendations. (Chapter 4 gives suggestions on using bulleted and numbered listings.)
- **Summary paragraph(s):** When a listing is not appropriate, you may want to write a concluding paragraph or two. Here you can leave readers with an important piece of information and make clear the next step to be taken.

Whichever alternative you choose, your goal is to return to the main concerns of the most important readers—decision makers. Both the abstract and conclusion, in slightly different ways, should respond to the needs of this primary audience.

>>> Tips for Organizing Sections and Paragraphs

First and foremost, the ABC format pertains to the organization of entire documents. Yet the same beginning-middle-end strategy applies to the smaller units of discourse— document sections and paragraphs. In fact, you can view the entire document as a series of interlocking units, each responding to reader expectations as viewed on the reader interest curve in Figure 3–3.

Common Patterns of Organization

As you are planning your writing projects, you may find it useful to use familiar patterns of organization to arrange the sections, and even paragraphs, of longer documents. Once you have clearly identified how information in the sections of your document will be organized, it will be easier for you to develop your ideas and to make the connections between your ideas clear. Commonly recognized patterns of organization also help your readers to find the information that they need and to understand that information.

There are several organizational patterns that you may already be familiar with from your previous writing classes. You should use the pattern that is most appropriate to the purpose of your document and to the topic that you are writing about.

Sequence

Documents that emphasize a sequence are usually organized chronologically or spatially. Chronological documents, such as instructions and process explanations, identify steps or stages and show how they are related in time. Documents that are organized spatially, such as technical descriptions, identify the parts of an object and show how they are related physically. For example, descriptions of machinery may be organized from front to back, top to bottom, or from the outside in. Whether a document uses time or space as an organizing principle, a sequential pattern moves from one section of the subject to the next in a clear, linear system.

Guidelines for sequential organization

- Identify clearly the steps or parts of the process or object you are describing.
- Follow a clear sequence from one step or part to the next.
- Use signals, whether transition words, bullets, or numbers, to help your reader identify the individual steps or parts.
- Make sure that connections between the steps or parts are clear.

Classification

Classification helps your reader make sense out of diverse but related items. In technical documents, you often group lists of items into categories. For example, a report on a department's activities may group them by client or project. Even a resume groups experience and knowledge into classifications such as *skills* and *education*.

Guidelines for classification

- Identify clearly the basis for your groupings.
- Create equivalent groupings.

- Avoid groups that overlap.
- Carefully classify the items based on the criteria and groups that you have defined.
- Limit the number of groups to something manageable.

Division

Division is often used to identify parts of an object, an organization, or a system. It begins with an entire item that must be broken down or partitioned into its components. Division is often used to describe mechanisms, but it can be applied to a variety of subjects. A company's organizational chart, like the one in Figure 2–2, is an example of division, with the company's personnel and responsibilities clearly divided into units.

Guidelines for division

- Identify clearly the basis for your divisions.
- Create equivalent and parallel divisions.
- Create divisions that include all of the parts, without overlapping.

Comparison/Contrast

Many writing projects obligate you to show similarities or differences between ideas or objects. (For our purposes, the word *comparison* emphasizes similarities, whereas the word *contrast* emphasizes differences.) Although you will emphasize one or the other, you will probably include both similarities and differences in your document. This technique especially applies to situations wherein readers are making buying decisions.

Guidelines for comparison/contrast

- Decide whether you want to emphasize the similarities or differences, and then organize your text so that the emphasis is clear.
- Use the whole-by-whole approach to emphasize similarities or for short comparison/contrasts. This strategy requires that you discuss one item in full, then another item in full, and so on.
- Use the part-by-part approach to emphasize differences or for long comparison/contrasts. This strategy requires that you discuss a parallel part of each item before moving on the next parallel part.
- Set clear standards of comparison and then apply them uniformly.
- Use parallel structure to discuss your sub-points.

General to Specific (or Vice Versa)

Often, documents are organized from general statements to supporting details in a process known as *deductive reasoning*. For example, a description of the function of a department may open with an overview of the department's responsibilities, and then explain how individuals in the department contribute to meeting those responsibilities. At other times, the details must be described first, before the results can be understood. This is known as *inductive reasoning.* For example, an analysis of a bridge failure may identify flaws in specific structural elements before explaining how they led to the failure of an entire section of the bridge.

Guidelines for general-to-specific organization (or vice versa)

- Identify clearly which pattern you are using.
- Use transitions and signal words to guide your reader.
- Conclude with an overview of the whole topic.

Cause and Effect

Workplace writing often examines the causes of events, or predicts the results of an action. For example, a report of an injury incident on a job site may identify unsafe practices that led to the incident and suggest changes that will prevent such incidents from happening in the future.

Guidelines for cause and effect

- Explain the connections in your document clearly. Connections that seem obvious to you may not be obvious to your readers.
- Identify the evidence for your claims clearly.
- Avoid making claims that your evidence doesn't support. Don't be afraid to use qualifiers (like *often* or *most*) when appropriate.

Problem/Solution

A large proportion of writing in organizations is designed to identify and analyze problems and offer solutions. Many companies gain clients through *proposals,* a formal way of presenting solutions to problems.

Guidelines for problem/solution organization

- Identify the problem clearly.
- Explain how you analyzed the problem.
- Make the connections between the problem and your solution clear.
- Explain why your solution is the best alternative under the circumstances.

These patterns can be used in a variety of combinations within a single document. An investigative report about an accident at a plant will probably include a narrative of the accident, organized sequentially; an analysis of the reasons for the accident, with cause-and-effect connections spelled out clearly; and a recommended solution to the problem that caused the original accident. Choose the patterns that help you organize information in the way that will be most useful to your readers.

Document Sections

As mentioned earlier, readers often move from the document abstract to the specific body sections they need to solve their problem or answer their immediate question. Just as they need abstracts and conclusions in the whole document, they need mini-abstracts and brief wrap-ups at the start and finish of each major section.

To see how a section abstract works, we must first understand the dilemma of readers. Refer to Model 3–2 on page 99, which contains one section from a long report. Some readers may read it from beginning to end, but others might not have the time or interest to do so in one sitting. Instead, they would look to a section beginning for an abstract, and then move around within that section at will. Thus the beginning must provide them with a map of what's ahead. Following are the two items that should be part of every section abstract:

1. **Interest grabber:** A sentence or more that captures the attention of the reader. Your grabber may be one sentence or an entire paragraph, depending on the overall length of the document.

2. **Lead-in:** A list, in sentence or bullet format, that indicates main topics to follow in the section. If the section contains subheadings, your lead-in may include the same wording as the subheadings and be in the same order.

The first part of the section gives readers everything they need to read on. First, you get their attention with a grabber. Then you give them an outline of the main points to follow so that they can move to the part of the section that interests them most. As in Model 3–2 on page 99, the section abstract immediately precedes the first subheading when subheads are used.

Sections also should end with some sort of closing thought, rather than just dropping off after the last supporting point has been stated. For example, you can (1) briefly restate the importance of the information in the section or (2) provide a transition to the section that follows. Model 3–2 takes the latter approach by suggesting the main topic for the next section. Whereas the section lead-in provides a map to help readers navigate through the section, the closing gives a sense of an ending so that readers are ready to move on.

Paragraphs

Paragraphs represent the basic building blocks of any document. Organizing them is not much different in technical writing than it is in nontechnical prose. Most paragraphs contain these elements:

1. **Topic sentence:** This sentence states the main idea to be developed in the paragraph. Usually it appears first. Do not delay or bury the main point, for busy readers may read only the beginnings of paragraphs. If you fail to put the main point there, they may miss it entirely.

2. **Development of main idea:** Sentences that follow the topic sentence develop the main idea with examples, narrative, explanation, or other details. Give the reader concrete supporting details, not generalizations.

3. **Transitional elements:** Structural transitions help the paragraph flow smoothly. Use transitions in the form of repeated nouns and pronouns, contrasting conjunctions, and introductory phrases.

4. **Closing sentence:** Most paragraphs, like sections and documents, need closure. Use the last sentence for a concluding point about the topic or a transitional point that links the paragraph with the one following it.

Model 3–3 on page 100 shows two paragraphs from an M-Global recommendation report that follow this pattern of organization. M-Global was hired to suggest ways for a hospital to modernize its physical plant. Each paragraph is a self-contained unit addressing a specific topic, while being linked to surrounding paragraphs (not shown) by theme and transitional elements.

This suggested format applies to many, but not to all, paragraphs included in technical documents. In one common exception, you may choose to delay statement of a topic sentence until you engage the reader's attention with the first few sentences. In other cases, the paragraph may be short and serve only as an attention grabber or a transitional device between several longer paragraphs. Yet for most paragraphs in technical communication, the beginning-middle-end model described here will serve you well. Remember these other points as well as you organize paragraphs:

- **Length:** Keep the typical length of paragraphs at around 6 to 10 lines. Many readers won't read long blocks of text, no matter how well organized the information may be. If you see that your topic requires more than 10 lines for its development, split the topic and develop it in two or more paragraphs.

- **Listings:** Use short listings of three or four items to break up long paragraphs. Readers lose patience when they realize information could have been more clearly presented in listings. Chapter 4 offers detailed suggestions on using lists.

- **Use of numbers:** Paragraphs are the worst format for presenting technical data of any kind, especially numbers that describe costs. Readers may ignore or miss data packed into paragraphs. Usually, tables or figures are a more clear and appropriate format. Also, be aware that some readers may think that cost data couched in paragraph form represent an attempt to hide important information.

>>> Modular Writing

Changes in technology have changed the way we write and read in the workplace. When readers access content through a web page, they do not read in a clearly identified sequence. Instead, they use links and tabs to pick and choose the topics that they want. (You learn more about writing for websites in chapter 11.) Another form of writing that is affected by technology is found in Help files in computer programs. The text in Help files is read in small, discrete parts, accessed as the reader needs it. Some texts are published in print and in digital form.

Today, many organizations are using *single sourcing* to create text for larger documents. In single sourcing, chunks of information are designed to be used in more than one document. For example, a passage that relates a company's history may be used in proposals, annual reports, promotional brochures, press releases, and on the company website.

This new technology poses new challenges for those writing technical documentation. Isolated sections of text must be clear, but they must also follow style and organizational specifications that cover a number of documents. Often, writers create only some of the sections that are assembled into larger documents; thus modular writing has become an important part of collaborative writing. This is discussed in greater length in chapter 13.

This chapter mostly concerns the ordering of ideas within paragraphs, sections, and whole documents. Good organization helps make your writing successful. Organization alone, however, will not win the day. Readers also expect a visually appealing document. Chapter 4 describes technical devices for creating the best possible design of your pages.

>>> Chapter Summary

Good technical communication calls on special skills, especially in organization. Writers should follow three guidelines for organizing information: (1) Write different parts of the document for different readers, (2) place important information at the beginnings and endings, and (3) repeat key points throughout the document.

This chapter recommends the ABC format for organizing technical documents. This format includes an *Abstract* (summary), a *Body* (supporting details), and a *Conclusion* (wrap-up and transition to next step). The abstract section is particularly important because most readers give special attention to the start of a document.

Individual sections and paragraphs also require attention to organization. Using familiar patterns of organization can be useful to both writers and readers. Sections need overviews and closing passages so that busy readers can find information quickly. Most paragraphs should contain a topic sentence, supporting details, transitional words and phrases, and a closing sentence that leads into the next paragraph. As technology for producing and reading documents changes, writers must create consistent and well-written shorter passages that can be read in a variety of formats and contexts.

>>> **Learning Portfolio**

Communication Challenge   Telecommuting: The Last Frontier?

Calling themselves the "Commute Group," five managers at M-Global's Boston office have been meeting to discuss telecommuting (i.e., permitting some or all employees to do part of their work at home). The branch manager, Richard DeLorio, expressed interest in the group's work and suggested that group members write a report proposing a pilot project at the branch. The report would be read by Richard and by members of the M-Global corporate staff in Baltimore—especially Karrie Camp, Vice President of Human Resources. It will probably also be read by Jeannie McDuff, vice president for domestic operations, Richard's boss. Any change in branch work schedules must be approved by corporate headquarters.

The Commute Group now must decide (1) what to include in its report to Richard DeLorio and (2) how to *organize* information for maximum impact. What follows are some details on the audience for the report, the group's reasons for favoring telecommuting, some problems discussed by the group, and questions that remain about the organization of the report. Although the group has made progress in discussing telecommuting, it has been unable to decide on a structure for its report.

This case study explains their approach for preparing the report, how to organize the information, which includes questions and comments for discussion, and an assignment for a written response to the Challenge.

Report Audience

The group has spent much time discussing what points would be most persuasive with the primary audience, Richard DeLorio and Karrie Camp. Richard has been open to new ideas since being chosen for the manager job a year ago. He meets often with all departments in the office and shows a genuine interest in creating a more comfortable workplace. For example, he recently accepted recommendations by department managers to purchase office chairs and desks that allow employees to work with less physical strain.

As Vice President of Human Resources, Karrie Camp sees part of her responsibility as protecting the assets of M-Global, and making sure that employees work effectively and efficiently. Indeed, Karrie, who has been with the organization for 30 years, has a master's degree in finance, and keeps a close eye on the bottom line of each branch. She is interested in exploring new work practices only if they may

improve employee productivity. More than likely, she will be the final decision maker about the pilot project, although she will inform Jeannie McDuff if there is a change of policy in the Boston branch.

Jeannie McDuff, Richard's boss, evaluates branch managers largely on the financial performance of the branches, but she is interested in innovation, and has been one of the main forces behind the organization's new image.

Rationale for Pilot Project

The Commute Group spent much time discussing two topics: branch jobs that would be best suited to telecommuting and specific arguments in support of a telecommute policy.

Group members agreed that employees who do much independent work, especially on the computer, would be the best candidates for a pilot project. In particular, members of the technical and scientific staff often spend half their days at personal computers, either performing technical calculations or drafting sections of reports and proposals.

Next the group discussed reasons for adopting a telecommute pilot project. The group first met to discuss the issue after a series of horrible rush hours over the holiday season in December. Bad weather forced most of the 125 branch employees either to miss some workdays during the period or to arrive up to two hours late several days. Most employees already have a one-way commute of at least one hour, because there is little affordable housing close to the office location in downtown Boston. Thus the heavy holiday traffic prompted the discussion about telecommuting.

In its deliberations, the Commute Group focused mostly on the kind of work that could be done by employees at home. What follows are some of the points discussed by the group, in random order.

- Telecommuting will save time either by eliminating commuting (on days the employee works exclusively at home) or by reducing commuting time (on days when the employee comes to the office for part of the day and thus avoids one or both rush-hour periods that day).
- Employees can write and edit reports and proposals at home for several hours at a time, without the usual office interruptions of meetings, phone calls, drop-in visitors, and so on. Some experts claim that writers are most productive during the drafting stage if they have uninterrupted blocks of writing time.

- Morale will improve as long as there is a clear rationale for adopting the policy and selecting participants for the pilot project. Employees chosen should be those who work well independently, whose jobs can be handled through telecommuting, and who have already made significant contributions to their departments.
- If M-Global adopts a telecommuting policy after the pilot project, the firm may attract an additional pool of excellent employees.
- Telecommuting will improve some employees' productivity by allowing them to work when they are recovering from an illness at home or when family members are ill—in other words, times when the employee would normally be on sick leave.
- The company would benefit from the increase in computer literacy among both the telecommuters and those who work with them back at the office. The firm would begin to take advantage of the considerable investment it already has made in computer technology—personal computers, laptops, networking, groupware, software for instant messaging, webcams, and so on. In particular, email and instant messaging would become a way of life. Until now, many employees have been reluctant to replace time-consuming meetings, phone calls, and memos with email and on-line discussions.
- If telecommuting were to become a regular way of doing business, it might reduce the amount of workspace needed at the office and thus reduce overhead. For example, several employees could share the same office workspace if much of their work time were spent at home.
- Even some non-computer tasks, such as phone calls to clients, could be done best in the quiet environment of the home, as opposed to the hectic environment of the office, where noise and interruptions are a part of doing business.
- If a telecommuting policy were adopted, M-Global would gain public support by showing that it is part of the solution to the central problems of traffic congestion and air pollution. Even some potential clients might be attracted by the firm's progressive policies.

Possible Problems with Telecommuting

The Commute Group also addressed problems that might arise with the pilot project and with telecommuting in general. Group members were unsure how or if the problems should be woven into the fabric of the report. Following are some concerns that were discussed:

- The right employees must be selected for the pilot project. Whereas some employees might improve their productivity at home, others might find it difficult to stay on task, either because of their own work habits or because of their home environment. Some kind of appropriate screening device would be in order.
- The branch must determine how to evaluate the success of the pilot project, perhaps by some combination of (1) self-evaluation by the employee, (2) performance evaluations by the employees' supervisors, (3) productivity assessment by the corporate office, or (4) opinions gathered by surveying employees who are not part of the pilot project but who interact regularly with the employees who are telecommuting.
- Good communication is central to the project. Employees must be involved in selecting participants, planning the study, conducting the project on a day-to-day basis, and evaluating its success.

Organization of the Report

The Commute Group has agreed on the audience for the report, the likely qualifications for participation in the pilot study, advantages of telecommuting, and some possible problems with the study and with telecommuting in general. However, the group has not resolved two main questions: (1) What part of the information assembled should be included in the report, and (2) what order this information should assume. In other words, the group must wrestle with matters of organization. Indeed, disagreements about these two issues created a stumbling block in the group's work.

Assume the role of a documentation specialist who is assigned to a standing proposal-writing team at the Boston branch. You have been called in by the Commute Group to help create an effective and persuasive argument. Answer the following questions, remembering that you are not to be concerned with specific report sections or headings described later in this book. Instead, this exercise concerns only the generic ABC (Abstract/Body/Conclusion) structure explained in this chapter.

1. Briefly, how would you describe an overall ABC structure that would work in this report? Your answer should take into account the intended purpose and audience of the report.
2. More specifically, what points would you suggest be included in the abstract component? What issues need not be addressed? Why?
3. What points would you suggest be included in the body? Why? In what order? Explain the rationale for the order you suggest. If you have excluded some information discussed by the committee, explain why.
4. Given what you've read so far, what one main purpose should be served by the conclusion component of the

group's report? To accomplish this purpose, what information should be included? Why?

5. What issues, if any, remain to be discussed by the Commute Group before it writes its recommendation report?

Write About It

Members of M-Global's Publications Development team are assigned to branches throughout the organization, but all members of The Pub share certain goals and responsibilities. One of the unofficial duties that they have agreed on is to help all employees at M-Global become better writers. Although you work in Boston, you are a member of The Pub. You take this opportunity to teach your fellow employees about organization. Create a general outline for the body of the report, identifying its key sections. Decide on patterns of organization for each section. Then write a memo to the members of the Commute Group. Include your suggested outline and organization patterns for each section and explain your rationale for your suggestions.

Collaboration at Work   Organizing the Catalog

General Instructions

Each Collaboration at Work exercise applies strategies for working in teams to chapter topics. The exercise assumes you (1) have been divided into teams of about three to six students, (2) will use team time inside or outside of class to complete the case, and (3) will produce an oral or written response. For guidelines about writing in teams, refer to pages 24–28.

Background for Assignment

Here, the term *organization* means the arrangement of information, such as purpose statements, supporting details, conclusions, and recommendations. As explained in this chapter, you should aim to choose patterns of organization that fit the context. In particular, they should respond to the needs of the specific readers.

Your college or university catalog is an example of a document that includes varied information, varied readers, and, in many cases, varied patterns of organization. Standard topics covered in catalogs often include the following:

- Accreditation organization, status, and guidelines
- Mission of the institution

- Admissions procedures
- Academic departments
- Degree programs
- Course descriptions
- Financial aid
- Extracurricular activities
- Academic regulations

To add to the complexity, different sections of the catalog may have been written by different writers. However, usually one or two people are responsible for coordinating and editing the entire document.

Team Assignment

Your team will either choose or be assigned one or more sections of your institution's catalog. Your task is to (1) describe the manner in which information is organized, (2) speculate about the rationale the writer had for the pattern(s) selected, and (3) develop an opinion as to whether the patterns meet the needs of the catalog's main readers.

Assignments

1. Overall Organization

Find an example of technical writing directed to more than one reader. Prepare a written or an oral report (your instructor's choice) that explains how well the excerpt follows this chapter's guidelines for organization.

2. Evaluating an Abstract

Read the following abstract and evaluate the degree to which it follows the guidelines in this chapter.

BMDO [The Ballistic Missile Defense Organization] is using a layered approach to protect U.S. forces and allies against ballistic missile attacks. This approach focuses on three priority areas: (1) theater missile defense (TMD), to address the short-range, widely dispersed threat from short-range ballistic missiles; (2) national missile defense (NMD), to address the long-range threat from intercontinental ballistic missiles; and (3) advanced technology, to continue advancing BMDO's capabilities to counter more complex future threats from ballistic missiles. Each priority area is discussed below.

Source: Ballistic Missile Defense Organization. (1998). *Ballistic Missile Defense Organization 1998 Technology Applications Report.* Alexandria, VA: National Technology Transfer Center, 8.

3. All Patterns of Organization— Recognition Exercise

For this group assignment, your instructor will provide each group with a different packet of "junk mail" (catalogs, sales letters, promotions, etc.) and perhaps other documents such as memos or product information sheets. Your group will search for and evaluate examples of various patterns of organization in the documents and then report its findings to the whole class.

4. Section Organization

As a graphics specialist at M-Global, you have written a recommendation report on ways to upgrade the graphics capabilities of the firm. One section of the report describes a new desktop publishing system, which you believe will make M-Global proposals and reports much more professional looking. Your report section describes technical features of the system, the free training that comes with purchase, and the cost.

Write a lead-in paragraph for this section of your report. If necessary, invent additional information for writing the paragraph.

5. Classification

Many websites offer advice to incoming freshmen about what to pack for their college dorm room. Using at least two lists as a starting place, create your own list of recommendations. You may include as many or as few of the recommended items as you feel worthwhile, and you can add your own items to the list. Then choose a principle for classifying the items on your list. Group the items and clearly identify the characteristics that helped you group the items. When you turn in your lists or share them with the class (as your teacher instructs), identify the Internet sites that you used as a starting point.

6. Classification: M-Global Projects

For this assignment you will use the main technical tasks listed on the eight projects at the end of chapter 2. Perform a classification exercise by finding a common basis, selecting an appropriate number of groups, and placing each of the technical tasks into one of the groups.

7. Division

Compare syllabi for several different courses. (Your teacher may ask you to work in groups, so that you have many examples to study.) How do they divide information about courses? Identify the divisions that appear most often in the syllabi. If you note significant differences between the syllabi—differences between the principle used for dividing information (e.g., by time or task) or divisions that are used in one syllabus but not in the others—analyze those differences. Do the differences seem to be connected to the topic of the course? To the particular type of course (whether it is a lecture or a lab course)? To the course level? Write a short essay that reports your findings.

8. Paragraph Organization: Analysis

Select a paragraph from each of four different articles taken from periodicals in your campus library. Choose one from a nationally known newspaper (like *The New York Times*), one from a popular magazine (like *Time* or *Sports Illustrated*), one from a business magazine (like *Forbes* or *Business Week*), and one from a technical journal (like *IEEE Transactions on Professional Communication*). Explain in writing how each of the paragraphs does or does not follow the top-down pattern of organization discussed in this chapter. If a paragraph does not follow the top-down pattern, indicate whether you believe the writer made the right or wrong decision in organizing the paragraph. In other words, was there a legitimate reason to depart from the ABC pattern? If so, what was the reason? If not, how would you revise the paragraph to make it fit the ABC model?

9. Paragraph Organization: Writing

With the following list of related information, write a paragraph that follows the organizational guidelines in this chapter. Use all the information, change any of the wording when necessary, and add appropriate transitions. Assume that the paragraph is part of an internal M-Global document suggesting ways to improve work schedules.

- Four-day weeks may lower job stress—employees have long weekends with families and may avoid worst part of rush hour.
- A four-day, 10-hour-a-day workweek may not work for some service firms, where projects and clients need five days of attention.

- Standard five-day, eight-hour-a-day workweeks increase on-the-job stress, especially considering commuter time and family obligations.
- M-Global is considering a pilot program for one office, whereby the office would depart from the standard 40-hour workweek.
- There are also other strategies M-Global is considering to improve work schedules of employees.
- The 40-hour workweek came into being when many more families had one parent at home while the other worked.
- Some firms have gone completely to a four-day week (with 10-hour days).
- M-Global's pilot program would be for one year, after which it would be evaluated.

10. Writing an Abstract

The short report that follows lacks an abstract that states the purpose and provides the main conclusion or recommendation from the body of the report. Write a brief abstract for this report.

DATE: June 13, 2008
TO: Ed Simpson
FROM: Jeff Radner
SUBJECT: Creation of an Operator Preventive Maintenance Program

The Problem

The lack of operator involvement in the equipment maintenance program has caused the reliability of equipment to decline. Here are a few examples:

- A tractor was operated without adequate oil in the crankcase, resulting in a $15,000 repair bill after the engine locked.
- Operators have received fines from police officers because safety lights were not operating. The bulbs were burned out and had not been replaced. Brake lights and turn-signal malfunctions have been cited as having caused rear-end collisions.
- A small grass fire erupted at a construction site. When the operator of the vehicle nearest to the fire attempted to extinguish the blaze, he discovered that the fire extinguisher had already been discharged.

When the operator fails to report deficiencies to the mechanics, dangerous consequences may result.

The Solution

The goal of any maintenance program is to maintain the company equipment so that the daily tasks can be performed safely and on schedule. Since the operator is using the equipment on a regular basis, he or she is in the position to spot potential problems before they become serious. For a successful maintenance program, the following recommendations should be implemented:

- Hold a mandatory four-hour equipment maintenance training class conducted by mechanics in the motor pool. This training would consist of a hands-on approach to preventive maintenance checks and services at the operator level.
- Require operators to perform certain checks on a vehicle before checking it out of the motor pool. A vehicle checklist would be turned in to maintenance personnel.

The attached checklist would require 5 to 10 minutes to complete.

Conclusion

I believe the cost of maintaining the vehicle fleet at M-Global will be reduced when potential problems are detected and corrected before they become serious. Operator training and the vehicle pretrip inspection checklist will ensure that preventable accidents are avoided. I will call you this week to answer any questions you may have about this proposal.

M-Global, Inc.
Fleet Maintenance Division
Vehicle Checklist
Pretrip Inspection

Inspected by: _____   Date: _____

Vehicle #: _____   Odometer: _____

Fluid Levels, Full/Low   Comments

_____ Engine Oil   _____

_____ Transmission Fluid   _____

_____ Brake Fluid   _____

_____ Power Steering   _____

_____ Radiator Level   _____

Before Cranking Vehicle

_____ Tire Condition   _____

_____ Battery Terminals   _____

_____ Fan Belts   _____

_____ Bumper and Hitch   _____

_____ Trailer Plug-in   _____

_____ Safety Chains   _____

After Cranking Vehicle

_____ Parking Brakes   _____

_____ Lights   _____

_____ All Gauges   _____

_____ Seat Belts   _____

_____ Mirrors/Windows/Wipers _____

_____ Clutch   _____

_____ Fire Ext. Mounted and Charged _____

_____ Two-Way Radio Working   _____

Additional Comments:_____

_____

_____

11. Ethics Assignment

Read this chapter's "Communication Challenge" section entitled "Telecommuting: The Last Frontier?" (You may also want to conduct some library or Internet research on telecommuting, or discuss the concept with someone who works full-time.) Then write a short essay or report that examines (a) any personal ethical dilemmas that may arise for M-Global employees if the Boston office adopts a telecommuting policy, and (b) possible solutions to the ethical problems you discuss.

12. International Communication Assignment

In the ABC pattern, the beginning of a document (A, or Abstract) includes a clear purpose statement, summary of main points for decision makers, and brief description or listing of information to follow in the document. Although this pattern works in most situations—especially in the United States and many Western countries—there may be contexts and cultures in which it is not the best choice. For example, the front-end location of important points may appear abrupt and even offensive in some cultures. Using a country or culture outside the United States, describe a context in which you would depart from the strict ABC pattern. (Possible countries to examine include China, Japan, Saudi Arabia, and Germany.) Be specific about purpose, readers, and preferred organizational pattern. Also, give the source upon which you base your conclusions—family experience, business experience, interviews, books, Internet research, and so on.

 ### 13. A.C.T. N.O.W Assignment (Applying Communication To Nurture Our World)

Find a document (report, article, letter to editor, poster with text, editorial, etc.) intended to alert readers to a health or safety issue. Depending on the instructions you are given, prepare an oral or a written report in which you (a) analyze the degree to which the document does or does not subscribe to the ABC pattern of organization and (b) offer suggestions for how the document might be reorganized to more effectively model the ABC format. For example, you may want to give specifics about the manner in which existing information may be rearranged or new information added.

MEMORANDUM

DATE: September 5, 2008
TO: Danielle Firestein
FROM: Barbara Ralston *BR*
SUBJECT: Recommendation for AIM 500 Fax

INTRODUCTORY SUMMARY

The purpose of this report is to present the results of the study you requested on the AIM 500 facsimile (fax) machine. I recommend purchase of additional AIM 500 machines, when needed, because they deliver fast, dependable service and include features we need most. This report includes the following sections: Ease of Operation, Features of the AIM 500, Dependability of the AIM 500, and Conclusion.

ABSTRACT
Identifies purpose of report and recommendations. Provides overview of structure.

EASE OF OPERATION

The AIM 500 is so easy to operate that a novice can learn to transmit a document to another location in about two minutes. Here's the basic procedure:

BODY
Headings and sub headings indicate structure.

1. Press the button marked TEL on the face of the fax machine. You then hear a dial tone.
2. Press the telephone number of the person receiving the fax on the number pad on the face of the machine.
3. Lay the document facedown on the tray at the back of the machine.

At this point, just wait for the document to be transmitted—about 18 seconds per page to transmit. The fax machine will even signal the user with a beep and a message on its LCD display when the document has been transmitted. Other more advanced operations are equally simple to use and require little training. Provided with the machine are two different charts that illustrate the machine's main functions.

The size of the AIM 500 makes it easy to set up almost anywhere in an office. The dimensions are 13 inches in width, 15 inches in length, and 9.5 inches in height. The narrow width, in particular, allows the machine to fit on most desks, file cabinets, or shelves.

FEATURES OF THE AIM 500

The AIM 500 has many features that will be beneficial to our employees. In the two years of use in our department, the following features were found to be most helpful:

Automatic redial
Last number redial memory
LCD display

M-Global Inc | 127 Rainbow Lane | Baltimore MD 21202 | 410.555.8175

■ **Model 3–1** ■ ABC format in whole document

Ralston to Firestein. 2

Preset dialing
Group dialing
Use as a phone

Automatic Redial. Often when sending a fax, the sender finds the receiving line busy. The redial feature will automatically redial the busy number at 30-second intervals until the busy line is reached, saving the sender considerable time.

Last Number Redial Memory. Occasionally there may be interference on the telephone line or some other technical problem with the transmissions. The last number memory feature allows the user to press one button to automatically trigger the machine to retry the number.

LCD Display. This display feature clearly shows pertinent information, such as error messages that tell a user exactly why a transmission was not completed.

Preset Dialing. The AIM 500 can store 16 preset numbers that can be engaged with onetouch dialing. This feature makes the unit as fast and efficient as a sophisticated telephone.

Group Dialing. Upon selecting two or more of the preset telephone numbers, the user can transmit a document to all of the preset numbers at once.

Use as a Phone. The AIM 500 can also be used as a telephone, providing the user with more flexibility and convenience.

DEPENDABILITY OF THE AIM 500

Over the entire two years our department has used this machine, there have been no complaints. We always receive clear copies from the machine, and we never hear complaints about the documents we send out. This record is all the more impressive in light of the fact that we average 32 outgoing and 15 incoming transmissions a day. Obviously, we depend heavily on this machine.

So far, the only required maintenance has been to change the paper and dust the cover.

CONCLUSION

CONCLUSION
Clear recommenda-
tion.
Invites contact.

The success our department has enjoyed with the AIM 500 compels me to recommend it highly for additional future purchases. The ease of operation, many exceptional features, and record of dependability are all good reasons to buy additional units. If you have further questions about the AIM 500, please contact me at extension 3646.

■ **Model 3–1** ■ continued

ADDITIONAL FEATURES OF MAGCAD

This report has presented two main advantages of the MagCad Drawing System: ease of correction and multiple use of drawings. However, there are two other features that make this system a wise purchase for M-Global's Boston office: the selective print feature and the cost.

Selective Printing

When printing a MagCad drawing, you can "turn off" specific objects that are in the drawing with a series of keystrokes. The excluded items will not appear in the printout of the drawing. That is, the printed drawing will reflect exactly what you have temporarily left on the screen, after the deletions. Yet the drawing that remains in the memory of the machine is complete and ready to be reconstructed for another printout.

The selective print feature is especially useful on jobs where different groups have different needs. For example, in a drawing of a construction project intended only for the builder, one drawing may contain only land contours and the building structures. If the same drawing is going to the paving company, we may need to include only land contours and parking lots. In each case, we will have used the selective print feature to tailor the drawing to the specific needs of each reader.

This feature improves our service to the client. In the past, we either had to complete several different drawings or we had to clutter one drawing with details sufficient for the needs of all clients.

Cost of MagCad

When we started this inquiry, we set a project cost limit of $12,000. The MagCad system stays well within this budget, even considering the five stations that we need to purchase.

The main cost savings occurs because we have to buy only one copy of the MagCad program. For additional work stations, we need pay only a $400 licensing fee per station. The complete costs quoted by the MagCad representative are listed below:

1.	MagCad Version 5	$ 5,000
2.	Licenses for five additional systems	2,000
3.	Plotter	2,000
4.	Installation	1,000
	TOTAL	$ 10,000

With the $2,000 difference between the budgeted amount and the projected cost of the system, we could purchase additional work stations or other peripheral equipment. The next section suggests some add-ons we might want to purchase later, once we see how the MagCad can improve our responsiveness to client needs.

ABSTRACT
Clearly identifies change in topic.

BODY
Headings identify structure.

CONCLUSION
Suggests benefits.

■ **Model 3–2** ■ ABC format in document section

Introduces topic.

Specifies advantages.

Summarizes result.

Conversion to a partial solar heating and cooling system would upgrade the hospital building considerably. In fact, the use of modern solar equipment could decrease your utility bills by up to 50%, using the formula explained in Appendix B. As you may know, state-of-the-art solar systems are much more efficient than earlier models. In addition, equipment now being installed around the country is much more pleasing to the eye than was the equipment of ten years ago. The overall effect will be to enhance the appearance of the building, as well as to save on utility costs.

We also believe that changes in landscaping would be a useful improvement to the hospital's physical plant. Specifically, planting shade trees in front of the windows on the eastern side of the complex would block sun and wind. The result would be a decrease in utility costs and enhancement of the appearance of the building. Of course, shade trees will have to grow for about five years before they begin to affect utility bills. Once they have reached adequate height, however, they will be a permanent change with low maintenance. In addition, your employees, visitors, and patients alike will notice the way that trees cut down on glare from the building walls and add "green space" to the hospital grounds.

■ **Model 3–3** ■ ABC format in paragraphs

Chapter | 4 | Page Design

>>> Chapter Outline

Elements of Page Design   102
   Grids   103
   White Space   105
   Headings   107
   Lists   111
   In-Text Emphasis   112

Fonts and Color   112
   Type Size   113
   Font Types   114
   Color   114

Computers in the Page Design Process   115
   Headers and Footers   115
   Templates   115
   Style Sheets   115

Using the Elements of Page Design   117

Chapter Summary   117

Learning Portfolio   119
   Communication Challenge   119
   Collaboration at Work   120
   Assignments   121

Model for Good Writing   124
   Model 4–1: Page design in memorandum   124

The Information Services (IS) Department at M-Global is preparing to introduce new security sign-on procedures for all employees in the organization, and Mark Merrill, one of the programmers, was assigned the task of writing the new sign-on procedures. He sent his draft as an email attachment to David Carlyle, a documentation specialist. Mark doesn't mind writing, but he is glad that David has been assigned to help make the information in IS documents more accessible.

When David opens the email from Mark, he knows that he won't have to do much editing to Mark's text. Mark is generally a good writer; his documents are usually written clearly and accurately. David's biggest challenge is to make sure that the instructions are easy to read and use by employees as they access the company intranet from their desks or from laptops and other electronic devices in the field. David decides to create signs that can be posted at shared terminals, and smaller laminated cards that can be folded to business-card size and kept in a desk, laptop case, or pocket. The next problem he faces is how to format and arrange the information so that it is easy to read and use. He must apply principles of good page design to the document.

As with the organizing principles discussed in chapter 3, good page design can help your readers find the information that they need. This chapter covers page design, another basic building block in technical communication. An operating definition is as follows:

Page design: A term that refers to creating clear, readable, and visually interesting documents through judicious use of white space, headings, lists, emphasis, and other design elements. Many firms combine these elements into style sheets that allow employees to produce documents in uniform and consistent formats.

This chapter presents guidelines and examples for page design, as well as commentary on the use of computers in the design process.

>>> Elements of Page Design

In the workplace, readers are busy, and few take the time to read a document from cover to cover. Some documents, such as manuals, are used as reference works, only consulted as a last resort. As one expert says, the subjects of technical documents do not invite casual reading. "Realize that people are lookers first and that they become readers only if you have revealed a good reason for them to want to [be]" (White, J. 2004. Building Blocks of Functional Design. *Technical Communication* 52, 37). Your challenge is to make your documents inviting by making pages interesting to the eye. A document may offer your reader great opportunities but never get read. Why? Because it doesn't look inviting.

Good organization, as pointed out in chapter 3, can fight readers' indifference by giving information when and where they want it. However, to get and keep readers interested, you must use effective page design—on each page of your document. Each page needs the right combination of visual elements to match the needs of your readers and the purpose of the document.

Good page design can work with good organization to help readers find information that they need in a document. Readers can recognize important information by its location on a page, by use of contrast, or by repetition of identically formatted elements such as guidelines or warning boxes. One way that readers can locate the information they need is by using navigational tools. You may be used to thinking of navigational tools in electronic texts such as web pages or even PDFs, but print documents also use navigational devices such as tables of contents, running headers and footers, headings, and even color coding.

Many firms know the benefits of visual imaging so they develop company style sheets for frequently used documents, including letters, memos, various types of reports, and proposals. Once developed, these style sheets are assembled into a style manual and distributed for general use. To make universal use easier, they are often loaded as templates or styles into the company's text editing software. More information about templates and styles appears on pages 115–117 at the end of this chapter.

Using style sheets saves time and reinforces the firm's corporate image. Their use makes it possible for members of writing teams to work independently while adhering to corporate formatting guidelines. Use of style sheets also ensures that readers see a clear relationship among ideas because headers and sub-headers are used consistently.

If the company you work for does not use a style manual, use the elements of page design shown in this chapter to develop your own style sheets: grids, white space, headings, lists, in-text emphasis, fonts, and color.

Grids

It is useful to approach page design by visualizing the elements on a page organized on a grid.[1] Planning your layout as a grid can help maintain a consistent unified appearance, especially in longer documents. This technique can also help you decide when elements should break the space, for example, to cover two columns or extend into the margin. When grids are used for page layout, blocks of text are usually represented with gray rectangles, and illustrations are usually represented with white boxes with a large "X" in them (Figure 4–1), allowing you to focus on the visual design of a page. As you design page layouts, focus on using the two basic page design elements that readers notice first:

1. **Text:** Long lines can be an obstacle to keeping the readers' attention. Eyes get weary of overly long lines, so some writers add double columns to their design options. This "book look" uses white space between columns to break up text and thus reduce line length. Line length can also be shortened by using wider margins with a single column of text. Figure 4–1 shows four different ways of arranging text blocks to affect line length.

---

[1]Some of the information in this section was taken from Christine Sevilla (2002, June), Page Design: Directing the Reader's Eye, *Intercom*, 49, 7–9; and Jan V. White (2005), Building Blocks of Functional Design, *Technical Communication*, 52, 37–41.

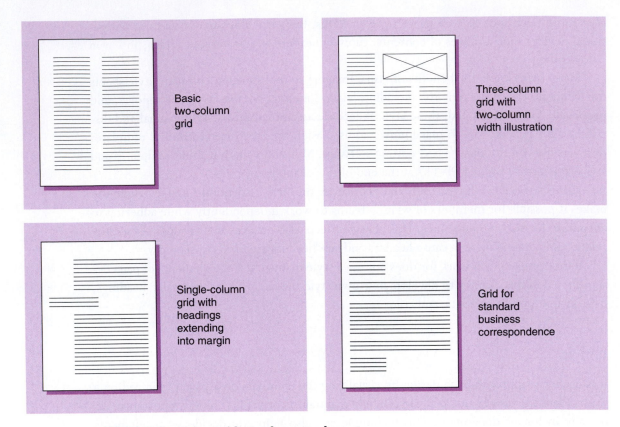

Basic
two-column
grid

Three-column
grid with
two-column
width illustration

Single-column
grid with
headings
extending
into margin

Grid for
standard
business
correspondence

■ **Figure 4–1** ■ Using grids to plan page layout

2. **Graphics:** Any illustration within the text needs special attention. Whereas chapter 12 provides a complete discussion of graphics, some pointers follow considering their placement for visual appeal:

- Make sure there is ample white space between any graphic and the text. If the figure is too large to permit adequate margins, reduce its size while maintaining readability.
- When you have the choice, place graphics near the top of the page, where they receive the most attention.
- When a graphic doesn't fit well on a page with text, place it on its own page to ensure adequate space and readability. Normally, it appears on the page following the first reference to it.
- Pay special attention to page balance when graphics are included on multicolumn pages, two-page spreads, or both.

To avoid confusing the reader, make sure that each page has no more than one dominant element. You can use more than one basic grid pattern within a document, but if you do, make sure that the patterns are related, and that there is a good reason to use an alternative grid. For example, you could design one grid for most of the pages in a long document, but a second grid for first pages of chapters or major sections within the document.

White Space

The term *white space* simply means the open places on the page with no text or graphics—literally, the white space (assuming you are using white paper). Experts have learned that readers are attracted to text because of the white space that surrounds it, as with a newspaper ad that includes a few lines of copy in the middle of a white page. Readers connect white space with important information.

In technical communication you should use white space in a way that (1) attracts attention, (2) guides the eye to important information on the page, (3) relieves the boredom of reading text, and (4) helps readers organize information. Here are some opportunities for using white space effectively:

1. **Margins:** Most readers appreciate generous use of white space around the edges of text. Marginal space tends to frame your document, so the text doesn't appear to push the boundaries of the page. Good practice is to use 1- to 1½-inch margins, with more space on the bottom margin. When the document is bound, the margin on the edge that is bound should be larger than the outside margin to account for the space taken by the binding (Figure 4–2.)

2. **Hanging Indents:** Some writers place headers and subheads at the left margin and indent the text block an additional inch or so, as shown in Figure 4–3. Headers and subheads force the readers' eyes and attention to the text block. Another common use of hanging indents is bulleted and numbered lists.

3. **Line Spacing:** When choosing single, double, or 1½ line spacing, consider the document's length and degree of formality. Letters, memos, short reports, and other documents read in one sitting are usually single-spaced. Longer documents, especially if they are formal, are usually 1½-space or double-spaced, sometimes with extra spacing between paragraphs. Manuscripts or documents that are typeset professionally are always double-spaced (Figure 4–4.)

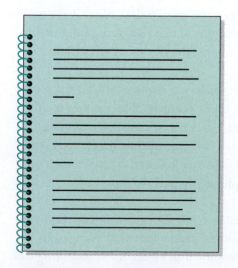

■ **Figure 4–2** ■ Use of white space: margins

■ **Figure 4–3** ■ Use of white space: hanging indents

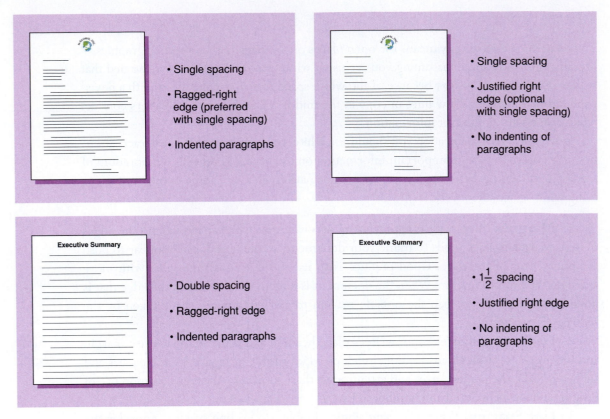

■ **Figure 4–4** ■ Use of white space: line spacing

4. **Justification:** The choice of justification should be based on line length and the formality of the document. In full-justified copy, all lines are the same length—as on this textbook page. In ragged-edge copy, lines are variable length. Some readers prefer ragged-edge copy because it adds variation to the page, making reading less predictable for the eye. Yet many readers like the professional appearance of full-justified lines, especially in formal documents or documents that use columns of text. However, full justification can sometimes result in odd spacing between letters as the computer tries to fill a line with a few words.

5. **Paragraph Length:** New paragraphs give readers a chance to regroup as one topic ends and another begins. These shifts also have a visual impact. The amount of white space produced by paragraph lengths can shape reader expectations. For example, two long paragraphs suggest a heavier reading burden than do three or four paragraphs of differing lengths. Thus, it is helpful to break complex information into shorter paragraphs. Most readers skip long paragraphs, so vary paragraph lengths and avoid putting more than 10 lines in any one paragraph (Figure 4–5).

6. **Paragraph Indenting:** Another design decision involves indenting the first lines of paragraphs. As with ragged-edge copy, most readers prefer indented paragraphs because the extra white space creates visual variety. As shown in Figure 4–4, indenting can be used in single- or double-spaced text. Reading text is hard work for the eye. You should take advantage of any opportunity to keep the reader's attention.

Poor Format: One long
paragraph on page

Better Format: Several
paragraphs on page

7. **Heading Space and Ruling:** White space helps the reader connect related information immediately. Always have slightly more space above a heading than below it. That extra space visually connects the heading with the material into which it leads. In a double-spaced document, for example, you would add a third line of space between the heading and the text that came before it. In addition, some writers add a horizontal line across the page above headings, to emphasize the visual break. The next section covers other aspects of headings.

8. **Running Headers and Footers:** Headers and footers help readers locate information in a document. They may be as simple as page numbers, or much more complex—using chapter titles, page numbers, and even colors in the top or bottom margins of the page. Your readers must be able to find pages and sections of a document easily; at the same time, the headers and footers should not clutter up the appearance of the document.

In summary, well-used white space can add to the persuasive power of your text. As with any design element, however, it can be overused and abused. Make sure there is a reason for every decision you make with regard to white space on your pages.

Headings

Headings are brief labels used to introduce each new section or subsection of text. They serve as (1) a signpost for the reader who wants to know the content, (2) a grabber to entice readers to read documents, and (3) a visual oasis of white space where the reader gets relief from text.

As a general rule, every page of any document over one page should have at least one

heading so that readers can find their way through your text. Models throughout this book show how headings can be used in short and long documents. Of course, heading formats differ greatly from company to company and even from writer to writer. With all the typographical possibilities of word processing, there is incredible variety in typeface, type size, and the use of bold, underlining, and capitals. Following are some general guidelines:

1. **Use your outline to create headings and subheadings.** A well-organized outline lists major and minor topics. With little or no change in wording, they can be converted to headings and subheadings within the document. As with outlines, you must follow basic principles of organization. The number of subheadings should be one indication of the relative length or importance of the section. Be consistent with your approach to headings throughout the document.

2. **Use substantive wording.** Headings give readers an overview of the content that follows. They entice readers into your document; they can determine whether readers—especially those who are hurried and impatient—read or skip over the text. Strive to use concrete rather than abstract nouns, even if the heading must be a bit longer. Note the improvements in the following revised headings:

Original: "Background"

Revised: "How the Simmons Road Project Got Started" or "Background on Simmons Road Project"

Original: "Discussion"

Revised: "Procedure for Measuring Toxicity" or "How to Measure Toxicity"

Original: "Costs"

Revised: "Production Costs of the FastCopy 800" or "Producing the FastCopy 800: How Much?"

3. **Maintain parallel form in wording.** Headings of equal value and degree should have the same grammatical form, as shown in the following:

A. *Headings That Lack Parallel Form*

Scope of Services

How Will Fieldwork Be Scheduled?

Establish Contract Conditions

B. *Revised Headings with Parallel Form*

Scope of Services

Schedule for Fieldwork

Conditions of Contract

You don't have to be a grammar expert to see that the three headings in option A are in different forms. The first is a noun phrase, the second is a question, and the third is an action phrase beginning with a verb. Because such inconsistencies distract the reader, you should make headings in each section uniform in wording, such as the headings in option B that each start with a noun.

4. **Establish clear hierarchy in headings.** Whatever typographical techniques you choose for headings, your readers must be able to distinguish one heading level from another. Visual features should be increasingly more striking as you move up the ranking of levels. Figures 4–6 and 4–7 show heading formats recommended by a professional organization and a professional publication.

Subheads: All subheads should be flush with the left margin, with one line space above.

FIRST-LEVEL SUBHEAD
(all capitals, boldface, on separate line)

Second-Level Subhead
(initial capitals, boldface, on separate line)

Third-Level Subhead
(initial capitals, italic, on separate line)

Fourth-Level Subhead (initial capitals, boldface, on same line as text, with extra letter space between subhead and text)

Fifth-Level Subhead (initial capitals, italic, on same line as text, with extra letter space between the subhead and text)

■ **Figure 4–6** ■
Required heading formats for Transportation Research Board publications and manuscripts

Source: Transportation Research Board of the National Academies (2007). *Information for Authors.* Retrieved Nov. 6, 2007, from http://www.TRB.org/Guidelines/Authors.pdf. Reproduced with permission of TRB.

Use up to three levels of headings and indicate them clearly.

FIRST-LEVEL HEADING
(all caps, bold, on a line by itself)

Second-level heading
(first word only capitalized, bold, on a line by itself)

Third-Level heading (first word only capitalized, bold, followed by two spaces, as part of the first line of the paragraph)

■ **Figure 4–7** ■
Required heading formats for manuscripts submitted to *Technical Communication*

Source: Society for Technical Communication (2007). *Author Guidelines for Technical Communication.* Retrieved Nov. 6, 2007, from http://www.stc.org/pubs/techcomGuidelines01.asp

Following are specific guidelines for using typographical distinctions:

■ **Use larger type size for higher-level headings.** You want readers to grasp quickly the relative importance of heading levels as they read your document. Type size fixes this relative importance in their minds so that they can easily find their way through your material both the first time and upon rereading it. The incremental upgrading of type size helps readers determine the relative importance of the information.

■ **Use heading position to show ranking.** In formal documents, your high-level headings can be centered. The next two or three levels of headings are at or off the left margin. Be sure these lower-level headings also use other typographical techniques, such as bolding, to help the reader distinguish levels.

■ **Use typographical techniques to accomplish your purpose.** Besides type size and position, as previously mentioned, you can vary heading type with such features as

Uppercase and lowercase

Bold type

Underlining

Changes in type font

With this embarrassment of riches, writers must be careful not to overdo it and create "busy" pages of print. Use only those features that provide an easy-to-grasp hierarchy of levels for the reader.

■ **Consider using decimal headings for long documents.** Decimal headings include a hierarchy of numbers for every heading and subheading listed in the table of contents. Many an argument has been waged over their use. People who like them say that they help readers find their way through documents and refer to subsections in later discussions. People who dislike them say that they are cumbersome and give the appearance of bureaucratic writing.

Unless decimal headings are expected by your reader, use them only with formal documents that are fairly long. Following is the normal progression of numbering in decimal headings for a three-level document:

1.0 xxxxxxxxxxxxx
   1.1 xxxxxxxxxx
      1.1.1 xxxxxxxxxx
      1.1.2 xxxxxxxxxx
   1.2 xxxxxxxxxx
      1.2.1 xxxxxxxxxx
      1.2.2 xxxxxxxxxx
2.0 xxxxxxxxxxxxx
   2.1 xxxxxxxxxx
      2.1.1 xxxxxxxxxx
      2.1.2 xxxxxxxxxx
   2.2 xxxxxxxxxx
3.0 xxxxxxxxxxxxx

Lists

Technical communication benefits from the use of lists. Readers welcome your efforts to cluster items into lists for easy reading. In fact, almost any group of three or more related points can be made into a bulleted or numbered listing. Following are some points to consider as you apply this important feature of page design:

1. **Typical uses:** Lists emphasize important points and provide a welcome change in format. Because they attract more attention than text surrounding them, they are usually reserved for these uses:

Examples

Reasons for a decision

Conclusions

Recommendations

Steps in a process

Cautions or warnings about a product

Limitations or restrictions on conclusions

2. **Number of items:** The best lists are those that subscribe to the rule of short-term memory; that is, people can retain no more than five to nine items in their short-term memory. A listing of more than nine items may confuse rather than clarify an issue. Consider placing 10 or more items in two or three groupings, or grouped lists, as you would in an outline. This format gives the reader a way to grasp information being presented.

3. **Use of bullets and numbers:** The most common visual clues for listings are numbers and *bullets* (enlarged dots or squares like those used in the following listing). Following are a few pointers for choosing one or the other:

■ **Bullets:** Best in lists of five or fewer items, unless there is a special reason for using numbers.

■ **Numbers:** Best in lists of more than five items or when needed to indicate an ordering of steps, procedures, or ranked alternatives. Remember that your readers sometimes infer sequence or ranking in a numbered list.

4. **Format on page:** Every listing should be easy to read and pleasing to the eye. The following specific guidelines cover practices preferred by most readers:

■ **Indent the listing.** Although there is no standard list format, readers prefer lists that are indented farther than the standard left margin. Five spaces is adequate.

■ **Hang your numbers and bullets.** Visual appeal is enhanced by placing numbers or bullets to the left of the margin used for the list, as done with the items here.

■ **Use line spaces for easier reading.** When one or more listed items contain over a line of text, an extra line space between listed items can enhance readability.

■ **Keep items as short as possible.** Depending on purpose and substance, lists can consist of words, phrases, or sentences—such as the list you are reading. Whichever format you choose, pare down the wording as much as possible to retain the impact of the list format.

5. **Parallelism and lead-ins:** Make the listing easy to read by keeping all points grammatically parallel and by including a smooth transition from the lead-in to the listing itself. (The term *lead-in* refers to the sentence or fragment preceding the listing.) *Parallel* means that each point in the list is in the same grammatical form, whether a complete sentence, verb phrase, or noun phrase. If you change form in the midst of a listing, you take the chance of upsetting the flow of information.

Example: To complete this project, we plan to do the following:

➤ Survey the site
➤ Take samples from the three boring locations
➤ Test selected samples in our lab
➤ Report the results of the study

The listed items are in verb form (note the introductory words *survey, take, test,* and *report*).

6. **Punctuation and capitalization:** Although there are acceptable variations on the punctuation of lists, preferred usage includes a colon before a listing, no punctuation after any of the items, and capitalization of the first letter of the first word of each item. Refer to the alphabetized Handbook under "Punctuation: Lists" for alternative ways to punctuate lists.

In-Text Emphasis

Sometimes you want to emphasize an important word or phrase within a sentence. Computers give you these options: underlining, boldface, italics, and caps. The least effective are FULL CAPS and <u>underlining</u>. Both are difficult to read within a paragraph and distracting to the eye. The most effective highlighting techniques are *italics* and **boldface**; they add emphasis without distracting the reader.

Whatever typographical techniques you select, use them sparingly. They can create a busy page that leaves the reader confused about what to read. Excessive in-text emphasis also detracts from the impact of headings and subheadings, which should be receiving significant attention.

>>> Fonts and Color

Besides page format, you have other choices to make: changes in the size and type of font you use in the text itself, as well as the use of color.

Type Size

Traditionally, type size has been measured in points, with 72 points to an inch. When you go to the font-selection menu, the sizes may be listed as such: 9, 10, 12, 14, 18, and 24.

Despite these many options, most technical writing is printed off the desktop in 10- or 12-point type. When you are choosing type size, however, be aware that the actual size of letters varies among font types. Some 12-point type appears larger than other 12-point type. Differences stem from the fact that your selection of a font affects (1) the thickness of the letters, (2) the size of lowercase letters, and (3) the length and style of the parts of letters that extend above and below the line. Figure 4–8 shows the differences in three common fonts. Note that the typeface used in setting the text of this book is 12-point Perpetua.

New Century Schoolbook
9 point
10 point
12 point
14 point
18 point
24 point

Times Roman
9 point
10 point
12 point
14 point
18 point
24 point

Helvetica
9 point
10 point
12 point
14 point
18 point
24 point

■ **Figure 4–8** ■ Type sizes

■ **Figure 4–9** ■ Font types

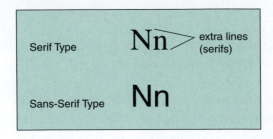

Before selecting your type size, run samples on your printer so that you are certain of how your copy will appear in final form.

Font Types

Your choice of fonts may be either prescribed by your employer or determined by you on the basis of (1) the purpose of the document, (2) the image you want to convey, and (3) your knowledge of the audience.

Font types are classified into two main groups:

- **Serif fonts:** Characters have "tails" at the ends of the letterlines.
- **Sans-serif fonts:** Characters do not have tails (Figure 4–9).

If you are able to choose your font, the obvious advice is to use the one that you know is preferred by your readers. A phone call or a look at documents generated by your reader may help you. If you have no reader-specific guidelines, following are three general rules:

1. **Use serif fonts for regular text in your documents.** The tails on letters make letters and entire words more visually interesting to the reader's eye, and they reduce eye fatigue. In this sense, they serve the same purpose as ragged-edge copy—helping your reader move smoothly through the document.

2. **Consider using another typeface—sans serif—for headings.** Headings benefit from a clean look that emphasizes the white space around letters. Sans-serif type helps attract attention to these elements of organization within your text.

3. **Avoid too many font variations in the same document.** There is a fine line between interesting font variations and busy and distracting text—but there is a line. Your rule of thumb might be to use no more than two fonts per document: one for text and another for headings and subheadings.

Because font selection is an important tool for developing graphics as well as page format, fonts are covered in more detail in chapter 12.

Color

Use of color, like fonts, is a graphic design tool that should reflect the tone, mood, and image of your documents. When used effectively, it focuses your reader's attention on important details. When used indiscriminately—inserted into a document just to show that color can be used—it is distracting.

Limit your use of color in routine documents for two reasons: (1) When you use professional printers, printing in color can be very expensive, and (2) when you are using desktop printers, printing color documents can be very slow. Use of color is discussed in detail in chapter 12.

>>> Computers in the Page Design Process

Most word-processing programs include tools to make page design easier and more consistent. They allow you to format running headers and footers, ensure consistency of elements such as headers and lists, change the appearance of tagged elements, and save time by using templates for the types of documents that you write most often. They even automatically generate tables of contents and indexes.

Headers and Footers

As noted earlier, running headers and footers are important navigation devices in documents, and most word-processing programs make the creation of headers and footers easy. In addition to being able to insert automatic page numbering, you can insert other information, such as short titles, your name, or your organization's name, on each page. You can decide where to position that information, and you can hide it on selected pages. Some organizations put information such as the computer filename or project identification number in document footers.

Templates

If you have a type of document that you must create often, such as progress reports, lab reports, memos, or even papers for school, you may find it useful to create a template for that type of document. Your word-processing program probably already has several templates preloaded for memos, letters, and reports. Although these templates are handy, they may not exactly fit your needs. If you need to include a company logo on a letterhead or alter the headings in the report template, you can modify existing templates or you can create your own. Some software publishers also make a large number of templates available for downloading from their websites. Templates include a catalog of styles for elements such as heading, lists, and even body text. They can also include passages of text or elements such as tables that are included in the same place in every document. For example, your teacher may use a template for class policy sheets that include the same information (e.g., office hours, contact information) or even the same text (e.g., absence policies, academic honesty statements).

Style Sheets

When you are writing a long document with many headings or other typographical elements, it may be difficult to remember how you formatted each element. For example, if it has been several pages since you used a third-level heading, you may have to scroll back to see what type size you used and whether you bolded or italicized it. This problem can be solved by using the styles in your word-processing software. A style sheet allows you to assign formatting to specific kinds of elements in your document, such as headings, body text, and lists. This formatting is done with *tags* or codes that your computer attaches to the elements. (If you are familiar with HTML coding, this tagging is similar.) You select the text, such as a first-level heading, select the appropriate style from a pull-down menu, and assign it to the selected text with a single mouse click. Figure 4–10 shows a style sheet, or *catalog,* that is part of a document template.

If you decide that you want to change the formatting, you can change it from the Styles menu and apply it to all of the tagged elements at once. For example, if you have set the font size for first-level headings at 12 points, and then decide that you want to change it to 14 points, you would open the Styles menu, find the Style for first-level headings, change the font size, and then apply it to all of the first-level headings. Instead of having to go through the document, find each heading, and change it, all first-level headings are changed at once. The heading tags that you created for your style sheet can also be used to automatically generate a Table of Contents. This process, and the one for creating indexes, can be a bit complicated, so consult your program's Help file or an after-market manual for instructions about how to do this.

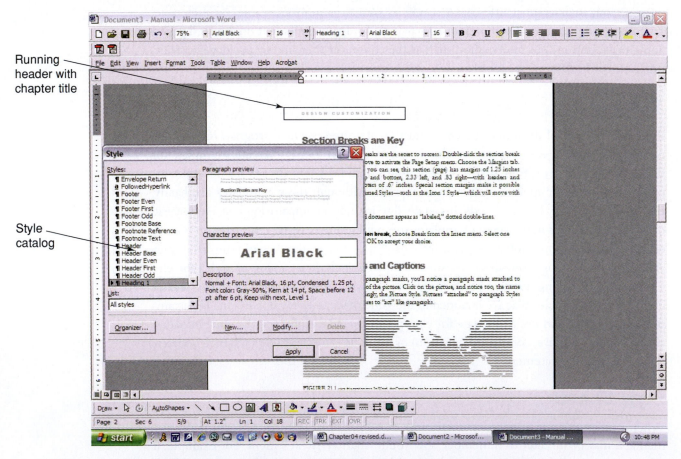

Running header with chapter title

Style catalog

■ **Figure 4–10** ■ Microsoft Word document template with style sheet

Source: "Microsoft product screen shot(s) reprinted with permission from Microsoft Corporation."

Learning to use the page design tools in your word-processing program can save you time and help you create consistent and professional-looking documents. However, these tools differ among the many word-processing programs (and sometimes from one version of a word-processing program to the next), so take the time to learn how to use the tools that are available in your word-processing software.

>>> Using the Elements of Page Design

Just as consistency and clarity are important in your writing, they are also important in your document design. A document that is cluttered with too many design elements can seem confusing and poorly thought out; however, consistent use of elements such as fonts and white space creates a sense of a unified document. At the same time, page design elements such as paragraphing and lists can help readers identify individual sections of a document. When you are designing your page layout, it is just as important to know your audience as it is when you are planning your document's text. Following are some questions to ask:

- Where will this document be used? At a desk? On a shop floor? In the field? The answers to these questions help you make decisions about page size and the number of pages.

- How will the document be used? Will it be kept on a desk as a reference? Read for background information for decision making? The answers to these questions help you make decisions about navigational devices.

- How do I want readers to perceive the document? As very businesslike? As friendly and easy to use? As complex and authoritative? The choices you make in font, use of color, headings, and graphics can influence how readers approach a document before they begin reading it.

Think of page design as a quick way of communicating the purpose and style of your document. A document that consists of densely packed text seems serious, even boring. A document that uses white space, headings, and lists seems clear and readable.

>>> Chapter Summary

This chapter shows you how to apply principles of page design to your assignments in this class and your on-the-job writing. The term *page design* refers to the array of design and layout options you can use to improve the visual effect of your document.

Effective page design requires that you use specific elements such as white space, headings, listings, and in-text emphasis. White space draws attention to adjacent items. Headings quickly lead the reader to important points and sub-points. Listings emphasize related groups of points, and conservative use of in-text emphasis, such as italics and bold-face print, can draw attention to items within sentences and paragraphs.

Another strategy for page design is to change the size and type of fonts in your documents. As with other strategies, this one must be used with care so that your document does not become too busy. Page design remains a technique for highlighting content, not a substitute for careful organization and editing.

The last several decades have seen an incredible change in the way documents are produced. Today, individual writers working at their personal computers write, edit, design, and print sophisticated documents. Writers who learn to use the tools available in word-processing programs find it easier to create consistent, professional-looking documents.

>>> Learning Portfolio

Communication Challenge "The St. Paul Style Guide: Trouble in the River City"

Frustrated by inconsistency in report styles, Elaine Johnson, a department manager at M-Global's St. Paul office, decided to take action. After collecting examples of office reports with diverse page designs, she met with her friend and branch manager Randall DiSalvo to complain.

"Enough is enough, Randall," Elaine said as the meeting started. "The technical staff produces all kinds of style, the administrative support staff doesn't know what designs are approved, and the clients get a fragmented image of the firm. Let's decide on one page design for reports and stay with it."

After an hour's talk, Elaine and Randall agreed that the office needed a style sheet to describe the required style for each document type written at the St. Paul branch. Busy with many other tasks, Randall told Elaine that he didn't have time to supervise the project, so he gave Elaine the authority to design what he wanted to be called the *Style Guide*. However, first she had to meet with all department managers and a few other employees about the project. Also, he asked that her first version of the guide be a modest one that covered only brief letter reports; later, the guide could be expanded.

What follows are details about (1) Elaine's process of gathering information, (2) some actual guidelines she decided to include in the *Style Guide*, and (3) some problems that arose with the project. This case study ends with questions and comments for discussion and an assignment for a written response to the Challenge.

Soliciting Opinions from Around the Office

The same morning she met with Randall, Elaine met with all five of her fellow department managers in the St. Paul office. Gathering in a meeting room overlooking the Mississippi River, they agreed immediately that the format problem needed to be solved.

The managers then concurred with the branch manager's idea about starting small—that is, covering only short letter reports now but later adding formal reports, proposals, manuals, letters, and memos, along with suggestions on style and grammar. Knowing that Elaine was one of the best writers and editors in the office, the managers said they were comfortable with her writing the manual herself. She could draw from whatever information she gathered from around the office and whatever guidelines she collected from her research on the subject. When the draft was com-

plete, she would run it by them for their comments. Then it would go to Randall for final approval before distribution to all branch employees.

Elaine could hardly believe it. In one morning, five department managers and the branch manager had reached consensus about the nature of the format problem and its solution. Buoyed by her success, Elaine was almost able to look beyond the fact that she had been given the job of writing the manual. Always the one to get a job done quickly, however, she moved to the following steps in the next few days:

- Notepad in hand, she interviewed all seven administrative assistants about their preferences in page design.
- She emailed all members of the professional staff, asking them to respond in writing in three days if they had specific preferences about the look they wanted to achieve in their letter reports.
- She called a local chapter of the Society for Technical Communication (STC), asking for references on page design.
- She located the four sources she received from STC and read them cover to cover.

By the end of the following week, Elaine was ready to begin writing the first draft of what would become the *Style Guide* for M-Global's St. Paul office.

Elaine's Format for Informal Reports

Elaine quickly developed a clear idea of what features should be part of M-Global's informal reports. To be sure, some of what she heard from department managers and other employees was at odds with her own views. For example, her preference for ragged-right margins in letter reports differed from that of many colleagues. When there were differences of opinion, she made decisions on the basis of her research and the level of response from those employees she surveyed. The following information summarizes some of the guidelines included in her draft:

1. Font choice should be 12-point New Century Schoolbook.
2. There should be 1.2-inch margins on the sides, a ½-inch margin on the top, and a 1-inch margin on the bottom (except for the first page, where letterhead requires the use of a top margin of 1 inch and a bottom margin of 1-1/2 inches).

3. Paragraphs should be block style without indenting the first line.

4. Text margins should be ragged-right, not fully justified.

5. Text should be single-spaced, with double-spacing between paragraphs.

6. Arrangement of date, inside address, and report title should be the same as that used in Model 8–1.

7. Every page after the first page should include a header. Placed in the top right corner (see margin guideline above), the header should include three single-spaced items: the company name (M-Global, Inc.—St. Paul), the project number (e.g., M-Global Project #134), and the date (e.g., July 29, 2009).

8. The heading system should follow this pattern:

Level 1 Heading

XXXXXXXXXXXXXXXXXXXXXXXXXXXXXXXXXXXX
XXXXXXXXXXXXXXXXXXXXXXXXXXXXXXXXXXX
(Futura 14-point bold with initial caps, on a separate line)

Level 2 Heading

XXXXXXXXXXXXXXXXXXXXXXXXXXXXXXXXXXXX
XXXXXXXXXXXXXXXXXXXXXXXXXXXXXXXXXXX
(Futura 12-point bold with initial caps, on a separate line)

Level 3 Heading.   XXXXXXXXXXXXXXXX

XXXXXXXXXXXXXXXXXXXXXXXXXXXXXXXXXXX
(Futura 12-point bold, in line with text)

9. Bulleted and numbered lists should be indented ½ inch from the left margin, with double spacing before and after the list and between items in the list.

Once her draft was approved by the branch manager, Elaine had 95 copies printed and distributed to all employees of the St. Paul office. She wrote a cover memo to accompany the *Style Guide,* explaining what it was and how it was to be used.

Questions and Comments for Discussion

Having finished her project by her deadline, Elaine was pleased. There would be clear guidelines for the staff, and the office would reap the rewards of a more efficient process of producing letter reports. As you look back on Elaine's activities and the guidelines she developed, consider the following questions and comments for discussion:

1. Elaine did a good job of seeking opinions of branch employees before she began her draft. Would it have been useful to consult with M-Global customers while the guide was being developed? Why or why not?

2. Should Elaine have tested the usefulness of the *Style Guide* before it was issued to all employees, or was her pilot draft approach adequate? If you think further testing was needed, what specifically would you have suggested?

3. Elaine chose to issue the final *Style Guide* through the office mail, with a cover memo. Was this strategy ideal? If so, why? If not, in what other way might she have introduced the manual?

4. Using Elaine's nine guidelines, edit the letter report included in Model 8–1 in this textbook. Do you think the revision is better designed than the original? Why or why not?

5. Elaine's manual provides a fairly rigid set of guidelines, as shown by the guidelines excerpt included in this case. Do you think a company should require employees to follow such a narrowly prescribed layout? Why or why not? Give the advantages and disadvantages of each point of view.

6. One engineer called Elaine to complain that the new guidelines did not allow him to use decimal numbered headings and subheadings in his short report. He said that he preferred such headings, and he suspected that his clients did as well. If you were in Elaine's position, how would you respond to this complaint?

7. If you were designing an M-Global *Style Guide* for short reports, what are some of the guidelines you would include, considering your own personal preferences?

Write About It

Taking the role of Elaine, write the cover memo for the style guide, indicating that (1) the guide should be considered the new model for all letter reports leaving the office, (2) it is a pilot draft that the office will review after six months, and (3) that the guide will be expanded later to include other documents. In your memo, explain how this style guide will benefit the St. Paul office and why employees should use it.

Collaboration at Work Design of the Campus Paper

General Instructions

Each Collaboration at Work exercise applies strategies for working in teams to chapter topics. The exercise assumes you (1) have been divided into teams of about three to six students, (2) will use team time inside or outside of class to complete the case, and (3) will produce an oral or written response. For guidelines about writing in teams, refer to pages 24–28.

Background for Assignment

Chapter 3 used the term *organization* to refer to the structure of ideas in your writing. This chapter introduces the term *page design* to refer to the arrangement of all visual elements—including text—on the page. Effective page design incorporates features such as white space, headings, lists, font size and type, and color. Choices in page design can greatly influence the interest readers show in a document.

This exercise concerns page design in a document with which you may be familiar—the campus print or online newspaper. Typical features of a college or university paper include the following:

- Campus news and events
- Local or regional news
- Updates on academic programs
- Updates on student organizations and activities
- Editorials
- Letters to the editor
- Advertisements

How these features are presented through effective (or ineffective) page design helps determine whether members of the campus community read the paper and take it seriously as journalism.

Team Assignment

The purpose of this assignment is to evaluate the page design of your campus newspaper. (If your campus has no paper, use a similar document available on or near campus.) Your team has two options for this assignment. Option 1 is to review a current issue of your campus newspaper, note elements of its page design, and determine whether these elements serve a useful purpose. Option 2 is to compare and contrast the effectiveness of the page design of two different campus newspapers or two different issues of the same paper.

Assignments

1. Team Evaluation of Page Design

Working in small teams, analyze the effectiveness of the page design of the document in Model 4–1 on pages 124–125. Your instructor will indicate whether you should prepare a written or an oral report of your findings. Give specific support for your praise or criticism. (Assignment 4 uses this same memorandum for a writing exercise.)

2. Individual Evaluation of Page Design

Option A Locate an example of technical writing, such as a user's manual or instructions. Use the guidelines in this chapter to analyze the document's page design. Your instructor will indicate whether your report should be oral or written.

Option B Use the guidelines in this chapter to evaluate the page design of one of the M-Global project sheets at the end of chapter 2 of this textbook. What works well? What could be improved? Be specific in your comments.

3. Individual Practice in Page Design

As a manager at M-Global, you have just finished a major report to a client. It gives recommendations for transporting a variety of hazardous materials by sea, land, and air. The body of your report contains a section that defines the term *stowage plan* and describes its use. Given your mixed technical and non-technical audience, this basic information is much needed. What follows is the text of that section. Revise the passage by applying any of this chapter's principles of page design that seem appropriate—such as adding headings, graphics, lists, and white space. If you wish, you may also make changes in organization and style. *Optional:* Share your version with another student to receive his or her response.

In the chemical shipping industry, a stowage plan is a kind of blueprint for a vessel. It lists all stowage tanks and provides information about tank volume, tank coating, stowed product, weight of product, loading port, and discharging port. A stowage plan is made out for each vessel on each voyage and records all chemicals loaded. The following information concerns cargo considerations (chemical properties and tank features) and some specific uses of the stowage plan in industry.

The three main cargo considerations in planning stowage are temperature, compatibility, and safety. Chemicals have physical properties that distinguish them from one another. To maintain the natural state of chemicals and to prevent alteration of their physical properties, a controlled environment becomes necessary. Some chemicals, for example, require firm temperature controls to maintain their physical characteristics and degree of viscosity (thickness) and to prevent contamination of the chemicals by any moisture in the tanks. In addition, some chemicals, like acids, react violently with each other and should not be stowed in adjoining, or even neighboring, tanks. In shipping, this relationship is known as chemical compatibility.

The controlled environment and compatibility of chemicals have resulted in safety regulations for the handling and transporting of these chemicals. These regulations originate with the federal government, which bases them on research done by the private manufacturers. Location and size of tanks also determine the placement of cargo. A ship's tanks are arranged with all smaller tanks around the periphery of the tank grouping and all larger tanks in the center. These tanks, made of heavy steel and coated with zinc or epoxy, are highly resistant to most chemicals, thereby reducing the chance of cargo contamination. Each tank has a maximum cargo capacity, and the amounts of each chemical are matched with the tanks. Often chemicals to be discharged at the same port are staggered in the stowage plan layout so that after they are discharged the ship maintains its equilibrium.

The stowage plan is finalized after considering the cargo and tank characteristics. In its final form, the plan is used as a reference document with all information relevant to the loading/discharging voyage recorded. If an accident occurs involving a ship, or when questions arise involving discharging operations, this document serves as a visual reference and brings about quick decisions.

4. Individual Practice in Word-Processing Tools

Review the summary of the guidelines in the Communication Challenge in this chapter (pp. 119–120). Use the Styles and Templates functions in your word processing program to create a template that follows the guidelines.

5. Team Practice in Page Design

Working in small teams, prepare a redesigned version of the memorandum in Model 4–1 (pp. 124–125). If your class is being held in a computer lab, present your team's version on-screen. If you are not using a lab, present your version on an overhead transparency.

6. Team Practice in Page Design: Using Computer Communication

This assignment is feasible only if you and your classmates have access to software that allows you to post messages to team members, edit on screen, and send edited copy back and forth. Your task is to add appropriate page design features to either (a) the *stowage plan* excerpt in assignment 3 or (b) any other piece of straight text permitted for use by your instructor. Choose a team leader who will collect and collate the individual edits. Choose another team member to type or scan the excerpt into the computer and then email the passage to other team members. Then each person should add the features desired and email the edited document to the team leader, who will collate the revisions and email the new version to team members for a final edit. Throughout this process, participants may conduct email conversations about the draft and resolve differences, if possible, before sending drafts to the leader. The team may need one or two short meetings in person, but most business should be conducted via the computer. The goal is to arrive at one final version for your team.

7. Individual or Team Practice in Organization and Page Design

The list that follows mostly includes exact wording or paraphrased excerpts from the National Center for Environmental Health Publication No. 01–0164—March 2001. (Some information has been slightly altered to accommodate this assignment, and much information in the publication has been left out.) Assume that the points are to be included as a *section* of a report you are producing. (NOTE: You are not being asked to produce a complete technical report, a subject covered in chapters 8 and 9.) First, arrange the information in an order that generally follows the ABC format described in chapter 3, eliminating any possible redundant information or any items that do not seem to fit the report section you are producing. Second, make adjustments in wording or style you consider appropriate. Third, add appropriate elements of page design that have been covered in this chapter.

A. This report will be followed by yearly updates. Future reports will attempt to answer the following questions:

1. Are exposure levels increasing or decreasing over time?
2. Are public health efforts to reduce exposure working?
3. Do certain teams of people have higher levels of exposure than others?

B. *Cotinine* is a metabolite of nicotine that tracks exposure to environmental tobacco smoke (ETS) among nonsmokers—higher levels reflect more exposure to ETS.

C. An environmental chemical is a chemical compound or chemical element in air, water, soil, dust, food, or other environmental media.

D. *Biomonitoring* is the assessment of human exposure to environmental chemicals by measuring the chemicals (or their breakdown products) in human specimens, such as blood or urine.

E. Because the sample size was relatively small and because the sampling was only conducted in 12 locations, the data cannot be considered conclusive. Additional studies should be conducted.

F. It should be noted that just because people have an environmental chemical in their blood or urine does not mean that the chemical causes disease. Research studies, separate from this report, are required to determine at what level the chemical may cause disease and what levels are of negligible health concern.

G. The reduction in cotinine levels reflected in the attached table indicates a dramatic reduction in exposure of the general population to environmental tobacco smoke since 1988–1991.

H. Special populations of children at high risk for lead exposure (e.g., those living in homes containing lead-based paint or lead-contaminated dust) remain a major public health concern.

I. The report provides new data on blood mercury levels among children ages 1–5 years and among women of childbearing age (16–49 years).

J. This report must be updated each year. The results from all samplings of the 27 chemicals at the 12 locations are included on the attached table. Most of the data will not be considered conclusive, and therefore will not be commented on until we have results for additional years, which will give more support for comments on trends.

K. This report measures the exposure (through biomonitoring) of a sample population to 27 environmental chemicals, which include 13 metals (antimony, barium, beryllium, cadmium, cesium, cobalt, lead, mercury, molybdenum, platinum, thallium, tungsten, and uranium), 6 organophosphate pesticide metabolites, 7 phthalate metabolites, and cotinine.

L. The results showed that levels of some metabolites in the sample population were considerably higher than other metabolites, indicating a need for further study.

M. Compared with an adult, the fetus and children are usually more vulnerable to the effects of metals, such as mercury. One goal of collecting such data is to better estimate health risks for the fetus, children, and women of childbearing age from potential exposures to mercury.

N. One noteworthy conclusion is that lead levels in the blood continue to decline among U.S. children when considered as a team, highlighting the success of public health efforts to decrease the exposure of children to lead.

O. Because more than half of American youth are still exposed to ETS, it remains a major public health concern.

P. Plans are to expand the list of measured chemicals from 27 to approximately 100.

⑧ 8. Ethics Assignment

Page design greatly influences the way people read documents no matter what message is being delivered. Even a product, a service, or an idea that could be considered harmful to the individual or public good can be made to seem more acceptable by a well-designed piece of writing. Find a well-designed document that promotes what is, in your opinion, a harmful product, service, or idea. Explain why you think the writers made the design decisions they did. Although your example may include graphics, focus mainly on page design covered in this chapter.

🌐 9. International Communication Assignment

Collect one or more samples of business or technical writing that originate in—or are designed for—cultures outside the United States. (Use either print examples or examples found on the Internet.) Comment on features of page design in the samples. If applicable, indicate how such features differ from those evident in business and technical writing designed for an audience within the United States.

🔵 10. A.C.T. N.O.W. Assignment (Applying Communication To Nurture Our World)

For this assignment, use the same context as that described in the A.C.T. N.O.W. exercise in chapter 3 (p. 96). Depending on the instructions you are given, prepare an oral or a written report in which you (a) analyze the degree to which the document subscribes to page design guidelines mentioned in this chapter and (b) offer suggestions about how the document's page design might be improved so that it accomplishes more effectively what you believe to be the purpose of the document.

MEMORANDUM

DATE: August 19, 2007
TO: Randall Demorest, Dean
FROM: Kenneth Payne, Professor and Head *KP*
SUBJECT: BSTC Advisory Board

What? Lunch meetings between Advisory Board members and me
Why?   To get more Board support for the BSTW degree program
Who?   Each individual member at a separate luncheon
When? Fall 2007
How?   Allocation of $360 to pay for the lunches

Rationale

When we seek support for the college, we have to (1) make people feel that they will get something in return and (2) make them feel comfortable about us and our organization. As businesses have demonstrated, one way we can accomplish these goals is by taking potential donors to lunch.

As you and I have discussed, the B.S. in Technical Communication degree program (BSTC) needs to strengthen ties to its Advisory Board. We must ask Board participants to provide tangible support for the program *and* give them meaningful involvement in the work we are doing.

Method

The immediate need is to involve members of the Advisory Board in the coming year's program. I want to do this in two ways:

1. Plan carefully for a fall Board meeting
2. Discuss with each of them individually what we want to accomplish this year

Cost

To do the second item mentioned, I request an allocation of $360 so that I can take each member to lunch for an extended one-on-one discussion. I plan to discuss the needs of our program and each member's capabilities to support it.

M-Global Inc | 127 Rainbow Lane | Baltimore MD 21202 | 410.555.8175

■ **Model 4–1** ■ Page design in memorandum

Payne to Demorest, 2

Specifics

Each member of the Board will be asked individually to consider the following ways to contribute:

1. Continuing support for the internship program
2. Participation in the research project we began a year ago
3. Cooperative work experiences for BSTC faculty, possibly during the summer of 2008. Financial support for the following items:
   The college's membership as a sponsoring organization in the Society for Technical Communication
   - Contributions——financial or otherwise——to library holdings in technical writing
   - Usability testing laboratory
   - A workshop series bringing to the campus some outstanding technical communicators (for example, Edward Tufte, expert in graphics; JoAnn Hackos, expert in quality management; and William Horton, expert in online documentation)

Benefits

What are Board members going to get from this?

Long range: A better BSTC program, which will produce better technical communicators for them to hire

Immediately: Meaningful involvement in the program

Specifically: Training opportunities for their personnel through the workshops mentioned

My tentative plan for those workshops is to provide a one-day seminar for our students and a second seminar for employees of Advisory Board members. (We will allow them a number of participants based on how much they contribute to the workshops.)

Response Needed

Please let me know as soon as possible if money is available for the lunches. I hope to begin scheduling meetings within a week.

■ **Model 4–1** ■ continued

Chapter | 5 | Letters, Memos, and Electronic Communication

>>> Chapter Outline

General Guidelines for Correspondence 127

Letters 133
   Positive Letters 134
   Negative Letters 135
   Neutral Letters 136
   Sales Letters 137

Memoranda 138

Email 140
   Guidelines for Email 140
   Appropriate Use and Style for Email 144

Memoranda versus Email 145

Chapter Summary 146

Learning Portfolio 147
   Communication Challenge 147
   Collaboration at Work 148
   Assignments 148

Models for Good Writing 154
   Model 5–1: M-Global sample letter 154
   Model 5–2: M-Global sample memo 155

Model 5–3: M-Global sample email 157

Model 5–4: Block style for letters 158

Model 5–5: Modified block style (with indented paragraphs) for letters 159

Model 5–6: Simplified style for letters 160

Model 5–7: Memo style 161

Model 5–8: Positive letter in block style 162

Model 5–9: Negative letter in modified block style (with indented paragraphs) 163

Model 5–10: Neutral letter (invitation) in block style 164

Model 5–11: Neutral letter (placing order) in simplified style 165

Model 5–12: Sales letter in simplified style 166

Model 5–13: Memorandum: changes in services 167

Model 5–14: Memorandum: changes in benefits 168

Model 5–15: Email: changes in procedure 169

Marie Stargill, M-Global's fire-science expert, just returned from a seminar that emphasized new techniques for preventing injuries from job-site fires. Within 24 hours of her return, she has already done three things:

1. Written her manager an email message over the office computer network

2. Sent a letter to an M-Global client suggesting use of fire-retardant gloves she learned about at the seminar

3. Sent the conference director a letter of appreciation about the meeting

Like Marie, you will write many letters, memos, and emails in your career. In fact, you probably will write more of this correspondence than any other type of document.

Letters, memos, and emails are short documents written to accomplish a limited purpose. Letters are directed outside your organization, memos are directed within your organization, and email can be directed to either an external or an internal audience. (Longer, more complicated letters and memos—called *letter reports* and *memo reports*—are covered in chapter 8.) Here are some working definitions:

Letter: A document that conveys information to a member of one organization from someone outside that same organization. Letters usually cover one major point and fit on one page. This chapter classifies letters into these four groups, according to type of message: (1) positive, (2) negative, (3) neutral, and (4) sales.

Memorandum: A document written from a member of an organization to one or more members of the same organization. Abbreviated *memo,* it usually covers just one main point and no more than a few. Readers prefer one-page memos.

Email: A document written often in an informal style either to members of one's own organization or to an external audience. Email messages usually cover one main point. Characterized by the speed with which it is written and delivered, an email can include more formal attachments to be read and possibly printed by the audience.

As with other forms of technical communication, your ability to write good memos, letters, and emails depends on a clear sense of purpose, thorough understanding of reader needs, and close attention to correct formats. This chapter prepares you for this challenge by presenting sections that cover (1) general rules that apply to all workplace correspondence; and (2) specific formats for positive letters, negative letters, neutral letters, sales letters, memoranda, and email. Job letters and resumes are discussed in a separate chapter on the job search (chapter 16).

>>> General Guidelines for Correspondence

Letters convey your message to readers outside your organization, just as memos are an effective way to get things done within your own organization, and email is a way to communicate quickly with readers inside and outside of your organization. By applying the guidelines in this chapter, you can master the craft of writing effective correspondence. You must plan, draft, and revise each letter, memo, and email as if your job depends on it—for it may.

Refer to Models 5–1, 5–2, and 5–3 on pages 154–157 for M-Global examples that demonstrate the guidelines that follow. Later examples in this chapter show specific types of letters and additional memos.

>> Correspondence Guideline 1: Know Your Purpose

Before beginning your draft, write down your purpose in one clear sentence. This approach forces you to sift through details to find a main reason for writing every letter or memo. This *purpose sentence* often becomes one of the first sentences in the document. Following are some samples:

- Letter purpose sentence: "As you requested yesterday, I'm sending samples of the new candy brands you are considering placing in M-Global's office vending machines."
- Memo purpose sentence: "This memo explains M-Global's new policy for selecting rental cars on business trips."
- Email purpose sentence: "I have attached the most recent draft of the proposal for the PI Corp. pipeline project."

Some purpose statements are implied; others are stated. An implied purpose statement occurs in the second paragraph of Model 5–1. That paragraph shows that the writer wishes both to respond to requests for M-Global brochures and, just as important, to seek the professor's help in soliciting good graduates for M-Global's Atlanta office. In a sense, one purpose leads into the other. Model 5–2 shows a more obvious purpose statement in the second sentence.

>> Correspondence Guideline 2: Know Your Readers

Who are you trying to inform or influence? The answer to this question affects the vocabulary you choose, the arguments you make, and the tone you adopt. Pay particular attention when correspondence will be read by more than one person. If these readers are from different technical levels or different administrative levels within an organization, the challenge increases. A complex audience compels you to either (1) reduce the level of technicality to that which can be understood by all readers or (2) write different parts of the document for different readers.

Model 5–1 is directed to a professor with whom the writer wants to develop a reciprocal relationship—that is, George Lux gives free guest lectures in civil-engineering classes, hoping the professor will in turn help him inform potential job applicants about M-Global. Model 5–2, directed to an in-house technical audience, contains fairly general information about the new technical editor. This information applies to, and should be understood by, all readers. Model 5–3 is typical of email between people who know each other professionally. It has a conversational tone not found in other forms of correspondence.

>> Correspondence Guideline 3: Follow Correct Format

Most organizations adopt letter and memo formats that must be used uniformly by all employees. Following are the basic guidelines:

■ **Letters:** There are three main letter formats—block, modified block, and simplified. Models 5–4, 5–5, and 5–6 on pages 158–160 show the basic page design of each; letter examples throughout the chapter use the three formats. As noted, you usually follow the preferred format of your own organization.

Addresses on envelopes and in letters should use the format recommended by the United States Postal Service. Addresses should include no more than four lines, and should not include punctuation such as commas or periods.

■ **Memos:** With minor variations, all memos look much the same. The obligatory "Date/To/From/Subject" information hangs at the top left margin, in whatever order your organization requires. Model 5–7 on page 161 shows one basic format. These four lines allow you to dispense with lengthy introductory passages seen in more formal documents. Note that the sender signs his or her initials after or above the typed name in the "From" line.

■ **Letters and Memos:** Some format conventions apply to both letters and memos. Three of the more important features are:

Facsimile reference: Readers often need to know—for convenience and for the record—when memos or letters have been sent by fax. Type FAX TRANSMISSION or FACSIMILE before the "Date/To/From/Subject" lines for a memo and between the date and inside address for a letter. This fax line also can be used for other similar notations such as CONFIDENTIAL, PERSONAL, or REGISTERED.

Reference initials: If the document has been typed by someone other than the writer, place the typist's initials two lines beneath the signature block for letters and below the last paragraph for memos (e.g., jt). Some organizations prefer that the initials of the writer also be included, followed by those of the typist (e.g., GTY/jt).

Enclosure notation: If attachments or enclosures accompany the letter or memo, type the singular or plural form of "Enclosure" or "Attachment" one or two lines beneath the reference initials. Some writers also list the item itself (e.g., Enclosure: Code of Ethics).

Multiple-page headings: Each page after the first page often has a heading that includes the name of the person or company receiving the letter or memo, the date, and the page number. Some organizations may prefer an abbreviated form such as "Jones to Bingham, 2," without the date.

■ **Email:** Computers and email systems handle formatting of texts and special characters differently, so you should format your email so that it can be read on any computer. Use your system's default font, and avoid highlighting, color, bold, italics, and underlining. Emails are generally short, no longer than what can be seen on a computer screen all at once, and paragraphs should be short. Some email systems can't translate tabs, so use lines of white space between paragraphs.

In memos and emails, give the subject line special attention, because it telegraphs meaning to the audience immediately. In fact, readers use it to decide when, or if, they will read the complete correspondence. Be brief, but also engage interest. For example, the subject line of the Model 5–2 memo could have been "Editing." Yet that brevity would have sacrificed reader interest. The actual subject line, "New Employee to Help with Technical Editing," conveys more information and shows readers that the contents of the memo will make their lives easier.

■ **Letters, memos, and email:** Some format conventions apply to all three forms of correspondence. Two of the more important features are:

Copy notation: If the correspondence has been sent to anyone other than the recipient, type "Copy" or "Copies" one or two lines beneath the enclosure notation, followed by the name(s) of the person or persons receiving copies (e.g., Copy: Preston Hinkley). Some organizations prefer the initials "c" (for *copy*), "cc" (for *carbon copy,* even though carbons rarely exist anymore), or "pc" (for *photocopy*). Email inserts this information automatically. If you are sending a copy but do not want the original letter or memo to include a reference to that copy, write "bc" (for *blind copy*) and the person's name only on the copy—not on the original (e.g., bc: Jim McDuff). (Note: Send blind copies only when you are certain it is appropriate and ethical to do so.)

Postscripts: Items marked "PS" or "P.S." appear occasionally in letters and rarely in memos. They are considered by many readers to be symbols of poor planning, so use them with caution. If used, they appear as the last item on the document (beneath the copy notation) and can be typed or written in longhand.

>> Correspondence Guideline 4: Follow the ABC Format for All Correspondence

Correspondence subscribes to the same three-part ABC (Abstract/Body/Conclusion) format used throughout this book. This approach responds to each reader's need to know "What does this document have to do with me?" According to the ABC format, your correspondence is composed of these three main sections:

■ **Abstract:** The abstract introduces the purpose and usually gives a summary of main points to follow. It includes one or two short paragraphs.

■ **Body:** The body contains supporting details and thus makes up the largest part of a letter or memo. You can help your readers by using such techniques as:

Deductive patterns for paragraphs: In this general-to-specific plan, your first sentence should state the point that helps the reader understand the rest of the paragraph. This pattern avoids burying important points in the middle or end of the paragraph, where they might be missed. Fast readers tend to focus on paragraph beginnings and expect to find crucial information there. Note how most paragraphs in Model 5–2 follow this format.

Personal names: If they know you, readers like to see their names in the body of the letter or memo, or in the salutation of an email. Your effort here shows concern for the reader's perspective, gives the correspondence a personal touch, and helps strengthen your personal relationship with the reader. (See the last paragraph in Model 5–1 and Model 5–3.) Of course, the same technique in direct mail can sometimes backfire, because it is an obvious ploy to create an artificially personal relationship.

Lists that break up the text: Listed points are a good strategy for highlighting details. Readers are especially attracted to groupings of three items, which create a certain rhythm, attract attention, and encourage recall. Use bullets, numbers, dashes, or other typographic techniques to signal the listed items. For example, the bulleted list in Model 5–1 draws attention to three important points about M-Global that the writer wants to emphasize. Because some email systems can't read special characters like bullets, use asterisks or dashes for lists in email.

Strongest points first or last: If your correspondence presents support or makes an argument, include the most important points at the beginning or at the end—not in the middle. For example, Model 5–2 begins and ends with two crucial issues: the effect of poor editing on company productivity and the need to decide on specific ways the new editor can improve writing at M-Global.

Headings to divide information: One-page letters and memos, and even email, sometimes benefit from the emphasis achieved by headings. The three headings in Model 5–2 quickly steer the reader to main parts of the document.

■ **Conclusion:** Readers remember first what they read last. The final paragraph of your correspondence should leave the reader with an important piece of information—for example, (1) a summary of the main idea or (2) a clear statement of what will happen next. The Model 5–1 letter makes an offer that helps continue the reader's association with the university, whereas the Model 5–2 memo gives readers a specific task to accomplish before the next meeting.

Your final paragraph in external correspondence should always continue the business relationship by encouraging future contact. Internal correspondence may also include a statement offering to answer questions or concerns. The final paragraph in the Model 5–1 letter encourages the reader to call, and the final paragraph in the original email in Model 5–3 specifically encourages the recipient to respond with questions or concerns.

>> Correspondence Guideline 5: Use the 3Cs Strategy for Persuasive Messages

The ABC format provides a way to organize all letters and memos. Another pattern of organization for you to use is the *3Cs strategy*—especially when your correspondence has a persuasive objective. This strategy has three main goals:

■ **Capture** the reader's interest with a good opener, which tells the reader what the letter, memo, or email can do for him or her.

■ **Convince** the reader with supporting points, all of which confirm the opening point that this document will make life easier.

■ **Contact** solidifies your relationship with the reader with an offer to follow up on the correspondence.

Although neither Model 5–1 nor Model 5–2 is overtly persuasive, each has an underlying persuasive purpose, as does the original message in Model 5–3. Note how each uses the 3Cs strategy.

>> Correspondence Guideline 6: Stress the "You" Attitude

As noted earlier, using the reader's name in the body helps convey interest. However, your efforts to see things from the reader's perspective must go deeper than a name reference. For example, you should perform the following tasks:

■ Anticipate questions your reader might raise and then answer these questions. You can even follow an actual question ("And how will our new testing lab help your firm?") with an answer ("Now M-Global's labs can process samples in 24 hours").

■ Replace the pronouns *I, me,* and *we* with *you* and *your.* Of course, you must use first-person pronouns at certain points in a letter, but many pronouns should be second person. The technique is quite simple. You can change almost any sentence from writer-focused prose ("We feel that this new service will . . .") to reader-focused prose ("You'll find that this new service will . . .").

Model 5–1 shows this *you* attitude by emphasizing what M-Global and the writer himself can do for the professor and his students. Model 5–2 shows it by emphasizing that the new editor will make the readers' jobs easier.

>> Correspondence Guideline 7: Use Attachments for Details

Keep text brief by placing details in attachments, which readers can examine later, rather than bogging down the middle of the letter or memo. This way, the supporting facts are available for future reference, without distracting readers from the main message. The memo in Model 5–2, for example, includes a list of possible job tasks for the new M-Global editor. The listing would only clutter the body of the memo, especially because its purpose is to stimulate discussion at the next meeting.

>> Correspondence Guideline 8: Be Diplomatic

Without a tactful tone, all your planning and drafting will be wasted. Choose words that persuade and cajole, not demand. Be especially careful of memos written to subordinates. If you sound too authoritarian, your message may be ignored—even if it is clear that what you are suggesting will help the readers. Generally speaking, negative (or "bad news") letters often use the passive voice, whereas positive (or "good news") letters often use the active voice.

For example, the letter in Model 5–1 would fail in its purpose if it sounded too pushy and one-sided about M-Global's interest in hiring graduates. Similarly, the editing memo in Model 5–2 would be poorly received if it used stuffy and condescending wording, such as, "Be advised that starting next month, you are to make use of proofreading services provided in-house by. . . ."

>> Correspondence Guideline 9: Edit Carefully

Because letters, memos, and email are short, editing errors may be obvious to readers. Take special care to avoid the following errors:

■ Mechanics
  • Misspelled words of any kind, but especially the reader's name
  • Wrong job title (call the reader's office to double-check, if necessary)
  • Old address (again, call the reader's office to check)

- Grammar
  - Subject–verb and pronoun nonagreement
  - Misused commas

- Style
  - Stuffy phrases, such as, "Per your request" and "Enclosed herewith"
  - Long sentences with more than one main and one dependent clause
  - Presumptuous phrases, such as, "Thanking you in advance for . . ."
  - Negative tone suggested by phrases such as "We cannot," "I won't," and "Please don't hesitate to"

The last point is crucial and gets more attention later in this chapter. Use the editing stage to rewrite any passage that could be phrased in a more positive tone. You must always keep the reader's goodwill, no matter what the message.

>> Correspondence Guideline 10: Respond Quickly

A letter, memo, or email that comes too late fails in its purpose, no matter how well written. Mail letters within 48 hours of your contact with, or request from, the reader. Send memos in plenty of time for your reader to make the appropriate adjustments in schedule, behavior, and so forth. Respond to emails the same day you receive them. This rule applies, for example, to correspondence written in response to the following situations:

- You want to write a follow-up letter after meeting or talking with a client.
- A customer requests information about a product or service.
- You discover that there will be a delay in your supply of a product or service to a customer.
- You select a candidate to interview for a position.
- You announce a change in company policy.
- You set the time for a company meeting.

The first sentence in the Model 5–1 letter, for example, shows that George Lux writes the day after his guest lecture. This responsiveness helps secure the goodwill of the professor.

>>> Letters

Letters are to your clients and vendors what memos are to your colleagues. They relay information quickly and keep business flowing. This section gives you specific guidelines for the following documents:

- Letters with a positive message
- Letters with a negative message
- Letters with a neutral message
- Letters with a sales message

To be sure, many documents are hybrid forms that combine these patterns. As a technical sales expert for M-Global, for example, you may be writing to answer a customer question about a new piece of equipment just purchased from M-Global's Equipment Development group. Your main task is to solve a problem caused by a confusing passage in the owner's manual. At the same time, however, your concern for the customer's satisfaction can pave the way for purchase of a second machine later in the year. Thus the letter has both a positive message and a sales message. This example also points to a common thread that weaves all four letter types together: the need to maintain the reader's goodwill toward you and your organization.

The next four sections present (1) a pattern for each type of correspondence, based on the ABC (Abstract/Body/Conclusion) format used throughout this text; and (2) one or more brief case studies in which the pattern might be used at M-Global.

Positive Letters

Everyone likes to give good news; fortunately, you will often be in the position of providing it when you write. Following are some sample situations:

- Replying to a question about products or services
- Acknowledging that an order has been received
- Recommending a colleague for a promotion or job
- Responding favorably to a routine request
- Responding favorably to a complaint or an adjustment
- Hiring an employee

The trick is to recognize the good-news potential of many situations. This section gives you an all-purpose format for positive letters, followed by a case study from M-Global.

ABC Format: **Positive Letter**

- **ABSTRACT:** Bridge between this letter and last communication with person
  - Clear statement of good news you have to report
- **BODY:** Supporting data for main point mentioned in abstract
  - Clarification of any questions reader may have
  - Qualification, if any, of the good news
- **CONCLUSION:** Statement of eagerness to continue relationship, complete project, etc.
  - Clear statement, if appropriate, of what step should come next

ABC Format for Positive Letters

All positive letters follow one overriding rule. You must always:

State good news immediately!

Any delay gives readers the chance to wonder whether the news will be good or bad, thus causing momentary confusion. On the left is a complete outline for positive letters that corresponds to the ABC format.

M-Global Case Study for a Positive Letter

As a project manager at M-Global's Houston office, Nancy Slade has agreed to complete a foundation investigation for a large church about 300 miles away. There are cracks in the basement floor slab and doors that do not close, so her crew needs a day to analyze the problem (observing the site, measuring walls, digging soil borings, taking samples, etc.). She took this small job on the condition that she could schedule it around several larger (and more profitable) projects in the same area during mid-August.

Yesterday, Nancy received a letter from the minister (speaking for the church committee), who requested that M-Global change the date. He was just asked by the regional headquarters to host a three-day conference at the church during the same time that M-Global was originally scheduled to complete the project.

After checking her project schedule, Nancy determines that she can reschedule the church job. Model 5–8 on page 162 shows her response to the minister.

Negative Letters

It would be nice if all your letters could be as positive as those just described. Unfortunately, the real world does not work like that. You will have many opportunities to display both tact and clarity in relating negative information. Following are a few cases:

- Explaining delays in projects or delivery of services
- Declining invitations or requests
- Registering complaints about products or services
- Refusing to make adjustments based on complaints
- Denying credit
- Giving bad news about employment or performance
- Explaining changes from original orders

This section gives you a format to follow in writing sensitive letters with negative information. Then it provides one application at M-Global.

ABC Format for Negative Letters

One main rule applies to all negative letters:

Buffer the bad news, but still be clear.

Despite the bad news, you want to keep the reader's goodwill. Spend time at the beginning building your relationship with the reader by introducing less controversial information—before you zero in on the main message. On the right is an overall pattern to apply in each negative letter.

ABC Format: **Negative Letter**

- **ABSTRACT:** Bridge between your letter and previous communication
  - General statement of purpose or appreciation—in an effort to find common bond or area of agreement
- **BODY:** Strong emphasis on what can be done, when possible
  - Buffered yet clear statement of what cannot be done, with clear statement of reasons for negative news
  - Facts that support your views
- **CONCLUSION:** Closing remarks that express interest in continued association
  - Statement, if appropriate, of what will happen next

M-Global Case Study for a Negative Letter

Reread the letter situation described in the section on positive letters. Now, assume that instead of being able to comply with the minister's request, the writer is unable to complete the work on another date without changing the fee. This change is necessary because Nancy must send a new crew 300 miles to the site, rather than using a crew already working on a nearby project.

Nancy knows the church is on a tight budget, but she also knows that M-Global would not be in business too long by working for free. Most importantly, because the church is asking for a change in the original agreement, she believes it is fair to request a change in the fee. Model 5–9 on page 163 is the letter she sends. Note her effort to buffer the negative news.

Neutral Letters

Some letters express neither positive nor negative news. They are simply the routine correspondence written every day to keep businesses and other organizations operating. Some situations follow:

- Requesting information about a product or service
- Inviting the reader to an event
- Responding to an invitation or a routine request
- Placing orders
- Providing a transmittal letter for fax transmissions
- Sending solicited or unsolicited items through the mail

Use the following outline in writing your neutral letters. Also, refer to the M-Global examples that follow the outline.

ABC Format for Neutral Letters

Because the reader usually has no personal stake in the news, neutral letters require less emphasis on tone and tact than other types, yet they still require careful planning. In particular, always abide by this main rule:

Be absolutely clear about your inquiry or response.

Neutral letters operate a bit like good-news letters. You must make your point early, without giving the reader time to wonder about your message. Neutral letters vary greatly in specific organization patterns. The *umbrella plan* suggested here emphasizes the main criterion of clarity.

ABC Format: **Neutral Letters**

- **ABSTRACT:** Bridge or transition between letter and previous communication, if any
  - Precise purpose of letter (e.g., request, invitation, response to invitation)
- **BODY:** Details that support the purpose statement (e.g., a description of item(s) requested, the requirements related to the invitation, a description of item(s) being sent)
- **CONCLUSION:** Statement of appreciation
  - Description of actions that should occur next

M-Global Case Studies for Neutral Letters

Letters with neutral messages get written by the hundreds each week at M-Global. Following are four situations that require a neutral letter; items 2 and 4 provide the context for the examples in Model 5–10 on page 164 and Model 5–11 on page 165:

1. Zach Bowers, a lab assistant, writes a laboratory supply company for information about a new unbreakable beaker to use in testing.
2. Faron Abdullah, president of the Student Government Association at River College, asks representatives of M-Global's St. Louis office to attend a career fair.
3. Donna Martinich, a geologist, responds to the request of a past client for a copy of a report done three years ago.
4. Farah Linkletter, a supply assistant with M-Global's San Francisco office, orders three new transits, making sure to emphasize that one is not to include a field case.

Sales Letters

On hearing the term *sales letter,* some people have visions of direct-mail requests for magazine subscriptions, vacation land, or diet plans. In this text, however, sales letters mean something quite different. They name all your correspondence with a customer—from the first contact letter through the last thank-you note. This list gives you some idea of the possibilities for sales letters:

- Starting a relationship ("I'll be calling you . . . ")
- Following a phone call ("Good talking to you . . . . Can we meet to discuss your needs regarding . . .")
- Following a meeting ("You mentioned that you could use more information . . . so here's a brochure on . . .")
- Following completion of sale or project ("We enjoyed working with you on . . .")
- Seeking repeat business ("I'd like to know how the new machinery has been working . . .")

Notice that sales letters almost always work together with personal contacts, such as meetings and phone calls. Your goal is to build a continuing relationship with the customer. Consult the following outline when writing sales letters for any context; the M-Global example shows the outline in action.

ABC Format for Sales Letters

The one main rule that governs all sales letters is as follows:

Help readers solve their problems.

Customers are interested in your product or service only insofar as it can assist them. You must engage the readers' interest by showing that you understand their needs and can help fulfill them. On the next page is a plan for writing a successful sales letter. Note

ABC Format: **Sales Letters**

- **ABSTRACT:** (Choose one or two to capture attention)
  - Cite a surprising fact
  - Announce a new product or service that client needs
  - Ask a question
  - Show understanding of a client's problem
  - Show potential for solving a client's problem
  - Present a testimonial
  - Make a challenging claim
  - Summarize results of a meeting
  - Answer a question reader previously asked
- **BODY:** (Choose one or two to convince the reader)
  - Stress one main problem reader has concern about
  - Stress one main selling point of your solution
  - Emphasize what is unique about your solution
  - Focus on value and quality, rather than price
  - Put details in enclosures
  - Briefly explain the value of any enclosures
- **CONCLUSION:** (Offer contact for the next step in sales process)
  - Leave the reader with one crucial point to remember
  - Offer to call (first choice) or ask reader to call (last choice)

reference to the 3Cs (Capture / Convince / Contact) strategy mentioned earlier in the chapter.

M-Global Case Study for a Sales Letter

M-Global provides customers with professional services and equipment, so sales letters have an important place in the firm. Benjamin Feinstein is one employee who writes them almost every day. As a first-year employee with a degree in industrial hygiene, Benjamin works in the newly formed asbestos-abatement group. Basically, he helps clients find out if there is any asbestos that must be removed from structures, recommends a plan for removal, and has the work done by another division of M-Global.

Following is one series of sales contacts that involves several letters. First, Benjamin sent *cold call* sales letters to 100 schools and small businesses in the St. Louis area, suggesting that they might want to have their structures checked for unsafe levels of asbestos. The letter contained a reply card. After calling and then meeting with a number of the respondents, he sent individualized follow-up letters that answered questions that came up in discussions and provided additional information. After another series of phone calls and meetings with some of the potential customers, he negotiated contracts with five of the businesses and completed the projects. Then within a few months of completion, he sent a final letter proposing additional M-Global services and began the cycle again. Model 5–12 on page 166 provides a sample sales letter that Benjamin used at the beginning of the cycle.

>>> Memoranda

Even though email has become common in the workplace, *memoranda* (the plural of *memorandum,* also called *memos*) are still important. You write them to peers, subordinates, and superiors in your organization—from the first days of your career until you retire. Even if you work in an organization that uses email extensively, you will still compose print messages that convey your point with brevity, clarity, and tact.

Memos can contain all four types of messages discussed with respect to letters—positive, negative, neutral, and sales. Following are some situations that require memos in an organization.

Positive

- Announcing high bonuses for the fiscal year
- Commending an employee for performance on a project
- Informing employees about improved fringe benefits

Negative

- Reporting decreased quarterly revenues for the year
- Requesting closer attention to filling out time sheets
- Asking for volunteers to work on a holiday

Neutral

- Announcing a meeting
- Summarizing the results of a meeting with a client
- Explaining a new laboratory procedure

Sales

- Requesting funding for a training seminar
- Recommending another staff member for the proposals unit
- Suggesting changes in the performance evaluation system

Below is an ABC format for memos, along with several case studies from M-Global.

ABC Format for Memoranda

Abide by this one main rule in every memo-writing situation:

Be clear, brief, and tactful.

Because many activities are competing for their time, readers expect information to be related as quickly and clearly as possible; however, you must be sure not to sacrifice tact and sensitivity as you strive to achieve conciseness. This ABC format helps you accomplish both goals.

M-Global Case Studies for Printed Memoranda

ABC Format: Memoranda

- **ABSTRACT:** Clear statement of memo's purpose
  - Outline of main parts of memo
- **BODY:** Supporting points, with strong points at the beginning or end
  - Frequent use of short paragraphs or listed items
  - Absolute clarity about what memo has to do with reader
  - Tactful presentation of any negative news
  - Reference to attachments, when much detail is required
- **CONCLUSION:** Clear statement of what step should occur next
  - Another effort to retain goodwill and cooperation of readers

At M-Global, memoranda are written to and from employees at all levels. To reflect this diversity, this section describes two different contexts for writing memoranda and shows accompanying examples.

In the first context, the lead secretary at M-Global's Baltimore office chaired an office committee to improve efficiency in using the centralized copy center. The committee was formed when the branch manager realized that many technical staff members had not been trained to use the center. This lack of training led to sloppy habits and loss of productivity. Rather than issue a "dictum" from his office, the branch manager established a small committee to review the problem and issue guidelines to the office staff. The memorandum in Model 5–13 resulted from the committee's meetings.

In the second M-Global case, the St. Louis personnel director, Timothy Fu, must announce several changes in benefits to the entire staff. Some changes are good news in that they expand employee benefits, and others are bad news in that they further limit benefits. Timothy has the difficult task of imparting both types of information in one memo to a broad audience. The memorandum in Model 5–14 is a result of his efforts.

>>> Email

Electronic communication (*email*) has become the preferred means of communication for many people in their professional lives and, often, in their personal lives. Some of us receive 100 or more messages a day. Any medium so widely used deserves special attention in a chapter on correspondence.

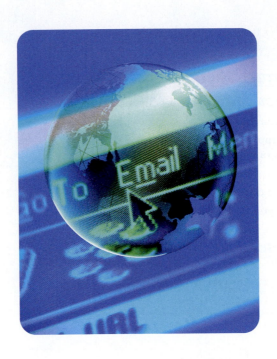

Guidelines for Email

When writing email, you should try to strike a balance between speed of delivery on one hand and quality of the communication to your reader on the other. In fact, the overriding rule for email is as follows:

Don't send it too quickly!

By taking an extra minute to check the style and tone of your message, you have the best chance of sending an email that will be well received.

>> Email Guideline 1: Use Style Appropriate to the Reader and Subject

Email sent early in a relationship with a client or other professional contact should be somewhat formal. It should be written more like a letter, with a salutation, closing, and complete sentences. Email written once a professional relationship has been established can use a more casual style. It can resemble conversation with the recipient on the phone. Sentence fragments and slang are acceptable, as long as they contribute to your objectives and are in good taste. Most important, avoid displaying a negative or angry tone. Don't push the *Send* button unless an email produces a constructive exchange.

>> Email Guideline 2: Be Sure Your Message Indicates the Context to Which It Applies

Tell your readers what the subject is and what prompts you to write your message. If you are replying to a message, be sure to include the previous message or summarize the message to which you are replying. Most email software packages include a copy of the message to which you are replying, as in Model 5–3. However, you should make sure that you include only the messages that provide the context for your reader. Long strings of forwarded email make it difficult to find the necessary information.

>> Email Guideline 3: Choose the Most Appropriate Method for Replying to a Message

Short email messages may only require that you include a brief response at the beginning or end of the email to which you are responding. For complex, multi-topic messages, however, you may wish to split your reply by commenting on each point individually (Figure 5–1).

>> Email Guideline 4: Format Your Message Carefully

Because email messages frequently replace more formal print-based documents, they should be organized and formatted so that the readers can locate the information you want to communicate easily.

- Use headings to identify important chunks of information.
- Use lists to display a series of information.
- Use sufficient white space to separate important chunks of information.
- Use separators to divide one piece of information from another.

Figure 5–2 illustrates an email message with headings, separators, and white space.

>> Email Guideline 5: Chunk Information for Easy Scanning

Break the information into coherent chunks dealing with one specific topic, including all the details that a reader needs to get all of the essential information. Depending on the nature of the information, include specific topic, time, date, location, and necessary prerequisites and details.

>> Email Guideline 6: When Writing to Groups, Give Readers a Method to Abstain from Receiving Future Notices

Email can easily become invasive and troublesome for recipients. You will gain favor—or at least not lose favor—if you are considerate and allow recipients to decide what email they wish to receive.

>> Email Guideline 7: When Writing to Groups, Suppress the Email Addresses of Recipients—Unless the Group Has Agreed to Let Addresses Be Known

It is inappropriate to reveal the email addresses of group members to other group members. Use the "bcc" line to suppress group members' addresses.

```
****************************************************
X-Sender: mckinley@mail2.m-global.com
Date: Tue, 11 Nov 2008 09:25:30 -0800
To: pcarmich@advantage.com
From: Mike McKinley <mckinley@mail2.m-global.com>
Subject: our recent visit
Mime-Version: 1.0
```

Dear Paul,

YOU WROTE:

>I hope that you had a good flight back home. I certainly enjoyed meeting you and look forward to the possibility of working with you this coming spring on the project that your firm, M-Global, may do for us.

REPLY:

The trip back was fine, but tiring. I enjoyed meeting you also and visiting with your staff. I particularly enjoyed meeting Harold Black, for he will be very valuable in developing the plans for the possible water purification plant.

YOU WROTE:

>If Advantage, Inc., does decide to build the water purification plant, we would be very interested in having M-Global's Mary Stevens as the project manager.

REPLY:

That certainly will be a possibility; Mary is one of our best managers.

YOU WROTE:

>After you left, I called the city administration here in Murrayville. M-Global does not need a business license for your work here, but, of course, you will need the necessary construction permits.

REPLY:

Thanks for taking care of this matter—I had not thought of that. We will supply the details to you for applying for the construction permits if you accept our proposal.

```
****************************************************
```

■ **Figure 5–1** ■ An email message that separates different topics for reply

A long email
message with use
of appropriate
headings,
separators, and
white space

```
**************************************************
Date: Tue, 7 Oct. 2008 09:25:30 -0800
To: Branch employees
From: Paul Carmichael <pcarmich@advantage.com>
Subject: October update
Mime-Version: 1.0
```

This is the October Electronic Update for Advantage, Inc. If you do not wish
to receive this electronic update, send a message to

pcarmich@advantage.com

With the message in the subject line: Unsubscribe.

```
*******************
UPCOMING EVENTS
*******************
```

Project managers' meeting

October 21—project managers meeting (notice the change of location):
Hereford building, room 209.

```
**********************
```
November department meetings

All departments will have their planning and reporting meetings on
November 18 at noon, with a joint lunch in the main dining room and
breakout sessions at 12:30. Meetings should conclude at 2 p.m.

```
**********************
```
December department meetings

NOTE CHANGE OF DATE: The December department meetings
will be held on December 10 (second Wednesday), NOT December 17
(third Wednesday).

```
*******************************************************
```

>> **Email Guideline 8: When Composing an Important Message,
Consider Composing It in Your Word Processor**

Important email messages should be not only clear in format but also correct in mechanics. Because email software may not have a spelling checker, compose important messages in your word processor and use your spelling checker to check accuracy. Then either cut and paste it into an email message or attach it as a file.

ABC Format for Email

Simply understanding that email *should* have a format puts you ahead of many writers, who consider email a license to ramble on without structure. Yes, email is casual and quick, but that does not make it formless. The three-part ABC format resembles that used for letters.

ABC Format: Email

- **ABSTRACT:** Casual, friendly greeting if justified by relationship
  - Short, clear statement of purpose for writing
  - List of main topics to be covered
- **BODY:** Supporting information for points mentioned in abstract
  - Use of short paragraphs that start with main ideas
  - Use of headings and lists
  - Use of abbreviations and jargon only when understood by all readers
- **CONCLUSION:** Summary of main point
  - Clarity about action that comes next

Remember—your reader is confronted with many emails during the day. Furthermore, the configurations of some computers make reading a screen harder on the eyes than reading print memos. So give each email a structure that makes it simple for your reader to find important information.

Appropriate Use and Style for Email

Email is an appropriate reflection of the speed at which we conduct business today. Indeed, it mirrors the pace of popular culture as well. Following are some of the obvious advantages that using email provides:

- It gets to the intended receiver quickly.
- Its arrival can be confirmed easily.
- Your reader can reply to your message quickly.
- It's cheap to use—once you have invested in the hardware and software.
- It permits cheap transmission of multiple copies and attachments.

Adding to the ease of transmission is the fact that email allows you to create mailing lists. One address label can be an umbrella for multiple recipients, saving you much time.

Of course, remember the flip side of this ease of use: Email is *not* private. Every time you send an email, remember that it may be archived or forwarded, and may end up being read by "the world." Either by mistake or design, many supposedly private emails often are received by unintended readers.

Email communication is often considered less formal and, therefore, less demanding in its format and structure than print-based messages such as memoranda and letters. However, because email messages have become so pervasive a means of communication, you should consider constructing them as carefully as you would a memorandum or a letter. Another reason to exercise great care is that email, like conventional documents, can be used in legal proceedings and other formal contexts.

Chapter 1, which mentions email in the context of team writing, shows how electronic mail helps you collaborate with others during the writing process—especially the planning stage. Interestingly, the email medium has produced a casual writing style similar to that of handwritten notes. It even has its own set of abbreviations and shortcut languages, which ranges so widely and changes so often that no list of abbreviations is included here. Following is an email message from one M-Global employee to another. Josh Bergen and Natalie Long are working together on a report in which they must offer suggestions for designing an operator's control panel at a large dam. Josh has just learned about another control panel that M-Global designed and installed for a Russian nuclear power plant (see the sixth project in the project sheets at the end of chapter 2). Josh wrote this email message to draw Natalie's attention to the related M-Global project:

DATE: September 15, 2008
TO: Natalie Long
FROM: Josh Bergen
SUBJECT: Zanger Dam Project
Natalie—
I've got an idea that might save us A LOT of time on the Zanger Dam project. Check out the company project sheet on the Russian nuclear plant job done last year.

Operators of hi-tech dams and nuke plants seem to face the same hassles:

- confusing displays
- need to respond fast
- distractions

When either a dam or nuke operator makes a mistake, there's often big trouble. I think we'd save time—and our client's money—if we could go right to some of the technical experts used in the nuke job. At least as a starting place. Maybe we'd even make our deadline on this project. That would be a change, considering the schedule delays this month on other jobs.

What do you think about this idea? Let me know today, if possible.

This message displays some of the most common stylistic features of electronic mail.

For everything you want to know about email, see http://www.learnthenet.com/english/section/email.html. Also, see Model 5–15 for another email example.

>>> Memoranda versus Email

Although email has become the most common form of internal correspondence in the workplace, there are times when a memo is a better option. Send a memo instead of an email in the following situations.

- The document is longer than can be viewed easily on a computer screen.
- The document must include symbols, special characters, or other formatting that may not be available through all computer systems.

- The document includes graphics.
- The document must be posted in print form.
- The document contains sensitive information, including information about clients, projects, or personnel.

>>> Chapter Summary

Correspondence keeps the machinery of business, industry, and government moving. Letters usually are sent to readers outside your organization, whereas memos are sent to readers inside. Email can be sent to internal or external audiences. In all types of correspondence, abide by these rules:

1. Know your purpose.
2. Know your readers.
3. Follow correct format.
4. Follow the ABC format for all letters and memos.
5. Use the 3Cs strategy for persuasive messages.
6. Stress the *you* attitude.
7. Use attachments for details.
8. Be diplomatic.
9. Edit carefully.

Besides following these basics, you must follow specific strategies for the four basic business letters, memos, and email. In letters with a positive message, the good news always goes first. In letters with a negative message, work on maintaining goodwill by placing a buffer before the bad news. Neutral letters, such as requests for information, should be absolutely clear in their message. Sales letters should show an interest in solving the readers' problems more than an eagerness to sell a product or service. Memos should strive for brevity, clarity, and tact. Your relationship with both superiors and subordinates can depend in part on how well you write memoranda. Email messages give you the additional flexibility of adopting an informal and more conversational style. Use email when speed and informality are desired.

>>> Learning Portfolio

Communication Challenge "Ethics and Sales Letters"

Technical firms like M-Global are known for taking a conservative approach to marketing. For the last six months, however, the M-Global corporate marketing department has been considering more imaginative marketing strategies. This case study explains their approach, presents the draft of a sales letter that is part of the proposed strategy, and ends with questions and comments for discussion and an assignment for a written response to the Challenge.

The Plan

Traditionally, technical consulting firms have done little if any advertising and sent few sales letters. Instead, they have just "been there" when clients needed services like the ones M-Global performs, such as foundation design, construction management, equipment development, environmental remediation, and training. M-Global has been successful because it has built up a stable of clients that return regularly for continuing services. Lately, however, a more competitive economy has reduced the amount of repeat business. Now some M-Global corporate managers want to introduce more aggressive marketing techniques into the firm's long-term plan.

Specifically, the marketing staff is proposing a three-stage plan to corporate management and branch managers. Their focus is government and private groups who own wetlands property being considered for commercial or recreational development. Eventually, owners will be told by federal agencies that the property must be evaluated rigorously for environmental impact. This analysis determines to what degree, if any, the land can be developed for parks, residential purposes, and so on. M-Global would like to write owners about this requirement now, before the owners request environmental studies officially. Thus, M-Global may beat competitors to the work. Following is the plan they propose:

1. Use public records to identify owners of wetland areas throughout the country.
2. Send a sales letter to all these owners.
3. Follow the letter with phone calls and meetings that provide a more detailed account of M-Global's services.

The marketing staff has decided the sales letter must depart from the predictable, boring letters sent in the past. In fact, the staff has devised a *grabber*: any recipient hiring M-Global to perform an environmental study before January 15 receives a fee reduction of 10%, up to a maximum of $1,000. Although the writers do not want their letter to look too commercial, they do want to depart from letters of the past. They believe something is needed to get readers to act; a financial incentive seems appropriate.

The Letter

The marketing staff just sent out the following letter to M-Global's corporate officers and branch managers for their comments. Included with the letter was a list of prospective clients and a reference to the three-point marketing strategy noted in the preceding list. Readers are to call or email the marketing staff with their response.

September 15, 2008

Drummond Recreational Development Inc
PO Box 303
Wilmington NC 20065

Reference: 100-Acre Parcel at Star-Gazer Swamp

Dear Mr. Drummond:

Would you like to save up to $1,000 on a study the government requires to evaluate the development potential of Star-Gazer Swamp? For a limited time, M-Global can save you this amount on an environmental evaluation of the site.

As you may know, the federal government requires that you complete an environmental assessment of property that may include wetlands areas important for biological diversity. M-Global has completed more than 500 such studies during the past 10 years. In each case, clients have received information they needed to determine

1. possible commercial or recreational development of their property and
2. applicable federal restrictions.

In making this evaluation, our team of technical experts has no equal among consulting firms throughout the country.

For a limited time, you have the opportunity to save 10%, up to a maximum of $1,000, on an environmental assessment that will cost you full price for any services contracted after January 15, 2009. Enclosed is a brochure that gives specifics about the features of our assessments and lists our previous clients.

I'll call you in a few days to see if your firm wants to take advantage of our offer for a significant reduction on an evaluation assessment. If you have questions before that time, please give me a call.

Sincerely,

Kathryn P. Phyzer

Kathryn P. Phyzer, Marketing Manager

Enclosure

Questions and Comments for Discussion

1. Explain how M-Global's letter does or does not satisfy this chapter's criteria for the content and style of an effective sales letter.
2. What kind of response do you think the marketing staff will get from the M-Global managers to whom they sent the draft?
3. Do you believe the financial incentive is ethical? Explain.

4. Do you think the financial incentive would be effective? Explain.
5. Generally speaking, what type of marketing strategy is either ethical or appropriate for technical firms like M-Global? Explain your answer.
6. In your view, should marketing techniques for firms like M-Global differ from techniques used by retail stores, credit card companies, and other similar organizations? Explain.

Write About It

Assume the role of one of the branch managers who has received this draft. Write the subject line and text of an email with your comments and suggestions for improving the sales letter. Remember that your branch depends on clients' goodwill, but it also depends on the clients brought to you by the sales staff. You also want to maintain M-Global's reputation for professional, quality work.

Collaboration at Work Choosing the Right Mode

General Instructions

Each Collaboration at Work exercise applies strategies for working in teams to chapter topics. The exercise assumes you (1) have been divided into teams of about three to six students, (2) use team time inside or outside of class to complete the case, and (3) produce an oral or written response. For guidelines about writing in teams, refer to pages 24–28.

Background for Assignment

A century ago, business professionals had few opportunities for communication beyond the formal letter or meeting; today, the range of options is incredibly broad. On one hand, we marvel at the choices for getting our message heard or read; on the other hand, the many ways to communicate present an embarrassment of riches that can be confusing.

In other words, when you have multiple communication options, you're challenged to match the right method with the right context—*right* in terms of what the reader wants and *right* in terms of the level of effort you should

exert to suit the purpose. You may think this challenge applies only to your working life. However, it also can influence your life in college, as this exercise shows.

Team Assignment

Brainstorm with your team to list every means you have used to communicate with your college and university, from the time you applied to the present. Then for each communication option that follows, provide two or three situations for which the option is the appropriate choice:

1. Letter that includes praise
2. Letter that describes a complaint
3. Letter that provides information
4. Letter that attempts to persuade
5. Telephone call
6. Fax transmission
7. Email
8. Memorandum
9. Personal meeting

Assignments

Follow these general guidelines for these assignments:

- Print or design a letterhead when necessary.
- Use whatever letter format your instructor requires.

- Invent addresses when necessary.
- Invent any extra information you may need for the correspondence, but do not change the information presented here.

Part 1: Letters

1. Positive Letter—Job Offer

Assume that you are the personnel director for M-Global's San Francisco office. Yesterday, you and your hiring committee decided to offer a job to Ashley Tasker, one of ten recent graduates you interviewed for an entry-level position as a lab technician. Write Ashley an offer letter and indicate a starting date (in two weeks), a specific salary, and the need for her to sign and return an acceptance letter immediately. In the interview, you outlined the company's benefit plan, but you are enclosing with your letter a detailed description of fringe benefits (e.g., health insurance, long-term disability insurance, retirement plan, vacation policy). Although Ashley is your first choice for this position, you are prepared to offer the job to another top candidate if Ashley is unable to start in two weeks at the salary you stated in the letter.

2. Positive Letter—Favorable Response to Complaint

The following letter was written in response to a complaint from the office manager at M-Global's Denver office. She wrote the manufacturer that the lunchroom microwave broke down just three days after the warranty expired. Although she did not ask for a specific monetary adjustment, she did make clear her extreme dissatisfaction with the product. The manufacturer responded with the following letter. Be prepared to discuss what is right and what is wrong with the letter. Also, rewrite it using this chapter's guidelines.

> This letter is in response to your August 3 complaint about the Justrite microwave oven you purchased about six months ago for your lunchroom at M-Global, Inc. We understand that the turntable in the microwave broke shortly after the warranty expired.
>
> Did you know that last year our microwave oven was rated "best in its class" and "most reliable" by *Consumers Count* magazine? Indeed, we have received so few complaints about the product that a recent survey of selected purchasers revealed that 98.5% of first-time purchasers of our microwave ovens are pleased that they chose our products and would buy another.
>
> Please double-check your microwave to make sure that the turntable is broken—it may just be temporarily stuck. We rarely have had customers make this specific complaint about our product. However, if the turntable is in need of repair, return the entire appliance to us, and we will have it repaired free of charge or have a new replacement sent to you. We stand behind our product, because the warranty period only recently expired.
>
> It is our sincere hope that you continue to be a satisfied customer of Justrite appliances.

3. Negative Letter—Explanation of Project Delay

You work for M-Global's Boston office. As project manager for the construction of a small strip shopping center, you have had delays about halfway through the project because of bad weather. Even worse, the forecast is for another week of heavy rain. Yesterday, just when you thought nothing else could go wrong, you discovered that your concrete supplier, Atlas Concrete, has a truck drivers' strike in progress. Because you still need half the concrete for the project, you have started searching for another supplier.

Your client, an investor/developer named Tanya Lee located in a city about 200 miles away, probably will be upset by any delays in construction, whether or not they are within your control. Write her a letter in which you explain weather and concrete problems. Try to ease her concern, especially because you want additional jobs from her in the future.

4. Negative Letter—Request for Prompt Payment

Recently, your M-Global office completed the project described in the fourth project sheet at the end of chapter 2. As indicated on the sheet, the city of Rondo was quite satisfied with your work. You billed the city within a week after completion and requested payment within 30 days of receipt of the bill, as you do for most clients. Forty-five days elapsed without payment being received, so you sent a second bill. Now it has been three months since completion of the project, and you still have not been paid. You suspect that your bill got lost in the paperwork at city hall, for there was a change of mayoral administrations shortly after you finished the project. Yet two phone calls to the city's business office have brought no satisfaction—two different assistants told you they could not find the bill and that you should rebill the city. You are steamed but would like to keep the client's goodwill, if possible. Write another letter requesting payment.

5. Negative Letter—Change in Project Scope and Schedule

As a marketing executive at the Cleveland office, you oversee many of the large accounts held by the office. One important account is a company that owns and operates a dozen radio and television stations throughout the Midwest. On one recent project, M-Global engineers and technicians did the foundation investigation for, and supervised construction of, a new transmitting tower for a television station in Toledo. First, your staff members completed a foundation investigation, at which time they examined the soils and rock below grade at the site. On the basis of what they learned, M-Global ordered the tower and the guy wires that connect it to the ground. Once the construction crew actually began excavating for the foundation, however, they found mud that could not support the foundation for the tower. Although unfortunate, it sometimes happens that actual soil conditions cannot be predicted by the preliminary study. Because of this discovery of mud, the tower must be shifted to another location on the site. As a result, the precut guy wires are the wrong length for the new site, requiring M-Global to order wire extenders. The extenders will arrive in two weeks, delaying placement of the tower by that much time. All other parts of the project are on schedule, so far.

Your client, Ms. Sharon West of Midwest Media Systems in Cleveland, doesn't understand much about soils and foundation work, but she does understand what construction delays mean to the profit margin of her firm's new television station as it attempts to compete with larger stations in Toledo. You must console this important client while informing her of this recent finding.

6. Neutral Letter—Response to Request for Information

As reservations clerk for the Best Central Inn in St. Louis, you just received a letter from Jerald Pelletier, an administrative secretary making arrangements for a meeting of M-Global managers from around the country. The group is considering holding its quarterly meeting in St. Louis in six months. Pelletier has asked you to send some brief information on hotel rates, conference facilities (meeting rooms), and availability. Send him some room rates for double and single rooms, and let him know that you have four conference rooms to rent out at $75 each per day. Also, tell him that at this time, the hotel rooms and conference rooms are available for the three days he mentioned.

7. Neutral Letter—Request for Information

For this assignment, choose one of the seven project sheets at the end of chapter 2. Assume you are a potential client of M-Global and have happened onto this project, which is similar to one your organization may need completed in the near future. Although you are not yet at the stage where you want to receive proposals, you would like additional information about firms that might be interested in your project. Using this context and information in the project description, write the marketing manager of M-Global for more information. Be specific about what you want to receive.

8. Sales Letter—to M-Global

Select a product or service with which you are familiar because of home, work, or school experience. Now assume that you are responsible for marketing this product or service to Ms. Janis Black, purchasing agent for M-Global, Inc. In a phone conversation earlier today, she showed some initial interest in purchasing the product or service for her firm. In fact, you managed to set up a meeting for two weeks from today, after she returns from a business trip. Now you need to write a follow-up letter in which you summarize the phone conversation, confirm the meeting date and time, and offer some additional information about your product or service that will keep her interest.

In selecting your product or service, you might want to review the information about M-Global, Inc., contained in chapter 2. Following are some sample products and services that Janis must evaluate routinely for use by the firm:

- Cleaning supplies, such as lavatory soap and paper towels
- Laboratory equipment, such as glass beakers and lab coats
- Office materials, such as notepads and pencils
- Office equipment, such as fax machines and pencil sharpeners
- Leased products, such as automobiles for managers and salespeople

9. Sales Letter—from M-Global

For this assignment, choose one of the seven project sheets included at the end of chapter 2. Although you will use the sheets as a source of information in writing your letter, you will not send the sheets as an attachment.

Option A Assume you are an M-Global employee writing to a new client. This client has shown initial interest in the service or product described in the project sheet you're using for this assignment. You've had one phone conversation with the potential client. Now you hope your letter will lay the foundation for a personal meeting. Write a sales letter that briefly describes (1) the service or product on the project sheet or (2) your success on the specific project.

Option B Assume you are an M-Global employee writing to the same client for whom you completed the project described on the sheet. You have heard there may be similar work available with this client, so you are seeking repeat business. Write your follow-up letter, referring to the success on the project and seeking information about possible future work.

Part 2: Memoranda

10. Memo—Positive News

Kevin Kehoe, an employee at the San Francisco office, is being considered for promotion to Manager of Technical

Services. He has asked you to write a memo to the San Francisco branch manager on his behalf. Although you now work as a marketing expert at the corporate office, several years ago you worked directly for Kevin on the Ocean Exploration Program in Cameroon (see the second project sheet at the end of chapter 2). Kevin has asked that your memo deal exclusively with his work on that program. Kevin was manager of the project; you believe that it was largely through his technical expertise, boundless energy, and organizational skills that the project was so successful. He developed the technical plan of work that led to the clear-cut set of findings. Write a memo that conveys this information to the branch manager considering Kevin for the promotion. Because the branch manager is new, he is not familiar with the project on which you and Kevin worked. Thus your memo may need to mention some details from the project sheet.

11. Memo—Negative News

You are project manager of the construction-management group at M-Global's St. Paul office. The current policy in your office states that employees must pass a pre-employment drug screening before being hired. After that, there are no tests unless you or one of your job supervisors has reason to suspect that an employee is under the influence of drugs on the job.

Lately, a number of clients have strongly suggested that you should have a random drug-screening policy for all employees in the construction-management group. They argue that the on-the-job risk to life and property is great enough to justify this periodic testing, without warning. You have consulted your branch manager, who likes the idea. You have also talked with the company's attorney, who assures you that such random testing should be legal, given the character of the group's work. After considerable thought, you decide to implement the policy in three weeks. Write a memo to all employees of your group and relate this news.

12. Memo—Negative News

A year ago, you introduced a pilot program to M-Global's London office, where you are branch manager. The program gave up to half the office employees the choice to work four 10-hour days each week, as opposed to five 8-hour days. You wanted to offer this flexibility to workers who, for whatever reason, desired longer weekends. As branch manager, you made it clear at the time that you would evaluate the program at the end of the one-year pilot.

Having completed your review, you've decided that employees need to return to the old schedule. Your main reason is that having the office short of staff on Friday (when the four-day employees are gone) has proved awkward in dealing with current and prospective clients. On many occasions, clients have called to find that their company contact is not working that day. In addition, the secretarial and clerical staff that does work on Friday cannot keep up with the end-of-week workload. Whereas you originally thought a split schedule in the office would work, now you know it causes more confusion than it's worth. People in the office are constantly forgetting who is working what schedule, although the schedule is published. So, your memo to office employees must inform them of your decision to discontinue the pilot program and to return to a five-day workweek for the whole office. The change will take place in one month.

13. Memo—Persuasive Message

For this assignment, choose either (1) a good reference book or textbook in your field of study or (2) an excellent periodical in your field. The book or periodical should be one that could be useful to someone working in a profession, preferably one that you may want to enter.

Now assume that you are an employee of an organization that would benefit by having this book or periodical in its staff library, customer waiting room, or perhaps as a reference book purchased for employees in your group. Write a one-page memo to your supervisor recommending the purchase. You might want to consider criteria such as:

- Relevance of information in the source to the job
- Level of material with respect to potential readers
- Cost of book or periodical as compared with its value
- Amount of probable use
- Important features of the book or periodical (such as bibliographies or special sections)

14. Memo—Neutral Message

Assume you work in the public relations area at M-Global's corporate office in Baltimore. Recently, your office began designing and producing company project sheets, each of which uses one page to describe a specific project completed by M-Global and provides an accompanying graphic on the page. Eight examples of these project sheets are included at the end of chapter 2. Write a memo to all 15 branch managers. Give them the information that follows: Beginning next month, project sheets will be written for every M-Global job that grosses $10,000 or more. Your office will have each sheet finished within 30 days of project completion and will then send each office the sheets for projects that were coordinated by that office. Then, as time permits, the public relations office will go back to significant previous projects, like the ones at the end of chapter 2, to do additional project sheets on previous work.

15. Memo—Persuasive Message

Assume you work at an M-Global office and have no undergraduate degree. You are not yet sure what degree program you want to enter, but you have decided to take one night course each term. Your M-Global office has agreed to pay 100 percent of your college expenses on two conditions. First, before taking each course, you must write a memo of request to your supervisor, justifying the value of the class to your specific job or to your future work with the company. Clearly, your boss wants to know that the course has specific application or that it will form the foundation for later courses. Second, you must receive a C or better in every class for which you want reimbursement.

Write the persuasive memo just described. For the purposes of this assignment, choose one course that you actually have taken or are now taking. Yet in your simulated role for the assignment, write as if you have not taken the course.

Part 3: Email

16. Email—Positive News

As branch manager of M-Global's Atlanta office, you just learned from your accounting firm that last year's profits were even higher than previously expected. Apparently, several large construction jobs had not been counted in the first reporting of profits. You and your managers had already announced individual raises before you learned this good news. Now you want to write an email that states that every branch employee will receive a $500 across-the-board bonus, in addition to whatever individual raises have been announced for next year. Include the subject line for the email.

17. Email—Declining a Request

Assume that you work at the M-Global office that completed the Sentry Dam (see the first project sheet at the end of chapter 2). Word of your good work has spread to the state director of dams. He has asked you, as manager of the Sentry project, to deliver a 20-minute speech on dam safety to the annual meeting of county engineers. Unfortunately, you have already agreed to be at a project site in another state on that day, and you cannot reschedule the site visit. Write an email, including the subject line, to the director of dams—who is both a former and, you hope, a future client—and decline the request. Although you know he expressly wanted you to speak, offer to send a substitute from your office.

18. Email—Neutral Message

As mailroom supervisor at M-Global's Baltimore office, you have a number of changes to announce to employees of the corporate office. Write an email, including the subject line, that clearly relates the following information: Deliveries and pickups of mail, which currently are at 8:30 A.M. and 3:00 P.M., will change to 9:00 A.M. and 3:30 P.M., starting in two weeks. Also, there will be an additional pickup at noon on Monday, Wednesday, and Friday. The mailroom will start picking up mail to go out by Federal Express or any other one-day carrier, rather than the sender having to wait for the carrier's representative to come to the sender's office. The sender must call the mailroom to request the pickup; and the carrier must be told by the sender to go to the mailroom to pick up the package. The memo should also remind employees that the mail does not go out on federal holidays, even though the mailroom continues to pick up mail from the offices on those days.

19. Email—Collaborative Project #1

Select one of the preceding memo assignments to complete as a team-writing project with two or three members of your class. Set up a plan of work that (1) involves the team in several face-to-face meetings and (2) requires each member of the team to send and receive at least one assignment-related email message to and from every other member of the team.

20. Email—Collaborative Project #2

Pair up with one or two members of your class. Assume your team has been asked to write an M-Global project sheet to add to the eight sheets at the end of chapter 2. Although you do not need to include a photo, hand in a description of what kind of photograph you would recommend for the sheet you are writing.

After your team has one or two face-to-face meetings to agree on a topic and research agenda, conduct all further team communication by email. One member is responsible for emailing a draft of the project sheet to the others. Then other members make all comments and suggestions by email. Once the email communication is complete, print and submit the project sheet along with hard copy of all email correspondence within the team.

? 21. Ethics Assignment

In the "Communication Challenge" section of this chapter, you were asked to examine the ethics of a particular sales letter. This assignment moves that ethical concern into the realm of electronic communication and assumes you have some familiarity with email. The project is best completed as a team assignment.

Pooling the experience that members of your team have had with email, focus specifically on inappropriate or unethical behavior. Possible topics include the content of messages, tone of language, and the use of distribution lists.

Now draft a simple Code of Ethics that could be distributed to members of any organization—such as M-Global, Inc.—whose members use email on a daily basis. See Assignment #6 in chapter 2 for an example of an actual code of ethics, or use "code of ethics" in an Internet search.

 ## 22. International Communication Assignment

Email messages can be sent around the world as easily as they can be sent to the next office. If you end up working for a company with international offices or clients, you probably will use email to conduct business.

Investigate the email conventions of one or more countries outside your own. Search for any ways that the format, content, or style of international email may differ from email in your country. Gather information by collecting hard copy of email messages sent from other countries, interviewing people who use international email, or consulting the library for information on international business communication. Write a memo to your instructor in which you (1) note differences you found and (2) explain why these differences exist. If possible, focus on any differences in culture that may affect email transactions.

 ## 23. A.C.T. N.O.W. Assignment (Applying Communication To Nurture Our World)

Whether you commute or live on campus, your everyday life at a college or university may be influenced by student government associations, usually made up of students elected to their positions. For example, such associations often receive fees (paid by students each term) to sponsor various cultural, intellectual, athletic, or entertainment events. Learn about what types of activities your student government association sponsors on campus. Then write a letter to your campus newspaper or to the student government association president in which you compliment or critique the use of funds—either related to a specific event or to the general use of the budget. If your campus has no student government, direct your letter to the campus administrator or office that does coordinate such events.

12 Peachtree Street
Atlanta GA 30056
404.555.7524

August 2, 2008

Professor Willard R Burton PhD
Department of Civil Engineering
Southern University of Technology
Paris GA 30007

Dear Professor Burton:

Expresses appreciation *and* provides lead-in to body. ▶ Thanks very much for your hospitality during my visit to your class yesterday. I appreciated the interest your students showed in my presentation on stress fractures in highway bridges. Their questions were very perceptive.

Responds to question that arose at class presentation. ▶ You may recall that several students requested further information on M-Global, so I have enclosed a dozen brochures for any students who may be interested. As you know, job openings for civil-engineering graduates have increased markedly in the last five years. Some of the best opportunities lie in these three areas of the discipline:

Uses bulleted list to emphasize information of value to professor's students.
▶ • Evaluation of environmental problems
• Renovation of the nation's infrastructure
• Management of construction projects

Adds unobtrusive reference to M-Global's needs. ▶ These areas are three of M-Global's main interests. As a result, we are always searching for top-notch graduates from solid departments like yours.

Closes with offer to visit class again. ▶ Again, I enjoyed my visit back to Southern last Friday, Professor Burton. Please call when you want additional guest lectures by me or other members of the M-Global's staff.

Sincerely,

George F. Lux, P.E.

George F Lux

Includes reference to enclosures. ▶ Enclosures

M-Global Inc | 127 Rainbow Lane | Baltimore MD 21202 | 410.555.8175

■ **Model 5–1** ■ M-Global sample letter

MEMORANDUM

DATE: December 4, 2008
TO: Technical Staff
FROM: Ralph Simmons, Technical Manager RS
SUBJECT: New Employee to Help with Technical Editing

Last week we hired an editor to help you produce top-quality reports, proposals, and other documents. This memorandum gives you some background on this change, highlights the credentials of our new editor, and explains what the change will mean to you.

BACKGROUND

At September's staff meeting, many technical staff members noted the excessive time spent editing and proofreading. For example, some of you said that this final stage of writing takes from 15-30 percent of the billable time on an average report. Most important, editing often ends up being done by project managers—the employees with the highest billable time.

Despite these editing efforts, many errors still show up in documents that go out the door. Last month I asked a professional association, the Engineers Professional Society (EPS), to evaluate M-Global-Boston documents for editorial correctness. (EPS performs this service for members on a confidential basis.) The resulting report showed that our final reports and proposals need considerable editing work. Given your comments at September's meeting and the results of the EPS peer review, I began searching for a solution.

SOLUTION: IN-HOUSE EDITOR

To come to grips with this editing problem, the office just hired Ron Perez, an experienced technical editor. He'll start work January 3. For the last six years, Ron has worked as an editor at Jones Technical Services, a Toronto firm that does work similar to ours. Before that he completed a master's degree in technical writing at Sage University in Buffalo.

At next week's staff meeting, we'll discuss the best way to use Ron's skills to help us out. For now, he will be getting to know our work by reviewing recent reports and proposals. Also, the attached list of possible activities can serve as a springboard for our discussion.

CONCLUSION

By working together with Ron, we'll be able to improve the editorial quality of our documents, free up more of your time for technical tasks, and save the client and ourselves some money.

I look forward to meeting with you next week to discuss the best use of Ron's services.

Enclosure
Copy: Ron Perez

M-Global Inc | 127 Rainbow Lane | Baltimore MD 21202 | 410.555.8175

Uses informative subject line.

Gives purpose of memo and highlights contents.

Uses side headings for easy reading.

Shows that the change arose from *their* concerns.

Adds evidence from outside observer.

Gives important information about Ron in *first* sentence.

Establishes his credibility.

Refers to attachment.

Focuses on *benefit* of change to reader.

Restates next action to occur.

■ **Model 5–2** ■ M-Global sample memo

POSSIBLE ACTIVITIES FOR IN-HOUSE EDITOR

1. Reviewing reports at all levels of production

2. Helping coordinate the writing of proposals

3. Preparing a format manual for the word-processing operations and secretaries

4. Preparing a report/proposal guide for the technical staff

5. Teaching luncheon sessions on editing

6. Teaching writing seminars for the technical staff

7. Working with the graphics department to improve the page design of our documents

8. Helping write and edit public-relations copy for the company

9. Visiting other offices to help produce consistency in the editing of documents throughout the company

■ **Model 5–2** ■ continued

From: "James Thuvenot" jthuvenot@rbirdarc.com
To: "Evelyn Dame" edame@m-global.com
Date: 3/14/2008 8:14 AM
Subject: RE: Riverview Shopping Center Project

Evelyn,

Thanks for the information. I just heard about the construction on the radio this morning and wondered if it would cause us problems.

So far, everything looks like it's going well.

Thanks for your good work.

Jim

James Thuvenot
Redbird Architects
335 River Ave.
Columbia, Illinois 62236

>Jim,
>
>We've just been informed that construction on the JB Bridge will start next
>month. We are adjusting our work schedule so that most of the material and
>equipment will be delivered to the Columbia worksite before the bridge
>construction begins.

>Although we may run into a few problems, we don't expect
>the bridge construction to delay the completion of the project by more than 2 or
>3 weeks.
>
>Please let me know if you have any questions or concerns.
>
>Evelyn Dame
>Project Manager
>M-Global St. Louis

■ **Model 5–3** ■ M-Global sample email

Letterhead of your organization

Two or more blank lines (adjust space to center letter on page)

Date of letter

Two or more blank lines (adjust space to center letter on page)

Address of reader

One blank line

Greeting

One blank line

Paragraph: single-spaced (indenting optional)

One blank line

Paragraph: single-spaced (indenting optional)

One blank line

Paragraph: single-spaced (indenting optional)

One blank line

Complimentary close

Three blank lines (for signature)

Typed name and title

One blank line

Typist's initials (optional: Writer's initials before typist's initials)

Computer file # (if applicable)

One blank line (optional)

Enclosure notation

One blank line (optional)

Copy notation

■ **Model 5–4** ■ Block style for letters

Letterhead of your organization

Two or more blank lines (adjust space to center letter on page)

Date of letter

Two or more blank lines (adjust space to center letter on page)

Address of reader

One blank line

Greeting

One blank line

Paragraph: single-spaced, with first line indented 5 spaces

One blank line

Paragraph: single-spaced, with first line indented 5 spaces

One blank line

Paragraph: single-spaced, with first line indented 5 spaces

One blank line

Complimentary close

Three blank lines (for signature)

Typed name and title

One blank line

Typist's initials (optional: Writer's initials before typist's initials)

Computer file # (if applicable)

One blank line (optional)

Enclosure notation

One blank line (optional)

Copy notation

■ **Model 5–5** ■ Modified block style (with indented paragraphs) for letters

Letterhead of organization

Two or more blank lines (adjust space to center letter on page)

Date of letter

Two or more blank lines (adjust space to center letter on page)

Address of reader

Three blank lines

Short subject line

Three blank lines

Paragraph: single-spaced, no indenting

One blank line

Paragraph: single-spaced, no indenting

One blank line

Paragraph: single-spaced, no indenting

Five blank lines (for signature)

Typed name and title

One blank line (optional)

Typist's initials (optional: Writer's initials before typist's initials)

Computer file # (if applicable)

One blank line (optional)

Enclosure notation

One blank line (optional)

Copy notation

■ **Model 5–6** ■ Simplified style for letters

Facsimile reference

One or more blank lines

Date of memo

Reader's name (and position, if appropriate)

Writer's name (and position, if appropriate)

Subject of memo

One or more blank lines

Paragraph: Single-spaced (optional–first line indented)

One blank line

Paragraph: Single-spaced (optional–first line indented)

One blank line

Paragraph: Single-spaced (optional–first line indented)

One blank line

Typist's initials (optional–writer's initials before typist's initials)

One blank line

Enclosure notation

One blank line

Copy notation

■ **Model 5–7** ■ Memo style

12 Post Street
Houston Texas 77000
713.555.9781

July 23, 2008

The Reverend Mr John C Davidson
Maxwell Street Church
Canyon Valley Texas 79195

Dear Reverend Davidson:

Mentions letter that prompted this response. Gives good news *immediately.* ▶ Thanks for your letter asking to reschedule the church project from mid-August to another, more convenient time. Yes, we'll be able to do the project on one of two possible dates in September, as explained below.

Reminds reader of rationale for original schedule—cost *savings.* ▶ As you know, M-Global originally planned to fit your foundation investigation between two other projects planned for the Canyon Valley area. In making every effort to lessen church costs, we would be saving money by having a crew already on site in your area—rather than having to charge you mobilization costs to and from Canyon Valley.

Offers two options—both save the church money. ▶ As it happens, we have just agreed to perform another large project in the Canyon Valley area beginning on September 18. We would be glad to schedule your project either before or after that job. Specifically, we

Shows M-Global's flexibility. ▶ could be at the church site for our one-day field investigation on either September 17 or September 25, whichever date you prefer.

Makes clear what should happen next. ▶ Please call me by September 2 to let me know your scheduling preference for the project. In the meantime, have a productive and enjoyable conference at the church next month.

Sincerely,

Nancy Slade

Nancy Slade, P.E.
Project Manager

NS/mh
File #34678

M-Global Inc | 127 Rainbow Lane | Baltimore MD 21202 | 410.555.8175

■ **Model 5–8** ■ Positive letter in block style

12 Post Street
Houston Texas 77000
713.555.1381

July 23, 2008

The Reverend Mr John C Davidson
Maxwell Street Church
Canyon Valley Texas 79195

Dear Reverend Davidson:

Thanks for your letter asking to reschedule the foundation project at your church from mid-August to late August, because of the regional conference. I am sure you are proud that Maxwell was chosen as the conference site.

One reason for our original schedule, as you may recall, was to save the travel costs for a project crew going back and forth between Houston and Canyon Valley. Because M-Global has several other jobs in the area, we had planned not to charge you for travel.

We can reschedule the project, as you request, to a more convenient date in late August, but the change will increase project costs from $1,500 to $1,800 to cover travel. At this point, we just don't have any other projects scheduled in your area in late August that would help defray the additional expenses. Given our low profit margin on such jobs, that additional $300 would make the difference between our firm making or losing money on the foundation investigation at your church.

I'll call you next week, Reverend Davidson, to select a new date that would be most suitable. M-Global welcomes its association with the Maxwell Street Church and looks forward to a successful project in late August.

Sincerely,

Nancy Slade

Nancy Slade, P.E.
Project Manager

NS/mh
File #34678

Provides "bridge" and compliments Davidson on conference.

Reminds him about original agreement— in tactful manner

Phrases negative message as positively as possible, giving rationale for necessary change.

Makes it clear what will happen next. Ends on positive note.

M-Global Inc | 127 Rainbow Lane | Baltimore MD 21202 | 410.555.8175

■ **Model 5–9** ■ Negative letter in modified block style (with indented paragraphs)

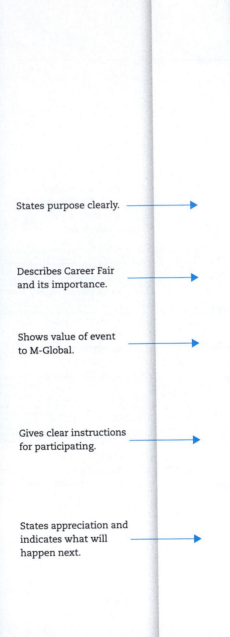

States purpose clearly. →

Describes Career Fair and its importance. →

Shows value of event to M-Global. →

Gives clear instructions for participating. →

States appreciation and indicates what will happen next. →

January 4, 2008

Mr Timothy Fu Personnel Director
M-Global Inc
211 River Front Circle
St Louis MO 63103

Dear Mr. Fu:

M-Global, Inc., has hired 35 graduates of River College since 1975. To help continue that tradition, we would like to invite you to the college's first Career Fair, to be held February 21, 2008, from 8 a.m. until noon.

Sponsored by the Student Government Association, the Career Fair gives juniors and seniors the opportunity to get to know more about a number of potential employers. We give special attention to organizations, like M-Global that have already had success in hiring River College graduates. Indeed, we have already had a number of inquiries about whether your firm will be represented at the fair.

Participating in the Career Fair is simple. We will provide you with a booth where one or two M-Global representatives can talk with students that come by to ask about your firm's career opportunities. Feel free to bring along brochures or other written information that would help our students learn more about M-Global's products and services.

I will call you next week, Mr. Fu, to give more details about the fair and offer a specific booth location. We at River College look forward to building on our already strong association with M-Global.

Sincerely,

Faron G. Abdullah

Faron G. Abdullah, President
Student Government Association

Copy: Gene Abrams, Placement Director

56 New Lane
Bolt Missouri
65101
314.555.0272

■ **Model 5–10** ■ Neutral letter (invitation) in block style

345 Underwood Street
Belforth California 90706
713.555.9781

April 2, 2008

Faraday Supply Company
34 State Street
San Francisco CA 94987

ORDER FOR FIELD TRANSITS

Yesterday I called Ms. Gayle Nichols to ask what transits you had in current
inventory. Having considered what you have in stock, I wish to order those
listed below.

Provides bridge to previous contact.
States purpose clearly.

Please send us these items:

1. One Jordan #456 Transit, with special field case
2. One Smith-Beasley #101FR, with special field case
3. One Riggins #6NMG, without special field case

Gives exact information needed by reader.

Note that we *do* want the special field cases with the Jordan and Smith-
Beasley units, but do *not* want the case with the Riggins unit.

Emphasizes important details about the order.

Please send the units and the bill to my attention. As always, we appreciate
doing business with Faraday.

States exactly what should happen next.

Farah Linkletter
Farah Linkletter
Supply Assistant

gh

M-Global Inc | 127 Rainbow Lane | Baltimore MD 21202 | 410.555.8175

■ **Model 5–11** ■ Neutral letter (placing order) in simplified style

211 River Front Circle
St Louis Missouri 63103
314.555.8175

August 21, 2008

Mr James Swartz Safety Director
Jessup County School System
1111 Clay Street
Smiley MO 64607

NEW ASBESTOS-ABATEMENT SERVICE NOW AVAILABLE

We enjoyed working with you last year, James, to update your entire fire alarm system. Given the current concern in the country about another safety issue, asbestos, we wanted you to know that our staff now does abatement work.

As you know, many of the state's school systems were constructed during years when asbestos was used as a primary insulator. No one knew then, of course, that the material can cause illness and even premature death for those who work in buildings where asbestos was used in construction. Now we know that just a small portion of asbestos produces a major health hazard.

Fortunately, there's a way to tell whether you have a problem: the asbestos survey. This procedure, done by our certified asbestos-abatement professionals, results in a report that tells whether your buildings are affected. And if we find asbestos, we can remove it for you.

Jessup showed real foresight in modernizing its alarm system last year, James. Your desire for a thorough job on that project was matched, as you know, by the approach we take to our business. Now we'd like to help give you the peace of mind that will come from knowing that either (1) there is no asbestos problem in your 35 structures or (2) you have removed the material.

The enclosed brochure outlines our asbestos services. I'll call you in a few days to see whether M-Global can help you out.

Barbara Feinstein
Barbara H. Feinstein
Certified Industrial Hygienist

BHF/sg

M-Global Inc | 127 Rainbow Lane | Baltimore MD 21202 | 410.555.8175

Simplified style eliminates salutation and closing.

Uses subject line to gain attention.

Refers to previous successful work.

Leads in naturally to letter's subject (asbestos abatement).

Comforts reader by showing how problem can be discovered and solved.

Reinforces relationship between writer's and reader's organizations.

Refers briefly to enclosures; stays in control by mentioning follow-up phone call.

■ **Model 5–12** ■ Sales letter in simplified style

MEMORANDUM

DATE: August 1, 2008
TO: All Employees
FROM: Gini Preston, Chair, Copy Services Committee
SUBJECT: Copy Center Changes

With the purchase of two new copiers and a folder, the Copy Center is able to expand its services. At the same time, we have had to reduce the paper stock that we keep on hand because of space limitations. This memo highlights the services and products now available at the Copy Center.

Gives brief purpose statement and overview of contents.

1. **Color copies:** With our new equipment, color copies do not require additional time to process. However, because color copies are expensive, please limit your use of them. If you have a document that includes both color and black-and-white pages, submit them as separate jobs so that the color copier is used only for color copies.

2. **Special stock:** The Copy Center now stocks only two colors of paper in addition to white paper: blue and goldenrod. Cover stock is available only in white and blue. We continue to stock transparencies. Although we are no longer stocking other kinds of paper, we are still able to meet requests for most special stock:

- **Stocks available with 24-hour notice:** We can purchase 11 x 17 inch paper, cover stock and regular stock in a variety of colors, and specialized paper such as certificates and NCR (carbonless copy) paper. Departments will be charged for all special stock.

Emphasizes need for special handling of requests for special paper.

- **Coated stock:** Our copiers do not produce quality copies on coated stock (paper or cover stock with a slick coating, like magazine paper). We will continue to outsource jobs that use coated stock to KDH Printing. Please allow at least one week for jobs that use coated stock.

3. **Bindery services:** With our new equipment, collating and stapling of large jobs no longer require additional time. The following bindery services are also available in house, but may require additional time:

- Perfect and spiral binding
- Folding
- Cutting and hole punching. (The paper cutter and paper drill can be used on up to 500 sheets at a time.)

The new equipment will be available August 15. Your efforts to make the most efficient use of Copy Center resources help improve the quality of your documents and the productivity of the company.

Makes it clear when changes will take place.

Feel free to call me at ext. 567 if you have any questions.

Invites contact.

M-Global Inc | 127 Rainbow Lane | Baltimore MD 21202 | 410.555.8175

■ **Model 5–13** ■ Memorandum: changes in services

Puts this memo in context of all benefits changes.

Emphasizes point of *agreement*—concerns about costs. Describes problem that led to need for change.

Gives overview of program.

Uses list to highlight three main elements of PAC.

Uses heading to focus on *main* concern of reader—*quality of care*.

Indicates that similar programs have worked well elsewhere.

MEMORANDUM

DATE: May 2, 2008
TO: All Employees of Cleveland Office
FROM: Timothy Fu, Personnel Director TF
SUBJECT: New Cost Containment Measure for Health Care

The next fiscal year will bring several changes in the company's fringe benefit plan. Later this month, you'll receive a complete report on all adjustments to go into effect July 1. For now, this memo will outline one major change in health care. Specifically, M-Global will adopt a cost-containment program called PAC—intended to help you and the company get more health care for the dollar.

WHAT IS PAC AND HOW DOES IT WORK?

Health costs have risen dramatically in the last 10 years. The immediate effect on M-Global has been major increases in insurance premiums. Both you and the company have shared this burden. This year M-Global will fight this inflationary trend by introducing a new cost-containment program called PAC—Pre-Admission Check.

Started by Healthco, our company medical supplier, PAC changes the procedure by which you and your dependents will be recommended for hospitalization. Except in emergencies, you or your physician will need to call the PAC hotline before admission to the hospital. The PAC medical staff will do the following:

1. Review the length of stay recommended by your physician, to make sure it conforms to general practice
2. Request a second opinion if the PAC staff believes that such an opinion is warranted
3. Approve final plans for hospitalization

If your physician recommends that you stay in the hospital beyond the length originally planned, he or she will call PAC for authorization.

WILL PAC AFFECT THE LEVEL OR QUALITY OF HEALTH CARE?

No. PAC will in no way restrict your health care or increase your personal costs. Quite the contrary, it may reduce total costs considerably, leading to a stabilization of the employee contributions to premiums next year. The goal is to make sure physicians give careful scrutiny to the lengths of hospital stays, staying within the norms associated with a particular illness unless there is good reason to do otherwise.

Programs like PAC have worked well for many other firms around the country; there is a track record of lowering costs and working efficiently with physicians and hospitals. Also, you will be glad to know that Healthco has the firm support of its member physicians on this program.

M-Global Inc | 127 Rainbow Lane | Baltimore MD 21202 | 410.555.8175

Leaves reader with clear sense of next step.

MEMO TO: all employees
May 2, 2008
Page 2

WHAT WILL HAPPEN NEXT?

As mentioned earlier, this change goes into effect with the beginning of the new fiscal year on July 1. Soon you will receive a report about this and other changes in benefits. If you have any questions before that time, please call the Corporate Benefits Department at ext. 678.

■ **Model 5–14** ■ Memorandum: changes in benefits

TO: Lab, Marketing, and Administrative Staff in U.S. Offices
FROM: Janice Simmons, Benefits Manager
SUBJECT: Training Funds for Fiscal Year 2008
DATE: January 2, 2008

Happy New Year to all of you! I hope you had a good break. I'm writing to announce some guidelines for approved training for the next twelve months—including an increased reimbursement. Please read on to see how these changes affect all lab, marketing, and administrative staff.

1. Lab Staff

   Maximum Reimbursement: $3,000 (up from $2,000)
   Approval Process: Discuss with your manager 21 days before tripTrip
   Purpose: To improve lab procedures

2. Marketing Staff

   Maximum Reimbursement: $4,000 (up from $3,500)
   Approval Process: Discuss with your manager 21 days before tripTrip
   Purpose: To learn new sales techniques

3. Administrative Staff

   Maximum Reimbursement: $4,500 (up from $4,000)
   Approval Process: Discuss with your manager 21 days before tripTrip
   Purpose: To improve productivity of office procedures

In the past most employees have failed to make use of their maximum training allotment. I encourage all of you to seek training opportunities that fit the guidelines listed above.

Please note the required 21-day lead time in the approval process!

Just send me an email if you have any questions about the procedure.

Janice

Annotations (right margin):

Begins with casual, friendly tone.

Includes clear purpose statement and three topics to be covered.

Supplies details about topics mentioned in first paragraph.

Uses list and parallel structure for easy reading.

Uses short paragraphs.

Concludes with reminder about an important part of the procedure.

Encourages them to contact her if there are questions.

■ **Model 5–15** ■ Email: changes in procedure

Chapter 6 | Definitions and Descriptions

>>> Chapter Outline

Definitions versus Descriptions   171
Technical Definitions at M-Global   171
Descriptions at M-Global   172

Guidelines for Writing Definitions   173
Example of an Expanded Definition   177

Guidelines for Writing Descriptions   177
Example of a Description   180

Chapter Summary   180

Learning Portfolio   182
Communication Challenge   182
Collaboration at Work   183
Assignments   183

Models for Good Writing   187
Model 6–1: Brief description (with formal definition included)   187
Model 6–2: Description from a user's manual   188
Model 6–3: Technical description (with definition included): Bunsen burner   190

Recently, M-Global has decided to change its insurance provider. Karrie Camp, Vice President of Human Resources, has decided to take this opportunity to revise the packet of information about benefits that is given to all employees in the United States. Although employees will continue to receive the detailed information about benefits from the insurance provider, Karrie knows that employees have asked for a quick, easy-to-read overview of the benefits available to them. She decides to create a set of information sheets that both define and describe each of the benefits in the new benefits package. (See Model 6–1 on page 187 for one of the sheets Karrie created for the packet.)

Definitions, descriptions, process explanations, and instructions are the types of writing that people often think of when they think of technical communication. This chapter and chapter 7 cover these four elements of technical communication. Definitions and descriptions are closely related; in fact, descriptions could be considered a type of definition. Process explanations and instructions are also closely related, with the difference being how the reader will use the documents.

>>> Definitions versus Descriptions

Definitions and descriptions can appear in any part of a document, from the introduction to the appendix. Like the example in Model 6–1, they may also be created as stand-alone documents. During your career, you will use technical terms known only to those in your profession. As a civil engineer, for example, you would know that a *triaxial compression test* helps determine the strength of soil samples. As a documentation specialist, you would know that *single-sourcing* allows the creation of multiple documents from the same original text. When writing to readers unfamiliar with these fields, however, you must define technical terms. You may also have to describe these technical objects, and the distinction between *definition* and *description* can sometimes be a bit confusing. In fact, you can consider a *description* to be a special type of definition that focuses on parts, functions, or other features. It emphasizes *physical* details. Descriptions often open with a sentence definition.

Technical Definitions at M-Global

Good definitions can support findings, conclusions, and recommendations throughout your document. They also keep readers interested. Conversely, the most organized and well-written report falls on deaf ears if it includes terms that readers do not grasp. "Define your terms!" is the frustrated exclamation of many a reader. For your reader's sake, then, you must be asking questions like these about definitions:

- How often do you use them?
- Where should they be placed?
- What format should they take?
- How much information is enough, and how much is too much?

To answer these questions, the following sections give guidelines for definitions and supply an annotated example. First, here are some typical contexts for definitions within M-Global, Inc.

- **Construction:** As an M-Global technician helping to build a power plant, you often use the term *turbine* in speech and writing. Obviously, your coworkers and clients understand the term. Now, however, you are using it in a plant brochure to be sent to consumers. For this general audience, you must define *turbine,* and you decide to accompany the definition with an illustration so that the unfamiliar audience can visualize how the turbine works.

- **Human Resources:** As a health and benefits specialist in M-Global's corporate office, you have been asked to introduce employees to a new organization-wide campaign to encourage healthy habits by employees. Your first project is a memorandum to all employees encouraging them to take advantage of free cholesterol screening being offered at all M-Global branches. Your memo must provide clear definitions for terms like *good cholesterol* and *bad cholesterol.*

- **Forestry:** As a forestry and agriculture expert with M-Global's Denver office, you have coordinated a major study for the state of Idaho. Your job has been to recommend ways that a major forested region can still be used for timber with little or no damage to the region's ecological balance. Although the report goes first to technical experts in Idaho's Department of Natural Resources, you have been told that it will also be made available to the public. Thus you have decided to include a glossary that defines terms such as *silvaculture, biodiversity, watershed management,* and *fuel reduction.*

In each case, you are including definitions to help readers with the least familiarity with the technical field understand the topic about which you are writing. When in doubt, insert definitions! Readers can always skip over ones they do not need.

Descriptions at M-Global

Descriptions are similar to definitions. In fact, they often open with a short definition, but they also emphasize the physical details of the object being described. Like definitions, descriptions often appear as supporting information in the document body or in appendices. Following are some typical contexts for descriptions within M-Global, Inc.

- **Site Recommendation:** M-Global's San Francisco office has been hired to recommend possible locations for a new swimming and surfing park in northern California. Written to a county commission (five laypersons who make the decision), your recommendation report gives three possible locations and the criteria for selecting them. The report includes appendices that give brief physical descriptions of the sites. Specifically, the appendices to the report describe (1) surface features, (2) current structures, (3) types of soils gathered from the surface, (4) water quality, and (5) aesthetic features, such as quality of the ocean views.

- **Sonar Equipment:** A potential M-Global client, Rebecca Stern, calls you in your capacity as a geologist at M-Global's Baltimore office. She wants information about the kind of sonar equipment M-Global uses to map geologic features on the seafloor. This client has a strong technical background, so you write a letter with a detailed technical description of the M-Global system. The body of the letter describes the locations and functions of (1) the seismic source (a device that sends the sound waves and is towed behind a boat) and (2) the receiver (a unit that receives the signals and is also towed behind the boat).

■ **Site Analysis:** M-Global's Cleveland office was hired to examine asbestos contamination in a large high school built in 1949. As a member of the investigating team, you found asbestos throughout the basement in old pipe coverings. Your final report to the school board provides conclusions about the level of contamination and recommendations for removal. An appendix gives a detailed technical description of the entire basement, including a map with a layout of the plumbing system.

■ **Office Equipment:** As the purchasing officer at M-Global's St. Paul office, you have been asked to provide information about printers and plotters that are used in the office. You gather the user's guides and manuals for the equipment and attach them to a cover memo that explains how often each piece of equipment is used and how well each piece of equipment has performed. Model 6–2 on pages 188–189 includes pages from the user's guide for the large document printer, which is used by a wide variety of employees.

>>> Guidelines for Writing Definitions

Once you know definitions are needed, you must decide on their format and location. Again, consider your readers. How much information do they need? Where is this information best placed within the document? To answer these and other questions, we offer five working guidelines for writing good definitions.

>> Definition Guideline 1:
Keep It Simple

Occasionally, the sole purpose of a report is to define a term; most often, however, a definition just clarifies a term in a document with a larger purpose. Your definitions should be as simple and unobtrusive as possible. Always present the simplest possible definition, with only that level of detail needed by the reader.

For example, in writing to a client on your land survey of her farm, you might briefly define a *transit* as "the instrument used by land surveyors to measure horizontal and vertical angles." The report's main purpose is to present property lines and total acreage, not to give a lesson in surveying, so this sentence definition is adequate. Choose from the following three main formats (listed from least to most complex) in deciding the form and length of definitions:

■ **Informal definition:** A word or brief phrase, often in parentheses, that gives only a synonym or other minimal information about the term.

- **Formal definition:** A full sentence that distinguishes the term from other similar terms and includes these three parts: the term itself, a class to which the term belongs, and distinguishing features of the term.
- **Expanded definition:** A lengthy explanation that begins with a formal definition and is developed into several paragraphs or more.

Guidelines 2–5 show you when to use these three options and where to put them in your document.

>> Definition Guideline 2: Use Informal Definitions for Simple Terms Most Readers Understand

Informal definitions appear right after the terms being defined, often as one-word synonyms in parentheses. They give just enough information to keep the reader moving quickly. As such, they are best used with simple terms that can be defined adequately without much detail.

One situation in which an informal definition would apply is as follows: M-Global has been hired to examine a possible shopping-mall site. The buyers, a group of physicians, want a list of previous owners and an opinion about the suitability of the site. As legal assistant at M-Global, you must assemble a list of owners in your part of the team-written report. You want your report to agree with court records, so you decide to include real-estate jargon such as *grantor* and *grantee*. For your nontechnical readers, you include parenthetical definitions such as these:

All *grantors* (persons from whom the property was obtained) and *grantees* (persons who purchased the property) are listed on the following chart, by year of ownership.

This same M-Global report has a section describing creosote pollution found at the site. The chemist writing the contamination section also uses an informal definition for the readers' benefit:

At the southwest corner of the mall site, we found 16 barrels of *creosote* (a coal tar derivative) buried under about three feet of sand.

The readers do not need a fancy chemical explanation of creosote. They need only enough information to keep them from getting lost in the terminology. Informal definitions perform this task nicely.

>> Definition Guideline 3: Use Formal Definitions for More Complex Terms

A formal definition appears in the form of a sentence that lists (1) the *term* to be defined, (2) the *class* to which it belongs, and (3) the *features* that distinguish the term from others in the same class. Use it when your reader needs more background than an informal definition provides. Formal definitions define in two stages:

- First, they place the term into a *class* (group) of similar items.
- Second, they list *features* (characteristics) of the term that separate it from all others in that same class.

In the list of sample definitions that follows, note that some terms are tangible (like *pumper*) and others are intangible (like *arrest*). Yet all can be defined by first choosing a class and then selecting features that distinguish the term from others in the same class.

Term	Class	Features
An *arrest* is	restraint of persons	that deprives them of freedom of movement and binds them to the will and control of the arresting officer.
A *financial statement* is	a historical report about a business	prepared by an accountant to provide information useful in making economic decisions, particularly for owners and creditors.
A *triaxial compression test* is	a soils lab test	that determines the amount of force needed to cause a shear failure in a soil sample.
A *pumper*	is a fire-fighting apparatus	used to provide adequate pressure to propel streams of water toward a fire.

This list demonstrates three important points about formal definitions. First, the definition itself must not contain terms that are confusing to your readers. The definition of *triaxial compression test,* for example, assumes readers understand the term *shear failure* that is used to describe features. If this assumption is incorrect, then the term must be defined. Second, formal definitions may be so long that they create a major distraction in the text. (See Guideline 5 for alternative locations.) Third, the class must be narrow enough so that you do not have to list too many distinguishing features.

>> Definition Guideline 4: Use the ABC Format for Expanded Definitions

Sometimes a parenthetical phrase or formal sentence definition is not enough. If readers need more information, use an expanded definition with this three-part structure:

■ The **Abstract** component provides an overview at the beginning, including a formal sentence definition and a description of the ways you will expand the definition

■ The **Body** component provides supporting information using headings and lists as helpful format devices for the reader

■ The **Conclusion** component should be brief, reminding the reader of the definition's relevance to the whole document

Following are seven ways to expand a definition, along with brief examples:

1. **Background or history of term**—expand the definition of *triaxial compression test* by giving a dictionary definition of *triaxial* and a brief history of the origin of the test

2. **Applications**—expand the definition of *financial statement* to include a description of the use of such a statement by a company about to purchase controlling interest in another

3. **List of parts**—expand the definition of *pumper* by listing the parts of the device, such as the compressor, the hose compartment, and the water tank

4. **Graphics**—expand the description of the *triaxial compression test* with an illustration showing the laboratory test apparatus

5. **Comparison/contrast**—expand the definition of a term like *management by objectives* (a technique for motivating and assessing the performance of employees) by pointing out similarities and differences between it and other management techniques

6. **Basic principle**—expand the definition of *ohm* (a unit of electrical resistance equal to that of a conductor in which a current of one ampere is produced by a potential of one volt across its terminals) by explaining the principle of Ohm's Law (that for any circuit, the electric current is directly proportional to the voltage and inversely proportional to the resistance)

7. **Illustration**—expand the definition of CAD/CAM (Computer-Aided Design/Computer-Aided Manufacturing—computerized techniques to automate the design and manufacture of products) by giving examples of how CAD/CAM is changing methods of manufacturing many items, from blue jeans to airplanes

Obviously, long definitions might seem unwieldy within the text of a report, or even within a footnote. For this reason, they often appear in appendices, as noted in the next guideline. Readers who want additional information can seek them out, whereas other readers are not distracted by digressions in the text.

>> Definition Guideline 5: Choose the Right Location for Your Definition

Short definitions are likely to be in the main text; long ones are often relegated to footnotes or appendices. However, length is not the main consideration. Think first about the *importance* of the definition to your reader. If you know that decision makers reading your report need the definition, then place it in the text—even if it is fairly lengthy. If the definition provides only supplementary information, then it can go elsewhere. You have these five choices for locating a definition:

1. **In the same sentence as the term,** as with an informal, parenthetical definition

2. **In a separate sentence,** as with a formal sentence definition occurring right after a term is mentioned

3. **In a footnote,** as with a formal or expanded definition listed at the bottom of the page on which the term is first mentioned

4. **In a glossary at the beginning or end of the document,** along with all other terms needing definition in that document

5. **In an appendix at the end of the document,** as with an expanded definition that would otherwise clutter the text of the document

Example of an Expanded Definition

Sometimes your readers may need a longer and more expanded definition. Expanded definitions are especially useful in reports from technical experts to nontechnical readers. M-Global's report writers, for example, often must explain environmental, structural, or geologic problems to concerned citizens or nontechnical decision makers. Figure 6–1 gives definitions of the same term from two reports about public health.

■ **Figure 6–1** ■
Definitions in two reports

BMI [Body Mass Index] is a practical measure that requires only two things: accurate measures of an individual's weight and height (Figure 1). BMI is a measure of weight in relation to height. BMI is calculated as weight in pounds divided by the square of the height in inches, multiplied by 703. Alternatively, BMI can be calculated as weight in kilograms divided by the square of the height in meters.

> Excerpted from *The Surgeon General's call to action to prevent and decrease overweight and obesity.* (2001). Office of Disease Prevention and Health Promotion. Centers for Disease Control and Preventions, National Institutes of Health. Rockville, MD: U.S. Dept. of Health and Human Services, Public Health Service, Office of the Surgeon General. Washington.

Body Mass Index measures weight in relation to height. It is a mathematical formula that divides a person's body weight in kilograms by the square of his or her height in meters. BMI is highly correlated with body fat, and can indicate that a person is overweight or obese. People with a Body Mass Index of 25 or higher are considered overweight, while those with a BMI of 30 or higher are considered obese. According to the National Institutes of Health, all adults who have a BMI of 25 or higher are considered at risk for premature death and disability as a consequence of being overweight.

> Excerpted from McCann, B.A. and Ewing, R. (Sept. 2003). *Measuring the Health Effects of Sprawl: A National Analysis of Physical Activity, Obesity and Chronic Disease.* Smart Growth America Surface Transportation Policy Project.

>>> Guidelines for Writing Descriptions

When your readers benefit from detailed information about parts, functions, or other elements, you should write a description. These five guidelines help you write accurate, detailed descriptions. Follow them carefully as you prepare assignments in this class and on the job.

>> Description Guideline 1: Remember Your Readers' Needs

The level of detail in a technical description depends on the purpose a description serves. Give readers precisely what they need—but no more. In the study of locations for a new swimming and surfing park, the commissioners do not want a detailed description of soil samples taken from borings. That level of detail is reserved for a few sites selected later for further study. Instead, they want only surface descriptions. Always know just how much detail will get the job done.

>> Description Guideline 2: Be Accurate and Objective

More than anything else, readers expect accuracy in descriptions. Pay close attention to details. (As noted previously, the *degree* of detail in a description depends on the *purpose* of the document.) In the asbestos analysis example, you should describe every possible location of asbestos in the school basement. Because the description becomes the basis for a cost proposal to remove the material, accuracy is crucial.

Along with accuracy should come *objectivity*. This term is more difficult to pin down, however. Some writers assume that an objective description leaves out all opinion. This is not the case. Instead, an objective description may very well include opinions that have these features:

- They are based on your professional background.
- They can be justified by the time you have had to complete the description.
- They can be supported by details from the site or object being described.

For example, your description of the basement pipes mentioned in the site analysis example might include a statement such as: "Because there is asbestos wrapping on the exposed pipes above the boiler, my experience suggests that asbestos wrapping probably also exists around the pipes above the ceiling—in areas that we were not able to view." This opinion does not reduce the objectivity of your description; it is simply a logical conclusion based on your experience.

>> Description Guideline 3: Choose an Overall Organization Plan

Like other patterns discussed in this chapter, technical descriptions usually make up only parts of documents. Nevertheless, they must have an organization plan that permits them to be read as self-contained, stand-alone sections. Indeed, a description may be excerpted later for separate use.

Following are three common ways to describe physical objects and events. In all three cases, a description should move from general to specific—that is, you begin with a view of the entire object or event, and in the rest of the description, you focus on specifics. Headings may be used, depending on the format of the larger document.

1. **Description of the parts:** For many physical objects, like sonar equipment, basement floor, and printers in the cases on pages 172–173, you simply organize the description by moving from part to part.

2. **Description of the functions:** Often the most appropriate overall plan relies on how things work, not on how they look. In the sonar example, the reader was more interested in the way that the sender and receiver worked together to provide a map of the seafloor. This function-oriented description should include only a brief description of the parts.

3. **Description of the sequence:** If your description involves events, as in a police officer's description of an accident investigation, you can organize ideas around the major actions that occurred, in their correct sequence. As with any list, it is best to place a series of many activities into just a few groups. It is much easier for readers to comprehend 4 groups of 5 events each than a single list of 20 events.

>> Description Guideline 4: Use "Helpers" Like Graphics and Analogies

The words of a technical description must come alive. Because your readers may be unfamiliar with the item, you must search for ways to connect with their experience and with their senses. Two effective tools are graphics and analogies.

Graphics respond to the desire of most readers to see pictures along with words. As readers move through your part-by-part or functional breakdown of a mechanism, they can refer to your graphic aid for assistance. The illustration helps you too, of course, in that you need not be as detailed in describing locations and dimensions of parts when you know the reader has easy access to a visual. Note how the diagrams in Models 6–2 and 6–3 on pages 188–192 give meaning to the technical details in the verbal descriptions.

Analogies, like illustrations, give readers a convenient handle for understanding your description. Put simply, an analogy allows you to describe something unknown or uncommon in terms of something that is known or more common. A brief analogy can sometimes save you hundreds of words of technical description. This paragraph description contains three analogies:

> M-Global, Inc. is equipped to help clean up oil spills with its patented product, SeaClean. This highly absorbent chemical is spread over the entire spill by means of a helicopter that makes passes over the spill, much like a lawnmower covers the complete surface area of a lawn. When the chemical contacts the oil, it acts like sawdust coming in contact with oil on a garage floor—that is, the oil is immediately absorbed into the chemical and physically transformed into a product that is easily collected. Then our nearby ship can collect the product, using a machine that operates much like a vacuum cleaner. This machine sucks the Sea-Clean (now full of oil) off the surface of the water and into a sealed container in the ship's hold.

>> Description Guideline 5: Give Your Description the "Visualizing Test"

After completing a description, test its effectiveness by reading it to someone unfamiliar with the material—someone with about the same level of knowledge as your intended

reader. If this person can draw a rough sketch of the object or events while listening to your description, then you have done a good job. If not, ask your listener for suggestions to improve the description. If you are too close to the subject yourself, sometimes an outside point of view helps refine your technical description.

Example of a Description

Fahdi Ahmad, the director of procurement at M-Global's office in Saudi Arabia, is changing his buying procedures. He has decided to purchase some basic lab supplies from companies in nearby developing nations rather than from firms in industrialized countries. For one thing, he thinks this move may save some money. For another, he believes it will help the company get more projects from these nations, for M-Global will become known as a firm that pumps back some of its profits into the local economies.

As a first step in this process, Fahdi is providing potential suppliers with descriptions of some basic lab equipment used at M-Global, such as pH meters, laboratory scales, glass beakers, and burners. Model 6–3 on pages 190–192 presents a moderately detailed description of one such piece of equipment—a Bunsen burner. Fahdi selected a burner model that is being successfully used at many M-Global offices in the United States. The Model 6–3 description can serve as a starting point for suppliers. However, M-Global and the suppliers realize that burners will have slightly different features, depending on the manufacturer.

>>> Chapter Summary

Definitions and descriptions help readers who are unfamiliar with technical terms understand your documents and use the documents you write to make informed decisions.

Definitions occur in technical communication in one of three forms: *informal* (in parentheses), *formal* (in sentence form with term, class, and features), and *expanded* (in a paragraph or more). The following main guidelines apply:

1. Keep it simple.
2. Use informal definitions for simple terms most readers understand.
3. Use formal definitions for more complex terms.
4. Use the ABC Format for expanded definitions.
5. Choose the right location for your definition.

Descriptions, like definitions, depend on detail and accuracy for their effect. Careful descriptions usually include a lengthy itemizing of the parts of a mechanism or the functions of a term. Follow these basic guidelines for producing effective descriptions:

1. Remember your readers' needs.
2. Be accurate and objective.
3. Choose an overall organization plan.
4. Use "helpers" like graphics and analogies.
5. Give your description the "visualizing test."

>>> Learning Portfolio

Communication Challenge "Biofuels Brainstorm: Describing New Technologies"

Sylvia Barnard, manager of the Denver branch of M-Global, has a special interest in the energy industry. As a geologist working in oil and gas exploration, she joined M-Global to contribute to its oil and gas industry construction projects, such as oil fields and refineries. Sylvia wants to see M-Global respond to changes in the energy industry by diversifying into work on biofuels projects. This case study explains her approach to the problem, and ends with questions and comments for discussion and an assignment for a written response to the Challenge.

Research

As a first step in developing a proposal for Jim McDuff, Sylvia wants to learn more about the biofuels industry and biofuels technology. Although she has read about biofuels in newspapers and general news magazines, she knows that to propose that M-Global enter the field, she must have more specialized knowledge about what biofuels are. With a better understanding of the technology, she will be able to focus her proposal on the areas in which M-Global's experience in the oil and gas industry can transfer to construction projects in the new biofuels industry. After her research, she decides to focus on the following types of fuels:

- Biodiesel
- Bioalcohols
- Biogas
- Cellulosic biofuels

The Report

Before she writes her proposal, Sylvia decides to create a report that compares refineries and refinery construction needs for biofuels to the oil and gas refineries that M-Global has worked on in the past. The report will be primarily descriptive. It must define biofuels and describe the equipment and site construction needs of biofuels refineries.

Sylvia knows that M-Global has a history of looking to environmental issues for business opportunities. In the 1970s, the company (then McDuff, Inc.) began work in hazardous waste disposal. (See pp. 48–49 in chapter 2.) At the time, however, it was clear that there was a need for such services, and that the technology was rapidly developing. Sylvia is concerned that her enthusiasm for biofuels may be premature. Although there are companies building biofuels refineries, many of them seem more focused on the environmental issues than on long-term profitability. Her research also suggests that the technology is in its earliest stages. She worries that it might be too early for M-Global to get into the biofuels industry, but decides to write the report anyway.

Questions and Comments for Discussion

1. How could Sylvia use her knowledge of M-Global's history, especially Rob McDuff's interest in environmental issues, to make her report appealing to Jim McDuff? Should Sylvia let Jim know that she plans on following this report with a proposal? If so, why, and what should she tell him?
2. What must Jim McDuff understand about biofuels before he can make a decision about exploring the opportunity further? What illustrations might help him make his decision?
3. What terms must Sylvia define? What kinds of definitions should she write, and where should they be included in the report?
4. Should Sylvia include her concerns about the fact that the biofuels industry is in its early stages? If so, what should she say? Should she even send the report, or should she save it until the biofuels industry is more well established?

Write About It

Sylvia has assigned you the task of writing a short description of biofuels that she can include in various documents related to her biofuels proposal. Write a one-page description of biofuels that could be used or adapted to a variety of documents related to the biofuels initiative at M-Global. You should define biofuels, and describe them. You may decide to describe the different types of biofuels (classification), or you may decide to compare them to oil and gas products (comparison/contrast). Use illustrations as appropriate. Include a list of references.

Collaboration at Work Analyzing the Core

General Instructions

Each Collaboration at Work exercise applies strategies for working in teams to chapter topics. The exercise assumes you (1) have been divided into teams of about three to six students, (2) will use team time inside or outside of class to complete the case, and (3) will produce an oral or written response. For guidelines about writing in teams, refer to pages 24–28.

Background for Assignment

Whereas some terms are easily defined, others, including abstract concepts, can be quite challenging. This also means that it is essential that abstract terms be clearly defined, because not all readers will understand the term the way that you do. This assignment asks your team to define the abstract concept of a college or university education.

Like other colleges and universities, your institution may require students to complete a core curriculum of required subjects. Some cores are virtually identical for students in all majors; others vary by major. Following is one example of a core curriculum (required for institutions in the University System of Georgia):

1. **Essential Skills** (9 semester hr): Includes two freshman composition courses and college algebra.
2. **Institutional Options** (4–5 semester hr): Includes courses of the institution's choosing, such as public speaking and interdisciplinary classes.

3. **Humanities/Fine Arts** (10–11 semester hr): Includes courses such as literature surveys, art appreciation, music appreciation, and foreign language.
4. **Science, Mathematics, and Technology** (10–11 semester hr): Includes laboratory and non-lab classes in fields such as physics, chemistry, biology, and calculus.
5. **Social Sciences** (12 semester hr): Includes courses such as American history, world history, political science, and religion.
6. **Courses Related to Student's Program of Study** (18 semester hr): Includes lower-level classes related to the student's specific major. For example, an engineering major may be required to take extra math, whereas a technical communication major may be required to take introductory technical communication.

Core curricula or general studies requirements like these suggest a definition of a college or university education. In this example, the required courses suggest that the university values developing intellectual curiosity and an understanding of communication, critical thinking, culture, and scientific reasoning.

Team Assignment

Examine the core curriculum at your institution and decide how it suggests that your school defines a university education. Write an extended definition that could be used on your school's website on materials that your school sends to potential students that help identify your school's philosophy and goals for its students.

Assignments

Part 1: Short Assignments

The following short assignments can be completed either orally or in writing. Unless a team project is specifically indicated, an assignment can be either a team or an individual effort. Your instructor will give you specific directions.

1. Definition

Using the guidelines in this chapter, discuss the relative effectiveness of the following short definitions. Speculate on the likely audience the definitions are addressing.

a. **Afforestation**—the process of establishing trees on land that has lacked forest cover for a very long period of time or has never been forested
b. **Carbon cycle**—the term used to describe the flow of carbon (in various forms such as carbon dioxide [CO_2], organic matter, and carbonates) through the atmosphere, ocean, terrestrial biosphere, and lithosphere
c. **Feebates**—systems of progressive vehicle taxes on purchases of less efficient new vehicles and subsidies for more efficient new vehicles

d. **Greenhouse gases**—gases including water vapor, CO_2, CH_4, nitrous oxide, and halocarbons that trap infrared heat, warming the air near the surface and in the lower levels of the atmosphere

e. **Mitigation**—a human intervention to reduce the sources of or to enhance the sinks of greenhouse gases

f. **Permafrost**—soils or rocks that remain below 0°C for at least two consecutive years

g. **Temperate zones**—regions of the Earth's surface located above 30° latitude and below 66.5° latitude

h. **Wet climates**—climates where the ratio of mean annual precipitation to potential evapotranspiration is greater than 1.0

Adapted from U.S. Climate Change Science Program. (November 2007.) The First State of the Carbon Cycle Report (SOCCR): *The North American Carbon Budget and Implications for the Global Carbon Cycle.* Synthesis and Assessment Product 2.2. http://www.climatescience.gov/Library/sap/sap2-2/final-report.

2. Definition

Create formal sentence definitions of the following terms. Remember to include the class and distinguishing features:

- Automated Teller Machine (ATM)
- Digital Video Disc (DVD)
- Website
- Job Interview

3. Definition

Reread the two definitions in Figure 6–1 on p. 177. Identify the formal sentence definition in each one. What other information is included in each definition? What information is found in both definitions? What audiences do you think these definitions were written for? Explain.

4. Definition—M-Global Context

Write definitions of the following words for the glossary mentioned in the "Forestry" example on p. 172.

- Silvaculture
- Biodiversity
- Watershed management
- Fuel reduction

5. Definition—M-Global Context

As part of its soil analysis work, M-Global works to identify brownfields. Write a one-paragraph definition of brownfields that could be used in all M-Global documents about the subject. Include a one-sentence formal definition.

6. Definition—M-Global Projects

This assignment can be completed as an individual or a team project. Select one of the projects located at the end of chapter 2. Using the outline of information on the project as a starting point, do the following:

- Conduct some research on the technical field reflected in the project.
- Select some terms related to the field.
- Write either short formal definitions or expanded definitions of the terms.

In assigning this project, your instructor will indicate (a) how many terms you should select and (b) whether you should write formal or expanded definitions.

7. Description

Write a description of a piece of equipment or furniture located in your classroom or brought to class by your instructor—for example, a classroom chair, an overhead projector, a screen, a three-hole punch, a mechanical pencil, or a computer mouse. Write the description for a reader totally unfamiliar with the item.

8. Description

Write a description of a piece of equipment that would be used in a hobby or activity in which you regularly take part. Write the description for someone who has just taken up the activity.

9. Description—M-Global Project

For this project put yourself in the position of someone who worked on the General Hospital project (third project in the project sheets in chapter 2). Write a description of a piece of equipment that either (a) could have been used in the construction of the hospital or (b) could be housed in any modern medical facility such as General Hospital. This assignment may require a visit to a local hospital, construction company, or library with resources in the fields of construction or medical care.

Part 2: Longer Assignments

These assignments test your ability to write the two patterns covered in this chapter—definitions and descriptions. Specifically, follow these guidelines:

- Write each exercise in the form of a letter report or memo report, as specified.
- Follow organization and design guidelines given in chapters 3 and 4, especially concerning the ABC format (**A**bstract/**B**ody/**C**onclusion) and the use of headings. Chapter 8 gives rules for short reports, but such detail is not necessary to complete the assignments here.

- Fill out a Planning Form (at the end of the book) for each assignment.

10. Technical Definitions in Your Field

Select a technical area in which you have taken course work or in which you have technical experience. Now assume that you are employed as an outside consulting expert, acting as a resource in your particular area to a M-Global manager not familiar with your specialty. For example, a food-science expert might provide information related to the dietary needs of oil workers working on an offshore rig for three months; a business or management expert might report on a new management technique; an electronics expert might explain the operation of some new piece of equipment that M-Global is considering buying; a computer programmer might explain some new piece of hardware that could provide supporting services to M-Global; and a legal expert might define *sexism in the workplace* for the benefit of M-Global's human resources professionals.

For the purpose of this report, develop a context in which you would have to define terms for an uninformed reader. Incorporate one expanded definition and at least one sentence definition into your report.

11. Description of Equipment in Your Field

Select a common piece of laboratory, office, or field equipment with which you are familiar. Now assume that you must write a short report to your M-Global supervisor, who wants it to contain a thorough physical description of the equipment. Later, he or she plans to incorporate your description into a training manual for those who must understand how to use, and perform minor repairs on, the equipment. For the body of your description, choose either a part-by-part physical description or a thorough description of functions.

12. Description of Position in Your Field

Interview a friend or colleague about the specific job that person holds. Make certain it is a job that you yourself have not had. On the basis of data collected in the interview, write a thorough description of the person's position—including major responsibilities, reporting relationships, educational preparation, and experience required.

Now place this description in the context of a letter report to Karrie Camp, the manager of human resources at M-Global. Assume she has hired you, a technical consultant to M-Global, to submit a letter report that contains the description. She is preparing to advertise such an opening at M-Global, but needs your report to write the job description and the advertisement. Because she has little firsthand knowledge of the position about which you are writing, you should avoid technical jargon.

13. Definition—M-Global Context

Model 6–1 is part of a packet explaining employee benefits. Write a definition of one of the following employee benefits that could be included in the same package:

- ESOP (Employee Stock Ownership Plan)
- 401K retirement plan
- HSA (Health Savings Plan)

14. Description—M-Global Context

As a website developer on the M-Global Publications Development Team, you are concerned with the accessibility of computers to those with visual disabilities. Write a description of one kind adaptive technology that can help those with visual disabilities use computers or access websites more easily.

Now place this description in the context of a memo to Karrie Camp, the manager of human resources at M-Global. Explain how the technology will make it possible for M-Global to hire qualified applicants with visual impairments.

15. Ethics Assignment

Although definitions and descriptions may appear neutral, they may be used to promote a point of view or to advance an argument on a controversial issue. Examine the following definitions of *global warming* from various sources on the Internet, and find and read each organization's home page. Can you see implied biases in the definition, or does the definition appear neutral? Does this bias or neutrality support the general goals of the organization that published the definition?

In a short essay, compare the definitions and identify the source of each one as well as any apparent bias in the original source. Discuss whether the definitions have been written to support their sources' points of view.

US Geological Service, National Wetlands Research Center
Global Warming—An increase of the earth's temperature by a few degrees resulting in an increase in the volume of water which contributes to sea-level rise.

"The Fragile Fringe: Glossary" <http://www.nwrc.usgs.gov/fringe/glossary.html>. Oct. 4, 2007.

Climate Change Central

Global Warming—Strictly speaking, global warming and global cooling refer to the natural warming and cooling trends that the earth has experienced all through its history. However, the term "global warming" has become a popular term encompassing all aspects of the global warming problem, including the potential climate changes that will be brought about by an increase in global temperatures.

"Glossary of Terms." Retrieved November 2007 from <http://www.climatechangecentral.com/default.asp?V_DOC_ID=849>, n.d.

Minnesota Pollution Control Agency

Global Warming—An increase in the Earth's temperature caused by human activities, such as burning coal, oil and natural gas. This releases carbon dioxide, methane, and other greenhouse gases into the atmosphere. Greenhouse gases form a blanket around the Earth, trapping heat and raising temperatures on the ground. This is steadily changing our climate.

"MPCA Glossary" <http://www.pca.state.mn.us/gloss/glossary.cfm?alpha=G&header=1&glossaryCat=0>, n.d.

Washington Council on International Trade

Global Warming—Heating that occurs when carbon dioxide traps the Sun's heat near Earth's surface, causing Earth's temperature to rise

"Trade is" <http://www.wcit.org/tradeis/glossary.htm>, n.d.

 ### 16. International Communication Assignment

In the global marketplace, companies are using illustrations and images to avoid expensive translation. Find examples of descriptions that use illustrations extensively. If possible, find descriptions in multiple languages, such as those in owner's manuals. (Focus on the descriptions of objects, not on instructions.) Analyze the illustrations for their effectiveness as descriptions. How important is text to the illustrations? Could the illustrations serve as descriptions without the text? If you have a document that is in multiple languages, do the illustrations differ from one version to the next? Write an essay that discusses the relationship of text and illustrations in descriptions. Include a discussion of whether you think companies should try to make their descriptions text-free.

 ### 17. A.C.T. N.O.W. Assignment (Applying Communication To Nurture Our World)

Colleges and universities work hard to foster good relationships with their surrounding communities. Sometimes referred to as *town–gown relations*, this connection between the institution and community is important for the obvious reason that both entities inhabit the same environment and depend on each other. For this assignment, select a project that you believe would improve or nurture town–gown relations in your community. Depending on the instructions you are given, prepare an oral or written report that describes the project. Your instructor may also ask you to use the description in the context of an argument for why the project would be useful.

Your M-Global Benefits

Flexible Spending Accounts (FSAs)

What is a Flexible Spending Account?
A Flexible Spending Account (FSA) is a pre-tax savings account that can be used for an employee's qualifying out-of-pocket expenses.

An FSA allows employees to set up an account that can be used for dependent care and health costs. FSAs are pre-tax benefits, meaning that they allow employees to designate an amount that will be withheld from their paychecks before taxes are figured. Taxes are then withheld based on the amount after the deduction for the FSA. As a result, participants can save as much as 35% on their federal income taxes.

At M-Global, Flexible Spending Accounts can be used for child care and for health costs that are not covered by insurance, including deductibles, co-insurance, and co-pays. Reimbursement is also available for out-of-pocket medical expenses such as prescriptions, orthodontia, and laboratory services. Details of employee benefits are available in the Employee Benefit Information packet distributed each November, in Employee Orientation materials, and on M-Global's Human Resources website.

When the employee has a reimbursable expense, he or she should submit the appropriate form to their branch Human Resources Office. Reimbursement forms are available through M-Global's Human Resources website. Reimbursement checks will generally be available one week after the form is submitted.

Are there any drawbacks to a Flexible Spending Account?
Contributions to an FSA will result in a decrease in take home pay, and child care costs are not eligible for child care credit on federal income tax if they are reimbursed through a Flexible Spending Account. In addition, FSAs are a "use it or lose it" plan. That is, any money that remains in the employee's account at the end of the plan year may be forfeited. Reimbursement requests must be submitted within 90 days of the end of the plan year.

How can I get the most benefit from a Flexible Spending Account?
It is important to estimate your deductions carefully. Figure your out-of-pocket expenses from the previous year. Then estimate any changes carefully. Keep track of your spending throughout the year to make sure that you use your Flexible Spending Account effectively.

Starts with a sentence definition

Emphasizes benefits to employees

Briefly explains how to use the account.

Points reader to documents and web site for further information

Meets legal requirement that limitations be explained

Closes with advice for using plan effectively

■ **Model 6–1** ■ Brief description (with formal definition included)

The front panel

Your printer's front panel is located on the front of the printer, on the right hand side. Use if for the following functions:

Starts with overview of important functions.

- Use it to perform certain operations, such as loading and unloading paper.

- View up-to-date information about the status of the printer, the ink cartridges, the printheads, the maintenance cartridge, the paper, the print jobs, and other parts and processes.

- Get guidance in using the printer.

- See warning and error messages, when appropriate.

- Use it to change the values of printer settings and the operation of the printer. However, settings in the Embedded Web Server or in the driver override changes made on the front panel.

Illustration focuses on parts being described

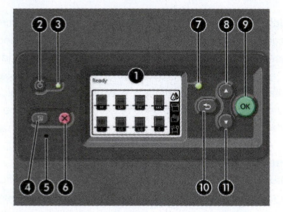

The front panel has the following components:

I. The display area, shows information, icons, and menus.

Numbers correspond to parts in illustration.

2. The Power button turns the printer on and off. If the printer is in sleep mode, this button will wake it up. (This is different from the hard power switch on the back of the printer. See Turn the printer on and off on page 21.)

3. The Power light is off when the printer is off. This light is amber when the printer is in sleep made, green when the printer is on, green and flashing when the printer is in transition between off and on.

4. The Form Feed and Cut button normally advances and cuts the roll. Here is a list of its other functions:

- If the printer is waiting for more pages to be nested, this button cancels the waiting time and prints the available pages immediately.

- If the printer is drying the ink after printing, this button cancels the waiting time and releases the page immediately.

Integrates description of parts with operating instructions.

- If the take-up reel is enabled, this button advances the paper 10 cm (3.9 inches), but does not cut the paper.

■ **Model 6–2** ■ Description from a user's manual
(Content courtesy of Hewlett-Packard Company)

5. The Reset button restarts the printer (as if it were switched off and switched on again). You will need a non-conductive implement with a narrow tip to operate the Reset button.

6. The Cancel button cancels the current operation. It is often used to stop the current print job.

7. The Status light is off when the printer is not ready to print: the printer is either off, or in sleep made. The Status light is green when the printer is ready and idle, green and flashing when the printer is busy, amber when a serious internal error has occurred, and amber and flashing when the printer is awaiting human attention.

8. The UP button moves to the previous item in a list, or increases a numerical value.

9. The OK button is used to select the item that is currently highlighted.

10. The Back button is used to return to the previous menu. If you press it repeatedly, or hold it down, you return to the main menu.

11. The Down button moves to the next item in a list, or decreases a numerical value.

To *highlight* an item on the front panel, press the Up or Down button until the item is highlighted.

To *select* an item on the front panel, first highlight it and then press the OK button.

The four front-panel icons are all found on the main menu. If you need to select or highlight an icon, and you do not see the icons in the front panel, press the Back button until you can see them.

Sometimes this guide shows a series of front panel items like this: **Item1** > **Item2** > **Item3**. A construction like this indicates that you should select **Item1,** select **Item2,** and then select **Item3.**

You will find information about specific uses of the front panel throughout this guide. ◄—— Refers user to more detailed information.

■ **Model 6–2** ■ continued

M-Global uses Bunsen burners in all its laboratories. Following some background information, this technical description provides details about three main parts of a typical burner:

Notes burner's main parts and sections that follow.

- Base
- Gas valve
- Pipe

The conclusion lists some standards for the burner's performance.

Background

Gives formal sentence definition and general information for non-technical readers.

The Bunsen burner is a basic piece of laboratory equipment used to produce a continuous flame at relatively low temperatures. Originally designed by Robert W. Bunsen in the late 1800s, the "Bunsen burner" has become a generic term for basic lab burners made by many firms.

Most burners look and perform alike, though burners from different companies do include slightly different features. What follows is a description of the Model 03–962 Bunsen-style burner manufactured by the Fisher Scientific Company. It runs on natural gas.

Base

The heavy die-cast base of the Fisher burner is very stable. It is made from nonferrous metal and has a nickel finish. Here are its main features and dimensions:

Uses bullets for technical detail.

- Hexagonal-shaped foundation that is 2 3/4″ in diameter and 2″ high

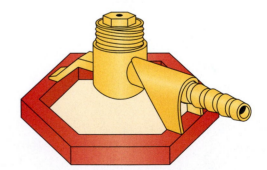

■ **Model 6–3** ■ Technical description (with definition included): Bunsen burner

- A 3/4″-diameter threaded cylinder at the top of the base
- A 1/2″-diameter hexagonal brass nut at the top of the cylinder, with a small hole in the center from which gas is emitted into the pipe

Jutting out from the side of the cylinder, parallel to the surface on which the base of the burner rests, is a tapered gas inlet 2 1/2″ long. The inlet has a grooved surface that holds the gas tube securely to the burner.

Gas Valve

A valve is threaded vertically up into the bottom center of the base of the burner. It allows the user to adjust the volume of gas that flows from the gas inlet up through the base cylinder.

The valve looks like a car axle with only one wheel at-tached. The 2″-long stem, or *axle*, rests on a round 3/4″-diameter base, or *wheel*. Actually, this base is about the diameter and thickness of a U.S. nickel coin. It has serrated edges so that it can be twisted with ease. The 1/8″-diameter stem is threaded and screws vertically into the base of the burner.

Uses analogies such as wheel, axle, *and* nickel coin.

When twisted clockwise, the valve closes and decreases gas flow. When twisted counterclockwise, it opens and increases gas flow.

Describes movement to help reader understand purpose of part.

Pipe

A 4 1/2″-long pipe extends straight up from the top of the base of the burner. Except for its flared ends, the pipe is 1/2″ in diameter. The combining of gas and air at the bottom of the pipe produces a flame that emerges at the top.

Includes common visual terms like flared.

The pipe threads on to the cylinder at the top of the burner base. The bottom of the pipe flares out to form an end piece with eight 3/16″ holes drilled around its circumference. The holes let in air that mixes with gas entering from the base. When the pipe is turned clockwise, the amount of air entering the holes is reduced and the temperature of the flame is lowered. When the pipe is turned counterclockwise, the amount of air entering the holes is increased and the temperature of the flame is raised.

Reveals purpose of parts through description.

The flared top end of the pipe looks much like a socket piece for a socket wrench. Called a *flame retainer* by the manufacturer, it helps keep the flame from going out. Viewed vertically from above the burner, the top of the retainer is shaped like a wagon wheel with short spokes. The *spokes* are actually eight ports that open to the pipe hole but close off before they reach the circumference of the pipe. Spaced evenly on the circumference, these ports help control the flame.

Employs two more analogies to help less technical readers.

■ **Model 6–3** ■ continued

Uses bullets for
technical detail.

Performance Specifications

The Fisher version of the Bunsen burner offers the following performance
standards:

- Produces flame that is adjustable from 3/4″ to 12″
- Has heat output of over 1465W (5000 BTU/hr.)
- Consumes 5 cu ft hr. of natural gas
- Meets U.S., NIST, USTM, and ASTM standards

■ **Model 6–3** ■ continued

Chapter 7 Process Explanations and Instructions

>>> Chapter Outline

Process Explanations versus Instructions 194

Process Explanations at M-Global 194

Instructions at M-Global 196

Guidelines for Process Explanations 197

Guidelines for Instructions 201

Chapter Summary 207

Learning Portfolio 209

Communication Challenge 209

Collaboration at Work 210

Assignments 211

Models for Good Writing 215

Model 7–1: M-Global process explanation: electronic mail 215

Model 7–2: M-Global instructions: electronic mail 216

Model 7–3: Process explanation 217

Model 7–4: M-Global process explanation with a flowchart 218

Model 7–5: Instructions for making travel arrangements 219

Model 7–6: M-Global memo containing how-to instructions for a scanner 221

Like other M-Global offices, M-Global's Denver office recently installed an updated electronic mail system. As office services manager, Jenny Vir met with a representative of the company that would install the system, shortly before the installation date. To help others who would be using this new system, Jenny wrote a memo to her boss, Leonard Schwartz, summarizing the installation process (Model 7–1 on page 215). She also wrote a memo to all employees, giving them instructions about how to read electronic-mail messages (Model 7–2 on page 216).

This brief M-Global case study demonstrates two types of technical communication you will often face—process explanations and instructions. Although technical communicators are creating a widening variety of documents, a survey of technical communication managers found that instructional materials such as manuals and online help remain the most common and most important documents in the field.[1] For this reason, instructions receive the most emphasis in this chapter.

Instructions and process explanations share an important common bond. Both must accurately describe a series of steps leading toward a specific result. Yet they differ in purpose, audience, and format. This chapter (1) explores these similarities and differences, with specific reference to M-Global applications, (2) gives specific guidelines for developing both types, and (3) provides models to use in your own writing.

---

[1]Rainey, K.T., Turner, R.K., and Dayton, D. (2005). "Do Curricula Correspond to Managerial Expectations? Core Competencies for Technical Communicators." *Technical Communication*, 52, 321–352.

>>> Process Explanations versus Instructions

In the M-Global example just given, Jenny's memo (Model 7–1) explained the process by which the vendor installed electronic mail. The other memo (Model 7–2) gave users the directions needed to read email. In other words, you write a process explanation to help readers *understand* what has been, is being, or will be done, whereas you write instructions to show readers how to perform the *process* themselves.

Process explanations are appropriate when the reader must be informed about the action but does not need to perform it. If you suspect a reader may in fact be a *user* (i.e., someone who uses your document to perform the process), always write instructions. Figure 7–1 provides a list of contrasting features of process explanations and instructions; the two subsections that follow give these features some realism by briefly describing some M-Global contexts.

Process Explanations at M-Global

Process explanations provide information for interested readers who do not need instructional details. At times, explaining a process may be the sole purpose of your document, as in Model 7–1. More often, however, you use process explanation only as a pattern of organization within a document with a larger purpose. The following M-Global examples (1) show the supporting purpose of process explanations and (2) reinforce the difference between process explanations and instructions.

- **Accounting:** As an accountant at M-Global's corporate office, you have just finished auditing the firm's books. Now you must write a report to the vice president for business and marketing on the firm's compliance with Generally Accepted Accounting

PROCESS DESCRIPTIONS

Purpose: Explain a sequence of steps in such a way that the reader understands a process

Format: Use paragraph descriptions, listed steps, or some combination of the two

Style: Use *objective* point of view ("2. The operator started the engine ..."), as opposed to *command* point of view ("2. Start the engine ...")

INSTRUCTIONS

Purpose: Describe a sequence of steps in such a way that the reader can *perform* the sequence of steps

Format: Employ numbered or bulleted lists, organized into subgroups of easily understandable units of information

Style: Use *command* point of view ("3. Plug the phone jack into the recorder unit"), as opposed to *objective* point of view ("3. The phone jack was plugged into the recorder unit")

Principles (GAAP). Along with your findings, the vice president wants an overview of the procedure you followed to arrive at your conclusions.

■ **Human Resources:** As a nurse in M-Global's Munich branch, you are preparing materials about a Wellness Fair that the Munich Human Resources office will distribute to all employees in the branch. This year, the fair will include a new ultrasound test for osteoporosis that will be conducted on-site. To encourage employees to take advantage of the test, you write an explanation of the process that emphasizes how quick and easy the test will be.

■ **Laboratory work:** As lab supervisor at the St. Louis office, you spent all day Saturday in the lab assembling a new gas chromatograph needed to analyze gases. To justify the overtime hours, you write a memo to your manager explaining the assembly process.

■ **Welding inspection:** As a nondestructive testing (NDT) expert at M-Global's San Francisco office, you were hired by a California state agency to x-ray all welds at a bridge damaged by an earthquake. The text of your report gives test results; the appendices explain the procedures you followed.

- **Marketing:** As M-Global's marketing manager, you devised a new procedure for tracking contacts with prospective clients (from first sales call to getting the job). You must write a memo to M-Global's two vice presidents for operations, briefly explaining the process. Their approval is needed before the new marketing technique can be introduced at the 16 branch offices.

In each case, you are writing for a reader who wants to know what has happened or will happen, but who does not need to perform the process.

Instructions at M-Global

Think of instructions this way: They must provide users with a road map to *do* the procedure, not just understand it—that is, someone must complete a task on the basis of words and pictures you provide. Clearly, instructions present you, the writer, with a much greater challenge and risk. The reader must be able to replicate the procedure without error and, most importantly, with full knowledge of any dangers. The M-Global situations that follow reflect this challenge. Note that they parallel the case studies presented for process explanations.

- **Accounting:** As M-Global's lead accountant for the past 20 years, you have always been responsible for auditing the firm's books. Because you developed the procedure yourself over many years, there is no comprehensive set of instructions for completing it. Now you want to record the steps so that other company accountants besides you can perform them.

- **Human Resources:** As a nurse in M-Global's Munich branch, you are preparing materials about a Wellness Fair that the Munich Human Resources office will distribute to all employees in the branch. The tests available to employees include cholesterol screening, which requires fasting before the test. You write instructions so that employees who will be having their cholesterol screened will prepare appropriately.

- **Laboratory work:** As lab supervisor for M-Global's St. Louis office, you have assembled one of the two new gas chromatographs just purchased by the company. You are supposed to send the other unit to the Tokyo branch, where it will be put together by Japanese technicians. Unfortunately, the manufacturer's instructions are poorly written, so you plan to rewrite them for the English-speaking technicians at the Tokyo office.

- **Welding inspection:** As M-Global's NDT expert at the San Francisco office, you have seen a large increase in NDT projects. Given California's aging bridges and constant earthquake activity, you persuaded your branch manager to hire several NDT technicians. Now you must write a training manual that instructs these new employees on methods for inspecting bridge welds.

- **Marketing:** As M-Global's marketing manager, you have suggested a new approach for tracking sales leads. Having had your proposal approved by the corporate staff, you must now explain the marketing procedure to technical professionals at all 16 offices. Your written set of instructions must be understood by technical experts in many fields and who have little if any marketing experience.

In each case, your instructions must explain steps so thoroughly that the reader will be able to replicate the process without having to speak in person with the writer of the instructions. The next two sections give rules for preparing both process explanations and sets of instructions.

>>> Guidelines for Process Explanations

You have already learned that process explanations are aimed at persons who must understand the process, not perform it. Process explanations often have the following purposes:

- Describing an experiment
- Explaining how a machine works
- Recording steps in developing a new product
- Describing procedures to ensure compliance with regulations
- Describing what will happen during a medical procedure

In each case, use the following guidelines to create first-rate process explanations:

>> Process Guideline 1: Know Your Purpose and Your Audience

Your intended purpose and expected audience influence every detail of your explanation. Following are some preliminary questions to answer before writing:

- Are you supposed to give just an overview, or are details needed?
- Do readers understand the technical subject, or are they laypersons?
- Do readers have mixed technical backgrounds?
- Does the process explanation supply supporting information (perhaps in an appendix), or is it the main part of the document?

Process explanations are most challenging when directed to a mixed audience. In this case, write for the lowest common denominator—that is, for your least technical readers. It is better to write beneath the level of your most technical readers than to write above the level of your nontechnical readers.

For example, the process explanation in Model 7–3 on page 217 is directed to a mixed audience of city officials—some technical staff and some nontechnical political officials. It is contained in an appendix to a long M-Global report that recommends immediate cleanup of a toxic-waste dump. Note that the writer either uses nontechnical language or defines any technical terms used.

>> Process Guideline 2: Follow the ABC Format

In chapter 3, you learned about the ABC format (**A**bstract/**B**ody/**C**onclusion) that applies to all documents. The abstract gives a summary, the body supplies details, and the conclusion provides a wrap-up or leads to the next step in the communication process.

Whether a process explanation forms all or part of a document, it usually subscribes to the following version of the three-part ABC plan:

- The **Abstract** component includes three background items:

  1. Purpose statement
  2. Overview or list of the main steps that follow
  3. List of equipment or materials used in the process

  Model 7–3 includes all three, with a separate heading for equipment. First, the purpose statement places the explanation in the context of the entire document. Then the list of main steps gives readers a framework for interpreting details that follow. Finally, the list of equipment or materials provides a central reference point as readers work through all the steps.

- The **Body** component of the process explanation moves logically through the steps of the process. By definition, all process explanations follow a *chronological,* or step-by-step, pattern of organization. These steps can be conveyed in two ways:

  1. **Paragraphs:** This approach weaves steps of the process into the fabric of typical paragraphs, with appropriate transitions between sentences. Use paragraphs when your readers would prefer a smooth explanation of the entire process, rather than emphasis on individual steps.
  2. **List of steps:** This approach includes a list of steps, usually with numbers or bullets. Much like instructions, a listing emphasizes the individual parts of the process. Readers prefer it when they must refer to specific steps later on.

  Both paragraph and list formats have their places in process explanations. In fact, most explanations can be written in either format. Figure 7–2 shows an M-Global

Steps of process are embedded in paragraph.

After brief lead-in, steps of process are placed in list format.

■ **Figure 7–2** ■ Two options for process explanation: (A) paragraph option; (B) list option

A. PARAGRAPH OPTION

The homeowner should select rough-grade 2 x 4s for building the wooden form for the patio. The form is just a box, with an open top and with the ground for the bottom, into which concrete will be poured. First the four sides are nailed together, and then the form is leveled with a standard carpenter's level. Finally, 2 x 4 stakes are driven into the ground about every 2 or 3 ft on the outside of the form to keep it in place during the pouring of the concrete.

B. LIST OPTION

Building a wooden form for a home concrete patio can be accomplished with some rough-grade 2 x 4s. This form is just a box with an open top and the ground for the bottom. Building involves three basic steps:

1. Nailing 2 x 4s into the intended shape of the patio
2. Leveling the box-shaped form with a standard carpenter's level
3. Nailing stakes (made from 2 x 4 lumber) every 2 or 3 ft at the outside edge of the form to keep it in place during the pouring of the concrete

example of both a paragraph and list explanation for the same process of laying a concrete patio. As a public-service gesture, M-Global, Inc., produced a pamphlet that briefly explains simple home improvements and is intended to help owners of small homes decide whether to complete renovations themselves or hire a contractor. If home owners are interested in one of the projects, they can find detailed instructions at the URL listed in the pamphlet.

Building a concrete patio is one project covered; the process explanation contains a subsection about constructing the wooden form into which concrete is poured.

■ The **Conclusion** component of a process explanation keeps the process from ending abruptly with the last step. Here you should help the reader put the steps together into a coherent whole. When the process explanation is part of a larger document, you can show how the process fits into a larger context (see Model 7–3).

>> Process Guideline 3: Use an Objective Point of View

Process explanations describe a process rather than direct how it is to be done. Thus they are written from an objective point of view—not from the personal *you* or *command* point of view common to instructions. Note the difference in these examples:

> *Process:* The concrete is poured into the two-by-four frame.
>
> *or*
>
> The technician pours the concrete into the two-by-four frame.
>
> *Instructions:* Pour the concrete into the two-by-four frame.

The process excerpts *explain* the steps, whereas the instructions excerpt *gives a command* for completing the instructions.

>> Process Guideline 4: Choose the Right Amount of Detail

Only a thorough audience analysis will tell you how much detail to include. Model 7–3 (p. 217), for example, could contain much more technical detail about the substeps for testing air quality at the site; however, the writer decided that the city officials would not need more scientific and technical detail.

In supplying specifics, be sure to subdivide information for easy reading. In paragraph format, headings and subheadings can be used to make the process easier to grasp. In list format, an outline arrangement of points and sub-points may be appropriate. When such detail is necessary, remember this general rule of thumb: *Place related steps into groups of from three to seven points.* Readers find it easier to remember several groupings with sub-points rather than one long list. Following are two rough outlines for a process explanation that was created to encourage consistency in the hiring process at all M-Global branches. The second is preferred in that it groups the many steps into three easily grasped categories.

Employment Interview Process

1. Interviewer reviews job description
2. Interviewer analyzes candidate's application
3. Candidate and interviewer engage in "small talk"

4. Interviewer asks open-ended questions related to candidate's resume and completed application form

5. Interviewer expands topic to include matters of personal interest and the candidate's long-term career plans

6. Interviewer provides candidate with information about the position (salary, benefits, location, etc.)

7. Candidate is encouraged to ask questions about the position

8. Interviewer asks candidate about her or his general interest, at this point, in the position

9. Interviewer informs candidate about next step in hiring process

Employment Interview Process

■ **Pre-interview Phase**

1. Interviewer reviews job description

2. Interviewer analyzes candidate's application

■ **Interview**

3. Candidate and interviewer engage in "small talk"

4. Interviewer asks open-ended questions related to candidate's resume and completed application form

5. Interviewer expands topic to include matters of personal interest and the candidate's long-term career plans

6. Interviewer provides candidate with information about the position (salary, benefits, location, etc.)

7. Candidate is encouraged to ask questions about the position

■ **Closure**

8. Interviewer asks candidate about his or her general interest, at this point, in the position

9. Interviewer informs candidate about next step in hiring process

>> Process Guideline 5: Use Flowcharts for Complex Processes

Some process explanations contain steps that are occurring at the same time. In this case, you may want to supplement a paragraph or list explanation with a flowchart. Such charts use boxes, circles, and other geometric shapes to show progression and relationships among various steps.

Model 7–4 on page 218, for example, shows a flowchart and an accompanying process explanation at M-Global. Both denote services that M-Global's London branch provides for oil companies in the North Sea. The chart helps to demonstrate that the geophysical study (mapping by sonar equipment) and the engineering study (securing and testing of seafloor samples) take place at the same time. Such simultaneous steps are difficult to show in a list of sequential steps.

>>> Guidelines for Instructions

Rules change considerably when moving from process explanations to instructions. Although both patterns are organized by time, the similarity stops there. Instructions walk readers through the process so that they can do it, not just understand it. It is one thing to explain the process by which a word-processing program works; it is quite another to write a set of instructions for using that word-processing program. This section explores the challenge of writing instructions by giving you some basic writing and design guidelines.

These guidelines for instructions also apply to complete operating *manuals,* a document type that many technical professionals will help to write during their careers. Those manuals include the instructions themselves, as well as related information such as (1) features, (2) physical parts, and (3) troubleshooting tips. In other words, manuals are complete documents, whereas instructions can be part of a larger piece.

>> Instructions Guideline 1: Select the Correct Technical Level

This guideline is just another way of saying you must know exactly who will read your instructions. Are your readers technicians, engineers, managers, general users, or some combination of these groups? Once you answer this question, select language that every reader can understand. If, for example, the instructions include technical terms or names of objects that may not be understood, use the techniques of definition and description discussed in chapter 6.

>> Instructions Guideline 2: Provide Introductory Information

Like process explanations, instructions follow the ABC format (**A**bstract/**B**ody/**C**onclusion) described in chapter 3. The introductory (or abstract) information should include (1) a purpose statement, (2) a summary of the main steps, and (3) a list or an illustration giving the equipment or materials needed (or a reference to an attachment with this information). These three items set the scene for the procedure itself.

Besides these three "musts," you should consider whether some additional items might help set the scene for your user:

- Pointers that help with installation
- Definitions of terms
- Theory of how something works
- Notes, cautions, warnings, or dangers that apply to all steps

>> Instructions Guideline 3: Use Numbered Lists in the Body

A simple format is crucial to the body of the instructions—that is, the steps themselves. Most users constantly go back and forth between these steps and the project to which they apply. Thus you should avoid paragraph format and instead use a simple numbering system. Model 7–5 on pages 219–220 shows a "before and after" example. The original version is written in paragraphs that are difficult to follow; the revised version includes nine separate numbered steps.

>> Instructions Guideline 4: Group Steps under Task Headings

Readers prefer that you group together related steps under headings, rather than present an uninterrupted "laundry list" of steps. Model 7–6 on pages 221–223 shows how this technique has been used in a fairly long set of instructions for operating a scanner. Given the number of steps in this case, the writer has used a separate numbering system within each grouping.

Groupings provide two main benefits. First, they divide fragmented information into manageable "chunks" that readers find easier to read. Second, they give readers a sense of accomplishment as they complete each task, on the way to finishing the whole activity.

>> Instructions Guideline 5: Place One Action in a Step

A common error is to bury several actions in a single step. This approach can confuse and irritate readers. Instead, break up complex steps into discrete units, as shown next:

■ **Original:**

Step 3:   Fill in your name and address on the coupon, send it to the manufacturer within two weeks, return to the retail merchant when your letter of approval arrives from the manufacturer, and pick up your free toaster oven.

■ **Revision:**

Step 3:   Fill in your name and address on the coupon.

Step 4:   Send the coupon to the manufacturer within two weeks.

Step 5:   Show your retail merchant the letter of approval after it arrives from the manufacturer.

Step 6:   Pick up your free toaster oven.

>> Instructions Guideline 6: Lead Off Each Action Step with a Verb

Instructions should include the *command* form of a verb at the start of each step. This style best conveys a sense of action to your readers. Models 7–5 and 7–6 on pages 219–223 use command verbs consistently for all steps throughout the procedures.

>> Instructions Guideline 7: Remove Extra Information from the Step

Sometimes you may want to follow the command sentence with an explanatory sentence or two. In this case, distinguish such helpful information from actions by giving it a label, such as *Note* or *Result* (e.g., see Model 7–2, page 216).

>> Instructions Guideline 8: Use Bullets or Letters for Emphasis

Sometimes you may need to highlight information, especially within a particular step. Avoid using numbers for this purpose, because you are already using them to signify steps.

Bullets work best if there are just a few items; letters are best if there are many, especially if they are in a sequence. The revised version in Model 7–5 shows the appropriate use of letters, and Model 7–6 shows the use of letters and bullets.

In particular, consider using bullets at any point at which users have an *option* as to how to respond. The following example uses bullets in this way; it also eliminates the problem of too many actions being embedded in one step.

Part of Procedure for Firing Clay in a Kiln

(Note: A *pyrometric cone* is a piece of test clay used in a *kiln,* an oven for baking pottery. The melting of the small cone helps the operator determine that the clay piece has completed the firing process.)

■ Original

Step 6: Check the cone frequently as the kiln reaches its maximum temperature of 1850°F. If the cone retains its shape, continue firing the clay and checking the cone frequently. When the cone begins to bend, turn off the kiln. Then let the kiln cool overnight before opening it and removing the pottery.

■ Revision

Step 6: Check the cone frequently as the kiln reaches its maximum temperature of 1850°F.

Step 7: Has the cone started to bend?

- If *no*, continue firing the piece of pottery and checking the cone frequently to see if it has bent.

- If *yes*, turn off the kiln.

Step 8: Let the kiln cool overnight after turning it off.

Step 9: Open the kiln and remove the pottery.

>> Instructions Guideline 9: Emphasize Cautions, Warnings, and Dangers

Instructions often require alerts that draw attention to risks in using products and equipment. Your most important obligation is to highlight such information. Unfortunately, professional associations and individual companies may differ in the way they use and define terms associated with risk, so you should make sure that the alerts in your document follow the appropriate guidelines. You must be certain to use language or graphics your reader understands. If you have no specific guidelines, however, the following definitions can serve as "red flags" to the reader. The level of risk increases as you move from 1 to 3:

1. **Caution:** possibility of damage to equipment or materials
2. **Warning:** possibility of injury to people
3. **Danger:** probability of injury or death to people

If you are not certain that these distinctions will be understood by your readers, define the terms *caution, warning,* and *danger* in a prominent place before you begin your instructions.

As for placement of the actual cautions, warnings, or danger messages, your options are as follows:

- **Option 1:** *In a separate section, right before the instructions begin.* This approach is most appropriate when you have a list of general warnings that apply to much of the procedure or when one special warning should be heeded throughout the instructions—for example: "WARNING: Keep main breaker on *off* during entire installation procedure." Figure 7–3

⚠**WARNING** Read all instructions. Failure to follow all instructions listed below may result in electric shock, fire, or serious injury.

INSTALLING THE SWITCH

⚠**WARNING** DISCONNECT MACHINE FROM POWER SOURCE.

1. Place switch (A) Fig. 11, behind the lip of extension wing (B). Insert M8x30 hex head screw (C) through wing and then switch support. Place an M8 flat washer and an M8 lock washer on the screw. Thread an M8 hex nut (D) onto screw and tighten nut securely.

2. Insert switch cord with female end through hole (F) Fig. 12 in upper left corner of the saw. Open motor cover and route the switch cord (F) Fig. 13 behind the cord guard (G) and then plug into motor cord (H), as shown in Fig. 13.

3. Make sure the slack is pulled down and rests on the dust chute as shown in Fig. 13.

⚠**WARNING** MAKE SURE CORD DOES NOT COME IN CONTACT WITH BLADE, BELT OR PULLEYS

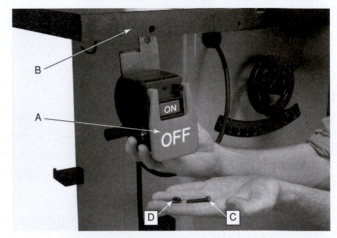

Fig. 11

Fig. 12

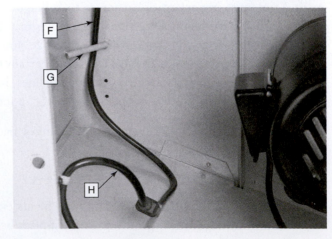

Fig. 13

- **Figure 7–3** -

Example of safety "warning"

(Courtesy of Delta International Machinery Co.)

> **CAUTION:**
> - Always make sure cords are NOT trapped or pinched between bed slats, mattress, springs, against wall, foot board or bed frame, or between furniture, walls or chairs.
> - Do not plug main power cord into the wall outlet until you have completely set up your warming product per the instructions.
> - Plug power cord into a 110–120 volt wall outlet. Do not attempt to force the plug. This appliance has a polarized plug (one blade is wider than the other). To reduce the risk of electrical shock, this plug is intended to fit only one way into an outlet. If the plug does not fit in the outlet, reverse the plug. If it still does not fit, contact a qualified electrician.

■ **Figure 7–4** ■
Example of a "caution"

shows both kinds of warnings. The first warning appears at the beginning of a manual for a portable table saw. The manual also includes warnings placed before and after each set of instructions.

■ **Option 2:** *In the text of the instructions.* This approach works best if the caution, warning, or danger message applies to the step that immediately follows it. Thus users are warned about a problem *before* they read the step to which it applies (Figure 7–4).

■ **Option 3:** *Repeatedly throughout the instructions.* This strategy is preferable with instructions that repeatedly pose risk to the user. For example, steps 4, 9, 12A, and 22—appearing on several different pages—may *all* include the hazard of fatal electrical shock. Your *danger* notice should appear in each step, as well as in the introduction to the document.

Give information about potential risks *before* the operator has the chance to make the mistake. Also, the caution, warning, or danger message can be made visually prominent by using one of the following techniques:

Underlining:	<u>Warning</u>
Bold:	**Warning**
Full Caps:	WARNING
Italics:	*Warning*
Oversized Print:	Warning
Boxing:	Warning
Color:	Warning
Combined Methods:	*Warning*
	WARNING
	WARNING

Color graphics are another effective indicator of risk. You have probably seen examples such as a red flame in a box for fire, a jagged line in a triangle for electrical shock, or an actual drawing of a risky behavior with an X through it.

The International Organization for Standardization (ISO) established international standards for safety alerts in ISO 3864, and the American National Standards Institute (ANSI) established domestic standards for safety alerts in ANSI Z535. If the organization you work for complies with ISO or ANSI, you should make sure that you are using the most recent version of the appropriate standards to reinforce the message in your text about cautions, warnings, and dangers.

>> Instructions Guideline 10: Keep a Simple Style

Perhaps more than any other type of technical communication, instructions must be easy to read. Readers expect a no-nonsense approach to writing that gives them required information without fanfare. Following are some useful techniques:

- Keep sentences short, with an average length of fewer than 10 words.
- Use informal definitions (parenthetical, like this one) to define any terms not understood by all readers.
- Never use a long word when a short one will do.
- Be specific and avoid words with multiple interpretations (*frequently, seldom, occasionally,* etc.).

>> Instructions Guideline 11: Use Graphics

Illustrations are essential for instructions that involve equipment. Place an illustration next to every major step when (1) the instructions or equipment is quite complicated or (2) the audience may contain poor readers or people who are in a hurry. Such word–picture associations create a page design that is easy to follow.

In other cases, just one or two diagrams may suffice for the entire set of instructions. The one reference illustration in Model 7–6 (pp. 221–223) helps the user of a scanner locate parts mentioned throughout the instructions.

Another useful graphic in instructions is the table. Sometimes within a step you must show correspondence between related data. For example, the instructions that follow would benefit from a list.

- **Original**

  Step 3: Use pyrometric cones to determine when a kiln has reached the proper temperature to fire pottery. Common cone ratings are as follows: a Cone 018 corresponds to 1200°F; a Cone 07 corresponds to 1814°F; a Cone 06 corresponds to 1859°F; and a Cone 04 corresponds to 1940°F.

- **Revision**

  Step 3: Use pyrometric cones to determine when a kiln has reached the proper temperature for firing pottery. Common cone ratings are as follows:

Cone 018	1200°F
Cone 07	1814°F
Cone 06	1859°F
Cone 04	1940°F

>> Instructions Guideline 12: Test Your Instructions for Usability

Testing instructions for *usability* ensures that your users are able to follow them easily. More information about usability and websites can be found in chapter 11, but understanding some of the basics of designing for usability will help you create effective instructions.[2] When you design for usability, you should be focused primarily on the user, not the product itself. This is true whether you are designing a document, software, a computer interface, or a piece of machinery. Products that are usable have the following qualities:

- Learning them is easy.
- Operating them requires the minimum number of steps.
- Remembering how to use them is easy.
- Using them satisfies the user's goals.

Usability does not happen automatically, but should be a concern from the earliest stages of the design of products and documentation.

Professional writers often test their instructions on potential users before completing the final draft. The most sophisticated technique for such testing involves a *usability laboratory* where test subjects are asked to use the instructions or manual to perform the process, often while speaking aloud their observations and frustrations (if any). The writers or lab personnel unobtrusively observe the process from behind a one-way mirror. Later, they may review audio- or videotaped observations of the test subjects, or they may interview these persons. This complex process helps writers anticipate and then eliminate many of the problems that users confront when they follow written instructions.

Of course, you probably will not have access to a usability laboratory to test your instructions. However, you can adapt the following user-based approach to testing assignments in this class and projects in your career. Specifically, follow these four steps:

1. Team up with another class member (or a colleague on the job). This person should be unfamiliar with the process and should approximate the technical level of your intended audience.

2. Give this person a draft of your instructions and provide any equipment or materials necessary to complete the process. Of course, for the purposes of a class assignment, this approach works only for a simple process with little equipment or few materials.

3. Observe your colleague following the instructions you provide. You should record both your observations and any responses this person makes while moving through the steps.

4. Revise your instructions to solve problems your user encountered during the test.

>>> Chapter Summary

Both process explanations and instructions share the same organization principle: time. That is, both relate a step-by-step description of events. Process explanations address an audience that wants to be informed but does not need to perform the process

---

[2]Adapted from Barnum, C.M. (2002). *Usability Testing and Research.* New York: Longman.

itself. Instructions are geared specifically for persons who must complete the procedure themselves.

In writing good process explanations, follow these basic guidelines:

1. Know your purpose and audience.
2. Follow the ABC format.
3. Use an objective point of view.
4. Choose the right amount of detail.
5. Use flowcharts for complex processes.

For instructions, follow these 12 rules:

1. Select the correct technical level.
2. Provide introductory information.
3. Use numbered lists in the body.
4. Group similar steps under heads.
5. Place one action in a step.
6. Lead off each action step with a verb.
7. Remove extra information from the step.
8. Use bullets or letters for emphasis.
9. Emphasize cautions, warnings, and dangers.
10. Keep a simple style.
11. Use graphics.
12. Test your instructions.

>>> Learning Portfolio

Communication Challenge "M-Global's Home of Hope: The Good, the Bad, and the Ugly?"

Recently M-Global's Atlanta office decided to change its approach to charitable giving at the branch. Instead of supporting various regional charities, employees could participate in a local project of their own—converting an abandoned building into a homeless shelter called "Home of Hope." The idea seemed to be a creative way to make a personal contribution to the community. What follows is a description of the stages of the project and some questions and comments for discussion followed by an assignment for a written response to the Challenge.

Project Planning

The process of making "Home of Hope" a reality began at M-Global–Atlanta's annual employee meeting last year. The human resources manager suggested that the office try a new approach to annual giving, and the office supervisors agreed to investigate. Eventually, the office decided to purchase an abandoned brick building on an acre lot in downtown Atlanta, in an area where homeless people often congregated.

M-Global conducted a preliminary study of the land and building, calculating that the project would cost about $150,000. Management developed a formula whereby the company would pay a 20% mortgage down payment from its savings and carry the monthly mortgage note. Then over a one-year period, the employees—through their annual financial contributions and personal labor—would renovate the house and add landscaping. The managers developed a suggested sliding scale for what money employees should contribute, based on their salaries. Managers and supervisors also were asked to meet individually with each employee to encourage contributions.

Once M-Global bought the land, the firm began benefiting from excellent publicity on local radio and in the papers. The media championed this effort by an Atlanta employer.

Site Problems

It appeared that nothing could go wrong—but something did. Ironically, considering that M-Global does environmental work, the firm found an environmental problem with the land that had not been detected before purchase. Apparently, part of the site had been used as a dumping ground for old car batteries and for chemicals from a nearby drycleaners. Both the batteries and a large portion of soil would have to be removed, adding $15,000 to the cost of the project.

Just as bad were the environmental surprises in the building itself. The company found some asbestos and lead paint that had not been detected before purchase. Removal would cost about $5,000. The increase in the total project cost irritated many employees, some of whom had been skeptical about the project from the start.

Employee Involvement

What did seem to go well were the weekend work groups that the company set up for the coming year. A group of five to ten employees worked a half day on each Saturday, meaning that most employees would end up working three or four Saturdays during the entire year-long project. Employees were strongly encouraged to participate, and about 85% of them signed up for the groups.

As noted previously, through meetings with managers and other means, employees were encouraged to contribute the amount suggested on the sliding scale. About 75% agreed to the amount suggested, 10% pledged more, 10% pledged less, and 5% pledged nothing. Pledges were drawn from paychecks over the one-year period.

Community Involvement

Once the lot was purchased and the renovation designed, M-Global worked with groups in the surrounding community, making sure that local people were informed about the project. One homeowners group from this working-class neighborhood raised questions about the project attracting even more homeless people to the area. The group worried that the possibility of increasing crime would lower the value of their homes. M-Global decided that an open community meeting was in order.

At the meeting at a local school, M-Global produced speakers who suggested that the home would actually help decrease crime by giving shelter, meals, and activities to people who otherwise would be vagrants. Although the answers seemed to satisfy many, M-Global officials were on the defensive and wished they had done more networking with local residents.

Final Preparations

Once the home and yard were finished, M-Global hired two permanent staff members and set up a group of volunteers from the community. Retired people were especially active as volunteers. The company also asked for, and received, an ongoing commitment of $17,000 a year from the city to pay half the salary of the Home of Hope director.

With these details handled, the home took in its first 25 residents several months ago. M-Global arranged for media coverage of the opening celebration, inviting a diverse group of company leaders. Of course, the company also made sure the event was covered in the M-Global corporate newsletter and by *EnviroNews*, a national news magazine in engineering and science.

Questions and Comments for Discussion

1. The M-Global corporate managers have expressed interest in the charity model developed by the Atlanta office. Specifically, they want the Atlanta human resources director to write a **process explanation** for the Home of Hope project. The explanation will be reviewed by all M-Global branch managers. What major points should be included in this process explanation? How should it differ from the way information is presented in the case just described?

2. Assume M-Global's corporate office has actually adopted a community-based charity option such as that reflected by the Home of Hope project. Now it wants to provide project **instructions** for other urban offices that may want to build shelters. What major points should be emphasized in the instructions and in what order? What particular problems did the Atlanta office encounter, and how can the instructions be written to help other offices avoid such problems? In other words, how should the *ideal* set of instructions differ from the *actual* process that was performed?

3. Answer these questions first with regard to M-Global employees and second with regard to the community surrounding Home of Hope. What tactical mistakes, if any, were made by M-Global management in the process of promoting, communicating, and running this project? How could the problems have been avoided?

4. Are there any ethical problems revealed in the process explained in this case? Specifically, how do you feel about the manner by which employees are encouraged to contribute to such causes?

5. Several large charity groups were disturbed that M-Global dropped them and instead involved employees in the Home of Hope. Give what you think would be the charities' point of view about the process explained in this case.

Write About It

Assume the role of the Atlanta human resources director. Write the process explanation for M-Global corporate managers that is described in question 1.

Collaboration at Work A Simple Test for Instructions

General Instructions

Each Collaboration at Work exercise applies strategies for working in teams to chapter topics. The exercise assumes you (1) have been divided into teams of about three to six students, (2) will use team time inside or outside of class to complete the case, and (3) will produce an oral or written response. For guidelines about writing in teams, refer to pages 24–28.

Background for Assignment

Writing instructions presents a challenge. The main problem is this: Although writers may have a good understanding of the procedure for which they are designing instructions, they have trouble adopting the perspective of a reader unfamiliar with the procedure. One way to test the effectiveness of instructions is to conduct your own usability test. The following exercise determines the clarity of instructions written by your team by asking another team to follow the instructions successfully.

Team Assignment

In this exercise, your team prepares a list of instructions for drawing a simple figure or object. The purpose is to write the list so clearly and completely that a classmate could draw the figure or object without knowing its identity. Following are instructions for completing the assignment:

1. Work with your team to choose a simple figure or object that requires only a relatively short set of instructions to draw. (Note: Use a maximum of 15 steps.)

2. Devise a list of instructions that your team believes cannot be misunderstood.

3. Test the instructions within your own team.
4. Exchange instructions with another team.
5. Attempt to draw the object for which the other team has written instructions. (Note: Perform this test without knowing the identity of the object.)
6. Talk with the other team about problems and suggestions related to the instructions.
7. Discuss general problems and suggestions with the entire class.

Assignments

Part 1: Short Assignments

Assignments 1–5 can be completed either as individual exercises or as team projects, depending on the instructions you are given in class.

1. Writing a Process Explanation—School Related

Your college or university has decided to evaluate the process by which students are advised about and registered for classes. As part of this evaluation, the registrar has asked a select team of students—you among them—to explain the actual process each of you went through individually during the last advising/registration cycle. These case studies collected from individual students—the customers—will be transmitted directly to a college-wide committee studying registration and advising problems.

Your job is to give a detailed account of the process. Remain as objective as possible without giving opinions. If you had problems during the process, the facts you relate will speak for themselves. Simply describe the process you personally experienced, and then let the committee members judge for themselves whether the steps you describe should be part of the process.

2. Writing a Process Explanation—M-Global Context

Choose one of the projects from the project sheets at the end of chapter 2. For this assignment, use (a) points listed in the "Main Technical Tasks" section of the project you have chosen; (b) related information you wish to supply from your own experience, reading, or imagination; or (c) both. Your assignment is to write a process explanation dealing with one or more of the bulleted points.

3. Writing Instructions—M-Global Context

Choose one of the projects from the project sheets at the end of chapter 2. For this assignment, conduct some research on either (a) one task or several related tasks in the "Main Technical Tasks" section or (b) a task that conceivably could be related to the project but is not specifically listed. Then write a set of numbered instructions for the task(s). Following are some sample tasks, with project references:

- Project #1: Conducting a lab test to evaluate a soil sample
- Project #2: Estimating the age of a geologic sample
- Project #3: Inspecting all or part of a new building
- Project #4: Running an effective meeting with subcontractors
- Project #5: Testing soil or water for signs of pollution
- Project #6: Evaluating the ergonometric features of a control panel
- Project #7: Evaluating the effectiveness of a technical report
- Project #8: Creating a printed guide to data security procedures

If you prefer to write about less-specific tasks, select a general topic related to a project. For example, here are four general topics derived from the specific ones just listed: conducting a lab test, running an effective meeting, evaluating the ergonometric features of any product, and evaluating the effectiveness of any document.

4. Writing Instructions—M-Global Context

As an employee at the corporate office of M-Global, you just received the job of writing a set of instructions for completing performance appraisal reviews (PARs). The instructions are included in a memo that goes to all supervisors at all branches of the firm, along with related forms. To help you get started on the instructions, you have been given a narrative explanation of the process (see the following). Your task is to convert this narrative into a simple set of instructions to go into the memorandum to supervisors.

> Performance Appraisal Reviews (PARs) are conducted annually for each employee during the anniversary month in which the employee was originally hired. Several days before the month in which the PARs are to be conducted, the corporate office sends each supervisor a list of employees in that supervisor's group who should receive PARs. The main portion of the PAR process is an interview between the supervisor and the employee receiving the PAR. Before this interview takes place, however, the supervisor should give the employee a copy of the *M-Global PAR Discussion Guide*,

which offers suggestions for the topics and tone of a PAR interview. The supervisor completes a *PAR Report Form* after each interview and then sends a copy to corporate and to the employee, and the original remains in the personnel files of that respective supervisor's branch. If for any reason a PAR interview and report form are not completed in the required month, the supervisor must send a memo of explanation to the corporate Human Resources Department, with a copy to the supervisor's branch manager.

5. Writing Instructions—School Related

In either outline or final written form, provide a set of instructions for completing assignments in this class. Consider your audience to be another student who has been ill and missed much of the term. You have agreed to provide her with an overview that will help her to plan and then write any papers she has missed.

Your instructions may include (1) highlights of the writing process from chapter 1 and (2) other assignment guidelines provided by your instructor in the syllabus or in class. Remember to present a generic procedure for all assignments in the class, not specific instructions for a particular assignment.

Part 2: Longer Assignments

These assignments test your ability to write and evaluate the two patterns covered in this chapter—process explanations and instructions. Specifically, follow these guidelines:

- Write each exercise in the form of a letter report or memo report, as specified.
- Follow organization and design guidelines given in chapter 3 and 4, especially concerning the ABC format (**A**bstract/**B**ody/**C**onclusion) and the use of headings. Chapter 8 gives rules for short reports, but such detail is not necessary to complete the assignments here.
- Fill out a Planning Form (at the end of the book) for each assignment.

6. Evaluating a Process Explanation

Using a textbook in a technical subject area, find an explanation of a process—for example, a physics text might explain the process of waves developing and then breaking at a beach, an anatomy text might explain the process of blood circulating, or a criminal justice text might explain the process of a criminal investigation.

Keeping in mind the author's purpose and audience, evaluate the effectiveness of the process explanation as presented in the textbook. Submit your evaluation in the form of a memo report to your instructor in this writing course, along with a copy of the textbook explanation.

For the purposes of this assignment, assume that your writing instructor has been asked by the publisher of the text you have chosen to review the book as an example of good or bad technical writing. Thus your instructor would incorporate comments from your memo report into his or her comprehensive evaluation.

7. Writing a Process Explanation— School Related

Conduct a brief research project in your campus library. Specifically, use company directories, annual reports, or other library sources to find information about a company or other organization that could hire students from your college.

In a memo report to your instructor, (1) explain the process you followed in conducting the search and (2) provide an outline or paragraph summary of the information you found concerning the company or organization. Assume that your report will become part of a volume your college is assembling for juniors and seniors who are beginning their job search. These students will benefit both from information about the specific organization you chose and from a explanation of the process that you followed in getting the information, because they may want to conduct research on other companies.

8. Writing a Process Explanation— M-Global Context

As a project manager for M-Global's Atlanta office, you just found out that your office has been selected as one of the firms to help renovate Kiddieworld, a large amusement park in the Southeast. Before Kiddieworld officials sign the contract, however, they want you to report on the process M-Global uses to report and investigate accidents (because the project involves some hazardous work). You found the following policy in your office manual, but you know it is not something you would want to send to a client. Take this stilted paragraph and convert it to a process explanation for your clients in the form of a letter report. Remember: The readers are not performing the process; they only want to understand it.

Accident reporting and investigation are an important phase of operations at M-Global, Inc. The main purpose of an accident investigation and report is to gain an objective insight into facts surrounding the accident in order to improve future accident control measures and activities as well as to activate the protection provided by our insurance policies. It is therefore imperative that all losses, no matter how minor, be reported as soon as possible, preferably within 48 hours, to the proper personnel. Specifically, all accidents must be reported orally to the immediate supervisor. For minor accidents that do not involve major loss of equipment or hospitalization, that supervisor has the responsibility of filling out an M-Global accident report form and then sending the form to the safety personnel at the appropriate branch office, who later sends it to the safety manager at the corporate office. For serious accidents that involve major loss of equipment or hospitalization of any individuals involved, the supervisor must call or fax the safety personnel at the appropriate branch office, who then should call or fax the safety manager at the corporate office. (A list of pertinent telephone numbers should be kept at every job site.) These oral reports are followed up with a written report.

9. Evaluation of Instructions

Find a set of operating or assembly instructions for a DVD player, microwave oven, CD player, computer, timing light, or other electronic device. Evaluate all or part of the document according to the criteria for instructions in this chapter.

Write a memo report on your findings and send it, along with a copy of the instructions, to Natalie Bern. As a technical writer at the company that produced the electronic device, Natalie wrote the set of instructions. In your position as Natalie's supervisor, you are responsible for evaluating her work. Use your memo report either to compliment her on the instructions or to suggest modifications.

10. User Test of Instructions

Find a relatively simple set of instructions. Then ask another person to follow the instructions from beginning to end. Observe the person's activity, keeping notes on any problems she or he encounters.

Use your notes to summarize the effectiveness of the instructions. Present your summary as a memo report to Natalie Bern, using the same situational context as described in assignment 9—that is, as Natalie's boss, you are to give her your evaluation of her efforts to produce the set of instructions.

11. Writing Simple Instructions

Choose a simple office procedure of 20 or fewer steps (e.g., changing a printer cartridge, filling a mechanical pencil, adding dry ink to a copy machine, adding paper to a laser printer). Then write a simple set of instructions for this process in the form of a memo report. Your readers are assistants at the many offices of a large national firm. Consider them to be new employees who have no background or experience in office work and no education beyond high school. You are responsible for their training.

12. Writing Complex Instructions, with Graphics—Team Project

Complete this assignment as a team project (see the guidelines for team work in chapter 1). Choose a process connected with college life or courses—for example, completing a lab experiment, doing a field test, designing a model, writing a research paper, getting a parking sticker, paying fees, or registering for classes.

Using memo report format, write a set of instructions for students who have never performed this task. Follow all the guidelines in this chapter. Include at least one illustration (along with warnings or cautions, if appropriate). If possible, conduct a user test before completing the final draft.

13. Writing Instructions—M-Global Context

M-Global does a good deal of environmental work around the country—cleaning up toxic-waste sites, building energy-efficient structures, removing asbestos from old buildings, and investigating construction sites to determine the most environmentally sound approach to design and construction. For business reasons—and also because of its sense of civic duty—the company encourages citizens to get directly involved in environmental action.

As public relations manager for M-Global, you have just received an interesting assignment from the president, Jim McDuff. He wants you to prepare a set of instructions that will go out to citizen and school groups in the Baltimore–Washington area. In the form of a memo report, this document should give readers specific directions for recycling one or more types of waste. Your instructions should be directed toward a broad audience, of course; moreover, they should give the kinds of details that allow someone to act without having to get more information.

To get information for this report, you might consider (1) calling individuals in the waste-management department of your local government, (2) reading relevant articles from recent periodicals, or (3) checking an environmental science textbook at your college.

14. Writing Instructions—Team Project with M-Global Context

M-Global's, increasing international work generated interest among the corporate staff in gaining ISO 9000 certification. (Based in Geneva, Switzerland, the International Organization for Standardization [ISO] helps organizations around the world develop standards in quality.) Your team will conduct some research on this topic of growing interest. Write a set of instructions for a company, like M-Global, that wishes to gain such certification. You may either (a) provide a generalized overview for completing the entire process or (b) focus on one limited, specific part of the process, such as the process for gaining certification for a particular product or service.

15. Evaluating Document with Embedded Instructions

An M-Global lab supervisor, Kerubo Awala, has very little time to train a new group of lab trainees who have little if any industrial lab experience. Although she has given the new recruits a detailed description of a Bunsen burner (see Model 6–3 in chapter 6), she also wants them to have a short, easy-to-read document that focuses on use and safety. For that purpose, she quickly assembled and distributed the following document. Evaluate its effectiveness for her intended purpose and audience. How could it be improved.

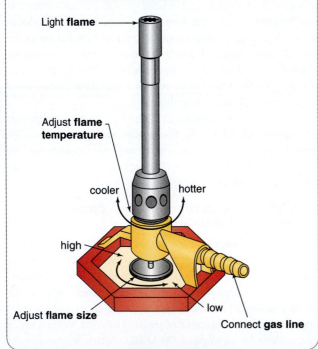

BUNSEN BURNER

Purpose The Bunsen burner is used in experiments that require you to heat solutions in a glass container (flask).

Warning The Bunsen burner produces an open flame that can burn you, burn other students, or set fire to flammable materials.
• Use it only when your lab instructor tells you.
• Apply the flame to glass or metal containers only.

Controls The diagram shows where to connect the gas line to the Bunsen burner and where to light and control the flame.

Light **flame**

Adjust **flame temperature**

cooler   hotter

high

Adjust **flame size**
low

Connect **gas line**

16. Ethics Assignment

Examine a set of instructions for a household or recreational device that—either in assembly or use—poses serious risk of injury or death. Evaluate the degree to which the manufacturer has fulfilled its ethical responsibility to inform

the user of such risk. You may want to consider the following questions:

a. Are risks adequately presented in text and/or graphic form?

b. Are risk notices appropriately placed in the document?

c. Is the document designed such that a user reading quickly could locate cautions, warnings, or dangers easily?

If you have highlighted any ethical problems, also suggest solutions to these problems.

17. International Communication Assignment

Sets of instructions may reflect *cultural bias* of a particular culture or country. Such bias may be acceptable if the audience for the instructions shares the same background. However, cultural bias presents a problem when (1) the audience represents diverse cultures and backgrounds or (2) the instructions must be translated into another language by someone not familiar with cultural cues in the instructions. Following are just a few categories of information that can present cultural bias and possibly cause confusion:

- Date formats
- Time zones
- Types of monetary currency
- Units of measurement
- Address and telephone formats
- Historical events
- Geographic references
- Popular culture references
- Acronyms
- Legal information
- Common objects in the home or office

[Adapted from a list on pp. 129–130 of Nancy L. Hoft's *International Technical Communication: How to Export Information About High Technology* (New York: John Wiley, 1995).]

Choose a set of instructions that reflects several types of cultural bias, such as those included on the previous list. Point out the examples of bias and explain why they might present problems to readers outside a particular culture.

18. A.C.T. N.O.W. Assignment (Applying Communication To Nurture Our World)

For this assignment, use the same context as that described in the ACT NOW exercise in chapter 6 (p. 186). Depending on the instructions given, prepare an oral or written report in which you either (a) explain the process for completing the town–gown project you proposed or (b) provide instructions for completing all or part of the project.

MEMORANDUM

DATE: November 7, 2008
TO: Leonard Schwartz
FROM: Jenny Vir
SUBJECT: New E-mail System

Yesterday I met with Jane Ansel, the installation manager at BHG Electronics, about our new e-mail system. She explained the process by which the system will be installed. As you requested, this memo summarizes what I learned about the setup process.

States purpose clearly.

BHG technicians will be at our offices on November 18 to complete the following tasks:

1. Removing old cable from the building conduits

2. Laying cable to link the remaining unconnected terminals with the central processing unit in the main frame

3. Installing software in the system that gives each terminal the capacity to operate the new e-mail system

Describes five main tasks, using parallel grammatical form.

4. Testing each terminal to make sure the system can operate from that location

5. Instructing selected managers on the use of the system

As you and I have agreed, when the installation is complete, I will send a memo to all office employees who will have access to e-mail. That memo will discuss setup procedures that each employee must complete before they are able to use their new e-mail accounts.

Confirms the follow-up activities they have already discussed.

Please let me know if you have further suggestions about how I can help make our transition to the new e-mail system as smooth as possible.

Gives reader opportunity to respond.

■ **Model 7–1** ■ M-Global process explanation: electronic mail

MEMORANDUM

DATE: November 20, 2008
TO: All Employees with Access to New E-mail System
FROM: Jenny Vir
SUBJECT: Instructions for Setting Up New E-mail Account

Gives clear purpose.

Earlier this month, we had a new e-mail system installed that will be used beginning December 1, 2008. This memo provides instructions on how to setup your new e-mail account and how to migrate all of your archived e-mail so that it will be ready for use when the new system goes into effect.

Identifies result of steps.

Please follow the step-by-step instructions below for proper setup of your e-mail and migration of your saved e-mail to the new system.

1. Double-click the **Email** icon.

Limits each step to one action.

2. Use the same **Username** and **Password** that you have used most recently with the old email system.
3. Select the **Accounts** menu.
4. Select the **Account Option**s sub-menu.

Separates results from actions.

RESULT: A window will open that prompts an **Account Name** and **Account Type.**

5. Enter a name (i.e., "Mail").
6. Use the drop-down menu to select **IMAP4** as the Account Type.
7. Click *Next.*

RESULT: You will be prompted to enter an **Incoming** and **Outgoing Mail Server**.

8. Enter as follows:

    Incoming: www.imap.mglobal.com

    Outgoing: www.smtp.mglobal.com

9. Click *Next.*

RESULT: You will be asked for your **email address**.

10. Use: *yourlastname*@mglobal.com
11. Click *Next.*
12. Click the radio button that reads: **Connect through my Local Area Network (LAN)**.
13. Click *Next.*
14. Name your "New Folder" (i.e., "Old Mail")
15. Click the **Finish** button.

Results if instructions have been followed correctly.

Your new account access should now be available, and your old e-mails will move to the new folder that you just named.

Shows reader how to get more information.

If you encounter any problems while performing the steps listed above, please contact a member of our IT staff for assistance.

■ **Model 7–2** ■ M-Global instructions: electronic mail

APPENDIX A: ON-SITE MONITORING

The purpose of monitoring the air is to determine the level of protective equipment needed for each day's work. This appendix gives an overview of the process for monitoring on-site air quality each day. Besides describing the main parts of the process, it notes other relevant information to be recorded and the manner in which data will be logged.

EQUIPMENT

This process requires the following equipment:

- Organic vapor analyzers (OVAs)
- Combustible-gas instruments
- Personal sampling devices

PROCESS

The project manager at the site is responsible for supervising the technician who performs the air-quality tests. At the start of every day, a technician uses an OVA to check the quality of air at selected locations around the site. Throughout the workday (at times specified by the project manager), the technician monitors the air with combustible-gas instruments and personal sampling devices. This monitoring takes place at the following locations:

1. Around the perimeter of the site
2. Downwind of the site (to determine the extent of migration of vapors and gases)
3. Generally throughout the site
4. At active work locations within the site

Then at the end of every workday, the technician uses the OVA to monitor the site for organic vapors and gases.

CONCLUSION

Besides the air-quality data, the following information is collected by the technician at each sampling time: percentage relative humidity, wind direction and speed, temperature, and atmospheric pressure. The project manager keeps records of air quality and weather conditions in dated entries in a bound log.

ABC format begins with abstract—with purpose statement and summary of appendix in this paragraph.

Abstract ends with list of equipment used in process that follows.

Body section of this process uses paragraph format and is aimed at non-technical audience.

Listing is used to highlight locations for sampling.

Conclusion part of ABC format puts this process in larger context.

■ **Model 7–3** ■ Process explanation

COMBINED SITE INVESTIGATION

In helping to select the site for an offshore oil platform, M-Global recommends a combined site investigation. This approach achieves the best results by integrating sophisticated geophysical work with traditional engineering activities.

As the accompanying flowchart shows, a combined site investigation consists of the following main steps:

Steps 1 and 2 are shown in top center portion of flowchart.

1. Planning the program, with M-Global's scientists and engineers and the client's representatives
2. Reviewing existing data

Steps 3 and 4 are shown in left and right portions of flowchart, respectively.

3. Completing a high-resolution geophysical survey of the site, followed by a preliminary analysis of the data
4. Collecting, testing, and analyzing soil samples

Step 5 is shown in bottom center portion of flowchart.

5. Combining geophysical and engineering information into one final report for the client

The report from this combined study will show how geological conditions at the site may affect the planned offshore oil platform.

Flowchart shows relationship among steps occurring at the same time.

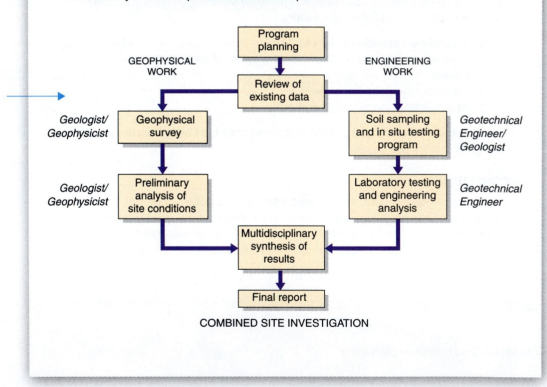

COMBINED SITE INVESTIGATION

■ **Model 7–4** ■ M-Global process explanation with a flowchart (both are included in an appendix to a report to a client)

MAKING TRAVEL ARRANGEMENTS
(Original Version)

When you're making travel arrangements, ask the person taking the trip to give you most of the details needed—dates, destinations, flight numbers, flight times, hotel requirements, rental car requirements, purpose of trip, and account number. Before proceeding, the first thing I do is confirm the flight information in the Official Airline Guide (OAG). You'll find the OAG on top of the credenza. The next step is to call Turner Travel (566-0998). Although I've had great luck with all the people there, ask for Bonnie or Charlie—these two are most familiar with our firm. Turner Travel will handle reservations for flights, hotels, and rental cars. Remind them that we always use Avis midsize cars.

After you have confirmed the reservations information, fill out the M-Global travel form. Here's where you need to know the purpose of the trip and the traveler's M-Global account number. Blank forms are in the top drawer of my file cabinet in the folder labeled "Travel Forms—Blank." Once the form is complete, file the original in my "Travel Forms—Completed" folder, also in the top drawer of the file cabinet. Give the copy to the person taking the trip.

When you get the ticket in the mail from Turner Travel, check the flight information against the completed travel form. If everything checks out, give the ticket to the traveler. If there are errors, call Turner.

Also, when making any reservations for visitors to our office, call either the Warner Inn (566-7888) or the Hasker Hotel (567-9000). We have company accounts there, and they will bill us directly.

Paragraph format makes it difficult for reader to locate individual steps.

■ **Model 7–5** ■ Instructions for making travel arrangements (original version)

MAKING TRAVEL ARRANGEMENTS
(Revised Version)

Arranging Travel for Employees

To make travel arrangements for employees, follow these instructions:

Step	Action
1.	Obtain the following information from the traveler:
	a. Dates
	b. Destinations
	c. Flight numbers
	d. Flight times
	e. Hotel requirements
	f. Rental car requirements
	g. Purpose of trip
	h. Account number
2.	Confirm flight information in the Official Airline Guide (OAG).
	Note: The OAG is on the credenza.
3.	Call Turner Travel (566-0998) to make reservations.
	Note: Ask for Bonnie or Charlie.
	Note: For car rental, use Avis midsize cars.
4.	Complete the M-Global travel form.
	Note: Blank forms are in the folder labeled "Travel Forms—Blank," in the top drawer of my file cabinet.
5.	Make one copy of the completed travel form.
6.	Place the original form in the folder labeled "Travel Forms—Completed," in the top drawer of my file cabinet.
7.	Send the copy to the person taking the trip.
8.	Check the ticket and the completed travel form after the ticket arrives from Turner Travel.
9.	Do the ticket and the completed travel form agree?
	a. If *yes*, give the ticket to the traveler.
	b. If *no*, call Turner Travel.

Arranging Hotel Reservations for Visitors

To make reservations for visitors, call the Warner Inn (566-7888) or the Hasker Hotel (567-9000). M-Global has company accounts there, and they will bill us.

Action steps all begin with *command* format verb.

Letters are used to show long list of subpoints for easy reference.

Notes are used to provide reader with *extra* information, separate from action of steps.

Although closely related, Steps 5–7 are best separated—for convenient reference by reader.

As noted in Guideline 8, two subpoints can show reader the *options* that exist.

■ **Model 7–5** ■ continued

MEMORANDUM

DATE: October 31, 2008
TO: Employees Receiving New Scanners
FROM: June Hier, Purchasing Agent
SUBJECT: Instructions for New Scanners

INTRODUCTORY SUMMARY

When we received our new flatbed scanners last week, it was brought to our attention that the accompanying instructions for setting up and operating the scanners had been lost. To help you begin using your scanner, I have written basic instructions. Before setting up your scanner, please make sure you have the following pieces:

Abstract information in ABC format places instructions in a context.

- Scanner
- Black connecting cable
- Software CD

The following illustration identifies your scanner's basic parts and the steps you need to install your software, set up your scanner, and begin scanning.

Reader receives purpose and overview information.

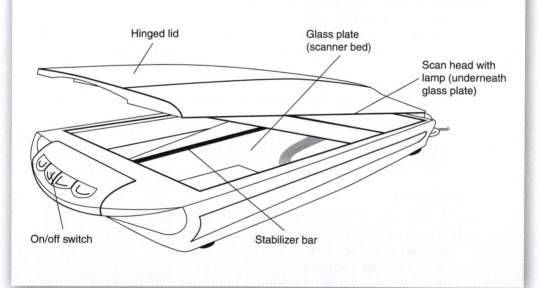

Hinged lid
Glass plate (scanner bed)
Scan head with lamp (underneath glass plate)
On/off switch
Stabilizer bar

■ **Model 7–6** ■ M-Global memo containing how-to instructions for a scanner

Main tasks are indicated in headings and subheads.

INSTALLING YOUR SCANNING SOFTWARE

Before you can use your scanner, you must install the appropriate software program. To install the software, insert the CD that came with the scanner into your CD drive. The installation wizard should automatically run (if it does not, go to Start > My Computer and double-click on MYSCNR). Follow these steps:

1. The installation wizard appears. Click *Begin*.
2. The wizard wants to know where it should install in the program.

Note provides troubleshooting help.

   Note: The default location should be Program Files. If not, click on the *Browse* button and go to My Computer > Local Disk (C:) > Program Files.

3. Click *OK*.
4. Click *Next*.
5. Make sure "Create Desktop Icon" is selected.
6. Click *Next*.
7. The wizard will install your program.
8. Click *Finish* when the installation is complete.

Identifies successful results

You have now successfully installed your scanner's software program. After installation, the program will run automatically. You can close it if you want to.

SETTING UP AND USING YOUR SCANNER

1. Hooking Up Your Scanner

Similar actions are separated into *two* different steps, to keep actions distinct.

   a. Plug the connecting cable into the back of the scanner.
   b. Plug the other end of the connecting cable into a wall socket.

2. Turning on Your Scanner

   a. Locate the On/Off switch on the front of the scanner.
   b. Switch to the On position.

   RESULT: Scanner's lamp will turn on and warm up. The scan head will move back and forth a few times.

3. Scanning

   a. Open the scanning program by double-clicking the desktop icon.
   b. Place a piece of paper on the scanner bed, in the upper right hand corner.
   c. Select *Scan Document*.
   d. Click *Preview Document*.

   NOTE: This will take 15–20 seconds.

   RESULT: A preview image will appear.

As with "notes," "results" should be *separated* from action in steps.

   e. Click and drag the edges of the crop box to fit the document.
   f. Click *Scan*.

   RESULT: The scanner will scan the selected area of the preview image.

▪ **Model 7–6** ▪ continued

4. Saving Scanned Files

    a. To save your scanned file(s), go to File > Save.

    b. Enter a name for your document.

    c. Choose to save it as either a JPG (for pictures) or PDF (for text).

    d. Find the file that you want to save the document in.

    e. Click *Save*.

5. Turning Off Your Scanner

    a. Locate the On/Off switch on the left side of the scanner.

    b. Switch to Off position.

Choosing Scanning Options

1. The scanner automatically scans in color. To scan in grayscale or black and white

    a. Follow steps a through C of #3.

    b. Look for the *Options* box above the *Preview Document* and *Scan* buttons.

    c. Select grayscale, black and white, or color from the drop-down menu.

2. Scanning Multiple-Page Documents

    a. To scan more than one page per document, open the scanning program.

    b. Select *Scan Multiple-Page Document*.

    c. Follow steps d through f of #3.

    RESULT: A dialogue box will pop up asking you if you want to add more pages to your document.

    d. Click *Yes*.

    e. Scan another page.

    f. When you are done adding pages, click *No* on the dialogue box.

    g. Save as a PDF.

CONCLUSION

If for any reason you have trouble following these instructions or do not have all the parts needed to set up and begin using your scanner, please contact Jerry (ext. 1781). If you encounter problems using the scanner, please report them to Jerry, especially if:

- The scanner cannot be detected.
- The scanning program freezes while saving.
- The scanner refuses to turn on.

Conclusion of ABC format wraps up memo by telling readers what to do if they encounter problems.

■ **Model 7–6** ■ continued

Chapter **8** Informal Reports

>>> Chapter Outline

When to Use Informal Reports   225
  Letter Reports at M-Global   225
  Memo Reports at M-Global   226

General Guidelines for Informal Reports   227

Specific Guidelines for Five Informal Reports   231
  Problem Analyses   232
  Recommendation Reports   233
  Equipment Evaluations   234
  Progress/Periodic Reports   235
  Lab Reports   236

Chapter Summary   237

Learning Portfolio   238
  Communication Challenge   238
  Collaboration at Work   239
  Assignments   239

Models for Good Writing   246
  Model 8–1: Recommendation report (letter format)   246
  Model 8–2: Equipment evaluation (memo format)   248
  Model 8–3: Problem analysis (memo format)   250
  Model 8–4: Progress report (memo format)   252
  Model 8–5: Periodic report (memo format)   254
  Model 8–6: Lab report (letter format)   256

Alan Murphy, a salesperson for M-Global's St. Paul office, has a full day ahead. Besides having to make some sales calls in the morning, he must complete two short reports back in the office. The first is a short progress report to Brasstown Bearings, a company that recently hired M-Global to train its technical staff in effective sales techniques. As manager of the project, Alan has overseen the efforts of three M-Global trainers for the last three weeks. According to the contract, he must send a progress report to Brasstown every three weeks during the project. Alan's second report is internal. His boss wants a short report recommending ways that M-Global can pursue more training projects like the Brasstown job.

Like Alan Murphy, you will spend much of your time writing informal reports in your career. Although short and easy to read like letters and memos, informal reports have more substance, are longer, and thus require more organization signals than everyday correspondence. A working definition follows:

Informal report: **A document that contains about two to five pages of text, not including**

attachments. It has more substance than a simple letter or memo, but is presented in letter or memo format. It can be directed to readers either outside or inside your organization. If outside, it may be called a *letter report;* **if inside, it may be called a** *memo report.* **In either case, its purpose can be** *informative* **(to clarify or explain),** *persuasive* **(to convince), or both.**

This chapter has three sections. The first shows you when to use informal reports in your career by describing some M-Global cases; the second provides 10 main writing guidelines that apply to both letter and memo reports; and the third focuses on specific suggestions for writing five common types of informal reports.

At the end of the chapter are examples with marginal annotations. They give you specific, real-life applications of the chapter's writing guidelines. As such, the models will help you complete chapter assignments and do actual reports on the job. During your career, you will write many types of informal reports other than those presented here. If you grasp this chapter's principles, however, you can adapt to other formats.

>>> When to Use Informal Reports

As noted previously, informal reports are clearly distinguished from both formal reports and routine letters and memos. Early in your career, however, you may have trouble deciding exactly where to draw the line. To help you decide, the two sections that follow describe briefly situations in which informal reports would be appropriate at M-Global.

Letter Reports at M-Global

Written to people outside your organization, letter reports use the format of a business letter because of their brevity; however, they include more detail than a simple business letter. Following are some sample projects at M-Global that would require letter reports:

- **Training recommendation:** M-Global's corporate training staff recommends changes in the training program of a large construction company. Courses that are recommended include technical writing, interpersonal communication, and quality management.

■ **Seafloor study:** M-Global's Nairobi staff writes a preliminary report on the stability of the seafloor where an oil rig might be located off the coast of Africa. This preliminary study includes only a survey of information on file about the site. The final report, involving fieldwork, will be longer and more formal.

■ **Marketing report:** A marketing specialist at the corporate office completes a study on "New Markets Beyond 2008." The report has been solicited by a professional marketing association to which M-Global belongs.

■ **Asbestos project:** M-Global's Houston staff reports to a suburban school board about possible asbestos contamination of an old elementary school. After two days on site, the crew of two technicians determined that the structure had no asbestos in its walls, plumbing, floors, or storerooms.

■ **Environmental study:** M-Global's San Francisco staff reports to the local Sierra Club chapter on possible environmental effects of an entertainment park, proposed for a rural area where eagles often nest. The project involved one site visit, interviews with a biologist, and some brief library research.

■ **Equipment design project:** M-Global's equipment development staff reports to a manufacturer on tentative designs for a computer-controlled device to cut plastic drainage pipe. The project involved several days' drafting work.

As these examples show, letter reports are the best format for projects with a limited scope. In addition, this informal format is a good sales strategy when dealing with customers greatly concerned about the cost of your work. When reading letter reports, they realize—consciously or subconsciously—that these documents cost them less money than formal reports. Your use of letter reports for small jobs shows sensitivity to their budget and may help gain their repeat work. See Model 8–1 on pages 246–247 for a letter report based on a small project at M-Global.

Memo Reports at M-Global

Memo reports are the informal reports that go back and forth among M-Global's own employees. Although in memorandum format, they include more technical detail and are longer than routine memos. These situations at M-Global show the varied contexts of memo reports:

■ **Need for testing equipment:** Juan Watson, a lab technician in the Denver office, evaluates a new piece of chemical testing equipment for his department manager, Wes Powell. Powell discusses the report with his manager.

■ **Personnel problem:** Werner Hoffman, a field engineer in the Munich office, writes to his project manager, Hans Schulman, about disciplinary problems with a field hand. Hoffman discusses the report with his manager and with the personnel manager.

■ **Need for laptop computers:** Susan Gindle, an equipment-development technician in Baltimore, writes a report to the equipment-development manager, Ralph Peak. Gindle recommends the company purchase five laptop computers from Simulon, Inc., as opposed to similar laptops from Sonet, Inc. Peak will discuss the report with the company's vice president for research and with the finance officer.

■ **Progress in hiring minorities:** Scott Sampson, personnel manager, reports to Karrie Camp, vice president of human resources, on the company's initial efforts to hire more minorities. Redmond will discuss Sampson's progress report with the company president and with all office managers.

■ **Report on training session:** Pamela Martin, a field engineer in St. Louis, reports to her project manager, Mel Baron, on a one-week course she took in Omaha on new techniques for removing asbestos from buildings. Baron circulates the report to his office manager, Ramsey Pitt, who then sends copies to the manager of every company office, because asbestos projects are becoming more common throughout the firm.

These five reports require enough detail to justify writing memo reports rather than simple memos. As for audience, each report goes directly to, or at least is discussed with, readers at high levels within the company. That means good memo reports can help advance your career. Model 8–2 on pages 248–249 provides an annotated example of a memo report about proposed computer software at M-Global.

>>> General Guidelines for Informal Reports

Following are 10 guidelines that focus mainly on report format.

>> Informal Report Guideline 1: Plan Well Before You Write

Like other chapters in this book, this section emphasizes the importance of the planning process. Complete the Planning Form at the end of the book for each assignment in this chapter, as well as for informal reports you write in your career. Before you begin writing a draft, use the Planning Form to record specific information about these points:

■ The document's purpose

■ The variety of readers who will receive the document

■ The needs and expectations of readers, particularly decision makers

■ An outline of the main points to be covered in the body
■ Strategies for writing an effective document

>> Informal Report Guideline 2: Use Letter or Memo Format

Model 8–1 shows that letter reports follow about the same format as typical business letters (see chapter 5). For example, both are produced on letterhead and both often include the reader's name, the date, and the page number on all pages after the first. Yet the format of letter reports differs from that of letters in the following respects:

- The greeting is sometimes left out or replaced by an attention line, especially when your letter report will go to many readers in an organization.
- A report title often comes immediately after the inside address. It identifies the specific project covered in the report. You may have to use several lines because the project title should be described fully, in the same words that the reader would use.
- Spacing between lines might be single, one-and-one-half, or double, depending on the reader's preference.

Model 8–2 shows the typical format for a memo report. Like most memos, it includes "Date/To/From/Subject" information at the top and has the reader's name, date, and page number on every page after the first. Also, both memos and memo reports have a subject line that should engage interest, give readers their first quick look at your topic, and be both specific and concise—for example, "Fracture Problems with Molds 43-D and 42-G" is preferable to "Problems with Molds." Because memo reports are usually longer than memos, they tend to contain more headings than routine memos.

>> Informal Report Guideline 3: Make Text Visually Appealing

Your letter or memo report must compete with other documents for each reader's attention. Following are three visual devices that help get attention, maintain interest, and highlight important information:

- Bulleted points for short lists (like this one)
- Numbered points for lists that are longer or that include a list of ordered steps
- Frequent use of headings and subheadings

Headings are particularly useful in memo and letter reports. As Models 8–1 and 8–2 (pp. 246–249) show, they give readers much-needed visual breaks. Because informal reports have no table of contents, headings also help readers locate information quickly. (Chapter 4 gives more detail on headings and other features of page design.)

>> Informal Report Guideline 4: Use the ABC Format for Organization

Headings and lists attract attention, but these alone do not keep readers interested. You must also organize information effectively. Most technical documents, including informal reports, follow what this book calls the *ABC format*. This approach to organization includes three parts: (1) **A**bstract, (2) **B**ody, and (3) **C**onclusion.

Abstract, Body, and *Conclusion* are only generic terms. They indicate the types of information included at the beginning, middle, and end of your reports—not necessarily the exact headings you will use. The next four guidelines give details on the ABC format as applied to memo and letter reports.

ABC Format: Organization

- **ABSTRACT:** Start with a capsule version of the information most needed by decision makers.
- **BODY:** Give details in the body of the report, where technical readers are most likely to linger a while to examine supporting evidence.
- **CONCLUSION:** Reserve the end of the report for a description or list of findings, conclusions, or recommendations.

>> Informal Report Guideline 5:
Create the Abstract as an Introductory Summary

Abstracts should give readers a summary, the "big picture." This text suggests that in informal reports, you label this overview *Introduction, Summary,* or *Introductory Summary*, terms that give the reader a good idea of what the section contains. (You also have the option of leaving off a heading label, in which case your first few paragraphs would contain the introductory summary information, followed by the first body heading of the report.)

In letter reports, the introductory summary comes immediately after the title; in memo reports, it comes after the subject line. Note that informal reports do not require long, drawn-out beginnings; just one or two paragraphs in this first section give readers three essential pieces of information:

1. **Purpose** for the report—why are you writing it?
2. **Scope** statement—what range of information does the report contain?
3. **Summary** of essentials—what main information does the reader most want or need to know?

>> Informal Report Guideline 6: Put Important Details in the Body

The body section provides details needed to expand on the outline presented in the introductory summary. If your report goes to a diverse audience, managers often read the quick overview in the introductory summary and then skip to conclusions and recommendations. Technical readers, however, may look first to the body section(s), where they expect to find supporting details presented in a logical fashion. In other words, here is your chance to make your case and to explain points thoroughly.

Yet the discussion section is no place to ramble. Details must be organized so well and put forth so logically that the reader feels compelled to read on. Following are three main suggestions for organization:

- **Use headings generously.** Each time you change a major or minor point, consider whether a heading change would help the reader. Informal reports should include at least one heading per page.
- **Precede subheadings with a lead-in passage.** Here you mention the subsections to follow, before you launch into the first subheading. (For example, "This section covers these three phases of the field study: clearing the site, collecting samples,

and classifying samples.") This passage does for the entire section exactly what the introductory summary does for the entire report—it sets the scene for what is to come by providing a "road map."

- **Move from general to specific in paragraphs.** Start each paragraph with a topic sentence that includes your main point, and then give supporting details. This approach always keeps your most important information at the beginnings of paragraphs, where readers tend to focus first while reading.

Another important consideration in organizing the report discussion is the way you handle facts versus opinions.

>> Informal Report Guideline 7: Separate Fact from Opinion

Some informal reports contain strong points of view. Others contain only subtle statements of opinion, if any. In either case, you must avoid any confusion about what constitutes fact or opinion. The safest approach in the report discussion is to move logically from findings to your conclusions and, finally, to your recommendations. Because these terms are often confused, some working definitions are as follows:

- **Findings:** Facts you uncover (e.g., you observed severe cracks in the foundations of two adjacent homes in a subdivision).
- **Conclusions:** Ideas or beliefs you develop based on your findings (e.g., you conclude that foundation cracks occurred because the two homes were built on soft fill, where original soil had been replaced by construction scraps). Opinion is clearly a part of conclusions.
- **Recommendations:** Suggestions or action items based on your conclusions (e.g., you recommend that the foundation slab be supported by adding concrete posts beneath it). Recommendations are almost exclusively made up of opinions, but recommendations should clearly be grounded in the facts presented in the report.

>> Informal Report Guideline 8: Focus Attention in Your Conclusion

Letter and memo reports end with a section labeled *Findings, Conclusions,* or *Conclusions and Recommendations.* Choose the wording that best fits the content of your report. In all cases, this section gives details about your major findings, your conclusions, and, if called for, your recommendations. People often remember best what they read last, so think hard about what you place at the end of a report.

The precise amount of detail in your conclusion depends on which of these two options you choose for your particular report:

Option 1: If your major conclusions or recommendations have already been stated in the discussion, then you only need to restate them briefly to reinforce their importance (see Model 8–2, pp. 248–249).

Option 2: If the discussion leads up to, but has not covered, these conclusions or recommendations, then you may want to give more detail in this final section (see Model 8–1, pp. 246–247).

As in Models 8–1 and 8–2, lists are often mixed with paragraphs in the conclusion. Use such lists if you believe they will help readers remember your main points.

>> Informal Report Guideline 9: Use Attachments for Less Important Details

The trend today is to avoid lengthy text in informal reports, yet technical detail is often needed for support. One solution to this dilemma is to replace as much report text as possible with clearly labeled attachments that could include the following items:

- **Tables and figures:** Illustrations in informal reports usually appear in attachments unless it is crucial that one is within the text. Memo and letter reports are so short that attached illustrations are easily accessible.

- **Costs:** It is best to list costs on a separate sheet. First, you do not want to bury important financial information within paragraphs. Second, readers must often circulate cost information, and a separate cost attachment is easy to photocopy and send.

>> Informal Report Guideline 10: Edit Carefully

Many readers judge you on how well you edit a report. A few spelling errors or some careless punctuation makes you seem unprofessional. Your career and your firm's future can depend on your ability to write final drafts carefully. Chapter 17 and the Handbook at the end of this text give detailed information about editing. For now, remember the following basic guidelines:

- Keep most sentences short and simple.
- Proofread several times for mechanical errors such as misspellings (particularly personal names).
- Triple-check all cost figures for accuracy.
- Make sure all attachments are included, are mentioned in the text, and are accurate.
- Check the format and wording of all headings and subheadings.
- Ask a colleague to check over the report.

These guidelines help memo and letter reports accomplish their objectives. Remember—both your supervisors and your clients will judge you as much on communication skills as they do on technical ability. Consider each report to be part of your resume.

>>> Specific Guidelines for Five Informal Reports

Report types vary from company to company. The ones described here are only a sampling of what you will be asked to write on the job. These five types were chosen because they are common and can be written as either memo reports or letter reports. Well-organized reports incorporate the writing patterns described in chapter 3 and often include the elements of technical communication discussed in chapters 6 and 7. If you master these five informal reports, you can probably handle other types that come your way.

The sections that follow include an ABC format for each report being discussed, some brief case studies from M-Global, and report models based on the cases. Remember to consult chapter 3 if you need to review general patterns of organization used in short and long reports, such as cause–effect; and chapters 6 and 7 if you need to review elements of technical communication, such as technical description.

Note: Informal and formal reports share common characteristics. Thus the following five report types also could be constructed as long, formal reports—assuming the report content was more complex and detailed than that included in an informal report.

Problem Analyses

Every organization faces both routine and complex problems. Routine problems are often handled without much paperwork; they are discussed and then solved. However, other problems must often be described in reports, particularly if they involve many people, are difficult to solve, or have been brewing for a long time. Use the following working definition of a *report* that analyzes a problem:

> **Problem analysis:** An informal report that presents readers with a detailed description of problems in areas such as personnel, equipment, products, and services. Its main goal is to provide objective information so that the readers can choose the next step. Any opinions must be well supported by facts.

Problem analyses, which can be either internal or external documents, should follow the pattern of organization described next.

ABC Format: **Problem Analysis**[1]

- **ABSTRACT:** Purpose of report
  - Capsule summary of problems covered in report discussion
- **BODY:** Background on source of problems
  - Well-organized description of the problems observed
  - Data that support your observations
  - Consequences of the problems
- **CONCLUSION:** Brief restatement of main problems (unless report is so short that such restatement would seem repetitious)
  - Degree of urgency required in handling problems
  - Suggested next step

ABC Format for Problem Analyses

Like other informal reports, *problem analyses* fit the simple ABC (**A**bstract/**B**ody/**C**onclusion) format recommended throughout this text. The three sections contain some or all of the following information, depending on the specific report. Note that solutions to problems are not mentioned; this chapter deals separately with (1) problem analyses, whose main focus is problems; and (2) recommendation reports, whose main focus is solutions. Of course, be aware that during your career, you will be called on to write reports that combine both types.

M-Global Case Study for a Problem Analysis

Model 8–3 on pages 250–251 presents a sample problem analysis that follows this chapter's guidelines. Harold Marshal, a longtime M-Global employee, supervises all technical work aboard the *Seeker II*, a boat that M-Global leases during the summer. Staffed with several technicians and engineers, the boat is used

---

[1]The content and order of items within each of the three sections can vary.

to collect and test soil samples from the ocean floor. Different clients purchase these data, such as oil companies that must place oilrigs safely, and telecommunications companies that must lay cable.

After a summer on the *Seeker II*, Harold has severe reservations about the safety and technical adequacy of the boat. Yet he knows that his supervisor, Jan Stillwright, will require detailed support of any complaints before she seriously considers negotiating a new boat contract next season. Given this critical audience, Harold focuses on specific problems that affect (1) the safety of the crew, (2) the accuracy of the technical work performed, and (3) the morale of the crew. He believes that this pragmatic approach, rather than an emotional appeal, will best persuade his boss that the problem is serious.

Most problem analyses contain both facts and opinions. As the writer, you must make special efforts to separate the two, for the following reason: Most readers want the opportunity to draw their own conclusions about the problem. Also, support all opinions with facts.

Recommendation Reports

Like problem analysis reports, recommendation reports include facts and opinions, because they include more personal views than other report types. Use the following working definition for recommendation reports:

Recommendation report: **An informal report that presents readers with specific suggestions that affect areas such as personnel, equipment, procedures, products, and services. Although the report's main purpose is to persuade, every recommendation must be supported by objective data.**

Recommendation reports can be either internal or external documents, both of which follow the ABC format.

ABC Format for Recommendation Reports

Problem analyses and recommendation reports sometimes overlap in content. You may recommend solutions in a problem analysis, just as you may analyze problems in a recommendation report. The ABC format assumes that you want to mention the problem briefly before proceeding to discuss solutions.

M-Global Case Study for a Recommendation Report

Model 8–1 (pp. 246–247) shows a typical recommendation report written at M-Global. The reader is a client oil firm about to place an oilrig at an offshore site in the Gulf of Mexico. Given the potential for risk to human life and to the environment, Big Muddy Oil wants to take every precaution. Therefore, it has hired M-Global to determine whether the preferred site is safe.

ABC Format: **Recommendation Report**[2]

- **A**BSTRACT: Purpose of report
  - Brief reference to problem to which recommendations respond
  - Capsule summary of recommendations covered in report discussion
- **B**ODY: Details about problem, if necessary
  - Description of recommendations (although some short reports place recommendations in the conclusion)
  - Data that support recommendations (with reference to attachments, if any)
  - Main benefits of recommendations
  - Any possible drawbacks
- **C**ONCLUSION: Brief statement or restatement of main recommendations (optional)
  - The main benefit of recommended change
  - Your offer to help with next step

This example presents an important problem you may face in writing recommendation reports. Occasionally, you may be asked to deliver recommendations sooner than you would prefer if you were working under ideal circumstances. In such cases, assume the cautious approach taken by Bartley Hopkins, the M-Global writer of Model 8–1. That is, make sure to state that your report is preliminary and based on incomplete data. This approach is even more important in situations like this case study in which there is risk to human life. Take pains to qualify your recommendations so that they cannot possibly be misunderstood by your audience.

Equipment Evaluations

Every organization uses some kind of equipment, and someone has to help buy, maintain, or replace it. Because companies put so much money into this part of their business, evaluating equipment has become an important activity. Following is a working definition of evaluation reports:

> **Equipment evaluation: An informal report that provides objective data about how equipment has, or has not, functioned. The report may cover topics such as machinery, tools, vehicles, office supplies, computer hardware, and computer software.**

ABC Format: **Equipment Evaluation**[2]

- **A**BSTRACT: Purpose of report
  - Capsule summary of what your report says about the equipment
- **B**ODY: Thorough description of the equipment being evaluated
  - Well-organized critique, either analyzing the parts of one piece of equipment or contrasting several pieces of similar equipment according to selected criteria
  - Additional supporting data, with reference to any attachments
- **C**ONCLUSION: Brief restatement of major findings, conclusions, or recommendations

Like a problem analysis, an equipment evaluation may focus only on problems; or like a recommendation report, it may go on to suggest a change in equipment. Whatever its focus, an equipment evaluation must provide a well-documented review of the exact manner in which equipment performed. Follow this ABC format in evaluating equipment.

ABC Format for Equipment Evaluations

Equipment evaluations that are informal reports should include some or all of the points listed here. Remember that in this type of report, the discussion must include evaluation criteria most important to the *readers*, not you.

---

[2]The content and order of items within each of the three sections can vary.

M-Global Case Study for an Equipment Evaluation

Like other firms, M-Global relies on word processing for almost all internal and external documents. Model 8–2 (pp. 248–249) contains an evaluation of a new word-processing package used on a trial basis. Melanie Frank, office manager in San Francisco, conducted the trial in her office and wrote the report to the branch manager, Hank Worley. Note that she analyzes each of the software's five main features, then ends with a recommendation, much like a recommendation report.

Pay special attention to the tone and argumentative structure of this example. Frank shows restraint in her enthusiasm, knowing that facts will be more convincing than opinions. Indeed, every claim about Best Choice software is supported either by evidence from her trial or by a logical explanation. For example, her praise of the file-management feature is supported by the experience of a field engineer who used the system for three days, and her statement about the well-written user's guide is supported by the few calls made to the Best Choice support center during the trial.

Progress/Periodic Reports

Some short reports are intended to cover activities that occurred during a specific period of time. They can be directed inside or outside your organization and are defined as follows:

Progress report: **An informal report that provides your manager or client with details about work on a specific project. Often you agree at the beginning of a project to submit a certain number of progress reports at certain intervals. The final progress report, submitted when a project is completed, is often called a** *project completion report.*

Periodic report: **An informal report, usually directed within your own organization, that summarizes your work on diverse tasks over a specific time period. For example, as supervisor of company publications, you may be asked to submit periodic reports each month on M-Global's new brochures, public-relations releases, and product flyers.**

Progress and periodic reports contain mostly objective data. Yet both of them, especially progress reports, sometimes may be written in a persuasive manner. After all, you are trying to put forth the best case for the work you have completed. The next section provides an ABC format for these two report types.

ABC Format for Progress/Periodic Reports

Whether internal or external, progress and periodic reports follow a basic ABC format. They contain some or all of the parts shown in the box on the next page.

M-Global Case Study for a Progress Report

As Model 8–4 on pages 252–253 indicates, Scott Sampson, M-Global's personnel manager, is in the midst of an internal project being conducted for Jeannie McDuff, vice president of domestic operations. Sampson's goal is to find ways to improve the company's training for technical employees. Having completed two of three phases, he is reporting his progress to McDuff. Note that Sampson organizes the body sections by task. This arrangement helps

ABC Format: **Progress/Periodic Report**[3]

- **ABSTRACT:** Purpose of report
  - Capsule summary of main project(s)
  - Main progress to date or since last report
- **BODY:** Description of work completed since last report, organized by task, by time, or by both
  - Clear reference to any dead ends that may have taken considerable time but yielded no results
  - Explanation of delays or incomplete work
  - Description of work remaining on project(s), organized by task, by time, or by both
  - Reference to attachments that may contain more specific information
- **CONCLUSION:** Brief restatement of work since last reporting period
  - Expression of confidence or concern about overall work on project(s)
  - Indication of your willingness to make any adjustments the reader may want to suggest

focus the reader's attention on the two main accomplishments—the successful phone interviews and the potentially useful survey. If, instead, Sampson had completed many smaller tasks, he may have wanted to organize the body of the report by time instead of by task.

Also note that Sampson adopts a persuasive tone at the end of the report—that is, he uses his solid progress as a way to emphasize the importance of the project. In this sense, he is "selling" the project to his "internal customer," Jeannie McDuff, who ultimately is in the position to make decisions about the future of technical training at M-Global.

M-Global Case Study for a Periodic Report

Model 8–5 on pages 254–255 shows the rather routine nature of most periodic reports. In this case, Nancy Fairbanks is simply submitting her usual monthly report. The greatest challenge in such reports is to classify, divide, and label information in such a way that readers can find what they need quickly. Fairbanks selected the kind of substantive headings that help the reader locate information (e.g., "Jones Fill Project," "Performance Reviews").

Lab Reports

College students write lab reports for courses in science, engineering, psychology, and other subjects, and this type of report also exists in technical organizations such as hospitals, engi-

neering firms, and computer companies. Perhaps more than any other type of informal report, the lab report varies in format from organization to organization (and from instructor to instructor, in the case of college courses). This chapter presents a format to use when no other instructions have been given. A working definition follows:

Lab report: An informal report that describes work done in any laboratory—with an emphasis on topics such as purpose of the work, procedures, equipment, problems, results, and implications. It may be directed to someone inside or outside your own organization. Also, it may stand on its own or it may become part of a larger report that uses the laboratory work as supporting detail.

[3]The content and order of items within each of the three sections can vary.

The next section shows a typical ABC format for lab reports, with the types of information that might appear in the three main sections.

ABC Format for Lab Reports

Whether simple or complicated, lab reports usually contain some or all of the parts shown on the right.

M-Global Case Study for a Lab Report

Model 8–6 on pages 256–257 shows an M-Global lab report that is not part of a larger document. In this case, the client sent M-Global some soils taken from borings made into the earth. M-Global analyzed the samples in its company laboratory and then drew some conclusions about the kind of rock from which the samples were taken. The report writer, a geologist named Joseph Rappaport, uses the body of the report to provide background information, lab materials, procedures, and problems encountered. Note that the report body uses process explanation, an element of technical communication covered in chapter 7.

ABC Format: **Lab Report**[4]

- **ABSTRACT:** Purpose of report
  - Capsule summary of results
- **BODY:** Purpose or hypothesis of lab work
  - Equipment needed
  - Procedures or methods used in the lab test
  - Unusual problems or occurrences
  - Results of the test with reference to your expectations (results may appear in conclusion, instead)
- **CONCLUSION:** Statement or restatement of main results
  - Implications of lab test for further work

>>> Chapter Summary

This chapter deals exclusively with the short and informal reports you write throughout your career. On the job, you write them for readers inside your organization (as *memo reports*) and outside your organization (as *letter reports*). In both cases, follow these basic guidelines:

1. Plan well before you write.
2. Use letter or memo format.
3. Make text visually appealing.
4. Use the ABC format for organizing information.
5. Start with an introductory summary.
6. Put detailed support in the body.
7. Separate fact from opinion.
8. Focus attention in your conclusion.
9. Use attachments for details.
10. Edit carefully.

Although letter and memo reports come in many varieties, this chapter conveys only five common types: problem analyses, recommendation reports, equipment evaluations, progress/periodic reports, and lab reports. Each follows its own type of three-part ABC format for organizing information.

---

[4]The content and order of items within each of the three sections can vary.

>>> Learning Portfolio

Communication Challenge "A Nonprofit Job: Good Deed or Questionable Ethics?"

It all started with an innocent conversation. Velora Nescon, a project manager at M-Global's Houston branch, had lunch with her old college friend Sibyl Sanders. As principal of Houston's Downtown Academy, Sibyl talked about her effort to keep the new private academy financially healthy and academically strong. Now she was busy with an expansion. What follows is a description of a problem she faced with the expansion, the help that Velora tried to offer, in addition to some ethical and procedural questions raised by the situation, and assignment for a written response to the Challenge.

A Working Lunch

Houston's Downtown Academy was established as a private elementary and middle school for bright inner-city kids who couldn't afford other private schools. Years ago, business leaders donated a renovated building for the new school and raised money for a scholarship account. Now the academy had a respectable enrollment, a good academic reputation, and excellent morale among faculty and staff. Overall, the future looked good for the school. Recently, a benefactor gave the school a piece of land adjacent to its campus where a recreational area would be built.

In her conversation with Velora, Sibyl noted that although the school had some money to begin construction, the budget would be tight. At this point, Velora reminded her friend that M-Global, Inc., offered some of the technical services the project might require. She added that she could ask her branch manager if M-Global might handle the job just for cost, as its way of contributing to the growth of the school. Velora genuinely wanted to help her friend with this worthwhile venture, but in truth, she also saw an opportunity to keep her technical staff busy during a slack period. By the end of lunch, Sibyl and Velora had reached a tentative agreement on M-Global doing the property study required before the land could be developed. Then, back at the office, Velora got her boss to agree to allowing the job to be done for cost.

Velora's Lucky Find

The project involved a soil and environmental study. M-Global was to drill borings to determine what foundations would be needed for small structures in the recreation area. In addition, soil samples would be taken from the site to check for contamination. These tasks were routine.

Before sending out her crew, Velora mentioned the project to a colleague, George Lightfoot, who thought he remembered doing some soil borings at the same location a couple of years ago. On checking his files, George found a report that included two borings paid for by Ace Enterprises, a firm that had considered buying the property. Later, Ace backed out of the purchase for reasons unrelated to the report. George loaned Velora the report and suggested she ask Ace for permission to use it, because she was trying to save the Downtown Academy money on the project. Velora thanked George for the report and said she would call Ace.

Velora wanted to get on this no-profit job right away, while her crew wasn't busy, so she tried all afternoon to contact Ace Enterprises. There was no listing for the firm in Houston, and the M-Global librarian found no address in a quick search of her regional files. Velora assumed the firm had gone out of business, changed names, or left town. That being the case, she decided to move ahead in using information in the report—thus saving the Downtown Academy money for soil borings that would have been done. She assumed that Ace, if it still existed, would not mind contributing information for a nonprofit job like this one.

M-Global's Fieldwork

The next day, the M-Global crew gathered soil samples from the surface and from shallow borings dug with a hand auger. Results of the lab tests on the samples showed there was an underground storage tank on the property. It had been used for kerosene, which had leaked into the surrounding soil. Both the tank and the soil would have to be removed.

Again trying to save her friend some money, Velora had the small tank and soil removed by an M-Global subcontractor working nearby later that week. The crew had been working for another M-Global client most of the day. Because that client had to pay for a full day's use of the crew and crane anyway, Velora didn't charge the Downtown Academy for the two hours it took to remove the tank and soil.

Questions and Comments for Discussion

Within a week, Velora handed Sibyl a complete report showing her fieldwork, lab tests, and conclusions. Sibyl was overjoyed that the study had come in even under the zero-profit budget she and Velora had first discussed. In thanking Velora, she assured her that M-Global would be a serious contender for the *profit* contracts that the Downtown Academy was sure to have in the future.

1. Do you think there are any ethical problems raised by this case study? If so, what are they and how would you have dealt with them? If not, explain your views.
2. Specifically, how do Sibyl's and Velora's actions either satisfy or violate the Equal Consideration of Interests principle described in chapter 2?

3. Putting aside the ethical issue, do you think Velora followed wise procedures in her handling of the Downtown Academy project? Why or why not?

Write About It

As Velora, write the project completion report (a final progress report) addressed to Sibyl and copied to Ralph Suarez, manager of the Houston branch. Identify all of the activities involved in completing the project, and address any ethical concerns that Ralph might have. Create whatever details necessary to make the report complete.

Collaboration at Work Suggestions for High School Students

General Instructions

Each Collaboration at Work exercise applies strategies for working in teams to chapter topics. The exercise assumes you (1) have been divided into teams of about three to six students, (2) will use team time inside or outside of class to complete the case, and (3) will produce an oral or written response. For guidelines about writing in teams, refer to pages 24–28.

Background for Assignment

This chapter introduced you to five types of informal reports, one of which is focused on recommendations. When you are writing a recommendation report with a team of colleagues, agreeing on content can be a challenge. Recommendations involve opinions, and opinions often vary about what should be presented to the reader. As with other collaborative efforts, first you share information in a nonjudgmental way; then you choose what should be included in the report based on your team discussions.

Team Assignment

Assume an association of colleges and universities has asked your team to help write a short report to be sent to high school students. The report's purpose is to assist students in selecting a college or university. Your team will prepare an outline for the body of the report by (1) choosing several headings that classify groupings of recommendations and (2) providing specific recommendations within each grouping. For example, one grouping might be "Support for Job Placement," with one recommendation in this grouping being "Request data on the job placement rates of graduates of the institution." After producing your outline, share the results with other teams in the class.

Assignments

This chapter includes both short and long assignments. The short assignments in Part 1 are designed to be used for in-class exercises and short homework assignments. The assignments in Part 2 generally require more time to complete.

Part 1: Short Assignments

1. Problem Analysis—Critiquing a Report

Using the guidelines in this chapter, analyze the level of effectiveness of the M-Global problem analysis that begins on the following pages.

2. Problem Analysis—Team Project

Divide into three- or four-person teams, as your instructor directs. In your team, share information about any problems that team members have encountered with services or facilities at the college or university you attend. Then select a problem substantive enough to be described in a short report. As a team, write a problem analysis in the format put forth in this chapter. Assume that your team represents an M-Global technical team that has been hired to investigate, and then write a series of reports on, problems at the school. Your report is one in the series. Select as your audience the appropriate administrators at the college or university.

April 16, 2008
Mr Jay Henderson
Christ Church
10 Smith Dr
Jar Georgia 30060

PROBLEM ANALYSIS:
NEW CHURCH BUILDING SITE

Introductory Summary

Last week, your church hired our firm to study problems caused by the recent incorporation of the church's new building site into the city limits. Having reviewed the city's planning and zoning requirements, we have found some problems with your original site design—which initially was designed to meet the county's requirements only. My report focuses on problems with four areas on the site:

1. Landscaping screen
2. Church sign
3. Detention pond
4. Emergency vehicle access

Attached to this report is a site plan to illustrate these problems as you review the report. The plan was drawn from an aerial viewpoint.

Landscaping Screen

The city zoning code requires a landscaping screen along the west property line, as shown on the attached site illustration sheet. The former design does not call for a screen in this area. The screen will act as a natural barrier between the church parking lot and the private residence adjoining the church property. The code requires that the trees for this screen be a minimum height of 8 feet with a height maturity level of at least 20 feet. The trees should be an aesthetically pleasing barrier for all parties, including the resident on the adjoining property.

Church Sign

After the site was incorporated into the city, the Department of Transportation decided to widen Woodstock Road and increase the setback to 50 feet, as illustrated on our site plan. With this change, the original location of the sign falls into the road setback. Its new location must be out of the setback and moved closer to the new church building.

Detention Pond

The city's civil engineers reviewed the original site drawing and found that the detention pond is too small. If the detention pond is not increased, rainwater may build up and overflow into the building, causing a considerable amount of flood damage to property in the building and to the building itself. There is a sufficient amount of land in the rear of the site to enlarge and deepen the pond to handle all expected rainfall.

Emergency Vehicle Access

On the original site plan, the slope of the ground along the back side of the new building is so steep that an ambulance or city firetruck will not be able to gain access to the rear of the building in the event of a fire. This area is shown on our site illustration around the north and east sides of the building. The zoning office enforces a code that is required by the fire marshal's office. This code states that all buildings within the city limits must provide a flat and unobstructed access path around the buildings. If the access is not provided, the safety of the church building and its members would be in jeopardy.

Conclusion

The just-stated problems are significant, yet they can be solved with minimal additional cost to the church. Once the problems are remedied and documented, the revised site plan must be approved by the zoning board before a building permit can be issued to the contractor.

I look forward to meeting with you and the church building committee next week to discuss any features of this study and its ramifications.

Sincerely,

Thomas K. Jones

Thomas K. Jones
Senior Landscape Engineer

Enclosure

3. Recommendation Report— Critiquing a Report

Using the guidelines in this chapter, analyze the level of effectiveness of the following M-Global recommendation report.

4. Writing a Recommendation Report

Divide into teams of three or four students, as your instructor directs. Consider your team to be a technical team from M-Global. Assume that the facilities director of your college or university has hired your team to recommend changes

---

April 21, 2008
Kenman Aircraft Company
76 Jonesboro Road
Sinman Colorado 87885

Attention: Mr. Ben Randall, Facilities Manager

EMERGENCY EXIT STUDY

Introductory Summary

As you requested, I have just completed a study of the emergency exits in your accounting office at the plant. My study indicates that you have two main problems: (1) easier access to exits is needed and (2) more exit signs and better visibility of these signs are needed. This report contains recommendations for rearranging the floor plan and improving signage.

Problems with Current Floor Plan

Two main problems cause the accounting office to fail to meet the county's guidelines for access to fire exits. First, the file cabinets on the north wall of the office are partially blocking the Reynolds Lane exit. Second, the office photocopier partially blocks the exit to the east hallway. In the first case, the file cabinets are so heavy that they cannot be moved by one person. In the second case, the photocopier can be rolled out of the way only by a very strong individual. Obviously, both situations are unacceptable and violate the current code.

The other problem is signage. The Reynolds Lane exit has an exit sign, but it is not easily seen. The east hallway exit has no sign at all. In addition, the rest of the office lacks any maps that show people the location of the two fire exits.

Recommendations for Solving Exit Problem

Fortunately, the existing problems can be corrected with only minor cost to the company. The following recommendations should be implemented immediately on your receipt of this report:

1. Move the file cabinets on the north wall to the east wall so that they no longer block the Reynolds Street exit.
2. Relocate the photocopier to the office supply room or the cubicle adjacent to it.
3. Remove the undersized exit sign from the Reynolds Street exit.
4. Purchase and install two county-approved exit signs above the two fire exits.
5. Draw up an emergency plan map and post a copy in every cubicle within the accounting office.

When you implement these recommendations, you will be in accordance with the county's current fire regulations.

continued

Conclusion

I strongly suggest that my recommendations be put into action as soon as possible. By doing so, you greatly reduce the risk to your employees and your associated liability.

If you have any questions or need additional information, please call me at your convenience.

Sincerely,

Howard B. Manwell

Howard B. Manwell
Field Engineer

---

that would improve your classroom. Write a team report that includes the recommendations agreed to by your team. For example, you may want to consider structural changes of any kind, additions of equipment, changes in the type and arrangement of seating, and so forth.

5. Writing an Equipment Evaluation

Assume you are a supervisor at M-Global's Equipment Development shop, located in Baltimore. The Procurement Office routinely asks you to write evaluations of new pieces of equipment being used in the shop. Such evaluations help the director of procurement, Brenda Seymour, decide on future purchases.

Write a brief memo to Seymour, evaluating the Brakoh cordless drills that your staff began using in the shop about a year ago. The Brakoh brand replaced a more expensive brand that the shop had used for the 10 previous years. In the past few months, your technicians have reported that the cheaper models have been falling apart after six or eight months of use. Information coming to you suggests that there are two main problems: (1) a grinding noise can be heard in the housing, resulting in the failure of the chuck (the piece that holds the drill bit) to rotate; and (2) drilling time between rechargings tends to decrease as the drill gets older. You believe that the manufacturer, in order to cut the cost of the drill, has substituted poorly made components in high-wear locations. For example, the gears responsible for turning the chuck are made of plastic. With a little wear, the gears tend to slip, which produces the grinding sound and the rotation failure. As for the recharging problem, the power cell just seems to hold less charge than the previous drill. In summary, the drill has broken down four times faster than the other model, causing many repair bills and a loss in productivity.

In writing your memo, remember that in this case, your main job is to provide information to the director of procure-

ment, not to make recommendations one way or the other. After receiving your memo, she probably will complete a cost analysis to determine if the problems with the cheaper drill outweigh the advantage of the initial cost savings.

Part 2: Longer Assignments—Individual or Team Work

While planning some of these assignments—especially assignment 6—you may need to review information in chapter 2 about M-Global, Inc. Also, for each assignment you should complete a copy of the Planning Form (included at the end of the book). These assignments can be completed as individual or team projects.

6. Report Based on Project Sheets

The project sheets at the end of chapter 2 summarize seven M-Global projects in various topics. These summaries were written for marketing purposes, after the jobs were completed.

Using the information on one of these sheets, write a brief informal report that summarizes the project for the client. If necessary, add details that are not on the sheet. Caution: Remember that marketing sheets may not be organized as reports are organized. Consider your purpose and audience carefully before writing.

7. Report Based on Internet "Surfing"

Use the Internet to collect actual information, or a list of sources that may contain information, about a topic that relates to your academic major. Then write an informal equipment evaluation in which you analyze (1) the ease with which the Internet allowed you to collect information on your topic and (2) the quality of the sources or information you received. Your audience is your instructor, who will let you know the degree of knowledge you can assume he or she has on this topic.

8. Problem Analysis

Assume you are an M-Global field engineer working at the construction site of a nuclear power plant in Jentsen, Missouri. For the past three weeks, your job has been to observe the construction of a water-cooling tower, a large cylindrical structure. As consultants to the plant's construction firm, you and your M-Global crew were hired to make sure that work proceeds properly and on schedule. As the field engineer, you are supposed to report any problems in writing to your project manager, John Raines, back at your St. Louis office. Then he will contact the construction firm's office, if necessary.

Write a short problem analysis in the form of a memo report to Raines. (Follow the guidelines in the "Problem Analyses" section of this chapter.) Take the following randomly organized information and present it in a clear, well-organized fashion. If you wish, add information of your own that might fit the context.

- Three cement pourings for the tower wall were delayed an hour each on April 21 because of light rain.
- Cement-truck drivers must slow down while driving through the site. Other workers complain about the excessive dust raised by the trucks.
- Mary Powell, an M-Global safety inspector on the crew, cited 12 workers for not wearing their hardhats.
- You just heard from one subcontractor, Allis Wire, Inc., that there will be a two-day delay in delivering some steel reinforcing wires that go into the concrete walls. That delay will throw off next week's schedule. Last Monday's hard rain and flooding kept everyone home that day.
- It is probably time once again to get all the subcontractors together to discuss safety at the tower site. Recently, two field hands had bad cuts from machinery.
- Although there have not been any major thefts at the site, some miscellaneous boards and masonry pieces are missing each day—probably because nearby residents (doing small home projects) think that whatever they find at the site has been discarded. Are additional "No trespassing" signs needed?
- Construction is only two days behind schedule, despite the problems that have occurred.

9. Problem Analysis

As a landscape engineer for M-Global, one of your jobs is to examine problems associated with the design of walkways, the location of trees and garden beds, the grading of land around buildings, and any other topographical features. Assume that you have been hired by a specific college, community, or company with which you are familiar. Your objective is to evaluate one or more landscaping problems at the site.

Write an informal report that describes the problem(s) in detail. (Follow the guidelines in the "Problem Analyses" section of this chapter.) Be specific about how the problem affects people—the employees, inhabitants, students, and so on. Following are some sample problems that could be evaluated:

- Poorly landscaped entrance to a major subdivision
- Muddy, unpaved walkway between dormitories and academic buildings on a college campus
- Unpaved parking lot far from main campus buildings
- Soil runoff into the streets from several steep, muddy subdivision lots that have not yet been sold
- City tennis courts with poor drainage
- Lack of adequate flowers or bushes around a new office building
- Need for a landscaped common area within a subdivision or campus
- Need to save some large trees that may be doomed because of proposed construction

10. Recommendation Report

For this document, choose a design problem at your college or company. Now put yourself in the position of an M-Global employee hired by your school or company to recommend solutions to the problem.

Your ideas must be in the form of a report that gives one or more recommendations resulting from your study. (Consult writing guidelines in the "Recommendation Reports" section of this chapter.) Assume that the problem is understood well enough to require only a brief summary, before launching into your recommendations. Because this is a short report, it may not contain many technical details for implementing your recommendations. Also, you must choose a topic that is specific enough to be covered in a short memo report. Following are some sample topics:

- Poor ventilation in an office or a classroom, such as one with sealed windows
- Inadequate space for quick exits during emergencies
- Poor visibility in a large auditorium
- Poor acoustics in a large classroom or training room
- Lack of, or improper placement of, lighting
- Energy inefficiency caused by structural flaws, such as poor insulation or high ceilings
- Rooms or walkways that are not handicapped accessible
- Failure to take advantage of solar heating
- Inefficient heating or air-conditioning systems

11. Recommendation Report

This project requires some research. Assume that your college plans either to embark on a major recycling effort or to expand a recycling program that has already started. Put yourself in the role of an M-Global environmental scientist or technician who has been asked to recommend these recycling changes.

First, do some research about recycling programs that have worked in other organizations. A good place to start is a periodical database such as *EBSCOhost* or *J-Stor*, which will lead you to some magazine articles of interest. Choose to discuss one or more recoverable resources such as paper, aluminum, cardboard, plastic, or glass bottles. Be specific about how your recommendations can be implemented by the organization or audience about which you are writing. (Consult the guidelines in the "Recommendation Reports" section of this chapter.)

12. Equipment Evaluation

For six months, you have driven a new Ford F-150 company truck at remote job sites. As lead field hand for M-Global's Boston office, you have been asked to write an evaluation of the vehicle for Brenda Seymour, director of procurement at the corporate office in Baltimore. Seymour will use your report to decide whether to recommend ordering five more F-150s for other offices. She has told you that you must discuss only major positive or negative features, not every detail. If she needs more information after reading your report, she will let you know.

Consider the following list to be your random notes. Use all this information to write a memo report that evaluates the truck. Make sure to follow the guidelines in this chapter.

- My 150 has been very reliable—it never failed to start, even during subzero ice storms last winter.
- The 4.6 liter small V-8 has provided plenty of power to handle any hauling I have done. No need to order the more expensive and less fuel-efficient 5.4 liter V-8.
- Have been to 18 job sites with the truck, from marshes in Maine to mountains in New Hampshire. Have put about 12,000 miles on it, on all kinds of roads and in all conditions.
- Tires that came with the truck did not work well in muddy locations, even with four-wheel drive. Suggest we buy all-terrain tires for future vehicles. Continue to order four-wheel drive—it is necessary at over half our job sites.
- The short bed (6 ft) did not provide enough hauling room, once I put my toolbox across the truck bed near the back window. Suggest company buy long-bed trucks with the added 2 ft of room.
- Given what I know now, I give the truck a good to excellent rating.
- Automatic transmission worked great. Am told by other owners that the automatic is better than the manual for construction jobs because the manual tends to burn out

clutches, especially when the truck needs to be "rocked" back and forth to get out of mud holes. My automatic has taken a lot of abuse without problems.
- Have had some problems with front-end handling on rough roads. Suggest that future trucks be ordered with special handling package, which includes two shock absorbers—not just one—on each front wheel.
- Have had no major repairs, just the regular maintenance checks at the dealer.
- There was one recall from the manufacturer concerning an exhaust pipe hanger that might bend, but the dealer fixed the problem in 20 minutes.
- Really need to have another six months to see how well truck holds up.

13. Equipment Evaluation

(For the purposes of this assignment, you may need to conduct research on the Internet or at your local electronics store.) M-Global, Inc., has decided to make a bulk purchase of 20 cell phones. The phones will go to a new department being set up in several months.

Assume that M-Global now uses five different types of cell phones. In the interests of a fair comparison/contrast, Brenda Seymour, director of M-Global's corporate Procurement Department, has asked you and several other employees to evaluate the effectiveness of your own phones. Write Seymour a memo report that includes your evaluation. (Consult guidelines in the "Equipment Evaluations" section of this chapter.) She will use the data and opinions in all the equipment evaluations she receives to make her choice for the bulk purchase. Your criteria for evaluation might include topics such as one or more of the following:

- Physical design of the equipment
- Ease with which system can be learned
- Quality of the written instructions
- Frequency and cost of maintenance
- Availability of appropriate features such as organizers
- Length of coverage of warranty
- Nearness to a service center
- Reputation of the manufacturer

14. Progress or Periodic Report

Assume that you have worked as a field hand at M-Global's Atlanta office for 15 years. Because of your reliability, good judgment, and intelligence, the company is paying for your enrollment at a local college. Also, you get half time off, with pay. Because of its investment in you, M-Global expects you to report periodically on your college work. Choose one of these two options for this assignment:

Progress report: Select a major project you are now completing in any college course. Following the guidelines in the

"Progress/Periodic Reports" section of this chapter, write a progress report on this project. Direct the memo report to the Atlanta office's manager of engineering, Wade Simkins. Sample topics might include a major paper, laboratory experiment, field project, or design studio.

Periodic report: Assume that M-Global requires that you submit periodic reports on your schooling every few weeks. Following the guidelines in the "Progress/Periodic Reports" section of this chapter, write a periodic report on your recent course work (completed or ongoing classes or both). Direct the memo report to the manager of engineering, Wade Simkins. Organize the report by class, and then give specific updates on each one.

15. Lab Report

For this assignment, you must be taking a lab course now or have taken such a course recently. As in assignment 14, assume you work as a field hand with M-Global's Atlanta office. The company is sponsoring your schooling and has requested that you report on a specific college lab.

Following the guidelines in this chapter's "Lab Reports" section, write a report to Wade Simkins, manager of engineering. The quality of your report may affect whether M-Global continues to fund your schooling. Be specific about the goals, procedures, and results of your laboratory—just as you would in an actual college lab report.

16. Ethics Assignment

Reread Model 8–3, a problem analysis concerning *Seeker II*, a ship leased by M-Global for its offshore jobs. Jan Stillwright receives the report just after she has signed a contract to complete a highly profitable one-week assignment for one of M-Global's best clients. This client urgently needs some geologic data from the floor of the Gulf of Mexico so that a bid

for a construction project can be submitted to the government. Jan immediately contacts other leasing operations but discovers that no other drilling ships are available when her project must begin—the day after tomorrow. At this point, Jan emails Harold Marshal that she will definitely address all concerns with the boat owner after this urgent one-week study. Moreover, she says that she will (1) give overtime pay to the crew for the upcoming one-week trip and (2) assign M-Global's top safety officer to the trip so that she can observe, record, and validate the points made in Harold's report. What are the ethical implications, if any, of Jan's response? If you were Harold, how would you respond?

17. Informal Report—International Context

Investigate features such as style, format, structure, and organization of short reports written in another country. For this assignment, it would be best to interview someone who does business in another country and, if possible, to get an actual report that you can submit. Write a memo report to your instructor that presents the results of your study.

 ## 18. A.C.T. N.O.W. Assignment (Applying Communication To Nurture Our World)

Interview a member of the campus staff responsible for adopting energy-saving measures at your college or university. Focus on one or more technical or social strategies such as alternative-energy vehicles, new technology for regulating energy systems, advanced insulation, variable work hours, and modification of human behaviors related to energy use. Then write a short report on the relative success of the strategies you have researched. If the report is well reviewed by your instructor, consider seeking wider distribution of the report by submitting it (or a version of it) to the campus—assuming you have the permission of the person interviewed.

12 Post Street
Houston Texas 77000
(713) 555-9781

April 22, 2008

Big Muddy Oil Company Inc
12 Rankin St
Abilene TX 79224

ATTENTION: Mr. James Smith, Engineering Manager

SHARK PASS STUDY
BLOCK 15, AREA 43-B
GULF OF MEXICO

Includes specific title.

INTRODUCTORY SUMMARY

Uses optional heading for abstract part of ABC format.

You recently asked our firm to complete a preliminary soils investigation at an off-shore rig site. This report presents the tentative results of our study, including major conclusions and recommendations. A longer, formal report will follow at the end of the project.

Draws attention to main point of report.

On the basis of what we have learned so far, it is our opinion that you can safely place an oil platform at the Shark Pass site. To limit the chance of a rig leg punching into the sea floor, however, we suggest you follow the recommendations in this report.

WORK AT THE PROJECT SITE

Gives on-site details of project—dates, location, tasks.

On April 15 and 16, 2008, M-Global's engineers and technicians worked at the Block 15 site in the Shark Pass region of the gulf. Using M-Global's leased drill ship, *Seeker II*, as a base of operations, our crew performed these main tasks:

- Seismic survey of the project study area
- Two soil borings of 40 feet each

Uses lead-in to subsections that follow.

Both seismic data and soil samples were brought to our Houston office the next day for laboratory analysis.

LABORATORY ANALYSIS

On April 17 and 18, our lab staff examined the soil samples, completed bearing capacity tests, and evaluated seismic data. Here are the results of that analysis.

Soil Layers

Highlights most important point about soil layer—that is the weak clay.

Our initial evaluation of the soil samples reveals a 7–9 ft layer of weak clay starting a few feet below the seafloor. Other than that layer, the composition of the soils seems fairly typical of other sites nearby.

M-Global Inc | 127 Rainbow Lane | Baltimore MD 21202 | 410.555.8175

■ **Model 8–1** ■ Recommendation report (letter format)

Bearing Capacity

We used the most reliable procedure available, the XYZ method, to determine the soil's bearing capacity (i.e., its ability to withstand the weight of a loaded oil rig). That method required that we apply the following formula:

Q = $cNv + tY$, where
Q = ultimate bearing capacity
c = average cohesive shear strength
Nv = the dimensionless bearing capacity factor
t = footing displacement
Y = weight of the soil unit

The final bearing capacity figure will be submitted in the final report, after we repeat the tests.

Notes why this method was chosen (i.e., reliability).

Seafloor Surface

By pulling our underwater seismometer back and forth across the project site, we developed a seismic "map" of the seafloor surface. That map seems typical of the flat floor expected in that area of the gulf. The only exception is the presence of what appears to be a small sunken boat. This wreck, however, is not in the immediate area of the proposed platform site.

Explains both how the mapping procedure was done and what results it produced.

CONCLUSIONS AND RECOMMENDATIONS

Based on our analysis, we conclude that there is only a slight risk of instability at the site. Although unlikely, it is possible that a rig leg could punch through the sea floor, either during or after loading. We base this opinion on (1) the existence of the weak clay layer, noted earlier; and (2) the marginal bearing capacity.

Leads off section with major conclusion, for emphasis.

Restates points (made in body) that support conclusion.

Nevertheless, we believe you can still place your platform if you follow careful rig-loading procedures. Specifically, take these precautions to reduce your risk:

1 Load the rig in 10-ton increments, waiting 1 hour between loadings.
2 Allow the rig to stand 24 hours after the loading and before placement of workers on board.
3 Have a soils specialist observe the entire loading process to assist with any emergency decisions if problems arise.

Uses list to emphasize recommendations to reduce risk.

As noted at the outset, these conclusions and recommendations are based on preliminary data and analysis. We will complete our final study in 3 weeks and submit a formal report shortly thereafter.

Again mentions tentative nature of information, to prevent misuse of report.

M-Global, Inc. enjoyed working once again for Big Muddy Oil at its Gulf of Mexico lease holdings. I will phone you this week to see if you have any questions about our study. If you need information before then, please give me a call.

Maintains contact and shows initiative by offering to call client.

Sincerely,

Bartley Hopkins

Bartley Hopkins, Project Manager
M-Global, Inc.

hg

■ **Model 8–1** ■ continued

Uses optional first heading for abstract section of ABC format. Gives background, main points, and scope statement.

Notes five main points to be covered.

Begins paragraph with most important point. Supports claim with evidence.

Uses specific example to document opinion.

Gives simple explanation of how spreadsheet works.

MEMORANDUM

DATE: July 25, 2008
TO: Hank Worley, Project Manager
FROM: Melanie Frank, Office Manager *MF*
SUBJECT: Evaluation of Best Choice Software

INTRODUCTORY SUMMARY

When the office purchased one copy of Best Choice Software last month, you suggested I send you an evaluation after 30 days' use. Having now used Best Choice for a month, I have concluded that it meets all our performance expectations. This memo presents our evaluation of the main features of Best Choice.

HOW BEST CHOICE HELPED US

Best Choice provides five primary features: word processing, file management, spreadsheet, graphics, and a user's guide. My critique of all five features is included here.

Word Processing

The system contains an excellent word-processing package that the engineers as well as the secretaries have been able to learn easily. This package can handle both our routine correspondence and the lengthy reports that our group generates. Of particular help is the system's 90,000-word dictionary, which can be updated at any time. The spelling correction feature has already saved much effort that was previously devoted to mechanical editing.

File Management

The file-manager function allows the user to enter information and then to manipulate it quickly. During one three-day site visit, for example, a field engineer recorded a series of problems observed in the field. Then she rearranged the data to highlight specific points I asked her to study, such as I-beam welds and concrete cracks.

Spreadsheet

Like the system's word-processing package, the spreadsheet is efficient and quickly learned. Because Best Choice is a multipurpose software package, spreadsheet data can be incorporated into letter or report format. In other words, spreadsheet information can be merged with our document format to create a final draft for submission to clients or supervisors, with a real savings in time. For example, the memo I sent you last week on budget projections for field equipment took me only an hour to complete; last quarter, the identical project took four hours.

M-Global Inc | 127 Rainbow Lane | Baltimore MD 21202 | 410.555.8175

■ **Model 8–2** ■ Equipment evaluation (memo format)

Hank Worley
July 25, 2008
Page 2

Graphics

The graphics package permits visuals to be drawn from the data contained in the spreadsheet. For example, a pie chart that shows the breakdown of a project budget can be created easily by merging spreadsheet data with the graphics software. With visuals becoming such an important part of reports, we have used this feature of Best Choice quite frequently.

Shows relevance of graphics to current work.

User's Guide

Eight employees in my group have now used the Best Choice user's guide. All have found it well laid out and thorough. Perhaps the best indication of this fact is that in 30 days of daily use, we have placed only three calls to the Best Choice customer-service number.

Supplies strong supporting statistic.

CONCLUSION

Best Choice seems to contain just the right combination of tools to help us do our job, both in the field and in the office. These are the system's main benefits:

Wraps up report by restating main points.

- Versatility—it has diverse functions
- Simplicity—it is easy to master

The people in our group have been very pleased with the package during this 30-day trial. If you like, we would be glad to evaluate Best Choice for a longer period.

Offers follow-up effort.

■ **Model 8–2** ■ continued

MEMORANDUM

DATE: October 15, 2008
TO: Jan Stillwright, Vice President of Research and Training
FROM: Harold Marshal, Technical Supervisor *HM*
SUBJECT: Boat Problems During Summer Season

Gives abstract (or summary) in first paragraph.

INTRODUCTORY SUMMARY

▶ We have just completed a one-month project aboard the leased ship, *Seeker II*, in the Pacific Ocean. All work went just about as planned, with very few delays caused by weather or equipment failure.

Provides capsule listing of problems discussed in report.

▶ However, there were some boat problems that need to be solved before we lease *Seeker II* again this season. This report highlights the problems so that they can be brought to the owner's attention. My comments focus on four areas of the boat: drill rig, engineering lab, main engine, and crew quarters.

DRILL RIG

Opens with most important point—then qualifies it. Explains problem in layperson's language, indicating possible consequences.

▶ Thus far, the rig has operated without incident. Yet on one occasion, I noticed that the elevator for lifting pipe up the derrick swung too close to the derrick itself. A quick gust of wind or a sudden increase in sea height caused these shifts. If the elevator were to hit the derrick, causing the elevator door to open, pipe sections might fall to the deck below.

I believe the whole rig assembly needs to be checked over by someone knowledgeable about its design. Before we put men near that rig again, we need to know that their safety would not be jeopardized by the possibility of falling pipe.

ENGINEERING LAB

Quite frankly, it is a tribute to our technicians that they were able to complete all lab tests with *Seeker II's* limited facilities. Several weeks into the voyage, these four main problems became apparent:

Uses listing to draw attention to four main lab problems on board.

1. Ceiling leaks
2. Poor water pressure in the cleanup sink
3. Leaks around the window near the electronics corner
4. Two broken outlet plugs

Although we were able to devise a solution to the window leaks, the other problems stayed with us for the entire trip.

M-Global Inc | 127 Rainbow Lane | Baltimore MD 21202 | 410.555.8175

■ **Model 8–3** ■ Problem analysis (memo format)

Jan Stillwright
October 15, 2008
Page 2

MAIN ENGINE

On this trip, we had three valve failures on three different cylinder heads. From our experience on other ships, it is very unusual to have one valve fail, let alone three. Fortunately for us, these failures occurred between projects, so we did not lose time on a job. And fortunately for the owner, the broken valve parts did not destroy the engine's expensive turbocharger.

Only an expert will be able to tell whether these engine problems were flukes or if the entire motor needs to be rebuilt. In my opinion, the most prudent course of action is to have the engine checked over carefully before the next voyage.

Uses simple language to describe technical problems.

Closes section with opinion that flows from facts presented.

CREW QUARTERS

When 15 men live in one room for three months, it is important that basic facilities work. On *Seeker II*, we experienced problems with the bedroom, bathroom, and laundry room that caused some tension.

Gives lead-in to three sections that follow.

Bedroom

Three of the top bunks had such poor springs that the occupants sank 6 to 12 in. toward the bottom bunks. More important, five of the bunks are not structurally sound enough to keep from swaying in medium to high seas. Finally, most of the locker handles are either broken or about to break.

Describes three problem areas in great detail—knowing the owner will want facts to support complaints.

Bathroom

Poor pressure in three of the commodes made them almost unusable during the last two weeks. Our amateur repairs did not solve the problem, so I think the plumbing leading to the holding tank might be defective.

LaundryRoom

We discovered early that the filtering system could not screen the large amount of rust in the old 10,000-gallon tank. Consequently, undergarments and other white clothes turned a yellow-red color and were ruined.

CONCLUSION

As noted at the outset, none of these problems kept us from accomplishing the major goals of this voyage, but they did make the trip much more uncomfortable than it had to be. Moreover, in the case of the rig and engine problems, we were fortunate that injuries and downtime did not occur.

I strongly urge that the owner be asked to correct these deficiencies before we consider using *Seeker II* for additional projects this season.

Briefly restates problem, with emphasis on safety and profits.

Ends with specific recommendation.

■ **Model 8–3** ■ continued

MEMORANDUM

DATE: June 11, 2008
TO: Jeannie McDuff, Vice President of Domestic Operations
FROM: Scott Sampson, Manager of Personnel SS
SUBJECT: Progress Report on Training Project

INTRODUCTORY SUMMARY

On May 21, you asked that I study ways our firm can improve training for technical employees in all domestic offices. We agreed that the project would take about six or seven weeks and involve three phases:

Phase 1: Make phone inquiries to competing firms
Phase 2: Send a survey to our technical people
Phase 3: Interview a cross section of our technical employees

I have now completed Phase 1 and part of Phase 2. My observation thus far is that the project will offer many new directions to consider for our technical training program.

WORK COMPLETED

In the first week of the project, I had extensive phone conversations with people at three competing firms about their training programs. Then in the second week, I wrote and sent out a training survey to all technical employees in M-Global's domestic offices.

Phone Interviews

I contacted three firms for whom we have done similar favors in the past: Simkins Consultants, Judd & Associates, and ABG Engineering. Here is a summary of my conversations:

1 Simkins Consultants
Talked with Harry Roland, training director, on May 22. Harry said that his firm has most success with internal training seminars. Each technical person completes several one- or two-day seminars every year. These courses are conducted by in-house experts or external consultants, depending on the specialty.

2 Judd & Associates
Talked with Jan Tyler, manager of engineering, on May 23. Jan said that Judd, like Simkins, depends mostly on internal seminars. But Judd spreads these seminars over one or two weeks, rather than teaching intensive courses in one or two days. Judd also offers short "technical awareness" sessions at the lunch hour every two weeks. In-house technical experts give informal presentations on some aspect of their research or fieldwork.

M-Global Inc | 127 Rainbow Lane | Baltimore MD 21202 | 410.555.8175

Notes in left margin:

Summarizes project, to refresh reader's memory and establish common ground.

Gives overview of report.

Summarizes two main tasks, as lead-in to subsections.

Organizes this section by the companies consulted.

Creates parallel form in organization of all three points.

■ **Model 8–4** ■ Progress report (memo format)

3. ABG Engineering

Talked with Newt Mosely, personnel coordinator, on May 27. According to Newt, ABG's training program is much as it was two decades ago. Most technical people at high levels go to one seminar a year, usually sponsored by professional societies or local colleges. Other technical people get little training beyond what is provided on the job. In-house training has not worked well, mainly because of schedule conflicts with engineering jobs.

Internal Survey

After completing the phone interviews noted, I began the survey phase of the project. Last week, I finished writing the survey, had it reproduced, and sent it with cover memo to all 450 technical employees in domestic offices. The deadline for returning it to me is June 17.

Gives important details about the survey.

WORK PLANNED

With phone interviews finished and the survey mailed, I foresee the following schedule for completing the project:

Organizes section chronologically, making sure to stay within a six- or seven-week schedule.

June 17:	Surveys returned
June 18–20:	Surveys evaluated
June 23–27:	Trips taken to all domestic offices to interview a cross section of technical employees
July 3:	Submission of final project report to you

CONCLUSION

My interviews with competitors gave me a good feel for what technical training might be appropriate for our staff. Now I am hoping for a high-percentage return on the internal survey. That phase will prepare a good foundation for my on-site interviews later this month. I believe this major corporate effort will upgrade our technical training considerably.

Looks to future tasks.

Emphasizes major benefit, to "sell" the project internally.

I would be glad to hear any suggestions you may have about my work on the rest of the project. For example, please call if you have any particular questions you want asked during the on-site interviews (ext. 348).

Indicates flexibility and encourages response from reader.

■ **Model 8–4** ■ continued

Begins with overview of entire report.

Gives summary of small projects.

Uses list to highlight main types.

Indicates reasons for delays.

Again, gives *reasons* for delay.

Supports section with *specifics*—for example, the exact number of meetings.

MEMORANDUM

DATE: August 1, 2008
TO: Ralph Buzby, Manager of Engineering
FROM: Nancy Fairbanks, Project Manager *NF*
SUBJECT: Activity Report for July 2008

   July has been a busy month in our group. Besides starting and finishing many smaller jobs, we completed the Jones Fill project. Also, the John Lewis Dam borings began just a week ago. Finally, I did some marketing work and several performance reviews.

SMALL PROJECTS

   Last month, my group completed nine small projects, each with a budget under $20,000 and each lasting only a few days. These jobs were in three main areas:

1   Surveying subdivisions—five jobs
2   Taking samples from toxic sites—two jobs
3   Doing nearby soil borings—two jobs

All nine were completed within budget. Eight of the nine projects were completed on time. The Campbell County survey, however, was delayed for a day because of storms on July 10.

JONES FILL PROJECT

   Our written report on this 12-month job was finally submitted to Trunk Engineering, Inc., on July 23. The delay was caused by Trunk's decision to change the scope of the project again. The firm wanted another soil boring, which we completed on July 22.

JOHN LEWIS DAM PROJECT

   As you know, we had hoped to start work at the dam site last month. However, the client decided to make many design changes that had to be approved by subcontractors. The final approval to start came just last week; thus our first day on site was July 28.

MARKETING

   During July, my main marketing effort was to meet with some previous clients, acquainting them with some of our new services. I met with eight different clients at their offices, with two meetings occurring on each of these dates: July 15, 16, 22, and 23. There's a good possibility that several of these meetings will lead to additional waste-management work in the next few months.

M-Global Inc | 127 Rainbow Lane | Baltimore MD 21202 | 410.555.8175

■ **Model 8–5** ■ Periodic report (memo format)

Ralph Buzby
August 1, 2008
Page 2

PERFORMANCE REVIEWS

As we discussed last month, I fell behind on my staff's performance reviews in June. In July, I completed the three delayed reviews, as well as the four that were due in July. Copies of the paperwork were sent to your office and to the Personnel Department on July 18. This brings us up to date on all performance reviews.

CONCLUSION

July was a busy month in almost all phases of my job. Because of this pace, I haven't had time to work on the in-house training course you asked me to develop. In fact, I'm concerned that time I devote to that project will take me away from my ongoing client jobs. At our next meeting, perhaps we should brainstorm about some solutions to this problem.

Lays foundation for *next* meeting.

■ **Model 8–5** ■ continued

105 Halsey Street
Baltimore Maryland 21212
(301) 555-7588

December 12, 2008

Mr Andrew Hawkes
Monson Coal Company
2139 Lasiter Dr
Baltimore MD 21222

LABORATORY REPORT
BOREHOLE FOSSIL SAMPLES
BRAINTREE CREEK SITE, WEST VIRGINIA

INTRODUCTORY SUMMARY

> Last week, you sent us six fossilized samples from the Braintree Creek site. Having analyzed the samples in our lab, we believe they suggest the presence of coal-bearing rock. As you requested, this report will give a summary of the materials and procedures we used in this project, along with any problems we had.

> As you know, our methodology in this kind of job is to identify microfossils in the samples, estimate the age of the rock by when the microfossils existed, and then make assumptions about whether the surrounding rock might contain coal.

LAB MATERIALS

> Our lab analysis relies on only one piece of specialized equipment: a Piketon electron microscope. Besides the Piketon, we use a simple 400-power manual microscope. Other equipment is similar to that included in any basic geology lab, such as filtering screens and burners.

LAB PROCEDURE

Once we receive a sample, we first try to identify the exact kinds of microfossils that the rocks contain. Our specific lab procedure for your samples consisted of these two steps:

Step 1

We used a 400-power microscope to visually classify the microfossils that were present. On inspection of the samples, we concluded that there were two main types of microfossils: nannoplankton and foraminifera.

M-Global Inc | 127 Rainbow Lane | Baltimore MD 21202 | 410.555.8175

Marginal notes:

Gives overview of results.

Outlines *procedure* to be detailed in following paragraph.

Describes main equipment, in layperson's language.

Breaks down procedure into easy-to-read "chunks."

■ **Model 8–6** ■ Lab report (letter format)

Andrew Hawkes
December 12, 2008
Page 2

Step 2

Next, we had to extract the microfossils from the core samples you provided. We used two different techniques:

Provides smooth transitions.

Nannoplankton Extraction Technique
a. Selected a pebble-size piece of the sample
b. Thoroughly crushed the piece under water
c. Used a dropper to remove some of the material that floats to the surface (it contains the nannoplankton)
d. Dried the nannoplankton-water combination
e. Placed the nannoplankton on a slide

Itemizes steps because of their importance in procedure.

Uses parallel form in describing this process.

Foraminifera Extraction Technique
a. Boiled a small portion of the sample
b. Used a microscreen to remove clay and other unwanted material
c. Dried remaining material (foraminifera)
d. Placed foraminifera on slide

PROBLEMS ENCOUNTERED

The entire lab procedure went as planned. The only problem was minor and occurred when we removed one of the samples from the container in which it was shipped. As the bag was taken from the shipping box, it broke open. The sample shattered when it fell onto the lab table. Fortunately, we had an extra sample from the same location.

Does not bury sampling error—gives it proper treatment.

CONCLUSION

Judging by the types of fossils present in the sample, they come from rock of an age that might contain coal. This conclusion is based on limited testing, so we suggest you test more samples at the site. We would be glad to help you with this additional sampling and testing.

Ends with wrap-up that reinforces main point of report.

I will call you this week to discuss our study and any possible follow-up you may wish us to do.

Offers follow-up services.

Sincerely,

Joseph Rappaport

Joseph Rappaport
Senior Geologist

■ **Model 8–6** ■ continued

Chapter 9 | Formal Reports

>>> Chapter Outline

When to Use Formal Reports   259

Strategy for Organizing Formal
   Reports   261

Guidelines for the Nine Parts of
   Formal Reports   262

   Cover/Title Page   263

   Letter/Memo of Transmittal   264

   Table of Contents   265

   List of Illustrations   266

   Executive Summary   267

   Introduction   268

   Discussion Sections   269

   Conclusions and
      Recommendations   270

   End Material   271

Formal Report Example   271

Chapter Summary   272

Learning Portfolio   273

   Communication Challenge   273

Collaboration at Work   274

Assignments   274

Models for Good Writing   278

   Model 9–1: Title page with
      illustration (see explanation
      on page 263)   278

   Model 9–2: Letter of transmittal   279

   Model 9–3: Memo of
      transmittal   280

   Model 9–4: Table of contents
      (all subheadings included)   281

   Model 9–5: Table of contents
      (third-level subheadings
      omitted)   282

   Model 9–6: List of illustrations—
      formal report   283

   Model 9–7: Executive summary—
      formal report   284

   Model 9–8. Introduction—formal
      report   285

   Model 9–9: Formal report   286

Toby West, M-Global's director of marketing, was given an interesting assignment 2 months ago. Jim McDuff asked him to take a long, hard look at the company's clients. Are they satisfied with the service they receive? Do they routinely reward M-Global with additional work? Are there any features of the company, its employees, or its services that frustrate them? What do they want to see changed? In other words, Toby was asked to step back from daily events and evaluate the company's level of service. He attacked the project in four stages:

1. He designed and sent out a survey to all recent and current clients.

2. He followed up on some of the returned surveys with phone and personal interviews.

3. He evaluated the data he collected.

4. He wrote a formal report on the results of his study. Besides going to all corporate and branch managers, West's report later served as a basis for some in-house training sessions called "Quality at M-Global."

Like Toby West, you will write a number of long, formal reports during your career. Most will be written collaboratively with colleagues; others will be your responsibility. All will require major efforts at planning, organizing, drafting, and revising. Although informal reports are the most common report in business writing, formal reports become a larger part of your writing as you move along in your career. This text uses the following working definition:

Formal report: **A formal report covers complex projects and is directed to readers at different technical levels. Although not defined by length, a formal report usually contains at least 6 to 10 pages of text, not including appendices. It can be directed to readers either inside or outside your organization. Often bound, it usually includes the following separate parts: (1) cover/title page, (2) letter/memo of transmittal, (3) table of contents, (4) list of illustrations, (5) executive summary, (6) introduction, (7) discussion sections, and (8) conclusions and recommendations. (Appendices often appear after the report text.)**

To prepare you to write excellent formal reports, this chapter includes three main sections. The first section briefly describes four situations that would require formal reports—two are in-house (to readers within M-Global) and two are external (to readers outside M-Global). The second section provides guidelines for writing the main parts of a long report. The third section includes a complete long report from M-Global that follows this chapter's guidelines.

>>> When to Use Formal Reports

Like most people, you probably associate formal reports with important projects. What else justifies all that time and effort? In comparison to informal reports, formal reports usually (1) cover more-complicated projects and (2) are longer than their informal counterparts.

Although complexity of subject matter and length are the main differences between formal and informal reports, sometimes there is another distinction—formal reports may have a more diverse set of readers. In this case, readers who want just a quick overview can turn to the executive summary at the beginning or the conclusions and recommendations at the end; technical readers who want to check your facts and figures can turn to discussion sections or appendices; and all readers can flip to the table of contents for a quick outline of what sections the report contains. You must consider the needs of all these readers as you plan and write your formal reports.

The intended audience for formal reports can be internal or external, although the latter is more common in formal reports. Following are four situations at M-Global for which formal reports are appropriate:

■ **Salary study and recommendations (internal):** Mary Kennelworth, a supervisor at M-Global's San Francisco office, has just completed a study of technicians' salaries among M-Global's competitors on the West Coast. What prompted the study was the problem she had hiring technicians to assist environmental engineers and geologists. Lately, some top applicants have been choosing other firms. Because the salary scales of her office are set by M-Global's corporate headquarters, she wants to give the main office some data showing that San Francisco starting salaries should be higher. Mary decides to submit a formal report, complete with data and recommendations for adjustments. Her main audience includes the branch manager and the vice president of human resources (the company's top decision maker about salaries and other personnel matters).

■ **Analysis of marketing problems (internal):** For several years, Jim Springer, engineering manager at M-Global's Houston office, has watched profits decline in onshore soils work. (In this type of work, engineers and technicians investigate the geologic and surface features of a construction site and then recommend foundation designs and construction practices.) One problem has been the "soft" construction market in parts of Texas. However, the slump in work has continued despite the recent surge in construction. In other words, some other company is getting the work. Jim and his staff have analyzed past marketing errors with a view toward developing a new strategy for gaining new clients and winning back old ones. He plans to present the problem analysis and preliminary suggestions in a formal report. The main readers are his manager, the corporate marketing manager in Baltimore, and managers at other domestic offices who have positions that correspond to his.

■ **Waste-management survey (external):** For the last several years, the city of Belton, Georgia, has noticed increased fish kills on the Channel River, which flows through the city and serves as the city's main source of drinking water. Pollution has always been fairly well monitored on the river, so city officials are puzzled by the kills. M-Global's Atlanta office was hired to analyze the problem and present its opinion about the cause. A team of chemists, environmental engineers, and field technicians has just completed a study and will present its formal report. The audience is quite diverse—from the technical experts in the city's water department to the members of a special citizens' panel representing the residents.

■ **Collapse of oil rig (external):** A 10-year-old rig in the North Sea recently collapsed during a mild storm. Several rig workers died, and several million dollars' worth of equipment was lost. Also, the accident created an oil spill that destroyed a significant amount of fish and wildlife before it was finally contained. M-Global's London office was hired to examine the cause of the collapse of this structure, which supposedly was able to withstand hurricanes. After 3 months of onsite analysis and laboratory work, M-Global's experts are ready to submit their report. It will be read by corporate managers of the firm, agencies of the Norwegian government, and members of several major wildlife organizations, and will be used as the basis for some articles in magazines and newspapers throughout the world.

As these four situations show, formal reports are among the most difficult on-the-job writing assignments you face in your career. Some you will write yourself; others you will write as a member of a team of technical and professional people. In all cases, you must (1) understand your purpose, (2) grasp the needs of your readers, and (3) design a report that responds to these needs. The guidelines in the next two sections help you meet these goals.

>>> Strategy for Organizing Formal Reports

You will encounter different report formats in your career, depending on your profession and your specific employer. Whatever format you choose, however, there is a universal approach to good organization that always applies. This approach is based on these main principles, discussed in detail in chapter 3:

Principle 1: Write different parts for different readers.

Principle 2: Place important information first.

Principle 3: Repeat key points when necessary.

These principles apply to formal reports even more than they do to short documents, for the following reasons:

1. A formal report often has a very mixed audience—from laypersons to highly technical specialists to executives.

2. The majority of readers of long reports focus on specific sections that interest them most, reading selectively each time they pick up the report.

3. Few readers have time to wade through a lot of introductory information before reaching the main point. They will get easily frustrated if you do not place important information first.

This chapter responds to these facts about readers of formal reports by following the ABC format (for Abstract, Body, Conclusion). As noted in chapter 3, the three main rules are that you should (1) start with an abstract for decision makers, (2) put supporting details in the body, and (3) use the conclusion to produce action. This simple ABC format should be evident in all formal reports, despite their complexity. The particular sections of formal reports fit within the ABC format as shown on the right:

Several features of this structure deserve special mention. First, note that the generic abstract section

ABC Format: **Formal Report**

- **ABSTRACT:**
  - Cover/title page
  - Letter or memo of transmittal
  - Table of contents
  - List of illustrations
  - Executive summary
  - Introduction
- **BODY:**
  - Discussion sections
  - [Appendices—appear after text but support Body section]
- **CONCLUSION:**
  - Conclusions
  - Recommendations

includes five different parts of the report that help give readers a capsule version of the entire report. As we discuss shortly, the executive summary is by far the most important section for providing this *big picture* of the report. Second, appendices are placed within the body part of the outline, even though sequentially they come at the end of the report. The reason for this outline placement is that both appendices and body sections provide supporting details for the report. Third, remember that the generic conclusion section in the ABC format can contain conclusions or recommendations or both, depending on the nature of the report.

Before moving to a discussion of the specific sections that make up the ABC format, take note of the use of main headings in complex formal reports. (See Figure 4–6, p. 109.) Much like chapter titles, these headings are often centered, in full caps, in bold type, and larger in font size than the rest of the text. They usually also begin on a new page. This way, each major section of the formal report seems to exist on its own. Then you have three remaining heading levels for use within each section.

>>> Guidelines for the Nine Parts of Formal Reports

The nine parts of the formal report are as follows:

1. Cover/title page
2. Letter or memo of transmittal
3. Table of contents
4. List of illustrations
5. Executive summary
6. Introduction
7. Discussion sections
8. Conclusions and recommendations
9. End material

Because formal reports may be longer and more complex than other forms of technical communication, it is important to help your readers navigate through the report. You may be used to thinking of navigation devices in web pages and electronic documents, but they are also important for long print documents. For example, consider what information you can include in the header and footer of your report to help your reader find appropriate sections quickly. Your organization may require standardized information in headers and footers, such as the company name, the date of the report, or an identifying code. In very long reports, it may be useful to include section headings in the header, in the same way that this textbook includes chapter number and chapter title information in its headers. Obviously, including pagination in the header or footer of your report makes it much easier to find information. Many styles of pagination abound. Following are some guidelines for one commonly used pattern that is acceptable unless you have been instructed to use another:

- Use lowercase roman numerals for some or all of the front matter that precedes—and includes—the table of contents.
- Use arabic numbers for items that follow the table of contents (all of which are listed in the table of contents).
- Continue the arabic numbering for appendices if they are relatively short. Long sets of appendices sometimes have their own internal numbering (A–1, A–2, A–3 . . .; B–1, B–2, B–3 . . .)

See Model 9–9 (pp. 286–301) for an example of this approach to pagination and for an example of guidelines that follow.

Dividers, colors on the edges of pages, or tabbed sheets are also good ways to help readers find the report sections that they are interested in. Consider starting each section with a tabbed sheet so that the reader can "thumb" to it easily.

Cover/Title Page

Formal reports are usually bound, often with a cover used for all reports in the writer's organization. (Reports prepared for college courses, however, are often placed in a simple report cover.) Because the cover is the first item seen by the reader, it should be attractive and informative. It usually contains the same four pieces of information mentioned in the following list with regard to the title page; sometimes it may have only one or two of these items.

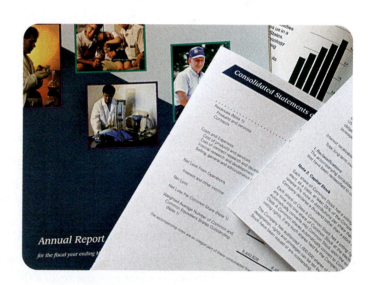

Inside the cover is the title page, which should include the following four pieces of information:

- Project title (exactly as it appears on the letter/memo of transmittal)
- Your client's name ("Prepared for . . .")
- Your name and/or the name of your organization ("Prepared by . . .")
- Date of submission

To make your title page or cover distinctive, you might want to place a simple illustration on it; however, do not clutter the page. Use a visual only if it reinforces a main point and if it can be done simply and tastefully. For example, assume that M-Global, Inc., submitted a formal report to a coastal city in California, concluding that an industrial park can be built near the city's bird sanctuary without harming the habitat—if stringent guidelines are followed. The report writer decides to place the picture of a bird on the title page, punctuating the report's point about the industrial park, as in Model 9–1 on page 278.

Letter/Memo of Transmittal

Letters or memos of transmittal are like an appetizer—they give the readers a taste of what is ahead. If your formal report is to readers outside your own organization, write a letter of transmittal. If it is to readers inside your organization, write a memo of transmittal. Models 9–2 and 9–3 on pages 279–280 show examples of both. Use the following guidelines for constructing this part of your report:

>> Transmittal Guideline 1: Place the Letter/Memo Immediately after the Title Page

This placement means that the letter/memo is bound with the document, to keep it from becoming separated. Some organizations paperclip this letter or memo to the front of the report, making it a cover letter or memo. In so doing, however, they risk having it become separated from the report.

>> Transmittal Guideline 2: Include a Major Point from Report

Remember that readers are heavily influenced by what they read first in reports. Therefore, take advantage of the position of this section by including a major finding, conclusion, or recommendation from the report—besides supplying necessary transmittal information.

>> Transmittal Guideline 3: Acknowledge Those Who Helped You

Recognizing those who have been particularly helpful with your project gives them recognition and identifies you as a team player. It reflects well on you and on your organization. Model 9–2 includes a thank you to members of the client's organization.

>> Transmittal Guideline 4: Follow Letter and Memo Conventions

Like other letters and memos, letters and memos of transmittal should be easy to read, inviting readers into the rest of the report. Keep introductory and concluding paragraphs relatively short—no more than three to five lines each. Also, write in a conversational style, free of technical jargon and stuffy phrases such as "per your request" or "enclosed herewith." See the models at the end of chapter 5 for more details concerning letter/memo format. For now, here are some highlights about the mechanics of format:

Letters and Memos
- Use single spacing and ragged-right-edge copy, even if the rest of the report is double spaced and fully justified.
- Use only one page.

Letters
- Include company project number with the letter date.
- Spell the reader's name correctly.

- Be sure the inside address includes the mailing address to appear on the envelope.
- Use the reader's last name ("Dear Mr. Jamison:") in the salutation or attention line because of the formality of the report—unless your close association with the reader would make it more appropriate to use first names ("Dear Bill:").
- Usually include a project title, as with letter reports. It is treated like a main heading. Use concise wording that matches wording on the title page.
- Use "Sincerely" as your closing.
- Include a line to indicate those who will receive copies of the report ("cc" for carbon copy, "pc" for photocopy, or just "c" or "copy" for copy).

Memos

- Give a clear description of the project in the subject line of the memo, including a project number if there is one.
- Include a distribution list to indicate those who will receive copies.

Table of Contents

Your contents page acts as an outline. Many readers go there right away to grasp the structure of the report, and then return repeatedly to locate report sections of most interest to them. Most word-processing programs allow you to generate a table of contents automatically from tagged headings, but tables of contents generated this way must often be edited, especially if you have decided to leave out lower-level headings (see guideline 3). Guidelines follow for assembling this important component of your report; see Model 9–4 on page 281 for an example.

>> Table of Contents Guideline 1: Make It Very Readable

The table of contents must be pleasing to the eye so that readers can find sections quickly and see their relationship to each other. Be sure to

■ Space items well on the page

■ Use indenting to draw attention to subheadings

■ Include page numbers for every heading and subheading, unless there are many headings in a relatively short report, in which case you can delete page numbers for all of the lowest level headings listed in the table of contents

>> Table of Contents Guideline 2: Use the Contents Page to Reveal Report Emphases

Choose the wording of headings and subheadings with care. Be specific yet concise so that each heading listed in the table of contents gives the reader a good indication of what the section contains.

Readers associate the importance of report sections with the number of headings and subheadings listed in the table of contents. If, for example, a discussion section called "Description of the Problem" contains many more heading breakdowns than other sections, you are telling the reader that the section is more important. When possible, it is best to have about the same number of breakdowns for report sections of about the same importance. In short, the table of contents should be balanced.

>> Table of Contents Guideline 3: Consider Leaving Out Low-Level Headings

In very long reports, you may want to declutter the table of contents by removing lower-level headings. As always, the needs of the readers are the most important criterion to use in making this decision. If you think readers need access to all levels of headings on the contents page, keep them there. If you think they would prefer a simple contents page instead of a comprehensive one, delete all the lowest-level headings from the table of contents (see Model 9–5 on page 282 compared to Model 9–4 on page 281).

>> Table of Contents Guideline 4: List Appendices

Appendices include items such as tables of data or descriptions of procedures that are inserted at the end of the report. Typically, they are listed at the end of the table of contents. They may be paged with arabic numerals, like the rest of the report. However, sometimes no page numbers are given in the table of contents, because many appendices contain *off-the-shelf* material such as resumes or project sheets and are thus individually paged (e.g., Appendix A might be paged A–1, A–2, A–3, etc.). Tabs on the edges of pages help the reader locate these sections.

>> Table of Contents Guideline 5: Use Parallel Form in All Entries

All headings in one section, and sometimes even all headings and subheadings in the report, have parallel grammatical form. Readers find mixed forms distracting. For example, "Subgrade Preparation" and "Fill Placement" are parallel, in that they are both the same type of phrase. However, if you switch the wording of the first item to "Preparing the Subgrade" or "How to Prepare the Subgrade," parallel structure is lost.

>> Table of Contents Guideline 6: Proofread Carefully

The table of contents is one of the last report sections to be assembled; thus it often contains errors. If you have used your word processor to generate a table of contents automatically, you may need to generate it again, after you have completed your final version of the report. Wrong page numbers and incorrect heading wording are two common mistakes. Another is the failure to show the correct relationship of headings and subheadings. Obviously, errors in the table of contents can confuse the reader and prove embarrassing to the writer. Proofread this section carefully.

List of Illustrations

Illustrations within the body of the report are usually listed on a separate page right after the table of contents. When there are few illustrations, another option is to list them at the bottom of the table of contents page rather than on a separate page. In either case, this list should include the number, title, and page number of every table and figure within the body of the report. If there are many illustrations, separate the list into tables and figures.

See the examples in Model 9–6 on page 283 and in Model 9–9 on pages 286–301. (For more information on illustrations, see chapter 12.)

Executive Summary

No formal report would be complete without an executive summary. This short section provides decision makers with a capsule version of the report. Consider it a stand-alone section that should be free of technical jargon. In some cases, a copy of the executive summary may be circulated and filed separate from the report (see, e.g., Model 9–7 on page 284). Follow these basic guidelines in preparing this important section of your formal reports:

>> Executive Summary Guideline 1: Put It on One Page

The best reason to hold the summary to one page is that most readers expect and prefer this length. It is a comfort to know that somewhere within a long report there is one page to which one can turn for an easy-to-read overview. Moreover, a one-page length permits easy distribution at meetings. When the executive summary begins to crowd your page, it is acceptable to switch to single spacing if such a change helps keep the summary on one page—even though the rest of the report may be space-and-a-half or double-spaced.

Some extremely long formal reports may require that you write an executive summary of several pages or longer. In this case, you must still provide the reader with a section that summarizes the report in less than a page. The answer to this dilemma is to write a brief *abstract,* a condensed version of the executive summary directed to the highest-level decision makers, that is placed right before the executive summary. (See chapter 15 for further discussion of abstracts.)

>> Executive Summary Guideline 2: Avoid Technical Jargon

Include only that level of technical language the decision makers comprehend. It makes no sense to talk over the heads of the most important readers.

>> Executive Summary Guideline 3: Include Only the Important Conclusions and Recommendations

The executive summary mentions only the major points of the report. An exhaustive list of findings, conclusions, and recommendations can come later at the end of the report. If you have trouble deciding what is most important, put yourself in the position of the readers. What information is most essential for them? If you want to leave them with one, two, or three points about the report, what would these points be? That is the information that belongs in the executive summary.

>> Executive Summary Guideline 4: Avoid References to the Report Body

Avoid the tendency to say that the report provides additional information. It is understood that the executive summary is only a generalized account of the report's contents. References to later sections do not provide the busy reader with further understanding.

An exception is those instances when you are discussing issues that involve danger or liability in which it may be necessary to add qualifiers in your summary—for example, "As noted in this report, further study will be necessary." Such statements protect you and the client in the event the executive summary is removed from the report and used as a separate stand-alone document.

>> Executive Summary Guideline 5: Use Paragraph Format

Whereas lists are often appropriate for body sections of a report, they can give executive summaries a fragmented effect. Instead, the best summaries create unity with a series of relatively short paragraphs that flow together well. Within a paragraph, there can be a short listing of a few points for emphasis (see Model 9–7 on page 284), but the listing should not be the main structural element of the summary.

Occasionally, you may be convinced that the paragraph approach is not desirable. For example, a project may involve a series of isolated topics that do not mesh into unified paragraphs. In this case, use a modified list. Start the summary with a brief introductory paragraph, followed by a numbered list of three to nine points. Each numbered point should include a brief explanation. For example,

1. Sewer Construction: We believe that seepage influx can be controlled by . . .
2. Geologic Fault Evaluation: We found no evidence of surficial . . .

>> Executive Summary Guideline 6: Write the Executive Summary Last

Only after finishing the report do you have the perspective to write a summary. Approach the task in a logical manner. First, sit back and review the report from beginning to end, and then ask yourself, "What would my readers really need to know if they had only a minute or two to read?" The answer to that question becomes the core of your executive summary.

Introduction

View this section as your chance to prepare both technical and non-technical readers for the discussion ahead. You do not need to summarize the report, because your executive summary has accomplished that goal. Instead, give information on the report's purpose, scope, and format, as well as a project description. Follow these basic guidelines, as reflected in Model 9–8 on page 285.

>> Introduction Guideline 1: State Your Purpose and Lead into Subsections

The purpose statement for the document should appear immediately after the main introduction heading (e.g., "This report presents M-Global's foundation design recommendations for the new Hilltop Building in Franklin, Maine"). Follow it with a sentence that mentions the introduction subdivisions to follow (e.g., "This introduction provides a description of the project site and explains the scope of activities we conducted").

>> Introduction Guideline 2: Include a Project Description

Here you must be precise about the project. Depending on the type of project, you may be describing a physical setting, a set of problems that prompted the report study, or some other data. The information may have been provided to you, or you may have collected it yourself. Accuracy in this section helps prevent any later misunderstandings between you and the reader. (When the project description is too long for the introduction, sometimes it is placed in the body of the report.)

>> Introduction Guideline 3: Include Scope Information

This section outlines the precise objectives of the study. Include all necessary details, using bulleted or numbered lists when appropriate. Your listing or description should parallel the order of the information presented in the body of the report. Like the project description, this subsection must be accurate in every detail. Careful and thorough writing here can prevent later misunderstandings about the tasks you were hired to perform.

>> Introduction Guideline 4: Consider Including Information on Report Format

Often, the scope section lists information as it is presented in the report. If this is not the case, end the introduction with a short subsection on the report format where you can give readers a brief preview of the main sections that follow. In effect, the section acts as a condensed table of contents and may list the report's major sections and appendices.

Discussion Sections

Discussion sections make up the longest part of formal reports. In general, they are written for the most technically oriented members of your audience. You can focus on facts and opinions, demonstrating the technical expertise that the reader expects from you. General guidelines for writing the report discussion are listed next. For a complete example of the discussion component, see the formal report example in Model 9–9 on pages 286–301.

>> Discussion Guideline 1: Move from Facts to Opinions

As you have learned, the ABC format requires that you start your formal report with a summary of the most important information—that is, you skip right to essential conclusions and recommendations the reader needs and wants to know. Once into the discussion section, however, you back up and adopt a strategy that parallels the stages of the technical project itself. You begin with hard data and move toward conclusions and recommendations (i.e., those parts that involve more opinion).

One way to view the discussion is that it should follow the order of a typical technical project, which usually involves the following stages:

First, collect data (e.g., samples, interviews, records).

Second, subject these data to verification or testing (e.g., lab tests, computer analyses).

Third, analyze all the information, using your professional experience and skills to form conclusions (or convictions based on the data).

Fourth, develop recommendations that flow directly from the conclusions you have formed.

Thus, the body of your report gives technical readers the same movement from fact toward opinion that you experience during the project itself. There are two reasons for this approach, one ethical and the other practical. First, as a professional, you are obligated to draw clear distinctions between what you observe and what you conclude or recommend. Second, reports are usually more persuasive if you give readers the chance to draw conclusions for themselves. If you move carefully through the four-stage process just described, readers are more likely to reach the same conclusions that you have drawn.

>> Discussion Guideline 2: Use Frequent Headings and Subheadings

Headings give readers handles by which to grasp the content of your report. They are especially needed in the report body, which presents technical details. Your readers view headings, collectively, as a sort of outline by which they can make their way easily through the report.

>> Discussion Guideline 3: Use Listings to Break Up Long Paragraphs

Long paragraphs full of technical details irritate readers. Use paragraphs for brief explanations, not for descriptions of processes or other details that could be listed.

>> Discussion Guideline 4: Use Illustrations for Clarification and Persuasion

A simple table or figure can sometimes be just the right complement to a technical discussion in the text. Incorporate illustrations into the report body to make technical information accessible and easier to digest.

>> Discussion Guideline 5: Place Excessive Detail in Appendices

Today's trend is to place cumbersome detail in appendices that are attached to formal reports, rather than weighing down the discussion with this detail. In other words, you give readers access to supporting information without cluttering up the text of the formal report. Of course, you must refer to appendices in the body of the report and label appendices clearly so that readers can locate them easily.

Conclusions and Recommendations

This section of the report gives readers a place to turn to for a comprehensive description—sometimes in the form of a listing—of all conclusions and recommendations. The points may or may not have been mentioned in the body of the report, depending on the length and complexity of the document. *Conclusions,* on the one hand, are convictions or beliefs based on the findings of your study. *Recommendations,* on the other hand, are actions you are suggesting based on your conclusions. For example, your conclusion may be that there is a dangerous level of toxic chemicals in a town's water supply, and your recommendation may be that the toxic site near the reservoir should be cleaned immediately.

What distinguishes this final section of the report text from the "Executive Summary" is the level of detail and the audience. The "Conclusions and Recommendations" section provides an exhaustive list of conclusions and recommendations for technical and management readers. The "Executive Summary" provides a selected list or description of the most important conclusions and recommendations for decision makers, who may not have technical knowledge.

In other words, view the "Conclusions and Recommendations" section as an expanded version of the "Executive Summary." It usually assumes one of these three headings, depending, of course, on the content:

1. Conclusions

2. Recommendations

3. Conclusions and Recommendations

Another option for reports that contain many conclusions and recommendations is to separate this last section into two sections: (1) "Conclusions" and (2) "Recommendations."

End Material

One kind of end material—appendices—is mentioned in the context of the discussion section. Note that formal reports may also contain works-cited pages or bibliographies, which should be included in the end materials. See chapter 14 for guidelines on formats. Finally, very long reports may include indexes.

>>> Formal Report Example

Model 9–9 (pp. 286–301) provides a long and formal technical report from M-Global, Inc. It contains the main sections discussed previously, including the list of illustrations. Marginal annotations indicate how the model reflects proper use of this chapter's guidelines for format and organization.

The report results from a study that M-Global completed for the city of Winslow, Georgia. Members of the audience come from both technical and non-technical backgrounds. Some are full-time professionals hired by the city, whereas others are part-time, unpaid citizens appointed by the mayor to explore environmental problems. The paid professionals include engineers, environmental specialists, accountants, city planners, managers, lawyers, real estate experts, and public relations specialists. The part-time appointees include citizens who work in a variety of blue-collar and white-collar professions or who are homemakers.

>>> Chapter Summary

In your career, you will write formal reports for large and complex projects, either inside or outside your organization. In both cases, you will send the report to people with different technical backgrounds. This complex audience will respond best to reports that subscribe to the ABC format, because it organizes information so that different readers can read different sections of the report. Although long report formats vary according to company and profession, most have these nine basic parts: (1) cover/title page; (2) letter/memo of transmittal; (3) table of contents; (4) list of illustrations; (5) executive summary; (6) introduction; (7) discussion sections; (8) conclusions and recommendations; and (9) end material. Follow the specific guidelines in this chapter for these sections. The annotated model can serve as your reference.

>>> Learning Portfolio

Communication Challenge   "The Ethics of Clients Reviewing Report Drafts"

Last week, Hank Wallace of M-Global, Inc.'s Kenya office completed the draft of an ocean exploration project for the Republic of Cameroon (see the second project sheet at the end of chapter 2). As is routine with major reports, Hank showed the client a draft before the final draft was submitted. For the first time in his career, he was asked by the client to make changes he thinks are difficult to justify by project data. What follows is an explanation of why M-Global shares report drafts with some clients, some background on the ocean exploration project, questions and comments for discussion, and an assignment for a written response to the Challenge.

Sharing Drafts with Clients

In M-Global's business, some of the firm's reports must be submitted both to the paying client and to regulatory agencies of the government. This dual audience has created a review procedure common in the industry. Client firms have an opportunity to review a draft and make suggestions before both they and the regulatory agencies are sent final drafts.

For example, a U.S. mining company hired M-Global to examine a Siberian site to determine if gold reserves could be mined without damaging the delicate permafrost surface of the tundra. Because the Russian government regulates development of the region, it received a final copy of the report. However, before the final copy was submitted to the government, M-Global shared a draft with engineers and executives from the mining company. These client representatives questioned several technical assumptions M-Global made about the site, but M-Global had adequate justifications for its work. In the end, M-Global made no change in its original draft recommendation—that is, that further study was needed before mining is permitted in the permafrost region.

In another case, however, a client's review of a report on a dam in the Midwestern United States prompted M-Global to adjust its report before submission of the final draft to the state's Department of Natural Resources, which regulates high-hazard dams. The owners of the dam—who paid for the study—convinced M-Global that the report should emphasize the fact that poor installation of a guardrail over the dam created a drainage problem. When heavy rains came, soil washed out an embankment near the dam's spillway. The first draft had failed to mention that the state's transportation group bore some responsibility for the dam's problems.

The Ocean Exploration Report Review

The draft review of the ocean exploration report for the Republic of Cameroon did not go as smoothly as the two reviews just described. Major differences of opinion were evident between the M-Global project manager and the client, Worldwide Energy, Inc.

As indicated on the project sheet at the end of Chapter 2, M-Global engineers developed conclusions and recommendations for the Cameroon coastal site. For the most part, they found that the offshore environment where they did the study would be too environmentally sensitive to drill offshore wells or run pipelines. There were two locations where a pipeline might be placed safely, but even in this case, some environmental damage was likely. When Worldwide Energy got the draft report, the company asked for a meeting.

At the meeting the following week, M-Global engineers reviewed their findings, conclusions, and recommendations with the client. Ultimately, M-Global managers were asked to change the wording in the report to present a more favorable view of oil exploration at the site because, in the client's opinion, M-Global was being too conservative in its conclusions. If M-Global would just adjust some wording so as not to emphasize what is, after all, only possible environmental damage, then the Cameroon government might be provided the support it needs to develop this potentially rich oil field. Cameroon, the client argued, needs oil revenues to improve its economy and assist poor farmers with the transition to a modern economy. M-Global is not being asked to alter the facts—only to adjust the tone of the language.

Back at the office, the M-Global project manager met with the branch manager and later with corporate staff via teleconference. The project manager presented the facts of the project and a summary of the meeting. To all present, it was clear that the relationship with a long-term client was at stake.

Questions and Comments for Discussion

1. How should M-Global, Inc., respond to the client's request?
2. Generally, do you think M-Global's procedure for reviewing report drafts with clients is ethically sound? Support your answer.

3. If you answered "yes" to question 2, do you have any suggestions to improve the procedure for this client report review? In other words, how might the process be adjusted to reduce the potential for misunderstandings and abuse?

4. If you answered "no" to question 2, is there any circumstance in which you would support the review of a report draft by a client before final submission to the client and its regulatory agency?

5. It is often said, in this text and elsewhere, that collaboration is essential in the workplace. Describe the kinds of on-the-job situations wherein you think collaboration between writer and reader would be useful, appropriate, and ethical.

Write About It

Assume the role of the project manager in this Communication Challenge. In preparing for your teleconference meeting, you have been thinking about whether M-Global's procedure for reviewing report drafts with clients is ethical (see question 2). Write a memo to Erik Schell, Vice President of International Operations, explaining your opinion. Respond to the issues raised in question 3 or question 4.

Collaboration at Work Critiquing an Annual Report

General Instructions

Each Collaboration at Work exercise applies strategies for working in teams to chapter topics. The exercise assumes you (1) have been divided into teams of about three to six students, (2) will use team time inside or outside of class to complete the case, and (3) will produce an oral or written response. For guidelines about writing in teams, refer to pages 24–28.

Background for Assignment

Many companies prepare annual reports that summarize activities during the previous year. Readers include stockholders, employees, and investment groups, among others. Besides providing useful data about the company, annual reports often provide good examples of the following:

- How complex and diverse information can be organized
- How words can be selected to accentuate the positive, even if negative information is being reported
- How page design can be used to create an engaging, readable document

Annual reports are sent to stockholders; however, they also may be available in libraries and on the Internet. For example, at http://www.pearson.com you will find the annual report for Pearson Education, the parent company of Prentice Hall, the publisher of this textbook.

Team Assignment

You may complete this assignment with a hard-copy annual report provided by your instructor or secured from your library, or you can complete it by locating an annual report on the World Wide Web, such as the Pearson Education annual report already mentioned. Try to find a report from an organization similar to one where you might work after graduation. Meet with your team to review the print or online version of the report. Develop responses to the following questions:

1. How is information organized? (Give examples.)
2. Does the report follow the ABC format? (Explain.)
3. Who is the audience for the report? (Support your conclusion.)

Assignments

Part 1 assignments ask you to evaluate a whole report, write an individual section, or evaluate an individual section. Part 2 assignments ask you to write complete formal reports. Remember to submit Planning Forms with the Part 2 assignments.

Part 1: Short Assignments

These assignments can be completed by individuals or by teams. If you are instructed to use teams, first review guidelines in chapter 1 on team writing.

1. Evaluation—A Formal Report

Use Model 9–9 on pages 286–301, the complete formal report example in this chapter, for this assignment. The audience for the report is described in the chapter section entitled "Formal Report Example." Although the writer directed the report to a mixed technical and non-technical audience, some sections clearly are more technical than others.

- Evaluate the likely audience for each section of the report.

- Discuss ways that the writer did or did not address the needs of specific audience types.
- Offer suggestions for improving the manner in which the report meets the needs of its intended audience.

2. Evaluation—A Formal Report

Locate a formal report written by a private firm or government agency, or use a long report provided by your instructor. Determine the degree to which the example follows the guidelines in this chapter. Depending on the instructions given by your teacher, choose between the following options:

- Present your findings orally or in writing.
- Select part of the report or all of the report.

3. Executive Summary

Choose one of the seven project sheets included at the end of chapter 2. Write a brief executive summary for the project. If necessary, provide additional information or transitional wording not included on the sheet, but do not change the nature of the information already provided.

4. Evaluation—An Introduction

Review the chapter guidelines for writing an effective introduction to a formal report. Then evaluate the degree to which the following example follows or does not follow the guidelines presented.

INTRODUCTION

M-Global, Inc., has completed a 3-week study of the manufacturing and servicing processes at King Radio Company. As requested, we have developed a blueprint for ways in which computer-aided testing (CAT) can be used to improve the company's productivity and quality.

Project Description

Mr. Dan Mahoney familiarized our project team with the problems that prompted this study of CAT. According to Mr. Mahoney, the main areas of concern are as follows:

- Too many units on the production line are failing post-production testing and thus returning to the repair line.
- Production bottlenecks are occurring throughout the plant because of the testing difficulties.
- Technicians in the servicing center are having trouble repairing faulty units because of their complexity.
- Customers' complaints have been increasing, both for new units under warranty and repaired units.

Scope

From May 3 through May 5, 2008, M-Global, Inc., had a three-person team of experts working at the King Radio Company plant. This team interviewed many personnel, observed all the production processes, and acquired data needed to develop recommendations. On returning to the M-Global office, team members met to share their observations and develop the master plan included in this report.

Report Format

This report is organized primarily around the two ways that CAT can improve operation at the King Radio Company plant. Based on the detailed examination of the plant's problems in this regard, the report covers two areas for improvement and ends with a section that lists main conclusions and recommendations. The main report sections are as follows:

- Production and Servicing Problems at King Radio
- CAT and the Manufacturing Process
- CAT and the Servicing Process
- Major Conclusions and Recommendations

The report ends with two appendices. Appendix A offers detailed information on several pieces of equipment we recommend that you purchase. Appendix B provides three recent articles from the journal *CAT Today*, all of which deal with the application of CAT to production and service problems similar to those you are experiencing.

Part 2: Longer Assignments

This section contains assignments for writing entire formal reports. Remember to complete the Planning Form for each assignment.

These assignments can be written by individual writers or by writing teams. If your instructor has made this a team assignment, review the guidelines on team writing in chapter 1.

5. Research-Based Formal Report

Complete the following procedure for writing a research-based report:

- Use library and Internet resources to research a general topic in a field that interests you. Do some preliminary reading to screen possible specific topics.
- Choose three to five specific topics that require further research and for which you can locate information.
- Work with your instructor to select the one topic that best fits this assignment, given your interests and the criteria set forth here.
- Develop a simulated context for the report topic, whereby you select a purpose for the report, a specific audience to whom it could be addressed (as if it were a real report), and a specific role for you as a writer.

For example, assume you have selected "Earth-Sheltered Homes" as your topic. You might be writing a report to the manager of a local design firm on the features and construction techniques of such structures. As a newly hired engineer or designer, you are presenting information so that your manager can decide whether the firm might want to begin building and marketing such homes. This report might present only data, or it could present data and recommendations.

- Write the report according to the format guidelines in this chapter and in consideration of the specific context you have chosen.
- Document your sources appropriately (see chapter 15).

6. Work-Based Formal Report

This assignment is based on the work experience that you may have had in the past or that you may be experiencing now.

- Choose 5 or 10 report topics that are based on your current or past work experience. For example, you could choose "Warehouse Design" if you stock parts, "Check-Out Procedure" if you work behind the counter at any retail store, "Report-Production Procedures" if you work as a secretary at an engineering firm, and so on. In other words, find a subject that you know about, or about which you can find more information, especially through interviews.
- Work with your instructor to select the one topic that holds out the best possibilities for a successful report on the basis of the criteria given here.
- Develop a context for the report in which you give yourself a role in the company where you work(ed). This role should be one in which you would actually write a formal in-house or external report about the topic you have chosen, but the role does not have to be the exact

one you had or have. Then select a precise purpose for which you might be writing the report, and finally a set of readers that might read such a report within or outside the organization. Your report can be a presentation of data and conclusions or a presentation of data, conclusions, and recommendations.

- Follow the guidelines included in this chapter for format and organization.

7. School-Based Formal Report

This assignment can be completed as an individual project or as a team project. As an individual project, it relies on observations you have made during the time you spent at a high school, college, or university—either the one where you are taking this course or another you attended previously. As a team project, it relies on either (1) team members from diverse majors using their varied backgrounds to examine a common campus problem, or (2) team members majoring in the same field or working in the same department exploring a problem they have in common.

Whether you write an individual or a team report, follow this general procedure:

- Assemble a list of 5 or 10 problems that you have observed at your school. These problems might concern (1) the physical campus (e.g., poor design of parking lots, inadequate lab space), (2) the curriculum (e.g., the need to update certain courses), (3) extracurricular activities (e.g., the need for more cultural or athletic events), or (4) difficulties with campus support services (e.g., red tape during registration).
- Work with your instructor to choose the one topic for which you can find the most information and for which you can develop the context described here.
- Collect information in whatever ways seem useful—for example, site observations, surveys to students, follow-up phone calls, or interviews.
- Submit progress reports at intervals requested by your instructor. (Consult guidelines in the "Progress/Periodic Reports" section of chapter 8.)
- Consider your role to be the one that you, in fact, have—a student or a group of students at the school. Then, select as your reader(s) the school officials who are actually in charge of solving the problem you identified. (You may or may not end up sending the report. Follow the advice of your instructor in this matter.) The purpose of this report is to explain, in great detail, all aspects of the problem and to form conclusions as to its cause. If it seems appropriate, you may take one further step to suggest recommendations for a solution—if your research has taken you this far. In any case, detail as well as tact are important criteria.

8. M-Global-Based Formal Report

For this assignment, place yourself in a role of your choosing at M-Global, Inc. Use the following procedure, which may be modified by your instructor:

- Review the section at the beginning of this chapter that lists M-Global cases for formal reports to get a sense of when formal reports are used at companies like M-Global.
- Review the M-Global information in chapter 2, especially with regard to the kinds of jobs people hold at the company and the kinds of projects that are undertaken.
- Choose a specific job that you could assume at M-Global, based on your academic background, your work experience, or your career interests.
- Choose a specific project that (a) could conceivably be completed at M-Global by someone in the role you have chosen, (b) would result in a formal report directed either inside or outside the company, and (c) would be addressed to a complex audience at two or three of the levels indicated on the Planning Form at the end of the book.
- Be sure you have access to information that will be used in this simulated report—for example, from work experience, from a term paper or class project in another course, or from your interviews of individuals already in the field. (For this assignment, you may want to talk with a professional, such as a recent graduate in your major.)
- Prepare a copy of the Planning Form at the end of the book for your instructor's approval before proceeding further with the project.
- Complete the formal report, following the guidelines in this chapter.

9. Ethics Assignment

Illustrations on cover pages of formal reports are one strategy for attracting the readers' attention to the document. Note the use of an illustration in Model 9–1, along with the rationale on page 263. Do you think Model 9–1 uses its graphics in an ethically sound way to engage the reader with the report? Why or why not? How do you determine whether a cover page illustration is an appropriate persuasive tool on the one hand, or an inappropriate attempt to manipulate the reader on the other? Give hypothetical examples, or find examples from reports available on the Internet.

10. International Communication Assignment

This assignment requires that you gain information about writing long technical reports designed for readers outside the United States (or outside the country where you are taking this course, if it is not the United States). The suggestions you develop can relate to either (1) reports written in English that will be read in English or (2) reports written in English that will be translated into another language.

Specifically, write a report that provides a wide range of recommendations for writing long technical documents to a specific international audience. Cover as many writing-related issues as possible—organization, format, page design, and style.

To gather information for this assignment, find someone who works for an international firm, deals with international clients, or has in some other way acquired information about the needs of international readers of technical documents. Possible sources include (1) your institution's alumni office, which may be able to provide names of graduates or employers of graduates; (2) friends or colleagues; (3) individuals contacted through websites of international organizations; and (4) local chambers of commerce and other organizations that promote international trade.

11. A.C.T. N.O.W. Assignment (Applying Communication To Nurture Our World)

This assignment is intended to be completed as a team project. Write a formal report in which you describe the efforts that have been made at an educational institution, a business, or a local government to promote diversity within the organization. Because you and your team members may not be experts in this subject, your report should be presented only with findings and conclusions—not with recommendations. In other words, your purpose is to describe strategies that have been used and—based on information gathered during your investigation—indicate their relative degree of success. On a campus, for example, you might cover topics such as enrollment issues and the resulting diversity among the student body, hiring issues and the resulting diversity among the faculty and staff, training and professional development of employees, and general atmosphere on campus.

Oceanside's New Industrial Park

Prepared for: City Council
                Oceanside, California

Prepared by: M-Global, Inc.
               San Francisco, California

Date: March 3, 2008

■ **Model 9–1** ■ Title page with illustration (see explanation on page 263)

12 Post Street
Houston Texas 77000
(713) 555-9781

Report #82-651
July 18, 2008

Belton Oil Corporation
PO Box 301
Huff Texas 77704

Attention: Mr. Paul A. Jones

GEOTECHNICAL INVESTIGATION
DREDGE DISPOSAL AREA F
BELTON OIL REFINERY
HUFF, TEXAS

This is the second volume of a three-volume report on our geotechnical investigation concerning dredge materials at your Huff refinery. This study was authorized by Term Contract No. 604 and Term Contract Release No. 20-6 dated May 6, 2008.

This report includes our findings and recommendations for Dredge Disposal Area F. Preliminary results were discussed with Mr. Jones on July 16, 2008. We consider the soil conditions at the site suitable for limited dike enlargements. However, we recommend that an embankment test section be constructed and monitored before dike design is finalized.

We appreciate the opportunity to work with you on this project, and we would like to thank Bob Berman and Cyndi Johnson for the help they provided on site. We look forward to assisting you with the final design and providing materials-testing services.

Sincerely,

George Fursten

George H. Fursten
Geotechnical/Environmental Engineer

GHF/dnn

M-Global Inc | 127 Rainbow Lane | Baltimore MD 21202 | 410.555.8175

■ **Model 9–2** ■ Letter of transmittal

MEMORANDUM

DATE: March 18, 2009
TO: Lynn Redmond, Vice President of Human Resources
FROM : Abe Andrews, Personnel Assistant   *aa*
SUBJECT: Report on Flextime Pilot Program at Boston Office

As you requested, I have examined the results of the six-month pilot program to introduce flextime to the Boston office. This report presents my data and conclusions about the use of flexible work schedules.

To determine the results of the pilot program, I asked all employees to complete a written survey. Then I followed up by interviewing every fifth person off an alphabetical list of office personnel. Overall, it appears that flextime has met with clear approval by employees at all levels. Productivity has increased and morale has soared. This report uses the survey and interview data to suggest why these results have occurred and where we might go from here.

I enjoyed working on this personnel study because of its potential impact on the way M-Global conducts business. Please give me a call if you would like additional details about the study.

M-Global Inc | 127 Rainbow Lane | Baltimore MD 21202 | 410.555.8175

■ **Model 9–3** ■ Memo of transmittal

CONTENTS

                                                                    PAGE
LIST OF ILLUSTRATIONS 1
EXECUTIVE SUMMARY 2

INTRODUCTION 3
    Project Description 3
    Scope of Study 4
    Report Format 4

FIELD INVESTIGATION 5

LABORATORY INVESTIGATION 7
    Classification Tests 7
    Strength Tests 8
    Consolidation Tests 10
        Boring 1 10
        Boring 2 10
        Boring 3 11
        Boring 4 11
        Boring 5 11

GENERAL SUBSURFACE CONDITIONS 12
    Site Description 12
    Site Geology 13
    Soil Stratigraphy 15
        Stratum 1 17
        Stratum 2 18
        Stratum 3 19
    Groundwater Level 21
    Variations in Subsurface Conditions 23

DIKE DESIGN 24
    Stability 25
    Settlement 27
    Construction 29
    Test Embankment 30
        Phase 1 31
        Phase 2 31
        Phase 3 32
        Phase 4 33

CONCLUSIONS AND RECOMMENDATIONS 34

■ **Model 9–4** ■ Table of contents (all subheadings included)

CONTENTS

PAGE

LIST OF ILLUSTRATIONS 1

EXECUTIVE SUMMARY 2

INTRODUCTION 3
    Project Description 3
    Scope of Study 4
    Report Format 4

FIELD INVESTIGATION 5

LABORATORY INVESTIGATION 7
    Classification Tests 7
    Strength Tests 8
    Consolidation Tests 10

GENERAL SUBSURFACE CONDITIONS 12
    Site Description 12
    Site Geology 13
    Soil Stratigraphy 15
    Groundwater Level 21
    Variations in Subsurface Conditions 23

DIKE DESIGN 24
    Stability 25
    Settlement 27
    Construction 29
    Test Embankment 30

CONCLUSIONS AND RECOMMENDATIONS 34

■ **Model 9–5** ■ Table of contents (third-level subheadings omitted)

ILLUSTRATIONS

TABLES   **PAGE**

1. Sampling Rates for 1990–2000 3
2. Pumping Rates after Start-up 9
3. Capital Costs Required for Vessel. 14

FIGURES

1. Rig Mounting Assembly 5
2. Dart System for Vessel 7
3. Exploded View of Extruder 8

■ **Model 9–6** ■ List of illustrations—formal report

EXECUTIVE SUMMARY

Gives brief background of project.

Quarterly monitoring of groundwater showed the presence of nickel in Well M–17 at the Hennessey Electric facility in Jones, Georgia. Nickel was not detected in any other wells on the site. Hennessey then retained M-Global's environmental group to determine the source of the nickel.

Keeps verbs in *active voice* for clarity and brevity.

The project consisted of four main parts. First, we collected and tested 20 soil samples within a 50-yard radius of the well. Second, we collected groundwater samples from the well itself. Third, we removed the stainless steel well screen and casing and submitted them for metallurgical analysis. Finally, we installed a replacement screen and casing built with teflon.

The findings from this project are as follows:

Uses short list to emphasize major findings.

- The soil samples contained *no* nickel.
- We found *significant* corrosion and pitting in the stainless steel screen and casing that we removed.
- We detected *no* nickel in water samples retrieved from the well after replacement of the screen and casing.

Emphasizes major conclusion in separate paragraph. (Note that this major point *could* have been placed after first paragraph, for a different effect.)

Our study concluded that the source of the nickel in the groundwater was corrosion on the stainless steel casing and screen.

■ **Model 9–7** ■ Executive summary—formal report

INTRODUCTION

This document examines the need for an M-Global, Inc., *Human Resources Manual*. Such a manual would apply to all U.S. offices of the firm. As background for your reading of this report, I have included (1) a brief description of the project, (2) the scope of my activities during the study, and (3) an overview of the report format.

Project Description

Three months ago, Jim McDuff met with the senior staff to discuss diverse human resources issues, such as performance appraisals and fringe benefits. After several meetings, the group agreed that the company greatly needed a manual to give guidance to managers and their employees. Shortly thereafter, I was asked to study and then report on three main topics: (1) the points that should be included in a manual, (2) the schedule for completing the document, and (3) the number of employees that should be involved in writing and reviewing policies.

Scope of Activities

This project involved seeking information from many M-Global employees and completing some outside research. Specifically, the project scope involved:

- Sending a survey to employees at every level at every domestic office
- Tabulating the results of the survey
- Interviewing some of the survey respondents
- Completing library research on the topic of human resource manuals
- Developing conclusions and recommendations that were based on the research completed

Report Format

To fulfill the report's purpose of examining the need for an M-Global *Human Resources Manual*, this report includes these main sections:

Section 1: Research Methods
Section 2: Findings of the Survey and Interviews
Section 3: Findings of the Library Research
Section 4: Conclusions and Recommendations

Appendices at the end of the text contain the survey form, interview questions, sample survey responses, and several journal articles of most use in my research.

Gives purpose of report and overview of introduction (as lead-in).

Describes the task the writer was given.

Denotes the major activities that were accomplished.

Provides reader with a preview of main sections to follow (as a sort of "mini" table of contents).

■ **Model 9–8** ■ Introduction—formal report

STUDY OF WILDWOOD CREEK

WINSLOW, GEORGIA

Prepared for:

The City of Winslow

Prepared by:

Christopher S. Rice, Hydro/Environmental Engineer
M-Global, Inc.

November 28, 2008

Uses graphic on title page to reinforce theme of environmental protection.

■ **Model 9–9** ■ Formal report

12 Peachtree Street
Atlanta GA 30056
(404) 555-7524

McDuff Project #99-119
November 28, 2008

Adopt-a-Stream Program
City of Winslow
300 Lawrence Street
Winslow Georgia 30000

Attention: Ms. Elaine Sykes, Director

STUDY OF WILDWOOD CREEK
WINSLOW, GEORGIA

We have completed our seven-month project on the pollution study of
Wildwood Creek. This project was authorized on May 16, 2008. We
performed the study in accordance with our original proposal No. 14-P72,
dated April 24, 2008.

This report mentions all completed tests and discusses the test results.
Wildwood Creek scored well on many of the tests, but we are concerned
about several problems—such as the level of phosphates in the stream. The
few problems we observed during our study have led us to recommend that
several additional tests should be completed.

Thank you for the opportunity to complete this project. We look forward to working
with you on further tests for Wildwood Creek and other waterways in Winslow.

Sincerely,

Christopher S. Rice

Christopher S. Rice, P.E.
Hydro/Environmental Engineer

M-Global Inc | 127 Rainbow Lane | Baltimore MD 21202 | 410.555.8175

> Lists project title as it appears on title page.

> Gives brief statement of project information.

> Provides major point from report.

■ **Model 9–9** ■ continued

CONTENTS

PAGE

LIST OF ILLUSTRATIONS 1
EXECUTIVE SUMMARY 2

INTRODUCTION 3
    Project Description 3
    Scope of Study 3
    Report Format 3

FIELD INVESTIGATION 4

    Physical Tests 4
        Air Temperature 4
        Water Temperature 4
        Water Flow 5
        Water Appearance 6
        Habitat Description 6
        Algae Appearance and Location 7
        Visible Litter 7
        Bug Count 7
    Chemical Tests 7
        pH 8
        Dissolved Oxygen (DO) 8
        Turbidity 8
        Phosphate 8

TEST COMPARISON 9

CONCLUSIONS AND RECOMMENDATIONS 10

    Conclusions 10
    Recommendations 10

APPENDICES

A. Background on Wildwood Creek 11
B. Water Quality Criteria for Georgia 12
C. Location of City of Winslow Parks and Recreation Facilities 13

Uses white space, indenting, and bold to accent organization of report.

■ **Model 9–9** ■ continued

ILLUSTRATIONS

FIGURES PAGE
1. Wildwood Creek—Normal Water Level 5
2. Wildwood Creek—Flash Flood Water Level 6

TABLES
1. Physical Tests 9
2. Chemical Tests 9

Includes illustration titles as they appear in text.

■ **Model 9–9** ■ continued

2

EXECUTIVE SUMMARY

Summarizes purpose
and scope of report.

The City of Winslow hired M-Global, Inc., to perform a pollution study of Wildwood Creek. The section of the creek that was studied is a one-mile-long area in Burns Nature Park, from Newell College to U.S. Highway 42. The study lasted seven months.

Describes major findings
and conclusions.

M-Global completed 13 tests on four different test dates. Wildwood scored fairly well on many of the tests, but there were some problem areas—for example, high levels of phosphates were uncovered in the water. The phosphates were derived either from fertilizer or from animal and plant matter and waste. Also uncovered were small amounts of undesirable water organisms that are tolerant to pollutants and can survive in harsh environments.

Includes main
recommendation
from report text.

M-Global recommends that (1) the tests done in this study be conducted two more times, through Spring 2009; (2) other environmental tests be conducted, as listed in the conclusions and recommendations section; and (3) a voluntary cleanup of the creek be scheduled. With these steps, we can better analyze the environmental integrity of Wildwood Creek.

■ **Model 9–9** ■ continued

3

INTRODUCTION

M-Global, Inc., has completed a follow-up to a study completed in 2000 by Ware County on the health of Wildwood Creek. This introduction describes the project site, scope of our study, and format for this report.

Gives lead-in to Introduction.

PROJECT DESCRIPTION

By law, all states must clean up their waterways. The State of Georgia shares this responsibility with its counties. Ware County has certain waterways that are threatened and must be cleaned. Wildwood Creek is one of the more endangered waterways. The portion of the creek that was studied for this report is a one-mile stretch in the Burns Nature Park between Newell College and U.S. Highway 42.

Briefly describes project.

SCOPE OF STUDY

The purpose of this project was to determine whether the health of the creek has changed since the previous study in 1996. Both physical and chemical tests were completed. The nine physical tests were as follows:

- Air temperature
- Water temperature
- Water flow
- Water appearance
- Habitat description
- Algae appearance
- Algae location
- Visible litter
- Bug count

Uses bulleted list to emphasize scope of activities.

The four chemical tests were as follows:

- pH
- Dissolved oxygen (DO)
- Turbidity
- Phosphate

REPORT FORMAT

This report includes three main sections:

1. Field Investigation: a complete discussion of all the tests that were performed for the project
2. Test Comparison: charts of the test results and comparisons
3. Conclusions and Recommendations

Provides "map" of main sections in report.

■ **Model 9–9** ■ continued

4

FIELD INVESTIGATION

Wildwood Creek has been cited repeatedly for environmental violations in the pollution of its water. Many factors can generate pollution and affect the overall health of the creek. In 1996, the creek was studied in the context of a study of all water systems in Ware County. Wildwood Creek was determined to be one of the more threatened creeks in the county.

The city needed to learn if much has changed in the past nine years, so M-Global was hired to perform a variety of tests on the creek. Our effort involved a more in-depth study than that done in 2000. Tests were conducted four times over a seven-month period. The 2000 study lasted only one day.

The field investigation included two categories of tests: physical tests and chemical tests.

PHYSICAL TESTS

The physical tests covered a broad range of environmental features. This section discusses the importance of the tests and some major findings. The Test Comparison section on page 9 includes a table that lists results of the tests and the completion dates. The test types were as follows: air temperature, water temperature, water flow, water appearance, habitat description, algae appearance, algae location, visible litter, and bug count.

Air Temperature

The temperature of the air surrounding the creek will affect life in the water. Unusual air temperature for the seasons will determine if life can grow in or out of the water.

Three of the four tests were performed in the warmer months. Only one was completed on a cool day. The difference in temperature from the warmest to coolest day was 10.5°C, an acceptable range.

Water Temperature

The temperature of the water determines which species will be present. Also affected are the feeding, reproduction, and metabolism of these species. If there are one or two weeks of high temperature, the stream is unsuitable for most species. If water temperature changes more than 1° to 2°C in 24 hours, thermal stress and shock can occur, killing much of the life in the creek.

During our study, the temperature of the water averaged 1°C cooler than the temperature of the air. The water temperature did not get above 23°C or below 13°C. These ranges are acceptable by law.

Amplifies information presented later in report.

■ **Model 9–9** ■ continued

5

Water Flow

The flow of the water influences the type of life in the stream. Periods of high flow can cause erosion to occur on the banks and sediment to cover the streambed. Low water flow can decrease the living space and deplete the oxygen supply.

The flow of water was at the correct level for the times of year the tests were done—except for June, which had a high rainfall. With continual rain and sudden flash floods, the creek was almost too dangerous for the study to be performed that month.

In fact, in June we witnessed the aftermath of one flash flood. Figure 1 shows the creek with an average flow of water, and Figure 2 shows the creek during the flood. The water's average depth is 10 inches. During the flash flood, the water level rose and fell 10 feet in about one hour. Much dirt and debris were washing into the creek, while some small fish were left on dry land as the water receded.

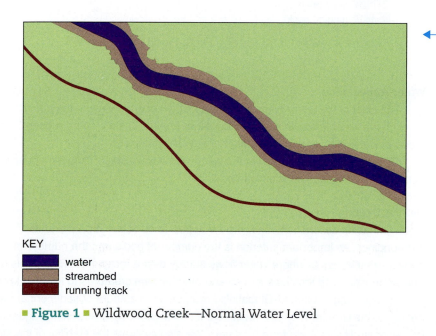

Incorporates graphic into page of text.

KEY

- water
- streambed
- running track

■ **Figure 1** ■ Wildwood Creek—Normal Water Level

6

KEY
- ■ water
- ■ streambed
- ■ running track

■ **Figure 2** ■ Wildwood Creek—Flash Flood Water Level

Water Appearance

The color of the water gives a quick but fairly accurate view of the health of the creek. If the water is brown or dirty, then silt or human waste may be present. Black areas of water may contain oil or other chemical products.

On each of the four test days, the water was always clear. Thus the appearance of the creek water was considered excellent.

Elaborates on importance of information shown in Table 1. Description parallels five items in table.

Habitat Description

The habitat description concerns the appearance of the stream and its surroundings. An important criterion is the number of pools and the number of ripples—that is, points where water flows quickly over a rocky area. Both pools and ripples provide good locations for fish and other stream creatures to live and breed.

In describing habitat, M-Global also evaluates the amount of sediment at the bottom of the stream. Too much sediment tends to cover up areas where aquatic life lays eggs and hides them from predators. We also evaluate the stability of the stream banks; a stable bank indicates that erosion has not damaged the habitat. Finally, we observe the amount of stream cover. Such vegetation helps keep soil in place on the banks.

■ **Model 9–9** ■ continued

7

Wildwood Creek tested fairly well for habitat. The number of pools and ripples was about average for such creeks. Stream deposits and stream bank stability were average to good, and stream cover was good to excellent. For more detail about test results, see the chart in the Test Comparison section on page 9.

Algae Appearance and Location

Algae is naturally present in any creek. The amount of algae can be a warning of pollution in the water. If algae is growing out of control, disproportionate amounts of nutrients such as nitrogen or phosphate could be present. These chemicals could come from fertilizer washed into the creek. Excessive amounts of algae cause the oxygen level to drop when they die and decompose.

During the four studies, algae was everywhere, but it was especially heavy on the rocks in the ripples of the creek. The algae was always brown and sometimes hairy.

Visible Litter

Litter can affect the habitat of a creek. Although some litter has chemicals that can pollute the water, other litter can cover nesting areas and suffocate small animals. Whether the litter is harmful or not, it is always an eyesore.

On all four test dates, the litter we saw was heavy and ranged from tires to plastic bags. Some of the same trash that was at the site on the first visit was still there seven months later.

Gives specific details that support the report's conclusions and recommendations, which come later.

Bug Count

The bug count is a procedure that begins by washing dirt and water onto a screen. As water drains, the dirt with organisms is left on the screen. The bugs are removed and classified. Generally, the lower the bug count, the higher the pollution levels. Bug counts were considered low to average.

Two types of aquatic worms were discovered every time during our count, but in relatively small amounts. In addition, the worms we observed are very tolerant of pollution and can live in most conditions. Finally, we observed only two crayfish, animals that are somewhat sensitive to pollution.

CHEMICAL TESTS

Although physical tests cover areas seen with the naked eye, chemical tests can uncover pollutants that are not so recognizable. Certain chemicals can wipe out all life in a creek. Other chemicals can cause an overabundance of one life-form, which in turn could kill more sensitive animals.

A chart of results of chemical tests is included in the Test Comparison section on page 9. The chemical tests that M-Global performed were pH, dissolved oxygen (DO), turbidity, and phosphate.

■ **Model 9–9** ■ continued

pH

The pH test is a measure of active hydrogen ions in a sample. The range of the pH test is 0–14. If the sample is in the range of 0–7.0, it is acidic; but if the sample is in the range of 7.0–14, it is basic. By law, the pH of a water sample must be within the range of 6.0–8.5.

For the tests we completed, the water sample was always 7.0, which is very good for a creek.

Dissolved Oxygen (DO)

Normally, oxygen dissolves readily into water from surface air. Once dissolved, it diffuses slowly in the water and is distributed throughout the creek. The amount of DO depends on different circumstances. Oxygen is always highest in choppy water, just after noon, and in cooler temperatures.

In many streams, the level of DO can become critically low during the summer months. When the temperature is warm, organisms are highly active and consume the oxygen supply. If the amount of DO drops below 3.0 ppm (parts per million), the area can become stressful for the organisms. An amount of oxygen that is 2.0 ppm or below will not support fish. DO that is 5.0 ppm to 6.0 ppm is usually required for growth and activity of organisms in the water.

According to the Water Quality Criteria for Georgia, average daily amounts of DO should be 5.0 ppm with a minimum of 4.0 ppm. Wildwood Creek scored well on this test. The average amount of DO in the water was 6.9 ppm, with the highest amount being 9.0 ppm on November 11, 2008.

Turbidity

Turbidity is the discoloration of water due to sediment, microscopic organisms, and other matter. One major factor of turbidity is the level of rainfall before a test.

Three of our tests were performed on clear days with little rainfall. On these dates, the turbidity of Wildwood Creek was always 1.0, the best that creek water can score on the test. The fourth test, which scored worse, occurred during a rainy period.

Phosphate

Phosphorus occurs naturally as phosphates—for example, orthophosphates and organically bound phosphates. Orthophosphates are phosphates that are formed in fertilizer, whereas organically bound phosphates can form in plant and animal matter and waste.

Phosphate levels higher than 0.03 ppm contribute to an increase in plant growth. If phosphate levels are above 0.1 ppm, plants may be stimulated to grow out of control. The phosphate level of Wildwood was always 0.5 ppm, considerably higher than is desirable.

■ **Model 9–9** ■ continued

9

TEST COMPARISON

There was little change from each of the four test dates. The only tests that varied greatly from one test to another were air temperature, water temperature, water flow, and DO. On the basis of these results, it would appear that Wildwood Creek is a relatively stable environment.

Table 1 Physical Tests ◄ —————————————— Brings together test results for easy reference.

TEST DATES	5/26/08	6/25/08	9/24/08	11/19/08
Air Temperature in °C	21.5	23.0	24.0	13.5
Water Temperature in °C	20.0	22.0	23.0	13.0
Water Flow	Normal	High	Normal	Normal
Water Appearance	Clear	Clear	Clear	Clear
Habitat Description				
Number of Pools	2.0	3.0	2.0	5.0
Number of Ripples	1.0	2.0	2.0	2.0
Amount of Sediment Deposit	Average	Average	Good	Average
Stream Bank Stability	Average	Good	Good	Good
Stream Cover	Excellent	Good	Excellent	Good
Algae Appearance	Brown	Brown/hairy	Brown	Brown
Algae Location	Everywhere	Everywhere	Attached	Everywhere
Visible Litter	Heavy	Heavy	Heavy	Heavy
Bug Count	Low	Average	Low	Average

Table 2 Chemical Tests

Test	5/26/08	6/25/08	9/24/08	11/19/08
pH	7.0	7.0	7.0	7.0
Dissolved Oxygen (DO)	6.8	6.0	5.6	9.0
Turbidity	1.0	3.0	1.0	1.0
Phosphate	0.50	0.50	0.50	0.50

■ **Model 9–9** ■ continued

CONCLUSIONS AND RECOMMENDATIONS

This section includes the major conclusions and recommendations from our study of Wildwood Creek.

CONCLUSIONS

Draws conclusions that flow from data in body of report.

Generally, we were pleased with the health of the stream bank and its floodplain. The area studied has large amounts of vegetation along the stream, and the banks seem to be sturdy. The floodplain has been turned into a park, which handles floods in a natural way. Floodwater in this area comes in contact with vegetation and some dirt. Floodwater also drains quickly, which keeps sediment from building up in the creek.

Uses paragraph format instead of lists because of lengthy explanations needed.

However, we are concerned with the number and types of animals uncovered in our bug counts. Only two bug types were discovered, and these were types quite tolerant to pollutants. The time of year these tests were performed could affect the discovery of some animals. However, the low count still should be considered a possible warning sign about water quality. Phosphate levels were also high and probably are the cause of the large amount of algae.

We believe something in the water is keeping sensitive animals from developing. One factor that affects the number of animals discovered is the pollutant problems in the past (see Appendix A). The creek may still be in a redevelopment stage, thus explaining the small numbers of animals.

RECOMMENDATIONS

On the basis of these conclusions, we recommend the following actions for Wildwood Creek:

Gives numbered list of recommendations for easy reference.

1. Conduct the current tests two more times, through Spring 2009. Spring is the time of year that most aquatic insects are hatched. If sensitive organisms are found then, the health of the creek could be considered to have improved.
2. Add testing for nitrogen. With the phosphate level being so high, nitrogen might also be present. If it is, then fertilizer could be in the water.
3. Add testing for human waste. Some contamination may still be occurring.
4. Add testing for metals, such as mercury, that can pollute the water.
5. Add testing for runoff water from drainage pipes that flow into the creek.
6. Schedule a volunteer cleanup of the creek.

With a full year of study and additional tests, the problems of Wildwood Creek can be better understood.

■ **Model 9–9** ■ continued

11

APPENDIX A

Background on Wildwood Creek

Wildwood Creek begins from tributaries on the northeast side of the city of Winslow. From this point, the creek flows southwest to the Chattahoochee River. Winslow Wastewater Treatment Plant has severely polluted the creek in the past with discharge of wastewater directly into the creek. Wildwood became so contaminated that signs warning of excessive pollution were posted along the creek to alert the public.

Today, all known wastewater discharge has been removed. The stream's condition has dramatically improved, but nonpoint contamination sources continue to lower the creek's water quality. Nonpoint contamination includes sewer breaks, chemical dumping, and storm sewers.

Another problem for Wildwood Creek is siltration. Rainfall combines with bank erosion and habitat destruction to wash excess dirt into the creek. This harsh action destroys most of the macroinvertebrates. At the present time, Wildwood Creek may be one of the more threatened creeks in Ware County.

■ **Model 9–9** ■ continued

APPENDIX B

Water Quality Criteria for Georgia

All waterways in Georgia are classified in one of the following categories: fishing, recreation, drinking, and wild and scenic. Different protection levels apply to the different uses. For example, the protection level for dissolved oxygen is stricter in drinking water than fishing water. All water is supposed to be free from all types of waste and sewage that can settle and form sludge deposits.

In Ware County, all waterways are classified as "fishing," according to Chapter 391-3-6.03 of "Water Use Classifications and Water Quality Standards" in the Georgia Department of Natural Resources *Rules and Regulations for Water Quality Control.* The only exception is the Chattahoochee River, which is classified as "drinking water supply" and "recreational."

■ **Model 9–9** ■ continued

APPENDIX C

Map 6
Location of City of Winslow
Parks and Recreation Facilities

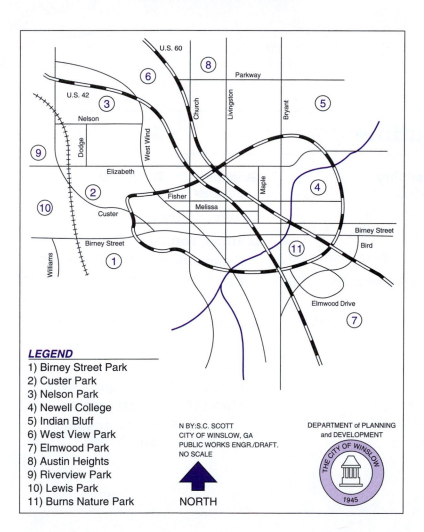

LEGEND
1) Birney Street Park
2) Custer Park
3) Nelson Park
4) Newell College
5) Indian Bluff
6) West View Park
7) Elmwood Park
8) Austin Heights
9) Riverview Park
10) Lewis Park
11) Burns Nature Park

N BY:S.C. SCOTT
CITY OF WINSLOW, GA
PUBLIC WORKS ENGR./DRAFT.
NO SCALE

NORTH

DEPARTMENT of PLANNING
and DEVELOPMENT

Chapter | 10 | Proposals and Feasibility Studies

>>> Chapter Outline

**Proposals and Feasibility
  Studies at M-Global**   303

  M-Global Proposals   304

  M-Global Feasibility Studies   305

Guidelines for Informal Proposals   306

Guidelines for Formal Proposals   311

  Cover/Title Page   311

  Letter/Memo of Transmittal   312

  Table of Contents   313

  List of Illustrations   313

  Executive Summary   313

  Introduction   314

  Discussion Sections   315

  Conclusion   315

  Appendices   316

**Guidelines for Feasibility
  Studies**   316

Chapter Summary   319

Learning Portfolio   321

  Communication Challenge   321

  Collaboration at Work   322

  Assignments   323

Models for Good Writing   329

  Model 10–1: Letter proposal   329

  Model 10–2: Memo proposal   332

  Model 10–3: Formal proposal   334

  Model 10–4: Formal proposal   343

  Model 10–5: Feasibility study
    (one alternative)   364

G ini Harris, a structural technician at M-Global's Houston office, has written many reports in her two years at the company. Last month, however, she had the chance to write her first major proposal. Anchor Productions, a Hollywood film company, asked M-Global for a proposal to build a full-scale replica of the Alamo. In an upcoming film about the Texas battle, most scenes will be filmed in this replica to be built on leased land west of San Antonio. To write her proposal for Anchor, Gini had to seek advice from a variety of technical specialists and architectural historians.

As it happens, Gini wrote a successful proposal that got the job. The work went so well that she asked her supervisor, Ken Blair, if M-Global could start a technical group to work just on historical restoration/replication projects around the country. Intrigued by the idea, Blair asked Harris to prepare a *feasibility study* that would focus on start-up costs, involvement of M-Global offices, hiring needs, and potential profit during the first five years. If the feasibility study showed promise, Blair would present the idea to M-Global's corporate management.

Although you probably won't write proposals to build an Alamo, you will write proposals and feasibility studies in your own field during your career. These modes of writing are crucial to most organizations— indeed, many companies rely on them, especially proposals, for their very survival. Proposals and feasibility studies are defined as follows:

Proposal: **A document written to convince your readers to adopt an idea, a product, or a service. They can be directed to colleagues inside your** own organization (*in-house proposals*), **to clients outside your organization (*sales proposals*), or to organizations that fund research and other activities (*grant proposals*).**

In all three cases, proposals can be presented in either a short, simple format (*informal proposal*) **or a longer, more complicated format (*formal proposal*). Also, proposals can be either requested by the reader (*solicited*) or submitted without a request (*unsolicited*).**

Feasibility study: **A document written to show the *practicality* of a proposed policy, product, service, or other change within an organization. Often prompted by ideas suggested in a *proposal*, they examine details such as costs, alternatives, and likely effects. Although they must reflect the objectivity of a report, most feasibility studies also try to convince readers either (1) to adopt or reject the one idea discussed or (2) to adopt one of several alternatives presented in the study.**

Feasibility studies can be *in-house* (written to decision makers in your own organization) or *external* (requested by clients from outside your organization).

There are four main sections in this chapter. The first gives specific situations in which you might write proposals and feasibility studies at M-Global; the second and third sections discuss informal and formal proposal formats while also denoting differences between in-house and sales versions of these formats; the fourth section covers feasibility studies. All chapter guidelines are followed by annotated examples to use as models for your own writing.

>>> Proposals and Feasibility Studies at M-Global

As the Alamo example shows, proposals and feasibility studies often work together. The proposal may suggest a topic on which a feasibility study is then written. The flowchart in Figure 10–1 shows another possible communication cycle that would involve both a proposal and a feasibility study. Note that the diagram includes the term *RFP*, which stands for *request for proposal.*

Request for proposal (RFP): A document sometimes sent out by organizations that want to receive proposals for a product or service. The RFP gives guidelines on (1) what the proposal should cover, (2) when it should be submitted, and (3) to whom it should be sent. As writer, you should follow the RFP religiously in planning and drafting your proposal.

RFPs generally are not used in the following situations:

- When the proposal is solicited from within your own organization
- When the proposal is requested less formally, as through a letter, phone call, or memo
- When the proposal is *unsolicited,* meaning that you are writing it without a request from the person who will read it

The sections that follow describe additional situations in which proposals and feasibility studies would be written at M-Global. Reading through these brief cases shows you the varied contexts for persuasive writing within just one company.

M-Global Proposals

Like many organizations, M-Global depends on (1) in-house proposals to breathe new life into its internal operations, (2) sales proposals to request work from clients, and (3) occasional grant proposals to seek research funds from outside organizations. Proposals are a main activity in healthy and growing organizations like M-Global.

■ **Figure 10–1** ■
Flowchart showing the main documents involved in an external, solicited proposal process.

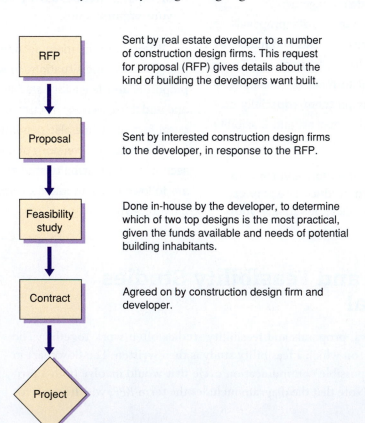

RFP — Sent by real estate developer to a number of construction design firms. This request for proposal (RFP) gives details about the kind of building the developers want built.

Proposal — Sent by interested construction design firms to the developer, in response to the RFP.

Feasibility study — Done in-house by the developer, to determine which of two top designs is the most practical, given the funds available and needs of potential building inhabitants.

Contract — Agreed on by construction design firm and developer.

Project

Of the five cases described here, the first and second are internal, requiring in-house proposals; the third and fourth are external, requiring sales proposals; and the fifth is external, requiring a grant proposal.

- **In-house proposal for structural design and analysis equipment:** Meg Stevens, a civil engineer at the Denver office, writes an in-house proposal to the construction manager, Elvin Lipkowsky, in which she proposes that the company purchase a new structural design and analysis system. Her proposal includes description of equipment that she recently saw demonstrated at a conference of civil engineers.

- **In-house proposal for retaining legal counsel:** Jake Washington, an employment specialist in the Human Resources Department in the Baltimore office, writes an in-house proposal to Karrie Camp, vice president of human resources. In it he proposes that the company retain legal counsel on a half-time basis (20 hours a week). In his position at M-Global, Jake uses outside legal advice in dealing with new hiring laws, unemployment compensation cases, affirmative action regulations, and occasional lawsuits by employees who have been fired. He is proposing that the firm retain regular half-time counsel, rather than dealing with different lawyers as is done now.

- **Sales proposal for asbestos removal:** Jane Wiltshire, asbestos department manager at M-Global's St. Paul office, regularly talks with owners of buildings that may contain asbestos. After an initial discussion with the head minister of First Street Church, she writes an informal sales proposal in which she offers M-Global's services in performing an asbestos survey of the church building. Specifically, she explains how M-Global will examine the structure for possible asbestos, gives a schedule for completing the survey and writing the final report, and proposes a lump-sum price for the project.

- **Sales proposal for work on wind turbine project:** A utility company in California plans to build 10 wind turbines in a desert valley in the southern part of the state. The "free" power that is generated will help offset the large increases in fuel costs for the company's other plants. Although the firm has selected a turbine design and purchased the units, it must decide where to place them and what kind of foundations to use. Thus it sent out a request for proposal to companies that have experience with foundation and environmental engineering. Louis Bergen, engineering manager at M-Global's San Francisco office, writes a proposal that offers to test the soils at the site, pinpoint the best locations for the heavy turbines, and design the most effective foundations.

- **Grant proposal for new equipment design:** Oilarus, Ltd., a British oil company, sometimes gives research and development funds to small companies. Such funding usually goes toward development of new technology or products in the field of petroleum engineering. Angela Issam, who works in M-Global's Equipment Development group, decides to apply for some of the funding. Her proposed project, if successful, would provide a new piece of oil drilling safety equipment that would reduce the chance of offshore oil spills at production sites.

M-Global Feasibility Studies

The next two examples show that feasibility studies often flow from proposals. They can be internal, to managers who need facts before making a decision, or they can be external, to clients who request a service.

■ **In-house feasibility study about legal counsel:** Karrie Camp, the vice president of human resources, recently received Jake Washington's proposal that M-Global retain half-time legal counsel (see second case in previous section). The idea interests her, but she is not convinced of its practicality. She calls a meeting with Jake Washington and Scott Sampson, the personnel manager, who also uses legal counsel on a part-time basis. Because Karrie will have to sell her boss on this idea, she asks Jake and Scott to write a feasibility study on it.

This study, unlike the proposal, must include a detailed comparison of the present mode of operating versus the proposed strategy of retaining a lawyer for 20 hours per week on a regular basis. The study must examine criteria such as current costs versus projected costs, current level of satisfaction versus projected level of satisfaction, and current level of services provided versus projected level of services provided. Lynn asks that the report include a clear recommendation, made on the basis of the data.

■ **External feasibility study on plant site:** Tarnak, Inc., a large furniture manufacturing company in North Carolina, has decided to build a new plant in northern Georgia. After getting proposals from many cities that want the plant, the company has narrowed its choices to three spots that are about equal in cost of living, access to workers, standard of living, construction costs, transportation facilities, and access to raw materials. Yet the firm has not studied any site with respect to waste management.

Specifically, Tarnak must decide which of the three cities is best prepared to handle the solid and chemical wastes from the plant in a safe and economical manner. It hires M-Global's Atlanta office to write a feasibility study. The first goal of the study is to determine what sites, if any, meet Tarnak's criteria for waste management. If the first objective yields more than one site, M-Global then must compare the sites and recommend the best one.

>>> Guidelines for Informal Proposals

Like informal reports, informal proposals are short documents that cover projects with a limited scope. But how short is short? And just what does *limited* mean? Following are some specific guidelines that you can use if your employer or client has not provided others:

Use Informal Proposals When

• The text of the proposal (excluding attachments) is no more than five pages
• The size of the proposed project is such that a long formal proposal appears to be inappropriate
• The client has expressed a preference for a leaner and less-formal document

Use Formal Proposals When

• The text of the proposal (excluding attachments) is more than five pages
• The size and importance of the project is such that a formal proposal is appropriate
• The client has expressed a preference for a more-formal document

These two formats can be used for proposals that are either *in-house* (to readers within your own organization) or *external* (to readers outside your organization). The rest of this section provides writing guidelines and an annotated model for informal proposals, the type of persuasive writing you will do most often in your career.

Informal proposals have two formats: (1) memos (for in-house proposals) and (2) letters (for external proposals). The guidelines recommended apply to both. With some variations, they are essentially the guidelines suggested in chapter 8 for informal reports. The formats are much the same, although the content and tone are different. Reports *explain,* whereas proposals *persuade.*

>> Informal Proposal Guideline 1: Plan Well Before You Write

Complete the Planning Form at the end of the book for all proposal assignments in class, as well as for proposals you write on the job. Carefully consider your purpose, audience, and organization. Two factors make this task especially difficult in sales-proposal writing:

1. You may know nothing more about the client than what is written on the RFP.
2. Proposals almost always are on a tight schedule, which limits your planning time.

Despite these limitations, try to find out exactly who will be making the decision about your proposal. Many clients will tell you if you give them a call. In fact, they may be pleased that you care enough about the project to target the audience. Once you identify the decision makers, spend time brainstorming about their needs before you begin writing. Proposals that betray an ignorance of client needs often do so because the writer began writing too soon about the product or service.

>> Informal Proposal Guideline 2: Use Letter or Memo Format

Letter proposals, such as the example shown in Model 10–1 on pages 329–331, basically follow the format of routine business letters (see chapter 5). This casual style gives readers the immediate impression that your document will be *approachable*—that is, easy to get through and limited in scope. Memo proposals, such as the example shown in Model 10–2 on pages 332–333, follow the format of an internal memorandum (see chapter 5). Following are a few highlights:

- Line spacing is usually single, but it may be one-and-a-half or double, depending on the reader's or company's preference.
- The recipient's name, date, and page number appear on sheets after the first.
- Most readers prefer an uneven or ragged-edge right margin, as opposed to an even or full-justified margin.

Your subject line in a memo proposal gives readers the first impression of the proposal's purpose. Choose concise yet accurate wording. Furthermore, the wording must match that used in the proposal text. See Model 10–2 for wording that gives the appropriate information and tries to engage the reader's interest.

>> Informal Proposal Guideline 3: Make Text Visually Appealing

The page design of informal proposals must draw readers into the document. Remember—you are trying to sell a product, a service, or an idea. If the layout is unappealing, then you will lose readers before they even get to your message. Also, remember that your proposal may be competing with others. Put yourself in the place of the reader who is wondering which one to pick up first. How the text looks on the page can make a big difference. Following are a few techniques to follow to help make your proposal visually appealing:

- Use lists (with bullets or numbered points) to highlight main ideas.
- Follow your readers' preferences as to font size, type, line spacing, and so forth. Proposals written in the preferred format of the reader gain a competitive edge.
- Use headings and subheadings to break up blocks of text.

These and other techniques help to reveal the proposal's structure and lead readers through the informal proposal. Given that there is no table of contents, you must take advantage of such strategies.

ABC Format: Informal Proposal

- **ABSTRACT:** Gives the summary or "big picture" for those who make decisions about your proposal. Usually includes a statement of the problem or other information that will entice the audience to read further.
- **BODY:** Gives the details about exactly what you are proposing to do.
- **CONCLUSION:** Drives home the main benefit and makes clear the next step.

>> Informal Proposal Guideline 4: Use the ABC Format for Organization

The ABC Format used throughout this book also applies, of course, to informal proposals.

Note: Beginning and ending sections should be easy to read and stress just a few points. They provide a short buffer on both ends of the longer and more technical body section in the middle. The next four guidelines give more specific advice for writing the main parts of an informal proposal.

>> Informal Proposal Guideline 5: Create the Abstract as an Introductory Summary

In an abstract, you capture the client's attention with a capsule summary of the entire proposal. This one- or two-paragraph starting section permits space only for what the reader really must know at the outset, such as the following:

- Purpose of proposal
- Reader's main need
- Main features you offer, as well as related benefits
- Overview of proposal sections to follow

As Models 10–1 and 10–2 show, the introductory summary appears immediately after the subject line of the memo proposal or salutation of the letter proposal. As with informal reports, you have the option of actually labeling the section "Introductory Summary" or simply leaving off the heading. In either case, keep this overview very brief. Answer the one question clients are thinking: "Why should we hire this firm instead of another?" If you find yourself starting to give too much detail, move background information into the first section of the discussion.

>> Informal Proposal Guideline 6: Put Important Details in the Body

The discussion of your proposal should address these basic questions:

1. What problem are you trying to solve, and why?
2. What are the technical details of your approach?
3. Who will do the work, and with what?
4. When will it be done?
5. How much will it cost?

Discussion formats vary from proposal to proposal, but here are some sections commonly used to respond to these questions:

1. **Description of problem or project and its significance.** Give a precise technical description, along with any assumptions that you have made on the basis of previous contact with the reader. Explain the importance or significance of the problem, especially to the reader of the proposal.

2. **Proposed solution or approach.** Describe the specific tasks you propose in a manner that is clear and well organized. If you are presenting several options, discuss each one separately—making it as easy as possible for the reader to compare and contrast information.

3. **Personnel.** If the proposal involves people performing tasks, it may be appropriate to explain qualifications of participants.

4. **Schedule.** Even the simplest proposals usually require some sort of information about the schedule for delivering goods, performing tasks, and so on. Be both clear and realistic in this portion of the proposal. Use graphics when appropriate (see chapter 13 for guidelines on creating Gantt and milestone charts).

5. **Costs.** Place complete cost information in the body of the proposal unless you have a table that would be more appropriately placed in an attachment. Above all, do not bury dollar figures in paragraph format. Instead, highlight these figures with indented or bulleted lists, or at least place them at the beginnings of paragraphs. Because your reader will be looking for cost data, it is to your advantage to make that information easy to find. Finally, be certain to include all costs—materials, equipment, personnel, salaries, and so on.

>> Informal Proposal Guideline 7: Give Special Attention to Establishing Need in the Body

A common complaint about proposals is that writers fail to establish the need for what is being proposed. As any good salesperson knows, customers must feel that they need your product, service, or idea before they can be convinced to purchase or support it. In other words, do not simply try to dazzle readers with the good sense and quality of what you are proposing. Instead, lay the groundwork for acceptance by first showing the readers that a strong need exists.

Establishing need is most crucial in unsolicited proposals, of course, when readers may not be psychologically prepared to accept a change that costs them money. Even in proposals that have been solicited, however, you should give some attention to restating the basic needs of the readers. If nothing else, this special attention shows your understanding of the problem.

>> Informal Proposal Guideline 8: Focus Attention in Your Conclusion

Called *conclusion* or *closing,* this section gives you the opportunity to control the readers' last impression. It also helps you avoid the awkwardness of ending proposals with the statement of costs, which is usually the last section in the discussion. In this closing section, you can

- Emphasize a main benefit or feature of your proposal
- Restate your interest in doing the work
- Indicate what should happen next

Regarding the last point, sometimes you may ask readers to call if they have questions. In other situations, however, it is appropriate to say that you will follow up the proposal with a phone call. This approach leaves you in control of the next step.

Incidentally, for informal sales proposals, there is a special technique that can push the proposal one step closer to approval. After the signature section, place an *acceptance block.* As shown here, this item makes it as easy as a signature for the reader to accept your proposal, rather than his or her having to write a return letter.

ACCEPTED BY LMN DEVELOPMENT, INC.

By: _____

Title: _____

Date: _____

>> Informal Proposal Guideline 9: Use Attachments for Less Important Details

Remember that the text of informal proposals is usually less than five pages. That being the case, you may have to put supporting data or illustrations in attachments that follow the conclusion. Cost and schedule information, in particular, is best placed at the end in well-labeled sections.

Make sure that the proposal text includes clear references to these visuals. If you have more than one attachment, give each one a letter and a title (for example, "Attachment A: Project Costs"). If you have only one attachment, include the title but no letter (for example, "Attachment: Resumes").

>> Informal Proposal Guideline 10: Edit Carefully

In the rush of completing proposals, some writers fail to edit carefully. That is a big mistake. Make sure to build in enough time for a series of editing passes, preferably by different readers. There are two reasons why proposals of all kinds deserve this special attention.

1. They can be considered contracts in a court of law. If you make editing mistakes that alter meaning (such as an incorrect price figure), you could be bound to the error.
2. Proposals often present readers with their first impression of you. If the document is sloppy, they can make assumptions about your professional abilities as well.

>>> Guidelines for Formal Proposals

Sometimes the complexity of the proposal may be such that a formal response is best. As always, the final decision about format should depend on the needs of your readers. Ask yourself questions such as the following in deciding whether to write an informal or a formal proposal:

- Is there too much detail for a letter or memo?
- Is a table of contents needed so that sections can be found quickly?
- Will the professional look of a formal document lend support to the cause?
- Are there so many attachments that a series of lengthy appendices would be useful?
- Are there many different readers with varying needs, such that there should be different sections for different people?

If you answer "yes" to one or more of these questions, give careful consideration to writing a formal proposal. This long format is most common in external sales proposals; however, important in-house proposals may sometimes require the same approach—especially in large organizations in which you may be writing to unknown persons in distant departments. Both in-house and sales examples follow the writing guidelines given next.

Formal proposals can be long and complex, so this part of the chapter treats each proposal section separately—from title page through conclusion. Two points become evident as you use these guidelines. First, formal reports and formal proposals are very much alike. A quick look at chapter 9 shows you the similarities in format. Second, a formal proposal—as with all technical writing described in this text—follows the basic ABC format described in chapter 3. Specifically, the parts of the formal proposal fit the pattern as shown in the following box.

As you read through and apply these guidelines, refer to Model 10–3 on pages 334–342 and Model 10–4 on pages 343–363 for annotated examples of the formal proposal. Note that each major section in the model proposals starts on a new page. Another alternative is to run most sections together, changing pages only at the end of the letter of transmittal and executive summary. Minor format variations abound, of course, but this chapter's guidelines will stand you in good stead throughout your career.

Cover/Title Page

Like formal reports, formal proposals are usually bound documents with a cover, which includes one or more of the items listed on the right for inclusion on the title page.

Just as important, the cover should be designed to attract the reader's interest—with good page layout and perhaps even a graphic. Remember—proposals are sales documents. No one has to read them.

ABC Format: Formal Proposal

- **ABSTRACT:**
  - Cover/Title Page
  - Letter of Transmittal
  - Table of Contents
  - List of Illustrations
  - Executive Summary
  - Introduction
- **BODY:**
  - Technical information
  - Management information
  - Cost information
  - (Appendices—appear after text, but support Body section)
- **CONCLUSION:**
  - Conclusion

Inside the cover is the title page, which contains these four pieces of information:

- **Project title,** sometimes preceded by *Proposal for* or similar wording
- **Your reader's name** (sometimes preceded by *Prepared for* . . .)
- **Your name or the name of your organization** spelled out in full (sometimes preceded by *Prepared by* . . .)
- **Date of submission**

The title page gives clients their first impression of you. For that reason, consider using some tasteful graphics to make the proposal stand out from those of your competitors. For sales proposals, a particularly persuasive technique is to place the logo of the client's company on the cover or title page. In this way, you imply your interest in linking up with that firm as well as your interest in satisfying its needs, rather than simply selling your products or services.

Letter/Memo of Transmittal

Internal proposals have *memos of transmittal;* external sales proposals have *letters of transmittal.* These letters or memos must grab the reader's interest. The guidelines for format and organization presented here help you write attention-getting prose. In particular, note that the letter or memo should be in single-spaced, ragged-right-edge format, even if the rest of the proposal is double-spaced copy with right-justified margins.

For details of letter and memo format, see chapter 5. The guidelines for the letter/memo of transmittal for formal reports also apply to formal proposals (see "Transmittal Guideline 4" in chapter 9). For now, some highlights of format and content that apply especially to letters and memos of transmittal are as follows:

1. Use short beginning and ending paragraphs (about three to five lines each).

2. Use a conversational style, with little or no technical jargon. Avoid stuffy phrases such as *per your request.*

3. Use the first paragraph for introductory information, mentioning what your proposal responds to (e.g., a formal RFP, a conversation with the client, your perception of a need).

4. Use the middle of the letter to emphasize one main benefit of your proposal, although the executive summary and proposal proper mention benefits in detail. Stress what you can do to solve a problem, using the words *you* and *your* as much as possible (rather than *I* and *we*).

5. Use the last paragraph to retain control by orchestrating the next step in the proposal process. When appropriate, indicate that you will call the client soon to follow up on the proposal.

6. Follow one of the letter formats described in chapter 5. Following are some exceptions, additions, or restrictions:

   - Use single-spaced, ragged-right-edge copy, which makes your letter stand out from the proposal proper.

- Keep the letter on one page—a two-page letter loses that crisp and concise impact you want to make.
- Place the company proposal number (if there is one) at the top, above the date. Exact placement of both number and date depends on your organization's letter style.
- Include the client's company name or personal name on the first line of the inside address, followed by the mailing address used on the envelope. If you use a company name, place an *Attention* line below the inside address. Include the full name (and title, if appropriate) of your contact person at the client firm. If you use a personal name, follow the last line of the inside address with a conventional greeting ("Dear Mr. Adams:").
- (Optional) Include the project title beneath the attention line, using the exact wording that appears on the title page.
- Close with "Sincerely" and your name at the bottom of the page. Also include your company affiliation.

Table of Contents

Create a very readable table of contents by spacing items well on the page. List all proposal sections, subsections, and their page references. At the end, list any appendices that may accompany the proposal.

Given the tight schedule on which most proposals are produced, errors can be introduced at the last minute because of additions or revisions. Therefore, take time to proofread the table of contents carefully. In particular, make sure to follow these guidelines:

- Wording of headings should match within the proposal text.
- Page references should be correct.
- All headings of the same order should be parallel in grammatical form.

List of Illustrations

When there are many illustrations, the list of illustrations appears on a separate page after the table of contents. When there are few entries, however, the illustrations may be listed at the end of the table of contents page. In either case, the list should include the number, title, and page number of every illustration appearing in the body of the text. (If there is only one illustration, a number need not be included.) You may divide the list into tables and figures if many of both appear in your proposal.

Executive Summary

Executive summaries are the most frequently read parts of proposals. That fact should govern the time and energy you put into their preparation. Often read by decision makers in an organization, the summary should present a concise one-page overview of the proposal's most important points. It should also accomplish the following objectives:

- Avoid technical language
- Be as self-contained as possible
- Make brief mention of the problem, proposed solution, and cost
- Emphasize the main benefits of your proposal

Start the summary with one or two sentences that command readers' attention and engage their interest, and then focus on just a few main selling points (three to five is best). You might even want to highlight these benefits with indented lead-ins such as *Benefit 1* and *Benefit 2*. When possible, use the statement of benefits to emphasize what is unique about your company or your approach so that your proposal attracts special attention. Finally, remember to write the summary after you have completed the rest of the proposal. Only at this point do you have the perspective to sit back and develop a reader-oriented overview.

Introduction

The introduction provides background information for both nontechnical and technical readers. Although the content varies from proposal to proposal, some general guidelines apply. Basically, you should include information on the (1) purpose, (2) description of the problem to which you are responding, (3) scope of the proposed study, and (4) format of the proposal. (A lengthy problem or project statement should be placed in the first discussion section of the proposal, not in the introduction.)

- Use subheadings if the introduction goes over a page. In this case, begin the section with a lead-in sentence or two that mention the sections to follow.
- Start with a purpose statement that concisely states the reason you are writing the proposal.
- Include a description of the problem or need to which your proposal is responding. Use language directly from the request for proposal or other document the reader may have given you so that there is no misunderstanding. For longer problem or need descriptions, adopt the alternative approach of including a separate needs section or problem description after the introduction.
- Include a scope section in which you briefly describe the range of proposed activities covered in the proposal, along with any research or preproposal tasks that have already been completed.
- Include a proposal format section if you believe the reader would benefit from a listing of the major proposal sections that follow.

Some formal sales proposals include information about the history, background, and expertise of an organization. This is especially appropriate in proposals that are responding to an RFP, where a number of organizations are competing to win the contract. This material is often *boilerplate,* or text that can be reused in all similar documents. M-Global has boilerplate passages about its history as well as about its experience in specific fields such as soils analysis, construction, or training. An M-Global proposal-writing team downloads the appropriate passages from where they are stored on the company's intranet, inserts them into a draft, and then polishes the entire passage so that the boilerplate is seamlessly integrated into the final proposal.

Discussion Sections

Aim the discussion or body toward readers who need supporting information. Traditionally, the discussion of a formal sales proposal contains three basic types of information: (1) technical, (2) management, and (3) cost. Following are some general guidelines for presenting each type. Remember that the exact wording of headings and subheadings varies depending on proposal content.

1. **Technical Sections**
   - Respond thoroughly to the client's concerns, as expressed in writing or meetings.
   - Follow whatever organization plan that can be inferred from the request for proposal.
   - Use frequent subheadings with specific wording.
   - Back up all claims with facts.

2. **Management Sections**
   - Describe who will do the work.
   - Explain when the work will be done.
   - Display schedule information graphically.
   - Highlight personnel qualifications (but put resumes in appendices).

3. **Cost Section**
   - Make costs extremely easy to find.
   - Use formal or informal tables when possible.
   - Emphasize value received for costs.
   - Be clear about add-on costs or options.
   - Always total your costs.

Conclusion

Formal proposals should always end with a section labeled "Conclusion" or "Closing." This final section of the text gives you the chance to restate a main benefit, summarize the work to be done, and assure clients that you plan to work with them closely to satisfy their needs. Just as important, this brief section helps you end on a positive note. You come back full circle to what you stressed at the beginning of the document—benefits to the client and the importance of a strong personal relationship. (Without the conclusion, the client's last impression would be made by the cost section in the discussion.)

Appendices

Because formal proposals are so long, readers sometimes have trouble locating information they need. Headings help, but they are not the whole answer. Another way you can help readers is by transferring technical details from the proposal text into appendices. The proposal still contains detail—for technical readers who want it—but details do not intrude into the text.

Proposals often include boilerplate such as the resumes of all major personnel who will be working on a project, or project sheets like the ones at the end of chapter 2. Creating boilerplate for this information can save you or your employer considerable time by permitting you to place the appropriate material appendices in all sales proposals.

Any supporting information can be placed in appendices, but following are some common items included there as well:

- Resumes
- Organization charts
- Company histories
- Detailed schedule charts
- Contracts
- Cost tables
- Detailed options for technical work
- Summaries of related projects already completed
- Questionnaire samples

This boilerplate is often printed from separate files, and thus is not paged in sequence with your text. Instead, it is best to use individual paging within each appendix. For example, pages in an Appendix B are numbered B-1, B-2, B-3, and so on.

>>> Guidelines for Feasibility Studies

Much like recommendation reports (see chapter 8), feasibility studies guide readers toward a certain line of action. Another similarity is that both report types can be either in-house or external. Yet most feasibility studies have the following five distinctive features that justify their being considered separately here:

1. They are always solicited by the reader, usually for the purpose of deciding on the best course of action.
2. They always assume one of these two patterns of organization:
   - An analysis of the advantages and disadvantages of one course of action, product, or idea
   - A comparison of two or more courses of action, products, or ideas
3. They are always intended to help managers and other decision makers vote for or against an idea or select among several alternatives.

4. They usually "nudge" (as opposed to "urge" or "push") the reader toward a decision—that is, they are supposed to be written in such a way that the facts speak for themselves.

5. They are often preceded in time by a proposal.

In some ways, feasibility studies could be viewed as a cross between technical reports and proposals. As a writer, you are expected to deal with the topic objectively and honestly, yet you are also expected to express your point of view. A feasibility study determines if some course of action is practical. For example, it may be desirable for a student to quit work and return to college full-time. Yet if that same student has hefty car and apartment payments, the only feasible alternative may be part-time course work.

The following guidelines help you prepare the kinds of feasibility studies requested by your boss (if the study is in-house) or by your client (if the study is external). In either case, your study may be used as the basis for a major decision. Refer to Model 10–5 on pages 364–365 as you read and apply these guidelines to your own writing.

>> Feasibility Study Guideline 1:
Choose Format Carefully

In deciding whether to use the format of an informal (letter or memo) or formal document, use the same criteria mentioned earlier in the chapter with regard to proposals. As always, the central questions concern your readers:

- What format gives them easiest access to the data, conclusions, and recommendations of your study?
- Are there enough pages to suggest the need for a table of contents (i.e., a formal report)?
- What is the format preference of your readers?
- What has been the format of previous feasibility studies written for the same organization?

>> Feasibility Study Guideline 2: Use the ABC Structure

Like other forms of technical writing, good feasibility studies have the basic three-part structure of **A**bstract, **B**ody, and **C**onclusion—although the exact headings you choose may vary from report to report. To the right is an overview of what the study includes.

ABC Format: **Feasibility Study**

- **ABSTRACT:** Capsule summary of information for the most important readers (i.e., the decision makers)
- **BODY:** Details that support whatever conclusions and recommendations the study contains, working logically from fact toward opinion
- **CONCLUSION:** Wrap-up in which you state conclusions and recommendations resulting from study

The following guidelines examine specific sections of feasibility studies, along with details of content and tone. The focus here is on what would be included in an informal document. (See chapter 9 for guidelines that could be applied to a formal feasibility study.)

>> Feasibility Study Guideline 3: Call Your Abstract an Introductory Summary

This section provides information that the most important readers would want if they were in a rush to read your study. With that criterion in mind, consider including these items:

- Brief statement about who has authorized the study and for what purpose
- Brief mention of the criteria used during the evaluation
- Brief reference to your recommendation

The last item is important, for it saves readers the frustration of having to wade through the whole document in search of the answers to the questions, "Is this a practical idea?" or "Which alternative is best?" It is best to mention the recommendation up front, giving readers a frame through which to see the entire report.

>> Feasibility Study Guideline 4: Organize the Body Well

More than anything else, readers of feasibility studies expect an unbiased presentation. That means the midsection of your report must clearly and logically work from facts toward recommendations. Following is one useful approach:

1. **Describe evaluation criteria used during your study,** if readers need more detail than was presented in the introductory summary.

2. **Describe exactly *what* was evaluated and *how*,** especially if you are comparing several items.

3. **Choose criteria that are most meaningful to the readers,** such as

   - Cost
   - Practicality of implementing idea
   - Changes that may be needed in personnel
   - Effect on growth of organization
   - Effects on day-to-day operations

   Of course, exact criteria depend on the precise topic you are investigating.

4. **Discuss both advantages and disadvantages** when you are evaluating just one item. Move from advantages to disadvantages. The conclusion allows you to come back around to supporting points.

5. **Follow organization guidelines for comparisons** when evaluating several alternatives (see chapter 3). You can discuss one item or criterion at a time.

>> Feasibility Study Guideline 5: Use the Conclusion for Detailed Conclusions and Recommendations

Here you get the opportunity to state (or restate) the conclusions evident from data you have presented in the discussion. First, state conclusions; then, state your recommendations. Use listings for three or more points to make this last section of the study as easy as possible to read.

>> Feasibility Study Guideline 6: Use Graphics for Comparisons

When comparing several items, you must consider most readers' preference for tabulated information. Tables can appear either in the discussion section or in attachments. In both cases, follow graphics guidelines explained in chapter 12.

>> Feasibility Study Guideline 7: Offer to Meet with the Readers

Most readers have many questions after reading a feasibility study, even if that study has been quite thorough. You score points for eagerness and professionalism if you anticipate needs and express your willingness to meet with readers later. Such meetings give you another opportunity to demonstrate your understanding of the topic.

>>> Chapter Summary

Proposals and feasibility studies stand out as documents that aim to convince readers. In the case of proposals, you are writing to convince someone inside or outside your organization to adopt an idea, a product, or a service. In the case of feasibility studies, you are marshalling facts to support the practicality of one approach to a problem—sometimes in comparison with other approaches. Both documents can be either informal or formal depending on length, complexity, or reader preference.

This chapter includes lists of writing guidelines for informal proposals, formal proposals, and feasibility studies. For informal proposals, follow these basic guidelines:

1. Plan well before you write.
2. Use letter or memo format.
3. Make text visually appealing.
4. Use the ABC format for organization.
5. Use the heading "Introductory Summary" for the generic abstract section.
6. Put important details in the body.
7. Give special attention to establishing need in the body.
8. Focus attention in your conclusion.
9. Use attachments for less important details.
10. Edit carefully.

In formal proposals, abide by the same general format presented in chapter 9 for formal reports. To be sure, formal proposals have a different tone and substance because of their more persuasive purpose. Yet they do have the same basic parts, with minor variations: cover/title page, letter/memo of transmittal, table of contents, list of illustrations, executive summary, introduction, discussion sections, conclusion, and appendices.

Feasibility studies demonstrate that an idea is or is not practical. They may also compare several alternatives. Follow these basic writing guidelines:

1. Choose format carefully.
2. Use the ABC structure.
3. Call your abstract an introductory summary.
4. Organize the body well.
5. Use the conclusion for detailed conclusions and recommendations.
6. Use graphics for comparisons.
7. Offer to meet with the readers.

>>> Learning Portfolio

Communication Challenge "The Black Forest Proposal: Good Marketing, or Bad Business?"

To strengthen its proposals, M-Global hired a new proposal writer at the corporate office in Baltimore. Ben Sadler came well recommended, having both advertising and marketing experience with technical firms. After he finished his first major proposal at M-Global, he had some disagreements during an internal review of his draft by one of the firm's technical experts. What follows is background information on the project, an overview of decisions Ben made in writing the proposal, some questions and comments related to the review, and an assignment for a written response to the Challenge.

Background on Black Forest Project

Jim McDuff and his staff hired Ben Sadler for a specific reason—although business was going fairly well, they decided the company needed new direction and energy in its marketing. Staff members seemed to be taking for granted that clients would always return and that word-of-mouth would keep new clients coming through the doors. Company leaders knew such an attitude was dangerous. Because they wanted to venture into new types of work, they believed the time was right for a new marketing expert. Ben Sadler seemed to be the catalyst the firm needed.

Just after arriving, Ben learned about a request for a proposal recently issued for a large construction job—building a new university campus in southern Germany. The project at Black Forest University included four buildings, for a total of $35 million in construction. It would become the centerpiece of the new campus. M-Global had thus far done no major construction work in Germany, nor had it done much work at colleges and universities anywhere in the world. Yet Ben felt the firm had the technical tools and the personnel to be a contender. After convincing his immediate boss, Kurt Fleisch, Vice President for Marketing, that the proposal was worth writing, Ben went to work.

Proposal Strategy

Although Ben had research assistance from M-Global employees in the United States and Germany, he wrote the draft himself. Following are parts of the writing and marketing strategy he planned to use:

- **Experience:** He emphasized the large construction jobs M-Global had done for other types of government-related agencies in Germany and around the world.

Although M-Global had done no major college or university construction, Ben believed that including work for government agencies would be an adequate substitute.

- **Technical Experts:** Although M-Global's London and Munich offices had no experts to coordinate large-scale construction, Ben knew the company could bring in experts from the United States. Admittedly, it would be more expensive to import talent—plus these high-salary individuals would be able to visit the site only periodically. Yet this arrangement satisfied the minimum technical requirements in the industry. Ben felt comfortable including the experts' resumes in the proposal and highlighting their experience, without mentioning the fact that they did not work out of the German office. That fact seemed to him to be an internal matter.

- **Costs:** When Ben calculated the tentative cost for the project, he was surprised that the figure was so high. The extra personnel costs previously noted were apparently part of the problem. But Ben also thought some costs may have been overstated because M-Global was not used to bidding on such jobs in Germany. (M-Global's corporate accounting manager had done the tentative cost estimate.) Believing the figures were inflated, Ben cut about 10 percent from personnel costs mentioned in the draft. He thought M-Global would perform more efficiently than the accounting manager had estimated.

- **Proposal Strategy:** Ben wanted to come on strong in the executive summary with what he saw as the benefits M-Global could offer. He focused on three main selling points: (1) availability of a nearby office and lab in Munich for project coordination, (2) experience with other large construction jobs in the United States, and (3) M-Global's history of good working relationships in Germany and the rest of Europe.

- **Personal Contacts:** Ben happened to have a close friend at a former firm (not a competitor of M-Global's) who went to college with an official now on the Black Forest University board of directors. Ben wanted to use this "friend of a friend" connection to get a meeting with the board member, perhaps to find out more about the project. He might then get the chance to give part of the "M-Global story" that was not revealed in the proposal.

Questions and Comments for Discussion

After Ben completed the draft, it was reviewed first by J. R. Link, one of the top technical experts at the Baltimore branch and an old-timer with M-Global. J. R. met with Ben and expressed reservations about the project and about the proposal. The questions that follow reflect their conversation, as well as some other concerns about the proposal:

1. J. R. first wondered why Sadler proposed on the job in the first place. Given that such proposals cost M-Global $10,000 or $15,000 to write, why did he bother with the Black Forest job when M-Global doesn't have experience on large construction jobs in Germany? Most projects out of the Munich office are environmental studies. Wouldn't it be a long shot to get the work? And shouldn't M-Global managers have some ethical concern about trying to get a job when they know they don't have the experience other competing firms probably have (or at least *should* have)?

   What is your view of the practical and ethical concerns raised by J. R.?

2. J. R. also questioned whether Ben was being deceptive about the way in which technical experts would be provided for the project. If the resumes were to be included in the proposal, shouldn't M-Global also mention that these experts reside in the United States? This matter didn't seem to be an "internal" one, as Ben stated.

   What's your view of the way Ben handled the issue of outside experts from the United States? How much of this sort of information must be put forth in a competitive proposal?

3. Ben had an honest disagreement about the calculation of costs by the accounting manager. He may have been right or wrong in his reservations about the accountant's estimate. Putting this point aside, was it procedurally correct for him, as project manager, to make changes in the costs submitted by an advisor? Why or why not? Was it ethical?

4. Ben chose a direct approach to proposal content by placing main selling points first (in the executive summary). Do you think this strategy is appropriate in all cultures? Why or why not? (If possible, do some research on technical communication in Germany, Japan, or China before answering the question.)

5. As noted earlier, Ben decided to pursue a connection he had on the board of directors of Black Forest University. Is this strategy ethical? Would it work? What are some possible results of such a strategy?

6. Have two students in the class conduct a role-play of the conversation between J. R. Link and Ben Sadler. The students can use the information just presented and any additional points that conceivably could be put forth by these two men, considering the sketches provided of them.

Write About It

Assume the role of J.R. Before your meeting with Ben, you were contacted informally by Kurt Fleisch about your concerns about the Black Forest proposal. Write a memo to Kurt explaining how you feel about the issues raised in the discussion questions. Now that you and Ben have had a chance to talk, write a memo to Kurt explaining how you feel about the issues raised in the discussion questions.

Collaboration at Work Proposing Changes in Security

General Instructions

Each Collaboration at Work exercise applies strategies for working in teams to chapter topics. The exercise assumes you (1) have been divided into teams of about three to six students, (2) will use team time inside or outside of class to complete the case, and (3) will produce an oral or written response. For guidelines about writing in teams, refer to pages 24–28.

Background for Assignment

Assume your school is reviewing all issues related to (1) the safety of students, faculty, and staff and (2) the security of equipment. This review has not been triggered by any particular event or problem; it is simply a periodic evaluation of conditions on campus. One step in the process has been to request that five consulting firms visit campus, spend a day on a field investigation, and then submit a proposal that lists specific work to be done and the cost of the work. Your school will choose one firm to do the job.

Team Assignment

Assume your team is one of five consulting firms submitting a proposal. As a preliminary step, team members toured the campus, recorded observations, and collected initial ideas to propose. After discussing your observations, agree on three to five main changes to propose to the school's administration. (As an alternative, your team could focus on proposing three to five changes in one particular activity, building, or area of campus.)

Assignments

The assignments in Part 1 and Part 2 can be completed either as individual projects or as team projects. If your instructor assigns team projects, review the information in chapter 1 on team writing.

Part 1: Short Assignments

These short assignments require either that you write parts of informal or formal proposals or that you evaluate the effectiveness of an informal proposal included here.

1. Introductory Summary

For this assignment, select one of the seven projects at the end of chapter 2. Now assume that you were responsible for writing the proposal that resulted in the project. In other words, work backward from the project to the informal proposal that M-Global used to get the work. Write a short introductory summary for the original proposal. Focus on the main reason you think the client would have for hiring M-Global. If necessary, invent additional information to complete this assignment successfully.

2. Needs Section

As this chapter suggests, informal proposals—especially those that are unsolicited—must make a special effort to establish the need for the product or service being proposed. Assume that you are writing an informal proposal to suggest a change in procedures or equipment at your college. Keep the proposal limited to a small change; you may even see a need in the classroom where you attend class (audiovisual equipment? lighting? heating or air systems? aes-

thetics? soundproofing?). Write the needs section that would appear in the body of the informal proposal.

3. Conclusion or Closing

For this assignment, as with assignment 1, select a project from the project sheets at the end of chapter 2. Assume that you were the M-Global employee responsible for writing the informal proposal that resulted in the work described in the project. Write an effective conclusion or closing for the proposal.

4. Boilerplate

For this assignment, review the information about the history of M-Global in chapter 2. Assume that you are a member of "The Pub," the Publications Development team at M-Global. Write a one paragraph history of M-Global that could be used in all formal M-Global proposals. Remember that this document will be boilerplate, and used in all of the company's proposals, whether they are for construction projects, environmental projects, training, equipment development, or some other kind of project. Thus, this history must be general enough that it could be used in any of these documents, but should emphasize M-Global's experience and expertise.

5. Evaluation—Informal Proposal

Review the informal proposal that follows, submitted by MainAlert Security Systems to the M-Global, Inc., office in Atlanta. Evaluate the effectiveness of every section of the proposal.

---

200 Roswell Road
Marietta Georgia 30062
(770) 555–2000

September 15, 2008

Mr Bob Montrose
Operations Manager
M-Global Inc
3295 Peachtree Road
Atlanta Georgia 30324

Dear Bob,

   Thank you for giving MainAlert Security Systems an opportunity to submit a proposal for installation of an alarm system at your new office. The tour of your nearly completed office in Atlanta last week showed me all I need to know to provide you with burglary and fire protection. After reading this proposal, I think you will agree with me that my plan for your security system is perfectly suited to your needs.

This proposal describes the burglary and fire protection system I've designed for you. This proposal also describes various features of the alarm system that should be of great value. To provide you with a comprehensive description of my plan, I have assembled this proposal in five main sections:

1. Burglary Protection System
2. Fire Protection System
3. Arm/Disarm Monitoring
4. Installation Schedule
5. Installation and Monitoring Costs

BURGLARY PROTECTION SYSTEM

The burglary protection system would consist of a 46-zone MainAlert alarm control set, perimeter protection devices, and interior protection devices. The alarm system would have a strobe light and a siren to alert anyone nearby of a burglary in progress. Our system also includes a two-line dialer to alert our central station personnel of alarm and trouble conditions.

Alarm Control Set

The MainAlert alarm control set offers many features that make it well suited for your purposes. Some of these features are as follows:

1. Customer-programmable keypad codes
2. Customer-programmable entry/exit delays
3. Zone bypass option
4. Automatic reset feature
5. Point-to-point annunciation

I would like to explain the point-to-point annunciation feature, because the terminology is not as self-explanatory as the other features are. Point-to-point annunciation is a feature that enables the keypad to display the zone number of the point of protection that caused the alarm. This feature also transmits alarm-point information to our central station. Having alarm-point information available for you and the police can help prevent an unexpected confrontation with a burglar.

Interior and Perimeter Protection

The alarm system I have designed for you uses both interior and perimeter protection. For the interior protection, I plan to use motion detectors in the hallways. The perimeter protection will use glass-break detectors on the windows and door contacts on the doors.

There are some good reasons for using both interior and perimeter protection:

1. Interior and perimeter protection used together provide you with two lines of defense against intrusion.
2. A temporarily bypassed point of protection will not leave your office vulnerable to an undetected intrusion.
3. An employee who may be working late can still enjoy the security of the perimeter protection while leaving the interior protection off.

Although some people select only perimeter protection, it is becoming more common to add interior protection for the reasons I have given. Interior motion detection, placed at carefully selected locations, is a wise investment.

Local Alarm Signaling

The local alarm-signaling equipment consists of a 40-watt siren and a powerful strobe light. The siren and strobe will get the attention of any passerby and unnerve the most brazen burglar.

Remote Alarm Signaling

Remote alarm signaling is performed by a two-line dialer that alerts our central station to alarm and trouble conditions. The dialer uses two telephone lines so that a second line is available if one of the lines is out. Any two existing phone lines in your office can be used for the alarm system. Phone lines dedicated for alarm use are not required.

FIRE PROTECTION SYSTEM

My plan for the fire protection system includes the following equipment:

1. Ten-zone fire alarm panel

2. Eight smoke detectors
3. Water flow switch
4. Water cutoff switch
5. Four Klaxon horns

The ten-zone fire alarm panel will monitor one detection device per zone. Because each smoke detector, the water flow switch, and the water cutoff switch have a separate zone, the source of a fire alarm can be determined immediately.

To provide adequate local fire alarm signaling, this system is designed with four horns. Remote signaling for the fire alarm system is provided by the MainAlert control panel. The fire alarm would report alarm and trouble conditions to the MainAlert control panel. The MainAlert alarm control panel would, in turn, report fire alarm and fire trouble signals to our central station. The MainAlert alarm panel would not have to be set to transmit fire alarm and fire trouble signals to our central station.

ARM/DISARM MONITORING

Because 20 of your employees would have alarm codes, it is important to keep track of who enters and leaves the office outside of office hours. When an employee arms or disarms the alarm system, the alarm sends a closing or opening signal to our central station. The central station keeps a record of the employee's identity and the time the signal was received. With the arm/disarm monitoring service, our central station sends you opening/closing reports on a semi-monthly basis.

INSTALLATION SCHEDULE

Given the size of your new office, our personnel could install your alarm in three days. We could start the day after we receive approval from you. The building is now complete enough for us to start anytime. If you would prefer for the construction to be completed before we start, that would not present any problems for us. To give you an idea of how the alarm system would be laid out, I have included an attachment to this proposal showing the locations of the alarm devices.

INSTALLATION AND MONITORING COSTS

Installation and monitoring costs for your burglary and fire alarm systems as I have described them in this proposal are as follows:

- $8,200 for installation of all equipment
- $75 a month for monitoring of burglary, fire, and opening/closing signals under a two-year monitoring agreement

The $8,200 figure covers the installation of all the equipment I have mentioned in this proposal. The $75-a-month monitoring fee also includes opening/closing reports.

CONCLUSION

The MainAlert control panel, as the heart of your alarm system, is an excellent electronic security value. The MainAlert control panel is unsurpassed in its ability to report alarm status information to our central station. The perimeter and interior protection offers complete building coverage that will give you peace of mind.

The fire alarm system monitors both sprinkler flow and smoke conditions. The fire alarm system I have designed for you can provide sufficient warning to allow the fire department to save your building from catastrophic damage.

The arm/disarm reporting can help you keep track of employees who come and go outside of office hours. It's not always apparent how valuable this service can be until you need the information it can provide.

I'll call you early next week, Bob, in case you have any questions about this proposal. We will be able to start the installation as soon as you return a copy of this letter with your signature in the acceptance block.

Sincerely,

Anne Rodriguez Evans

Anne Rodriguez Evans
Commercial Sales
Enc.

ACCEPTED by M-Global, Inc.

By: _____

Title: _____

Date: _____

ALARM SYSTEM LAYOUT FOR
M-GLOBAL INC.–ATLANTA, GA

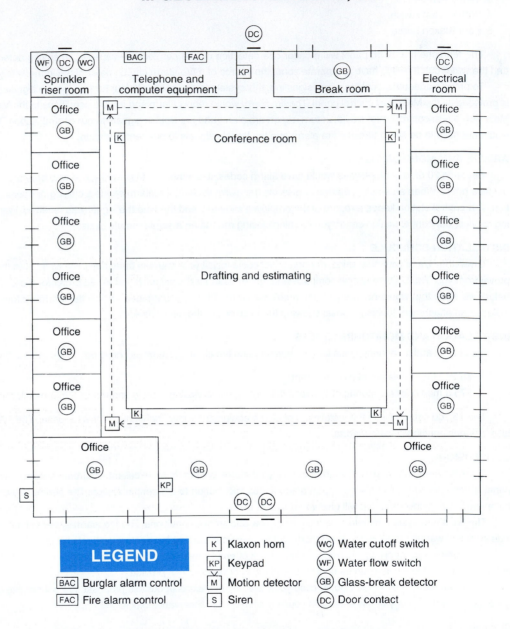

LEGEND

BAC	Burglar alarm control
FAC	Fire alarm control
K	Klaxon horn
KP	Keypad
M	Motion detector
S	Siren
WC	Water cutoff switch
WF	Water flow switch
GB	Glass-break detector
DC	Door contact

Part 2: Longer Assignments

For each of these assignments (except number 9), complete a copy of the Planning Form included at the end of the book.

6. Informal Proposal—M-Global

Choose Option A or Option B. Remember that informal proposals should be fairly limited in scope, given their length and format.

Option A: In-House

- Use your past or present work experience to write a memo proposal suggesting a change at M-Global, Inc. Possible topic areas include changes in operating procedures, revisions to company policies, additions to the work force, alterations of the physical plant, or purchase of products or services.

- Place yourself in the role of an employee of M-Global. The proposal may be solicited or unsolicited—whichever best fits your situation.
- Make sure that your proposal topic is limited enough in scope to be covered fully in an informal proposal with memo format.
- Choose at least two levels of readers who could conceivably be decision makers about a proposal such as the one you are writing—for example, branch or corporate managers. Review chapter 2 if necessary.

Option B: Sales

- Select a product or service (1) with which you are reasonably familiar (on the basis of your work experience, research, or other interests) and (2) that could conceivably be purchased by a company like M-Global.
- Put yourself in the role of someone representing the company that makes the product or provides the service.
- Write an informal sales proposal in which you propose purchase of the product or service by a representative of M-Global.

7. Formal Proposal—M-Global

Choose Option A, B, or C. Make sure that your topic is more complex than the one you would choose for the preceding informal-proposal assignments.

Option A: Community-Related

- Write a formal proposal in which you propose a change in (1) the services offered by a city or town (e.g., mass transit, waste management) or (2) the structure or design of a building, garden, parking lot, shopping area, school, or other civic property.
- Select a topic that is reasonably complex and yet one about which you can locate information.
- Place yourself in the role of an outside consultant with a division of M-Global, Inc., who is proposing the change.
- Choose either an unsolicited or a solicited context.
- Write to an audience that could actually be the readers. Do enough research to identify at least two levels of audience.

Option B: School Related

- Write a proposal in which you propose a change in some feature of a school you attend or have attended.
- Choose from topics such as operating procedures, personnel, curricula, activities, and physical plant.
- Select an audience that would actually make decisions on such a proposal.
- Give yourself the role of an outside consultant working for M-Global, Inc.

Option C: Work-Related

- Write a proposal in which you, as a representative of M-Global, propose purchase of a product or service by another firm.
- Choose a topic about which you have work experience, research knowledge, or keen interest—and one that could conceivably be offered in one of M-Global's project areas. Make sure you have good sources of information.
- Choose either a solicited or an unsolicited context.

8. Feasibility Study—M-Global

- Choose any one of the proposal assignments that you completed as part of the preceding assignments. (Or for this assignment, you can use a proposal completed by one of your classmates.)
- Take yourself out of the role of proposal writer. Instead, consider yourself to be someone assigned (or hired) to complete the task of evaluating the practicality of the proposal after it has been received.
- If appropriate, choose several alternatives to evaluate.

⑨ 9. Ethics Assignment

Reread this chapter's "Communication Challenge" and its "Questions and Comments for Discussion." Put yourself in the position of Jim McDuff, company president. You've just had a phone call from J. R. Link, your old friend and colleague, who proceeds to relate his concerns about Ben Sadler. Put bluntly, J. R. thinks you should fire Sadler immediately before he does serious damage to the firm. You indicate you'll think about it, talk to Sadler, and then call J. R. to tell him your decision. If you were Jim McDuff, how would you handle the conversation with Sadler? (Remember why you hired him.) Would you fire him, praise him, or seek to moderate or redirect his efforts? Explain your position in detail in an essay.

10. Informal or Formal Proposal—International Context

- Assume you are a consultant asked to propose a one-week training course to one of M-Global's offices outside the United States. (See the map with the project sheets at the end of chapter 2 for a list of these seven locations.) Most or all seminar participants are residents native to the country you choose—not U.S. citizens working overseas.
- Choose a seminar topic familiar to you—for example, from college courses, work experience, or hobbies—or one that you are willing to learn about quickly through some study.

- Research work habits, learning preferences, social customs, and other relevant topics concerning the country where the M-Global office you have chosen is located.
- Write M-Global, Inc., an informal or a formal proposal that reflects your understanding of the topic, your study of the country, and your grasp of the proposal-writing techniques presented in this chapter.

Optional Team Approach: If this assignment is done by teams within your class, assume that members of your team work for a company proposing training seminars at M-Global offices around the world. Each team member has responsibility for one of M-Global's non-U.S. offices.

Different sections of the proposal will be written by different team members, who may be proposing the same seminar for all offices or different seminars. Whatever the case, the document as a whole should be unified in structure, format, and tone. It will be read by (1) the vice president for international operations at the corporate office, (2) the vice president for research and training at the corporate office, and (3) all six branch managers in Venezuela, England, Saudi Arabia, Kenya, Germany, and Japan.

 11. A.C.T. N.O.W. Assignment (Applying Communication To Nurture Our World)

Many educational institutions sponsor credit or noncredit service projects that permit participating students to learn new skills while also helping people in a near or distant location. Determine the type of service learning activities, if any, that are available through your college or university. Then write a proposal for a new service-learning project that you believe would fit your campus and provide an important service to your local community, other domestic location, or an international setting. Direct your proposal to individuals who actually would consider such a proposal on your campus, and cover as much information as possible that would help them move to the next step in the decision-making process.

Professional
Documentation, Inc.

3450 Jones Mill Road
Neming Georgia 30092
(404) 555-8438

January 15, 2008

Mr David Barker
Technical Communication Manager
Real Big Professional Software
PO Box 123456
Atlanta Georgia 30339

Dear David:

I enjoyed meeting with you and learning about your new General Ledger software product. Because you require a March release, I can understand why you want to choose an approach to documentation and get the project started.

This proposal describes a strategy for completing the documentation in the 10 weeks between now and your March deadline. Included are these main sections:

1. Selection of the Best Format
2. Adoption of a Publication Plan
3. Control of Costs
4. Conclusion

SELECTION OF THE BEST FORMAT

I think your customers will be best served by a combined installation and user's guide. It uses a functional approach to show how General Ledger works. My assessment results from these completed steps:

- Interviews with support staffers responsible for providing technical support to customers using the company's other accounting products
- Interviews with programmers developing General Ledger, who have an intimate knowledge of how it works
- Conversations with you that clarified your organization's general expectations for the documentation

The assessment is also based on my experience developing documentation for other products. I strive to use clear, concise prose and ample white space to provide a visually appealing text. The text will be enhanced and supplemented with graphics depicting General Ledger's feature screens. The screens themselves will be captured directly from the program and inserted into the text by your staff using your in-house publishing system.

Shows understanding of client's main concern—scheduling.

Asserts ability to meet scheduling need.

Gives helpful overview of sections to follow.

Uses list to itemize important points—that is, the basis for his assessment.

■ **Model 10–1** ■ Letter proposal

David Barker
January 15, 2008
Page 2

Continues emphasis on benefits to reader.

This approach will yield a thorough and easy-to-use document that will allow your customers to take full advantage of General Ledger's many innovative features.

Leads in smoothly to next section.

As you know, writing documentation is a cooperative effort. Each member of the General Ledger product team will play key roles during the development process. To keep us all on track, I have put together a publication plan that shows how the project will progress from beginning to end.

ADOPTION OF A PUBLICATION PLAN

Starts with *overview* of sections to follow.

The publication plan shows how we can have the documentation ready for General Ledger's March unveiling. The four major steps are described here.

Define the Project

Organizes paragraph around three *main points*.

Much of this work has already been accomplished as a result of doing the research for this proposal. As a preliminary step, we will meet and review the project's scope and priority within the organization. We will detail the resources that will be available to complete the project. Most important, we will look at expectations: management's, yours, and the customers'.

Develop a Schedule

This step is the key to the publication process and ensures a common understanding of what has to be done and in what period of time. It has three basic steps:

Uses bulleted list for primary *steps* in project.

- We define the tasks that are part of the project.
- We define the resources we have available to deal with the identified task.
- We assign tasks to the most appropriate individuals.

Manage the Project

Introduces section with *question* to attract attention to passage.

What is good project management? In this plan, good management is essentially good communications. In the first three steps, we define the information that project members must have to understand how the document will be produced and their roles in that process. Ongoing management of the project will be a matter of keeping the channels of communication open.

Perform a Postmortem

The last step is an evaluation of the effectiveness of the publication plan. It provides the opportunity for us to learn how to do future documentation better. It is important to look back at what went right and what went wrong during a project and to share this information with the others. You will get a complete postmortem report from me after the project is completed.

■ **Model 10–1** ■ continued

David Barker
January 15, 2008
Page 3

Throughout the project, this management system will guide us in completing General Ledger's documentation on time and within budget.

CONTROL OF COSTS

Good documentation helps to sell software. By working smart, we can develop documentation that will enhance General Ledger's appeal, and we can do it at a reasonable cost.

My experience in this area and the management system described here will reduce waste and duplication of effort, two factors that affect cost. This savings means I can bring the project in within the 200-hour cap you mentioned.

This estimate assumes that three of the program's four main features are in a complete, or "fixed," state and that the fourth main feature is about 50 percent complete. This estimate also assumes that all programming will be finished by March 5, which will allow time to put the guide through final review and production.

CONCLUSION

The functional approach, which describes a product in terms of its operations, is the documentation format that will best serve General Ledger customers. Your goal of having the documentation ready by March will be aided by adopting a four-step publication plan. The plan will define the strategy for writing the documentation and will help keep costs down.

I'll call you in a few days, David, to answer any questions you might have about this proposal. I can begin work on the documentation as soon as you sign the acceptance block and return a copy of this letter to me.

Sincerely,

Steven Nickels

Steven Nickels
Documentation Specialist

Enclosure

ACCEPTED by Real Big Professional Software

By: _____

Title: _____

Date: _____

Shows interest in following through.

Places benefit in heading.

Shows he can meet project criteria—but also clarifies the assumptions he is making.

Returns to main concern of reader—scheduling.

Retains control of next step.

Includes acceptance block to simplify approval process.

■ **Model 10–1** ■ continued

Gives concise view of problem—and his proposed solution.

DATE: October 3, 2008
TO: Gary Lane
FROM: Jeff Bilstrom *JB*
SUBJECT: Creation of Logo for Montrose Service Center

Part of my job as director of public relations is to get the Montrose name firmly entrenched in the minds of metro Atlanta residents. Having recently reviewed the contacts we have with the public, I believe we are sending a confusing message about the many services we offer retired citizens in this area.

To remedy the problem, I propose we adopt a logo to serve as an umbrella for all services and agencies supported by the Montrose Service Center. This proposal gives details about the problem and the proposed solution, including costs.

The Problem

Includes effective lead-in.

The lack of a logo presents a number of problems related to marketing the center's services and informing the public. Here are a few:

Uses bulleted list to highlight main difficulties posed by current situation.

- The letterhead mentions the organization's name in small type, with none of the impact that an accompanying logo would have.
- The current brochure needs the flair that could be provided by a logo on the cover page, rather than just the page of text and headings that we now have.
- Our 14 vehicles are difficult to identify because there is only the lettered organization name on the sides without any readily identifiable graphic.
- The sign in front of our campus, a main piece of free advertising, could better spread the word about Montrose if it contained a catchy logo.
- Other signs around campus could display the logo, as a way of reinforcing our identity and labeling buildings.

Ends section with good transition to next section.

It's clear that without a logo, the Montrose Service Center misses an excellent opportunity to educate the public about its services.

The Solution

Starts with main point—need for logo.

I believe a professionally designed logo could give the Montrose Service Center a more distinct identity. Helping to tie together all branches of our operation, it would give the public an easy-to-recognize symbol. As a result, there would be a stronger awareness of the center on the part of potential users and financial contributors.

■ **Model 10–2** ■ Memo proposal

Gary Lane
October 3, 2008
Page 2

The new logo could be used immediately to do the following:

- Design and print letterhead, envelopes, business cards, and a new brochure.
- Develop a decal for all company vehicles that would identify them as belonging to Montrose.
- Develop new signs for the entire campus, to include a new sign for the entrance to the campus, one sign at the entrance to the Blane Workshop, and one sign at the entrance to the Administration Building.

Focuses on benefits *of proposed change.*

Cost

Developing a new logo can be quite expensive. However, I have been able to get the name of a well-respected graphic artist in Atlanta who is willing to donate his services in the creation of a new logo. All that we must do is give him some general guidelines to follow and then choose among eight to ten rough sketches. Once a decision is made, the artist will provide a camera-ready copy of the new logo.

Emphasizes benefit *of possible price break.*

- Design charge $0.00
- Charge for new letterhead, envelopes,
  business cards, and brochures
  (min. order)   545.65
- Decal for vehicles 14 @ $50.00 + 4%   728.00
- Signs for campus   415.28

Total Cost $1,688.93

Uses listing *to clarify costs.*

Conclusion

As the retirement population of Atlanta increases in the next few years, there will be a much greater need for the services of the Montrose Service Center. Because of that need, it's in our best interests to keep this growing market informed about the organization.

I'll stop by later this week to discuss any questions you might have about this proposal.

Closes with major benefit to reader and urge to action.

Keeps control of next step.

■ **Model 10–2** ■ continued

**PROPOSAL FOR SUPPLYING
TEAK CAM CLEAT SPACERS**

Prepared by
Totally Teak, Inc.

Prepared for
John L. Riggini
Bosun's Locker Marine Supply

August 22, 2008

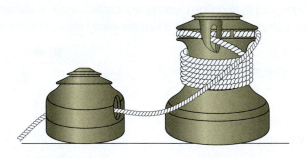

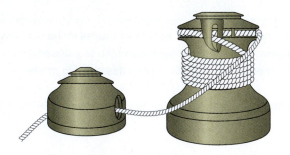

■ **Model 10–3** ■ Formal proposal

Totally Teak Inc
6543 Amster Avenue NW
Atlanta Georgia 30308
404.555.9425

August 22, 2008

John L Riggini President
Bosun's Locker Marine Supply
38 Oakdale Parkway
Norcross OH 43293

Dear Mr. Riggini:

I enjoyed talking with you last week about inventory needs at the 10 Bosun's stores. In response to your interest in our products, I'm submitting this proposal to supply your store with our Teak Cam Cleat Spacers.

This proposal outlines the benefits of adding Teak Cam Cleat Spacers to your line of sailing accessories. The potential for high sales volume stems from the fact that the product satisfies two main criteria for any boat owner:

1. It enhances the appearance of the boat.
2. It makes the boat easier to handle.

Your store managers will share my enthusiasm for this product when they see the response of their customers.

I'll give you a call next week to answer any questions you have about this proposal.

Sincerely,

William G. Rugg

William G. Rugg
President
Totally Teak Inc

WR/rr

In this example, the letter of transmittal appears immediately following the title page. It can also appear before the title page (see Model 10–4).

Establishes *link* with previous client contact.

Stresses two main benefits.

Says he will *call* (rather than asking client to call).

CONTENTS

Organizes entire proposal around *benefits*.

	PAGE
EXECUTIVE SUMMARY	1
INTRODUCTION	2
Background	2
Proposal Scope and Format	2
FEATURES AND BENEFITS	3
Practicality	3
Suitability for a Variety of Sailors	4
High-Quality Construction and Appearance	4
Dealer Benefits	4
Sizable Potential Market	5
Affordable Price	5
CONCLUSION	6

ILLUSTRATIONS

Figure: Side Views of Cleating Arrangement	3

■ **Model 10–3** ■ continued

EXECUTIVE SUMMARY

This proposal outlines features of a custom-made accessory designed for today's sailors—whether they be racers, cruisers, or single-handed skippers. The product, Teak Cam Cleat Spacers, has been developed for use primarily on the Catalina 22, a boat owned by many customers of the 10 Bosun's stores. However, it can also be used on other sailboats in the same class.

The predictable success of Teak Cam Cleat Spacers is based on two important questions asked by today's sailboat owners:

- Will the accessory enhance the boat's appearance?
- Will it make the boat easier to handle and, therefore, more enjoyable to sail?

This proposal answers both questions with a resounding affirmative by describing the benefits of teak spacers to thousands of people in your territory who own boats for which the product is designed. This potential market, along with the product's high profit margin, will make Teak Cam Cleat Spacers a good addition to your line of sailing accessories.

Briefly mentions main *need* to which proposal responds.

Reinforces main points mentioned in *letter* (*selective repetition* of crucial information is acceptable).

1

■ **Model 10–3** ■ continued

INTRODUCTION

Makes clear the proposal's purpose.

The purpose of this proposal is to show that Teak Cam Cleat Spacers will be a practical addition to the product line at the Bosun's Locker Marine Supply stores. This introduction highlights the need for the product, as well as the scope and format of the proposal.

Gives lead-in about section to follow.

Background

Sailing has gained much popularity in recent years. The high number of inland impoundment lakes, as well as the vitality of boating on the Great Lakes, has spread the popularity of the sport. With this increased interest, more and more sailors have become customers for a variety of boating accessories.

Establishes need for product.

What kinds of accessories will these sailors be looking for? Accessories that (1) enhance the appearance of their sailboats and (2) make their sailboats easier to handle and, consequently, more enjoyable to sail. With these customer criteria in mind, it is easy to understand the running joke among boat owners (and a profitable joke among marine supply dealers): "A boat is just a hole in the water that you pour your money into."

Shows his understanding of need (personal experience of designing owner survey).

The development of this particular product originated from our designers' first-hand sailing experiences on the Catalina 22 and knowledge obtained during manufacture (and testing) of the first prototype. In addition, we conducted a survey of owners of boats in this general class. The results showed that winch and cam cleat designs are major concerns.

Proposal Scope and Format

The proposal focuses on the main advantages that Teak Cam Cleat Spacers will provide your customers. These six sections follow:

Gives list of sections to follow, to reinforce organization of proposal.

1. Practicality
2. Suitability for a Variety of Sailors
3. High-Quality Construction and Appearance
4. Dealer Benefits
5. Sizable Potential Market
6. Affordable Price

2

■ **Model 10–3** ■ continued

FEATURES AND BENEFITS

Uses main heading that *engages reader's interest.*

Teak Cam Cleat Spacers offer Bosun's Locker Marine Supply the best of both worlds. On the one hand, the product solves a nagging problem for sailors. On the other hand, it offers your store managers a good opportunity for profitability. Described here are six main benefits for you to consider.

Practicality

Phrases each side heading in "benefit-centered" language.

This product is both functional and practical. When installed in the typical arrangement shown in the figure below, the Teak Cam Cleat Spacer raises the height of the cam cleat, thereby reducing the angle between the deck and the sheet as it feeds downward from the winch. As a result of this increased height, a crewmember is able to cleat a sheet with one hand instead of two.

SIDE VIEWS OF CLEATING ARRANGEMENT

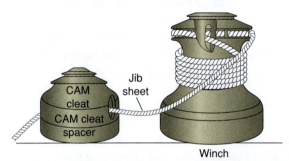

With Spacer

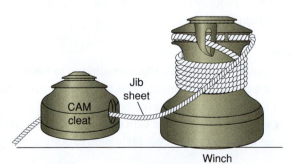

Without Spacer

3

Explains exactly how product will work.

Such an arrangement allows a skipper to maintain steerage of the boat, keeping one hand on the helm while cleating the sheet with the other. Securing a sheet in this manner can be done more quickly and securely. Also, this installation reduces the likelihood of a sheet "popping out" of the cam cleat during a sudden gust of wind.

Suitability for a Variety of Sailors

Moves logically from racing to cruising to single-handed sailing— all to show usefulness of product.

For the racer, cruiser, and single-handed sailor, sailing enjoyment is increased as sheets and lines become easier to handle and more secure. In a tight racing situation, these benefits can be a deciding factor. The sudden loss of sail tension at the wrong moment as a result of a sheet popping out of the cam cleat could make the difference in a close race.

A cruising sailor is primarily concerned with relaxation and pleasure. A skipper in this situation wants to reduce his or her workload as much as possible. In the instance of a sheet popping loose, the sudden chaos of a sail flapping wildly interrupts an otherwise tranquil atmosphere. Teak Cam Cleat Spacers reduce the chance of this happening.

A cruising sailor often has guests abroad. In this situation, as well as in a race, the skipper wants to maintain a high level of seamanship, especially where the control of the boat and the trim of its sails are concerned.

The single-handed sailor derives the greatest benefit from installing Teak Cam Cleat Spacers. Without crew nearby to assist with handling lines or sheets, anything that makes work easier for the skipper is welcome.

High-Quality Construction and Appearance

Stresses quality.

The teakwood frame from which this product is manufactured is well suited for use around water, since teak will not rot. It also looks nice when oiled or varnished.

The deck of most sailboats is made primarily of fiberglass. The appearance of such a boat can be significantly enhanced by the addition of some teak brightwork.

Each spacer is individually handcrafted by Totally Teak, Inc., to guarantee a consistent level of high quality.

Dealer Benefits

Appeals to self-interest of individual Bosun's dealers.

Teak Cam Cleat Spacers make a valuable addition to the dealer's product line. They complement existing sailing accessories as well as provide the customer with the convenience of a readily available prefabricated product.

A customer who comes in to buy a cam cleat is a ready prospect for the companion spacer. Such a customer will likely want to buy mounting hardware as well.

With this unique teak product readily available, a dealer can save the customer the time and trouble of fabricating makeshift spacers.

4

■ **Model 10–3** ■ continued

Sizable Potential Market

These Teak Cam Cleat Spacers are designed with a large and growing potential market in mind. They are custom-made for the Catalina 22, one of the most popular sailboats in use today. More than 13,000 of these sailboats have been manufactured to date. These spacers are also well suited for other similar-class sailboats.

Includes number of owners to emphasize potential sales.

Affordable Price

The Teak Cam Cleat Spacers made by Totally Teak, Inc., wholesale for $3.95/pair. Suggested retail is $6.95/pair. This low price is easy on the skipper's wallet and should help this product move well. And, of course, the obviously high profit margin should provide an incentive to your store managers.

Keeps price information short *and* clear.

■ **Model 10–3** ■ continued

CONCLUSION

Why should a marine supply dealer consider carrying Teak Cam Cleat Spacers? This product satisfies two common criteria of sailboat owners today: it enhances the appearance of any sailboat, and it makes the boat easier to handle. The potential success of this product is based on its ability to meet these criteria and the following features and benefits:

1. It is practical, allowing quick, one-handed cleating.
2. It is ideally suited for a variety of sailors, whether they are racing, cruising, or sailing single-handedly.
3. It is a high-quality, handcrafted product that enhances the appearance of any sailboat.
4. It is a product that benefits the dealer by making a valuable addition to her or his product. It complements existing sail accessories and satisfies a customer need.
5. It is geared toward a sizable potential market. Today there are thousands of sailboats in the class for which this accessory is designed.
6. It is affordably priced and provides a good profit margin.

Links list of benefits with order of same in discussion—drives home advantages of product to user and dealer.

6

■ **Model 10–3** ■ continued

Hydrotech Diving and Salvage Inc
Industrial Complex
Belle Chase Louisiana 70433

February 25, 2008

Peter Hancock
Purchasing Manager
M-Global Inc.
127 Rainbow Lane
Baltimore MD 21201

Dear Mr. Hancock:

Your February 1, 2008, RFP details the need for improved hull maintenance on your excelsior-class drillships. Traditional approaches to hull cleaning, as you know, have become costly and inefficient. This proposal offers Hydrotech's innovative and affordable solution to M-Global's maintenance needs.

Recent technology has spurred development of mobile hull maintenance systems that can be used on anchored vessels almost anywhere. This equipment, used regularly, will greatly reduce the accumulation of marine growth. As a result, your ships will cruise faster and more efficiently.

I'll give you a call next week to answer any questions you may have about how Hydrotech can improve hull maintenance at M-Global.

Sincerely,

Stephen B. Wilson

Stephen B. Wilson
Underwater Maintenance Supervisor

sf

Refers briefly to problem.

Emphasizes Hydrotech's main selling point: an innovative cleaning system.

Stays in control of proposal process by saying he will call.

■ **Model 10–4** ■ Formal proposal

Title mentions
benefit— increased
efficiency.

**INCREASED EFFICIENCY FOR
M-GLOBAL, INC.**

Prepared by

Hydrotech Diving and Salvage, Inc.

for

Peter Hancock

Purchasing Manager

M-Global, Inc.

February 25, 2008

■ **Model 10–4** ■ continued

CONTENTS

Uses white space and indentation to make contents page highly readable.

ILLUSTRATIONS 1

EXECUTIVE SUMMARY 2

INTRODUCTION 3

    Purpose

    Description of Hull Maintenance Problem

    Scope

    Proposal Format

SEASLED: HOW IT WILL WORK FOR M-GLOBAL, INC. 6

    Time-Saving Equipment

    Mobility That Makes Sense

    Diving Crew and Procedure

    Diver Rotation

    Completion Time

SCHEDULE AND QUALIFICATIONS 12

    Schedules That Save Money

    Variety of Services Available to You

    Professionals That Make the Difference

REDUCED COSTS USING SEASLED 16

CONCLUSION 18

■ **Model 10–4** ■ continued

ILLUSTRATIONS

TABLES

1. Job Completion Time: Cleaning the Galaxy Hull with Seasled .. 12

2. Hull Maintenance and Inspection Charges per Ship.................. 17

FIGURES

1. Seasled: Top View... 7

2. Seasled: Bottom View 8

3. Seasled: Side View 9

■ **Model 10–4** ■ continued

2

EXECUTIVE SUMMARY

Marine growth on hulls can reduce the speed of ships, increase fuel consumption, and cause more frequent hull repairs. Excelsior-class drillships have a particularly large hull surface area below the waterline. If marine growth is allowed to accumulate, excessive drag will occur between the hull and water.

The removal of marine growth usually is expensive and time-consuming. Hydrotech, however, offers a hull maintenance system that provides a more convenient and economical way to clean and preserve oil tanker hulls. The Seasled Hull Maintenance System uses a diver-operated, self-propelled, scrubbing and painting device that can be brought directly to your anchored vessels. The primary benefits to your company include the following:

1. Bottom cleaning and preservation completed in about 100 hours, 30 hours less than most conventional cleaning systems; and
2. Hull maintenance done at times and locations convenient for you.

Because the Seasled System reduces the cost of each cleaning, you can now afford annual hull maintenance and semiannual inspections. This regular attention will contribute significantly to the average cruising speed and fuel efficiency of your ships.

Provides overview of problem . . .

. . . and solution.

Focuses on main benefits to client.

Uses style that appeals to busy management readers— short paragraphs, numbered points, no technical jargon.

Summarizes entire proposal.

■ **Model 10–4** ■ continued

3

INTRODUCTION

Repeats background information for readers who start with introduction.

Recent technology has made many contributions to the petroleum shipping industry. A new method of hull cleaning and repair has been developed that offers many advantages to oil tanker operators. We believe this new method will contribute significantly to the success and prosperity of M-Global. Maintaining clean hulls increases ship cruising speed and fuel efficiency.

Purpose

This proposal describes the benefits you receive using the new Seasled Hull Maintenance System. Your company will save time and money by using this service regularly for all excelsior-class drillships.

Description of Hull Maintenance Problem

Shows understanding of problem mentioned in RFP but doesn't belabor it, since client already knows problem exists.

M-Global operates a fleet of six excelsior-class drillships engaged in worldwide exploration for petroleum. Your vessels travel to oceans and ports around the world, where many forms of marine growth collect on ships' hulls. Because the excelsior-class drillship has a large hull surface area below the waterline, even light marine growth causes excessive drag between the hull and water.

Furthermore, if marine growth is allowed to accumulate for an extended period, the hull will deteriorate. Marine growth results in slower ship speed, increased fuel consumption, and more frequent major hull repair.

Scope

Shows that Hydrotech has done its homework in preparing proposal.

This proposal reflects our thorough research and more than ten years' experience on hull maintenance and cleaning for tankers. Besides relying on our own experience, we have consulted experts at firms such as the following:

1. Yamamoto Shipbuilding Service Consultants
   Kure, Japan

■ **Model 10–4** ■ continued

4

2. Marine Corrosion Consultants
   Belle Chase, Louisiana

3. Ocean Science Center
   Key West, Florida

The result has been the development of the Seasled System. This proposal explains how using Seasled reduces cleaning time while maintaining safety-conscious crew procedures. We also cover features of scheduling as well as the backgrounds of key people who will work on the program. Then you will find a cost estimate, per ship, for both maintenance and inspection procedures.

■ **Model 10–4** ■ continued

5

Proposal Format

   For ease of reference, this proposal is divided into these three major sections:

Refers to remaining sections so readers don't have to consult contents.

1. Seasled: How It Will Work for M-Global, Inc.—which describes the equipment, crew, and procedures
2. Schedule and Qualifications—which provides information on schedules, services, and personnel

Gives brief overview of subsections to follow.

3. Reduced Costs Using Seasled—which provides information on the cost of our hull-cleaning service

■ **Model 10–4** ■ continued

6

SEASLED: HOW IT WILL WORK
FOR M-GLOBAL, INC.

This section describes the equipment and procedure used for Seasled hull maintenance of the excelsior-class drillship. The main services proposed are semiannual inspections and annual cleanings (which include spot painting of bare or badly worn sections).

Time-Saving Equipment

Recent advances in underwater technology have yielded specialized equipment that has revolutionized the ship-maintenance industry. The Seasled Hull Maintenance System is a self-propelled, diver-operated device capable of cleaning 400 square feet of hull surface per hour. (See Figures 1, 2, and 3 for top, bottom, and side views of this device.)

Shows that Hydrotech has done its homework in preparing proposal.

■ **Model 10–4** ■ continued

7

Comments and figures show that Seasled incorporates latest technology and that Hydrotech helped develop it.

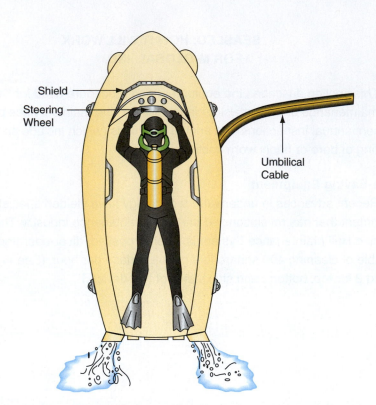

Shield

Steering Wheel

Umbilical Cable

■ **Figure 1** ■ Seasled: Top view

Hydrotech Diving and Salvage, Inc., helped develop the Seasled System and has used it successfully for the past two years. The system is quite reliable, as well as extremely efficient in removing marine growth.

■ **Model 10–4** ■ continued

8

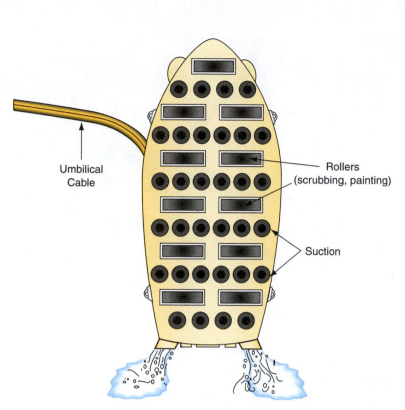

Umbilical
Cable

Rollers
(scrubbing, painting)

Suction

■ **Figure 2** ■ Seasled: Bottom view

As Figures 1 and 2 show, Seasled is designed to scrape and then paint the worn portions of the hull. All power and materials are transmitted to the sled through an umbilical to the tanker. It's simply a matter of lowering Seasled and the operator into position and then letting the device do its work.

■ **Model 10–4** ■ continued

9

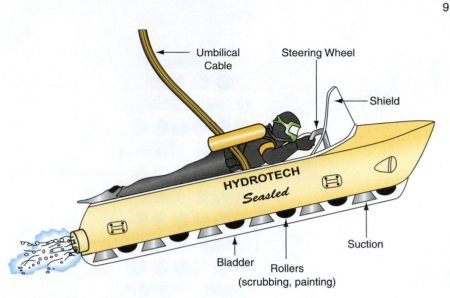

■ **Figure 3** ■ Seasled: Side view

Mobility That Makes Sense

Mobility is the real advantage of Seasled. Any scheduled layover of your drillships can be transformed from lost shipping time to convenient maintenance time.

In the past, drillship operators were often required to take vessels hundreds of miles to deep-water ports where expensive pier space was needed to perform regular hull maintenance. Now our mobile hull-cleaning system can be transported to anchored vessels at locations that are more convenient for you.

You'll also be glad to know that the equipment is easily transferred from small craft to tanker deck, using the cargo lifting davits you already have on board.

Explicitly states difference between results from old and new systems.

■ **Model 10–4** ■ continued

10

Furthermore, the hull cleaning and paint repair can be conducted while the ship is at anchor in calm water.

Diving Crew and Procedure

Our diving crews consist of highly trained professionals who ensure high standards of safety and productivity. A typical diving crew for hull maintenance includes the following members:

1. Two experienced diving supervisors
2. Six certified commercial divers
3. Four vocationally trained tenders
4. Two equipment technicians

These workers are on call 24 hours a day and will supervise the transport of the equipment to the desired location. Once aboard your ship, the system is assembled and operable in less than 12 hours.

Our personnel work 12-hour shifts around the clock with a diving supervisor on duty at all times. The hull-cleaning and paint-repair operation is conducted as follows:

1. Two divers, operating the sleds, start on the port and starboard sides of the bow and work toward the stern.
2. Each diver is monitored and assisted from topside by a tender.
3. Two standby divers are on deck and ready to assist the working diver in case of an emergency.
4. An equipment technician monitors and maintains the system.

Describes procedures with short paragraphs and lists.

■ **Model 10–4** ■ continued

11

During the entire operation, a surface-supplied life-support system and standard diving equipment are used by the diving crew.

Diver Rotation

The working divers are relieved every 4 hours, and the rotation proceeds as follows:

Emphasizes use of safe techniques.

1. The off-going divers are off duty for 12 hours following their dive.
2. The on-coming divers are on call 8 hours before their dive.
3. The on-coming divers act as standby divers for the 4 hours immediately preceding their dive.

Completion Time

Table 1 lists individual tasks in the left column and the maximum completion time for each task in the right column. As you can see, the maximum completion time for an average excelsior-class drillship is 101 hours—far less than other methods.

■ **Model 10–4** ■ continued

16

REDUCED COSTS USING SEASLED

As noted earlier, the Seasled System introduces increased efficiency, and thus reduced costs, to hull-cleaning projects. The charge for hull cleaning and paint repair for your vessels will vary by location, due to transportation costs. For your convenience, we have assigned flat rates for five geographic areas:

- Region 1—Southeastern U.S. (includes Georgia, Florida, Louisiana, and Texas)
- Region 2—the northern quadrant of the Western hemisphere (except that area designated Southeastern U.S.)
- Region 3—the southern quadrant of the Western hemisphere
- Region 4—the northern quadrant of the Eastern hemisphere
- Region 5—the southern quadrant of the Eastern hemisphere

Table 2 lists the charges for hull maintenance and semiannual inspection *per ship*. These costs are guaranteed through 2010 and include all transportation charges. As mentioned in your RFP, M-Global, Inc. will provide room and board for our crews on site.

Introduces innovative "flat rate" fee structure to simplify cost estimating for client.

Displays maintenance and inspection costs in table.

Guarantees costs through specific date to make budget more predictable.

■ **Model 10–4** ■ continued

TABLE 2

HULL MAINTENANCE AND INSPECTION CHARGES PER SHIP

	Regions				
Service	*Region 1*	*Region 2*	*Region 3*	*Region 4*	*Region 5*
Maintenance	$25,000	$30,000	$35,000	$40,000	$40,000
Inspection	$ 5,000	$ 7,000	$ 9,000	$11,000	$11,000

Our portable hull-maintenance system rates apply only to excelsior-class drillships up to 40 feet in length. Special rates can be quoted for loaded tankers on an individual basis.

■ **Model 10–4** ■ continued

18

CONCLUSION

Hull cleaning on drillships used to be expensive and time-consuming. Ships had to be taken hundreds of miles to deep-water ports, where expensive pier mooring was needed to perform underwater hull maintenance.

A new hull-maintenance system, which can be brought to your anchored vessels, is now available. The Seasled Hull Maintenance System uses a diver-operated, self-propelled, scrubbing and painting device. This portable system can clean the bottom of an excelsior-class drillship and paint damaged areas in about four days.

Our diving crew can rendezvous with your ships at times and locations that fit your schedule, and turn lost time into money-saving maintenance time. To keep your ships operating at maximum efficiency, we recommend annual hull cleaning and semiannual inspections of all surfaces beneath the waterline.

We look forward to tailoring a hull-maintenance schedule to fulfill your individual requirements.

Returns to client's major concerns—regular schedules and cost savings.

■ **Model 10–4** ■ continued

MEMORANDUM

DATE: July 22, 2008
TO: Greg Bass
FROM: Mike Tran *MT*
SUBJECT: Replacement of In-House File Server

INTRODUCTORY SUMMARY

Gives context for feasibility study.

The purpose of this feasibility study is to determine if the NTR PC905 would make a practical replacement for our in-house file server. As we agreed in our weekly staff meeting, our current file-serving computer is damaged beyond repair and must be replaced by the end of the week. This

Summarizes conclusion of report.

study shows that the NTR PC905 is a suitable replacement that we can purchase within our budget and install by Friday afternoon.

FEASIBILITY CRITERIA

Pinpoints three criteria to be discussed.

There are three major criteria that I addressed. First, the computer we buy must be able to perform the tasks of a file-serving computer on our in-house network. Second, it must be priced within our $4,000 budget for the project. Third, it must be delivered and installed by Friday afternoon.

Performance

As a file server, the computer we buy must be able to satisfy these criteria:
- Store all programs used by network computers
- Store the source code and customer-specific files for Xtracheck
- Provide fast transfer of files between computers while serving as host to the network
- Serve as the printing station for the network laser printer

Shows how NTR PC905 will fulfill performance criteria.

The NTR PC905 comes with a 240GB hard drive. This capacity will provide an adequate amount of storage for all programs that will reside on the file server. Our requirements are for 60GB of storage for programs used by network computers and 70GB of storage for source code and customer-specific programs. The 240GB drive will leave us with 110GB of storage for future growth and work space.

Only covers advantages because there are no disadvantages to buying PC905.

The PC905 can transfer files and execute programs across our network. It can perform these tasks at speeds up to five times faster than our current file server. Productivity should increase because the time spent waiting for transfer will decrease.

M-Global Inc | 127 Rainbow Lane | Baltimore MD 21202 | 410.555.8175

■ **Model 10–5** ■ Feasibility study (one alternative)

Greg Bass
July 22, 2008
Page 2

The computer we choose as the file server must also serve as the printing station for our network laser printer. The PC905 is compatible with our Hewy Packer laser printer. It also has 4.0GB more memory than our current server. As a result, it can store larger documents in memory and print them with greater speed.

Budget

The budget for the new file server is $4,000. The cost of the PC905 is as follows:

PC905 with 120MB Hard Drive	$2,910
Keyboard	112
Monitor	159
Total	$3,181

> Makes costs easy to find with simple table.

No new network boards need to be purchased because we can use those that are in the current server. We also have all additional hardware and cables that will be required for installation. Thus the PC905 can be purchased for $800 under budget.

Time Frame

Our sales representative at NTR guarantees that we can have delivery of the system by Friday morning. Given this assurance, we can have the system in operation by Friday afternoon.

> Highlights major goal—quick installation.

Additional Benefits

We are currently using NTR PCs at our customer sites. I am very familiar with the setup and installation of these machines. By purchasing a brand of computer currently in use, we will not have to worry about additional time spent learning new installation and operation procedures. In addition, we know that all our software is fully compatible with NTR products.

> Ends with "extras"— that is, benefits not among major criteria but still useful.

The warranty on the PC905 is for one year. After the warranty period, the equipment is covered by the service plan that we have for all our other computers and printers.

CONCLUSION

I recommend that we purchase the NTR PC905 as the replacement computer for our file server. It meets or exceeds all criteria for performance, price, and installation.

> Restates significant point already noted in introductory summary.

Chapter | 11 | Web Pages and Writing for the Web

Craig Baehr,
Texas Tech University

>>> Chapter Outline

Your Role in Developing Websites and Content 367

Planning 368

Content Development 371

Content Chunking 371

Guidelines for Writing Web Content 372

Adapting Content for the Web 372

Scripting Languages and Software Authoring Tools 373

Document Conversion Issues and Common File Formats 374

Structure 374

Site Structures and Types 375

Process of Developing a Structure 376

Navigation Design 379

Guidelines for Labeling 379

Grouping and Arrangement Strategies 381

Design 381

Design Conventions and Principles 382

Finding a Theme and Developing Graphic Content 384

File Formats and Graphics 384

Interface Layouts 385

Usability and Publication 388

Testing Your Site for Your User Base 389

Performing Usability Reviews 389

Quick Usability Checks and System Settings 389

Accessibility Guidelines 391

Publishing Your Site 393

Chapter Summary 393

Learning Portfolio 395

Communication Challenge 395

Collaboration at Work 395

Assignments 396

Models for Good Writing 398

Model 11–1: Original M-Global Website 398

Model 11–2: Proposed M-Global Website 399

Web pages have become an important form of online publication on the Internet for companies, institutions, and individuals. Most organizations use websites to communicate externally with a global audience and internally through *intranets,* or internal company networks. Your college or university probably includes websites in its array of communications with its audience. Thus it is probable that you will be using web pages throughout your career, as you may do now.

Some aspects of web pages—such as navigation, visual elements, and interactivity—differentiate them from printed pages and documents. Websites often have unique structures that define the arrangement and forms of navigation used to browse and search. They rely more on visual content and are interactive, providing feedback to users. They allow users to post comments, buy products, search databases, and perform other useful functions. Because of these characteristics, the nature of developing technical documentation for the web is different from its print-based counterparts. However, the core principles of good technical communication—as discussed throughout this book—can be applied.

This chapter provides you with an overview of developing websites and web content. This five-phase process includes planning, content development, structure, design, and usability. The first section helps you examine your role in website development and provides an overview of the five-step process. The subsequent sections describe each of the five phases. Finally, this chapter provides you with basic knowledge to publish your own finished site.

>>> Your Role in Developing Websites and Content

Depending on the team you are working with and your organization's structure, your role in developing websites may range from being a subject-matter expert who provides content to a development team to creating template-driven documents and completing site development. Much of your role will be determined by the development resources available, the scope of the project, and your own individual expertise. Some common roles on a web development team include project manager, programmer, graphic artist, writer/editor, content provider, and usability tester (Figure 11–1).

Each role may be defined slightly differently by your team or organization; however, these typical roles are found on most teams. You may have multiple persons serving in one role, or you may even have one person performing multiple roles. Other roles may be necessary depending on your project's scope and the available resources. In the preliminary planning of your project, defining roles and responsibilities will be important. (Review the "Guidelines for Team Writing" in chapter 1, pages 24–28.) Regardless of your role, you should understand the basic development process and elements of a website to familiarize yourself with the scope of a web project.

Keep in mind that the process of web development is an *iterative* one—that is, a change in one phase may require you to go back to a previous phase to make adjustments to your work. For example, after you have a site structure, you may need to modify the organization of the site structure to accommodate a navigation problem introduced in a previous phase. Formal and informal reviews throughout the process may require you to revisit other decisions. An iterative process allows errors to be corrected as they are discovered and greater flexibility in project development.

■ **Figure 11–1** ■
Typical web-
development
team roles

Role	Function
Project manager	Serves as the team leader, who establishes and manages the timeline, finances, and resources. May serve as the client's primary point of contact.
Programmer	Oversees the scripting, programming, publishing, and other technical issues.
Graphic artist	Acquires and/or develops graphic content.
Writer/editor	Writes and edits content for the Web project and any formal reports required.
Content provider	Provides content to the writer/editor and team to be included in the site. May serve as a reviewer of the project.
Usability tester	Tests the site for usability and accessibility guidelines.

■ **Figure 11–2** ■
The five-step
process of website
development

Planning	Conceptualizing the site, including analyzing the audience and defining the purpose, scope, and context.
Content Development	Analyzing, writing, editing, and adapting content.
Structure	Developing a site structure and navigation systems.
Design	Designing graphic content and interface layouts.
Usability	Testing the project using usability and accessibility checks and guidelines.

The remainder of this chapter outlines the five major phases of the web development process, which are summarized in Figure 11–2.

>>> Planning

The first phase of web development is the *planning* phase, during which you will make initial decisions to conceptualize the site. For example, you define the site's purpose, analyze your audience, outline the scope of the site the with clear goals, identify constraints that must be considered, and identify methods to incorporate user-centered design into the development process.

First, you must identify the site's purpose and project scope before undertaking its development. Websites have a variety of

Common Purpose	Description	Example
Search Portals	Searchable databases or indexes of websites and web content that provide links to other sites.	Google http://www.google.com
Sales or E-commerce	Sites devoted to selling products and services over the Internet.	Amazon http://www.amazon.com
Informational	Sites that provide information on specific subjects, such as news, government, or other general information.	Internal Revenue Service http://www.irs.gov
Educational	Websites that provide training, courses, tutorials, or supplementary instructional materials.	howtoons http://www.instructables.com/group/howtoons
Entertainment	Sites that provide games or online entertainment.	Comics.Com http://www.comics.com
Personal	Personal web pages that allow users to share information on the web with family and friends.	Any personal home page

■ **Figure 11–3** ■ Common website purposes

purposes, most of which fit into one of six categories: search portals, sales, informational, educational, entertainment, or personal. Figure 11–3 summarizes these common website purposes.

Figure 11–4, the Web Planning Form, is a modified version of the Planning Form from chapter 1. Use this form to help plan your site, and begin by asking and answering these two questions:

■ Why am I developing this site?

■ What response do I want from users of the site?

Next, you must define your audience. Use the Web Planning Form (Figure 11–4) to identify the types of users you expect at your site. Users will have different technical levels (managers, experts, operators, and general readers) as well as different decision-making levels (decision makers, advisers, and receivers). (See chapter 1, pages 9–17, for more information.)

As part of the audience analysis, you should identify all legal, ethical, cultural, social, usability, accessibility, or technical constraints, as well as any obstacles your users face that pertain to the site. For example, a financial consulting firm's site would be ethically constrained to provide accurate information of services and prices. A computer business that sold new systems would be constrained by legal limitations with regard to licensing software on the new systems. If either site had international customers, you should be

WEB PLANNING FORM

Name: _____ Assignment _____

I. Purpose: Answer each question in one or two sentences.

   A. Why are you developing this site?

   B. What response do you want from users?

II. Audience

   A. User Matrix: Fill in names and positions of people who may read the document

	Decision Makers	Advisers	Receivers
Managers			
Experts			
Operators			
General Readers			

   B. Information on Individual Users: Answer these questions about the selected members of your audience. Attach additional sheets as is necessary.

   1. What is this user's technical or educational background?

   2. What main question does this person need answered?

   3. What main action do you want this person to take?

   4. What features of this person's personality might affect his or her use of the site?

III. Website

   1. What content do you want to include on the website?

   2. What graphic choices will present a professional image for me and the organization I represent?

   3. What site structure is appropriate to the subject and purpose of the website? (Attach a site structure sketch.)

   4. What navigation tools are appropriate to the subject and purpose of the website?

■ **Figure 11–4** ■ Web Planning Form

conscious of cultural issues that require the site to be developed in different languages or with alternate content. You might also face competitive issues, such as not disclosing internal proprietary data on a public site.

Usability and accessibility issues are especially important, because you must provide equal access to all of your users, including those with limited access or disabilities. For instance, many sites provide text equivalents for graphic content that may not be viewable on the site by all its users. You must also consider the technological limitations of some computer systems that may be used to access your site. They may require you to consider system settings and features such as screen resolution, software plug-ins, Internet connection speed, and security settings required for users to view and access specific site content. Make a list of all important contextual issues to use as a guide when you are developing content, graphics, and other aspects of your website.

The information you gather during planning helps you design a website tailored to your users' unique needs and specifications. Focusing design on the human user throughout the planning and development process is known as *user-centered design*. The goal of user-centered design is to create a product that is both usable and accessible.

>>> Content Development

The second phase in creating a website is *content development,* which includes writing, editing, and adapting both new and existing content for a web environment. It also involves converting documents into readable formats for the web. Because of fundamental differences between printed documents and websites, usually you won't be able to drop content into a web page without some editing. This section explains content chunking, offers some guidelines for developing web content, shows methods of adapting content for the web, and explains some document conversion issues, common file formats, scripting languages, and software authoring tools.

Content Chunking

The basic written unit of a web page is a content chunk. A *content chunk* is a stand-alone unit of text and graphics. An individual content chunk is often separated from other content chunks by spacing on a page. Content chunks can vary in size, depending on their purpose, screen layout, or specific project style guidelines. Generally, content chunks tend to average around four to seven lines of text, depending on the audience and subject. If a chunk includes graphics, it could be as large as what fits on the user's screen.

The goal of content chunking is not to reduce everything to a short description, but rather to write content so it is more readable and more likely to be read. One of the advantages of the web is that individual chunks and entire sites can be interconnected by links, which may, in fact, accommodate more detailed and lengthy descriptions than most printed books.

Guidelines for Writing Web Content

Research suggests that web readers tend to scan, skim, and "raid" sites for specific content. They read pages in a nonsequential order, based on association, and tend to dislike excessive scrolling in documents. Because users read content differently on screen than in print, the guidelines for writing web content differ from writing printed materials. Following are some other general writing guidelines based on web reading habits:

- Provide a summary of important points first, followed by specifics. Web readers tend to look for results up front, and then look for more details.
- Organize content chunks into main content areas or categories that might serve as navigation links or sections of the website.
- Identify each content chunk or sections of chunks with meaningful headers to match users' habits of scanning documents.
- Rewrite and reformat paragraphs that use lists into bulleted or numbered lists, which are visually more readable on the web.
- Reduce the size of individual chunks to fit on a single screen when possible, because users tend to prefer less scrolling and shorter documents.
- Break up larger chunks into smaller ones that can be linked using hyperlinks. Providing "more information" links with summaries is a good method.
- Provide links to individual sections on longer pages with horizontal scrolling to aid users in searching and navigating.
- Develop a style sheet that establishes the sizes, font faces, colors, and spacing for headers, chunks, graphics, and text. Consistency in the visual style of text chunks helps establish credibility with users.
- Edit your content with a careful eye, because every mistake can compromise the credibility of your work.

Although not exhaustive, this list provides some general writing guidelines to help you write content chunks that best fit the users' reading habits. Be certain to follow proper use of grammar and punctuation, as you would in writing any document. Once the textual content is written and edited, the next step is to adapt it properly for the web.

Adapting Content for the Web

Before a document is converted, you should check for internal consistency to ensure that it is formatted according to your style sheet. Style sheets are composed of a list of rules that govern the consistent markup, format, and display of textual and graphic content for a website. They should include any internal style guidelines unique to your organization or project. For example, such guidelines may relate to fonts, colors, alignments, spacing, size, and format. Using consistent styles throughout your document makes it appear more professional, readable, and usable. After you determine your style guidelines and apply them, take a detailed approach to editing your work, just as you would for any printed document. Read through each chunk to ensure it is well written and edited, and conforms to your style guidelines.

Part of adapting content for the web involves deciding which electronic formats to use for documents. Initially, converting content into electronic format makes it easier to cut and paste your work into web-page templates and existing layouts. Most web content is formatted in *Hypertext Markup Language (HTML),* a basic markup language used to structure and notate content so it can be read by a web browser. However, if your website uses a large number of checklists, worksheets, fact sheets, or other such documents, it may be easier to convert these into file formats that are easily downloadable or printable. These may include Adobe Portable Document Format (.pdf), Microsoft Word (.doc), and Rich Text Format (.rtf). If you have larger printed documents, such as long reports, you may find it easier to use web authoring software or a word-processing program to convert them directly to HTML files or web pages.

Scripting Languages and Software Authoring Tools

Every web developer involved in developing web-page content should have at least a basic knowledge of HTML. HTML includes tags used to mark up structural elements, such as headers, titles, body text, hyperlinks, and graphic content. Other scripting languages can be used with HTML to allow you to add global style sheets, create interactive graphics and forms, and retrieve information from a database for display in a web browser, such as XHTML, Cascading Style Sheets (CSS), JavaScript, and Hypertext Preprocessor (PHP). If you have the time and interest, you can learn HTML and other scripting languages to create more dynamic web pages.

In many cases, you may want to use web authoring software that does the markup and scripting of content for you. If your preference is to use a web-authoring tool, such as Microsoft FrontPage or Adobe Dreamweaver, you may benefit from learning a little HTML to help customize design and fix minor problems. A basic understanding of these elements is sufficient to help solve problems common to most designers. Learning HTML helps familiarize you with the internal structure of web documents. You can find many useful books that will help you learn HTML in any computer section of most bookstores.

There are also many useful online references, such as the World Wide Web Consortium's (W3C) site (http://www.w3.org), which has many reference guides on a variety of scripting languages. Other useful references can be found by doing a quick search for "HTML reference guide" on any major search portal site (such as Yahoo! or Google). Many computer training centers and continuing-education programs offer introductory courses in HTML. One good book, reference website, or introductory course is usually enough to give you a basic understanding of the syntax, structure, and scripting tags used in scripting a website.

Just as no developers should be without some knowledge of markup and scripting, they should also have knowledge of at least one web software authoring tool, such as Microsoft FrontPage or Adobe Dreamweaver. Today, most web developers use software authoring tools or web software development programs. These software programs can make the development process much easier because they provide graphical tools and buttons that enable you to develop page layouts, format text, create navigation toolbars, and add interactive graphics, forms, and features. Learning a little about authoring software gives you an idea of its limitations and also gives you a clearer sense of what can and can't be done easily. Such knowledge may help your team refine goals and choice of development tools and methods.

Document Conversion Issues and Common File Formats

You may have reference documents, product descriptions, definitions, forms, and documents that must be posted in their native format rather than to be converted to a web-page format. It may not be feasible to convert all of your site's documents to HTML files, or you may need to convert them to more printer-friendly formats. For example, a site that sells personal computers might provide printable system configurations in Adobe pdf files because they tend to print more legibly than simply printing what appears on the screen.

You should try to select the file formats that are most commonly used, or provide documents in multiple formats to accommodate the widest user base. You may need to do some research to see what software most of your users have on their systems. If they have Microsoft Word, then they can view documents in that format (.doc files). Using file formats that are readable with free document viewers, such as the Adobe pdf, can save users the additional expense of purchasing software. You can put a link to the free downloads on your site to assist users.

If you are unable to use a single format for your reference documents, you can also provide multiple formats to accommodate users. Some sites provide one version in HTML format and an alternate version for printing. You can also save documents as Rich Text Format files (.rtf) or as Text files (.txt) that are viewable by most basic word-processing and text-editing programs. These file types are good for basic text documents that don't require complex formatting, tables, or graphics. Alternatively, you can scan documents and save them as images, or graphic formats, which are viewable by most web browsers, such as .jpg, .gif, or .png files. Although typically used for graphic images, these file types can also be used to create quick snapshots of documents as images. However, they have some limitations; for example, they can be difficult to modify and can result in larger file sizes, longer download times, and poorer readability. Providing two to three different formats of documents is a good rule of thumb to accommodate most users.

>>> Structure

The third phase in developing a website is to develop a *structure* for the site. The structure of books is both linear and hierarchical, in that they are ordered or outlined using sequential page numbers, chapters, and tables of contents. Websites tend to have more flexible structures—they are composed of content chunks interconnected by hyperlinks. Their structures may be linear, hierarchical, hypertextual, or custom. The structure helps define how individual pages are arranged and organized into a complete website.

Structure also encompasses navigation, because the various toolbars, menus, and hyperlinks are the means by which we search and browse a site's contents. It is important to help users understand the site structure so they can browse and search your site more efficiently. Our discussion of the structure phase includes types of structures, the structure development process, navigation design, labeling, and guidelines for grouping and arranging pages.

Site Structures and Types

This section describes the features and uses of four main web structures: linear, hierarchical, hypertextual, and customized. See Figure 11–5 for a summary of these types.

Linear structures follow a designated order for arranging pages. They are similar to printed books, in that pages are arranged so that one follows another in a sequence (i.e., page 1, page 2). Such structures offer less flexibility to readers because they must follow a rigid sequence. Linear structures require every user to read the same information in the same sequence. As such, they are used most commonly in web-based training sites and online tutorials, where users must learn a specific process or sequence. Linear structures may also be used in sites that sell products, in which users deposit items in a virtual shopping cart and check out using a specific process or sequence of paying and confirming the order.

Hierarchical structures organize pages into a hierarchy of categories and subcategories that most resemble an outline or a table of contents in a book. These structures are commonly used in education, business, and news websites (in the latter, individual articles may be organized into a hierarchy of categories, such as World, Science, Law, and Education). In hierarchical structures, major categories are provided as navigation options on the home page, and subcategories and related content are found in subsequent pages in

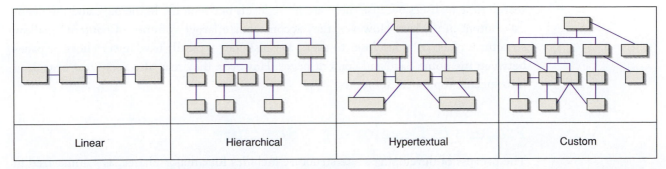

Linear	Hierarchical	Hypertextual	Custom

■ **Figure 11–5** ■ Four types of website structures

the site. Some news sites use a more complex structure that allows users to search and browse more dynamically, by jumping from one subcategory to another, but the basic overall arrangement of the site may still resemble a hierarchy.

Hypertextual structures are typically nonlinear structures in which any content chunk or page can link one or multiple pages in the site. Hypertextual structures work well for sites with large databases of information that permit flexible searching or browsing. These sites customize content by piecing together individual content chunks from a database. Many web-search portal sites, such as Yahoo! and Google, are hypertextual because the results of any keyword search could provide links to any number of pages. A hypertextual structure works well for sites that organize information associatively, where one page can link to any number of other pages. As such, hypertextual structures typically provide the most flexible navigation options for users. Other sites that use hypertextual structures are online reference guides, knowledge bases, and troubleshooting guides that link related content chunks to help users solve a problem.

Custom structures can combine multiple structural types and are usually tailored to the specific subject or purpose of the site. A custom structure might work best in a site that has multiple purposes or a specific need for a unique structural type. For example, an online university may require a hierarchical structure for its administration materials, such as policy statements, application guidelines, degree plans, and forms. Yet the same site might benefit from using a hypertextual structure for its course descriptions, which allows users to search a database of descriptions and titles in order to find those that best match their own academic interests. Many sites that sell products provide a hierarchical organization of products, a hypertextual search feature, and a linear process when purchasing items. If your site has more than one function or purpose, it may benefit from the use of a custom structure. Figure 11–5 summarizes the four types of structures.

Two important factors to consider in choosing a site structure are breadth and depth. *Breadth* is the number of choices or content areas at any given level in the site's structure. For example, a site with eight major content areas would be considered broad because it provides more categories or initial navigation pathways for users. A site with four major content areas is narrower because it offers fewer pathways. The breadth influences both the navigation and organization of your site.

Depth can be determined by the number of mouse clicks required to reach content at the lowest level in the structure. Deep structures have more levels in each content area or pathway, whereas shallow structures have fewer levels. Deep structures can require more search time by users because there are more levels between the home page and the deepest content in the site. However, they accommodate larger volumes of content. Shallow structures place content closer to users, because they typically have fewer clicks or pages between the home page and content pages. However, they may also seem less organized into specific topics and subtopics.

Process of Developing a Structure

The process of developing a site structure includes four major phases, as summarized in Figure 11–6.

Analysis	Determining major content areas, content types, and file formats.
Labeling	Selecting meaningful labels for pages and major content areas.
Layout	Sketching a site map or blueprint of the site's structure.
Editing	Making adjustments to the site structure based on user-testing and reviews.

■ **Figure 11–6** ■

The four steps of developing a site structure

In the analysis phase, you determine the major content areas and the types of documents to include. If you already have content chunks written, this step requires you to organize them into specific content areas. One method of determining major content areas is to create an index of your content chunks. Place the following information on an index card:

1. A one-sentence description of the chunk

2. The content type—for example, description, definition, form, or information graphic

3. The file format of each content type, such as web page, pdf document, .gif graphic, and so on

As you work, try to identify patterns or relationships between individual chunks and make note of them. After you have finished, sort your cards into stacks based on their similarities in subject or purpose.

During the labeling phase, your purpose is to select meaningful labels for all the pages in your site. First, write a label on each card that might serve as its title. Then, come up with a label for each card stack to serve as its group name. Be sure to choose a word or short phrase for each stack that represents the overall subject, purpose, or function of all of the content chunks in that stack. Typically, most labels are either nouns or verbs, but in some cases can be phrased as answers to questions (e.g., Who We Are, What We Do). Your labels might be organized using one of many methods: alphabetically, by function, using a metaphor or theme, by question, by task or topic, or by another method. Using a scheme can make the task of labeling much easier. Be sure to select concise and meaningful labels that help users scan and browse pages more easily.

To this point, your stacks of cards should organize content into groups of topics and subtopics. Now you must decide what type of structure best fits your site's subject and purpose. Figure 11–7 summarizes the advantages and disadvantages of site structures.

The next step toward developing a site structure is layout. The goal of the layout phase is to select and sketch a site structure that best fits the nature of the content and communicates the site structure to users. You should proceed to (1) sketch your site structure as a flowchart on a piece of paper, (2) use boldface rectangles to represent your major content area pages, (3) draw lines to indicate hyperlinks that link pages together, and (4) write your labels on each rectangle. As you develop your sketch, you might end up making changes to labels or the organization of content areas. Naturally, as you try to create a unified

Structural Type	Advantages	Disadvantages
Linear	• Is easy for users to learn the structure. • Lets users read all content in the same order. • Is good for process descriptions, instructions, and training.	• Has less flexible navigation options. • Links pages only in a specific sequence.
Hierarchical	• Organizes information into categories and subcategories. • Is relatively easy for users to learn the site structure.	• Can be complex and difficult to navigate larger structures. • Can be difficult to navigate hierarchies, causing users to back up in order to browse a different category.
Hypertextual	• Accommodates more flexible forms of searching and browsing. • Organizes content by association or relevance. • Works well for searching large volumes of content.	• May be difficult for users to discern site structure. • Can become complex to map the site structure in larger sites.
Custom	• Can accommodate multiple structural types. • Is the most flexible of structures.	• May be difficult for users to discern site structure. • Is time-consuming to design an effective structure.

■ **Figure 11–7** ■ Advantages and disadvantages of site structural types

structure, you will find better ways of arranging and labeling your pages as part of the process. Figure 11–8 shows a sample site-structure sketch.

The sample structure in Figure 11–8 is hierarchical, broad, and shallow. It has six major content areas, also known as *nodes,* which make it relatively broad. The structure has three levels at its deepest level, making it shallow. This type of structure offers users many

■ **Figure 11–8** ■
Sample site
structure
sketch

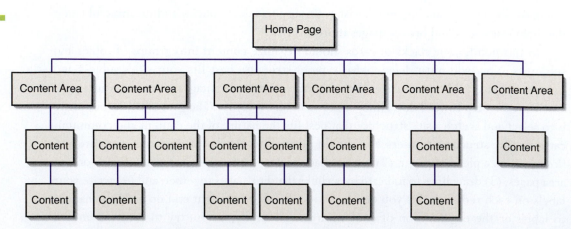

content areas from which to choose and places content pages relatively close to the home page. Although this structure offers many initial choices, or *pathways,* for users to follow, it makes finding content in the deepest levels a quicker task for users.

The purpose of the editing phase is to test the site structure to make sure it is organized properly and demonstrates good usability. One way to test your site's structure is with a small group of typical users (four or five) who are not part of the development team. You can even use your set of cards for this basic test, as follows:

1. Provide your testers with an overview of your site's subject, purpose, and general function.

2. Give them your stack of cards.

3. Have them organize the cards into stacks they think represent the best organization. To limit the time of the test, you could provide a reduced set of your cards to get a general idea of how they might organize the stacks.

This test examines how typical users might perceive the arrangement of pages in a site to help you design a more usable site structure. Once each tester finishes the test, record the information and compare it to your site structure sketch to see how well they match. You may or may not decide to make changes based on your findings.

This exercise should give you a general idea of how users might expect content to be organized. As a final step, consider having members of your development team review the work and comment on alternate ways of arranging the site. Then, with the information from your testers and team review, you can decide what changes in the structure are necessary before you start planning the site's navigation.

Navigation Design

Site structure and navigation are interrelated in that the structure dictates the navigation options, or pathways, users have in searching and browsing the site. The site structure and labeling system can be used as an outline or a blueprint for developing the site's navigation, but first you must decide what types of navigation to provide for users. Some types commonly used are hyperlinks, toolbar menus, site maps, and search fields. Figure 11–9 lists each type and describes its function.

Many sites use many types of navigation to provide users with multiple means to navigate the site. Some users may prefer toolbar menus and site maps, whereas other users prefer search fields. Because we all have different preferences, it is important to provide more than one way of searching and browsing your site. For example, if one of your navigation systems is a graphic toolbar menu, you might also provide a duplicate set of text links for users who may have trouble loading or using the graphics. Let the purpose of the site dictate the types of navigation you select, and choose at least one other type to accommodate users' individual preferences.

Guidelines for Labeling

Once the navigation systems have been selected, the next step is to provide text labels for your links, buttons, and toolbar menu options. You can use the same labels you

Navigation Type	Function
Hyperlinks	Hyperlinks can be individual words, phrases, or images that when clicked, link to another page or chunk of related content. Users can click on these words or images to access the related page or chunk. Hyperlinks are used in all site structures and are the foundation of other navigation types.
Toolbar menus	Toolbar menus are groups of navigation choices that show the major content areas or functions of the site. They can be drop-down, pullout, graphic, or basic text labels. Each item in the menu links to a related page or section of the site. In larger sites, submenus can provide links to more specific content. Toolbars work well in all sites, but are less effective in search portal sites, which have large databases of searchable content.
Site maps	Site maps provide a structural layout of the site and links to all pages in the structure. They can be a graphic map or textual outline of your site structure map. They explicitly show the organization of pages, like an index or a table of contents. They help users plan paths through pages to find information. Site maps work well in sites with custom or complex site structures.
Search fields	Search fields allow users to type in key words or phrases in a search box and press a button to search the entire content of a site. Based on the results of the search, a list of possible pages that match is displayed and links are provided. Search fields typically work best in sites with large amounts of searchable content.

■ **Figure 11–9** ■ Navigation types and function

selected for your site structure and simply create a toolbar menu, site map, or other navigation tools. Sometimes these labels may need to be different to make them more visible to users.

Consider using a scheme for labeling your navigation links and toolbars. Some commonly used labeling methods include alphabetic, by function, using a metaphor or theme, by question, by task or topic, or other means based on the site's subject and purpose. One advantage of using a scheme is that it follows an organized pattern that users recognize. Once users see a few links in the toolbar menu or site map, they can determine how each relates and have a better idea of how the site as a whole is arranged. Select a scheme and appropriate labels for your navigation toolbars, menus, and hyperlinks to help users understand their meaning and function.

Another issue to consider is how to communicate the site structure effectively to users. There are a variety of means to communicate the structure to users, such as the following:

■ Site maps and indexes act as interactive maps of the site structure, showing the arrangement of pages and providing links to each page in the site.

■ Toolbar menus show users how information is organized into major content areas.

- Headers and titles throughout the site can help users understand how information is arranged.

- "Breadcrumb" links show the trail of links a user has followed in the site.

- Visual cues and elements, such as colors, icons, or graphics, can represent themes, concepts, categories, or subcategories to users. These visual cues can quickly suggest which pages go together as a collection.

Whichever methods you select, think carefully about how you show users the arrangement of your site.

Grouping and Arrangement Strategies

After you have decided on the types of navigation and the labeling schemes, you should consider placement. The methods used to group hyperlinks and their placement on the page are both equally important tasks.

You have probably noticed some consistencies in the grouping and arrangement of elements in websites you have visited. These consistencies can be considered to be loose conventions, which many users have come to expect and look for in site layouts. Some common guidelines for placing navigation links and toolbar menus found in many site designs are as follows:

- Position toolbar menus and main site navigation links in the user's initial screen viewing area on every page. Users should be able to see the navigation when the page first loads, without having to scroll to find it.

- Place toolbar menus and main site navigation links in consistent locations throughout the site so users can find them easily. Most site navigation is placed in one or more of three areas: top margin beneath the site's title banner, left margin, or page footer.

- Group main site navigation links together so that users can easily find the site's major content areas.

- Place appropriate navigation tools in each new window or frame, if your site uses multiple windows or frames, in case users get separated from the main site and must return to a previous location.

- Provide links to your site map, search, and help systems near or with your main site navigation links so users can always find them when needed.

Once you have devised the structure and navigation, you have a structural blueprint of the site that is ready for the next step—the design of graphics and the interface.

>>> Design

The fourth phase, the *design phase,* involves developing graphic content, page layouts, and design of the *interface* (what the user sees on the screen) into a whole site. Design involves much of the actual production work of the website. Your work involves arranging,

formatting, and perhaps even redesigning some of your content to fit the site design. It includes the design of site maps, navigation tools, buttons, headers, backgrounds, and other elements developed in previous phases. You will create an effective interface design and individual page layouts for the site.

This section discusses design conventions and principles, development of graphic identity and content, different file formats, graphics, and guidelines for designing effective interface layouts.

Design Conventions and Principles

Effective design is much more than good aesthetics and instincts. Few people have the innate ability to design without some type of guidance or rules to assist them. Whether designing a graphic logo or an advertising banner or laying out the interface, it is important to use established design conventions and principles to guide your design work. Some of these commonly observed web-design conventions include the following guidelines:

- Place a hyperlink on the title banner or logo to the home page.
- Place the site's main navigation in the left or top margin.
- Use consistent font faces, sizes, and colors on pages.
- Use descriptive titles and headers for each graphic and on every page.
- Provide contextual cues that provide hints as to the function, concept, or arrangement of specific pages or graphics.
- Maintain a consistent graphic identity, or *look,* using consistent colors and visual elements.
- Provide redundant navigation links to supplement graphic links.
- Provide alternate descriptions of graphic content.
- Use no more than three font faces for text in your site.
- Use colors that contrast well.

Because web development is fairly new, no definitive list of design rules applies to all websites. Although this list may prove useful in your design work, no single set of conventions applies to all websites or documents. Your design team determines if there is a good reason for a convention to be followed or broken, considering your project scope, purpose, and context. To begin your design work, make a list of design conventions to use, then do some benchmarking research by viewing how conventions are used in sites that have a similar purpose and function. Add to your list any conventions that seem to apply to your site, ultimately arriving at a list that guides your design on a specific project.

Design principles are also based on theories of design. Although not prescriptive, they provide you with broad guidance to assist in the design of graphics and page layouts. For example, Gestalt theory provides a foundation for much contemporary visual design and suggests that users actively engage in organizing and making sense out of visual stimuli in their field of vision. Because of the visual and interactive nature of websites and web content, Gestalt is particularly useful for design work in those media. The following design principles based on the Gestalt approach can be used to guide your design work:

- Consistency in the use of elements in repetitive and similar ways on a page creates a unified look. Maintaining a graphic identity through the repeated use of specific logos, icons, and colors in a website is one way to demonstrate consistency.

- Contrast in the use of visual elements, such as colors, lines, or shaded regions, draws the eye to those elements and creates visual emphasis. Placing white text on a black background creates good visual contrast because the text stands out more and is easier to read.

- Group elements together through the use of space, color, or other graphic elements. This includes placing elements together to show a relationship between them. Placing all elements of a search feature, such as the text boxes, buttons, and instructions, in a shaded region creates conceptual grouping, telling users which elements belong to the feature.

The Internet Archive: Live Music Archive page (Figure 11–10) demonstrates good use of these three design principles. The page demonstrates consistency through its use of consistent font faces, sizes, colors, headers, and logos. It shows good use of grouping, using shaded headers, boxes, and white space to group related content together. The page also shows good contrast, using colors that make text easy to read and that emphasize headers and hyperlinks on a variety of background colors.

■ **Figure 11–10** ■ Example of design principles used in a web page design

Source: Internet Archive: Live Music Archive, http://www.archive.org/details/etree.

Finding a Theme and Developing Graphic Content

Once you have a set of design conventions and design principles, you can start designing your graphic content. First, you should establish a graphic identity or design theme for your site. The site's *graphic identity,* defined by the use of colors, fonts, lines, boxes, shapes, graphics, animation, and other visual information, defines the site's unique brand. Rather than selecting your choices at random, it is a good idea to select a specific theme or metaphor to guide your selections. You should consider the tone and impression you want the site to have. Many e-commerce sites that sell products use a shopping-cart theme. Select a theme that is appropriate for your users, purpose, content, and the tone you want to convey, and then use the defined theme to help you develop graphics for your site's design.

Many development teams use a variety of methods in developing graphic content for websites. If you have existing graphic content to use in your site, it may need to be converted or improved for a web environment. Graphic content for your website can come from a variety of sources. If you have printed material, such as sketches or photos, you can use a document scanner to convert them to graphic file formats for use on the web. Also, you can create your own graphics or hire a graphic design professional to create them using graphic design software.

If you don't have the time or resources to develop your own content, you can search the web's vast number of sites for graphic libraries (both free and for purchase). By typing in a search for "free web graphics" in any search portal site, you will get access to thousands of sites that offer graphic libraries and utilities for generating custom graphics and animations, some for free and some for a fee. In some cases, you might be able to get some good recommendations from colleagues or friends. Choose the methods that best fit your budget, time, and project scope.

File Formats and Graphics

For static images, the three most common types of file formats used in websites are the Graphic Interchange Format (.gif), the Joint Photographic Experts Group format (.jpg), and the Portable Network Graphic format (.png). Each type has relatively good *file compression capabilities,* which means that they typically have smaller file sizes compared to other formats. Smaller file sizes mean shorter user download times on the web. Commonly used formats that tend to have less compression, include Windows Bitmap (.bmp), Encapsulated PostScript (.eps), and Tagged Image File Format (.tif). Use these other formats only if you have a specific reason to do so.

When deciding which format to use, you should consider the following information about available features:

- *Color depths* affect both the quality and file size of the image. True color images display up to 16.7 million colors and produce images of photographic quality. However, not all graphics require this amount of color depth.

- *Transparency* allows you to set a single color in a graphic to be transparent, which allows graphics to blend well against different backgrounds.

■ **Figure 11–11** ■
Graphic file
formats

File Format	Graphic Interchange Format (GIF)	Joint Photographic Experts Group (JPG)	Portable Network Graphic (PNG)
Color Depth	• 256 colors	• 16.7 million colors	• 16.7 million colors
Transparency	• transparency	• no transparency	• transparency
Animation	• animation	• no animation	• no animation
Interlacing	• interlaced	• non-interlaced	• interlaced

■ Some types of animation don't require special software plug-ins or programs to view them.

■ *Interlacing* allows users to see pieces of the graphic on the screen as they load. Noninterlaced graphics remain unseen until the entire image is downloaded.

Figure 11–11 summarizes the capabilities of the three main graphics file formats.

Many other types of graphic formats are unique to different graphic design software and programs, such as animations, 3-D modeling, drafting files, and video clips. Most require specific software programs or plug-ins to view. If your site uses these types of files, make sure you instruct users on how to download and view them properly. Provide file formats that do not require them to purchase additional software or spend a lot of time downloading appropriate viewers. Be sure to consider all your audiences when making the choice. In designing a site for a general audience, consider using file formats that can be viewed with free viewers or software plug-ins, or provide users with alternate versions. Sometimes simple scanned images, saved as jpg files, are good supplements that allow users to view static images or parts of your dynamic content. If members of your team lack the expertise to address some of these issues, it might be best to hire a graphic design consultant to help you solve some of them.

Interface Layouts

Once the graphic content has been developed, the next task is to begin laying out the interface—what the user sees on the screen. The interface serves as the user's control panel for browsing, searching, and interacting with the site. The typical web interface includes the following five elements:

1. **Header**—includes the logo and a title banner that identifies the site's title and/or company name

2. **Navigation**—includes hyperlinks, site maps, toolbar menus, and search features, which are the main navigation tools for the user to search and browse the site

3. **Content window**—includes most content chunks and graphic content displayed on the screen and is placed in consistent locations throughout the site

4. **Graphic identity**—includes visual information (logo, icons, graphics, colors, or other visual elements) found throughout the site to give it a unique brand

5. **Contextual clues**—includes information that helps users understand the nature or organization of information on pages or in the site

Before you begin designing the interface, make a list of the graphics, content chunks, and objects that will be used in each of the basic interface elements.

The Internal Revenue Service website, shown in Figure 11–12, uses the five interface elements on this excerpt from its home page. The header displays the title of the site and the organization's logo. The site's main navigation is placed under the header and in the left margin. The content window is in the center of the page and features links and information by topic. Elements used in the site's graphic identity include a photo that suggests the average working taxpayer, graphic links for services such as E-file, and blue and white colors consistent with the logo. The page also uses contextual clues such as headers, boldface text, menus, toolbars, and graphic file folder tabs to help users understand how the site's contents are organized.

Designing the interface also includes arranging interface elements, such as content chunks, navigation tools, and graphics, as well as the interactive features of the site, including

■ **Figure 11–12** ■ Interface elements used in a web page

Source: Internal Revenue Service http://www.irs.gov

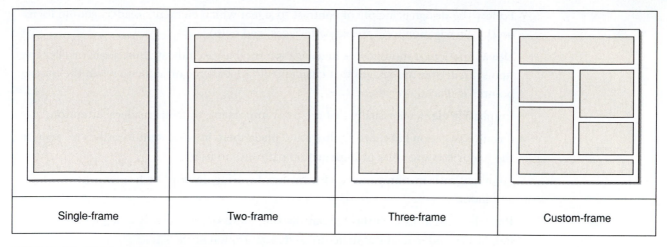

| Single-frame | Two-frame | Three-frame | Custom-frame |

■ **Figure 11–13** ■ Interface layout types

buttons, links, and forms. Most sites use the same interface design, or layout, for all the pages in the site. Some sites may find it necessary to use two or more layouts, depending on the function and purpose of other pages or sections of the site. For example, a *splash page,* or an introduction page with a short animated movie, might have a simpler layout for that page and a separate one for the rest of the site. A site that sells music CDs might use one layout for the home page, a second layout for all pages with product information, and a third layout for the shopping cart or purchasing page. You should decide if there are certain pages or sections of your site that require different interface layouts and make a list of the characteristics each layout should have.

Some typical types of interface layouts are single-frame, two-frame, three-frame, and custom-frame. Figure 11–13 shows a sample of each of the four interface layouts. There are many other ways of arranging layouts in the two-frame, three-frame, and custom-frame layouts, which you may devise on your own. One advantage of multiple-frame layouts is that you can anchor headers, navigation, and content in consistent locations on the screen to help users find each more easily as they move around in your site. Custom layouts usually are more complex and can be difficult to navigate or understand. Generally, the more complex the layout, the more contextual clues are needed to help users understand the arrangement of information in your site.

In drafting the layout, begin with a sheet of paper or workspace and sketch out a few interface layouts. Identify where to place each interface element, including the header, navigation tools, content window, graphic elements, and contextual clues. You might try a couple of different layouts to see which seems to best fit your site's needs. Sometimes the true test of an interface layout is to take your paper sketches and set them up in your web-authoring software program to see how they look on the screen. When designing your interface layout, remember to use the three design principles: consistency, contrast, and grouping. You may want to consider some of the following guidelines based on these principles:

■ Use site maps, indexes, breadcrumb links, and consistent graphics to help users identify the site's structure and major content areas.

- Follow the design principle of contrast to signal which elements readers should focus on (i.e., navigation tools, search interfaces, and help).

- Use animation or mouseovers in navigation toolbars to indicate functional or clickable items. *Mouseovers* are images that change their appearance or animate when the mouse pointer is placed over them.

- Emphasize elements visually that are most important, to focus readers' attention.

- Group navigation links into toolbars and place them in consistent locations to suggest the major content areas or information pathways to users.

- Highlight information pathways for readers by using descriptive headers, site maps, and indexes.

- Provide contextual cues throughout the site, but more frequently at higher levels, to help users understand the structure and organization of the site.

- Use familiar shapes, icons, and other visuals to suggest concepts to users.

- Pair graphics used in unfamiliar contexts with text descriptions to help users understand them.

- Group related visual and textual content using visual shapes or space so readers can understand their relationship or function in the whole.

Your main goal should be to select a layout that organizes your interface elements consistently throughout the site. Consistency makes it easier for users to learn the organization, layout, and functions of your site. Select backgrounds, colors, and other elements that create good contrast on the screen to maximize the readability, clarity, and legibility of your content. Use white space, lines, or shaded regions to group or set apart individual elements in the interface in order to demonstrate how they relate. You may have to go through several iterations before you decide on a final version to test. Once your team has devised an interface layout, you can begin the task of adding content, navigation tools, graphics, and other elements onto individual pages; then you can link pages together following your site structure map to create the finished product, or whole site.

>>> Usability and Publication

The fifth and final phase of developing websites is *usability*. It involves testing and editing the site so that it is readable, accessible, and useful to your audience. Resources for ensuring the usability of websites are available at www.usability.gov. Usability should be considered throughout the development process, not just at the end. Simply put, it means you must focus on the needs and expectations of your audience. When you make decisions on navigation, graphics, content chunks, site structure, and interface layout, always make choices that demonstrate good usability.

This section discusses testing your site using a usability review, quick usability checks, and accessibility checks. We also cover how to publish your finished site.

Testing Your Site for Your User Base

Design team members should participate in the usability process so that the site is organized and functioning properly. This section describes the most feasible methods for most developers—usability reviews and quick checks—while also covering the topic of accessibility and formal laboratory testing.

Although not described in this chapter, formal usability testing in a laboratory environment may be useful if you have the equipment, time, and financial resources available. It typically involves setting goals, selecting criteria, developing test materials, soliciting participants, setting up the testing environment, conducting the test, and writing a results report. For more information on formal usability testing, three useful books are Jakob Nielsen's *Usability Engineering,* Mark Pearrow's *Web Usability Handbook,* and Carol Barnum's *Usability Testing and Research.*

Performing Usability Reviews

Usability reviews function as a design review of a document or project. You can perform a simple usability review of your site using a limited number of participants (typically four or five) and a set of evaluation criteria to analyze your site. The review is composed of two parts: an internal review, using members of your design team; and an external review, using a small group of typical users.

To begin a usability review, select a set of evaluation criteria and write short definitions of each. You might choose to evaluate the site's navigation system, use of graphics, readability of content, or function. Once you have selected criteria, set up a checklist to use in the review. The checklist should define each criterion and any items reviewers should look for in their review. As part of your checklist, you may want to devise a ranking system for each item so they can assess the level of compliance. Provide adequate space for reviewers to record comments, observations, and notes. Figure 11–14 is a sample usability checklist that lists and defines evaluation criteria and identifies specific related items to look for in a review. You can create your own checklist using other evaluation criteria, items, and ranking system.

To begin the test, provide your reviewers with background information on the site, including its subject, purpose, and location. Explain the instructions of the test, including the allotted time, evaluation criteria, and any forms or checklists they will use. Instruct reviewers to identify items that violate any of the evaluation criteria and have them distinguish specific problems from general comments. After they complete their reviews, compile the responses and meet with your team to devise solutions for each of the problems noted by your reviewers. Be sure to take a close look at specific problems identified and any comments provided. You probably want to act on most of the problems identified, but may choose whether to act on general comments.

Quick Usability Checks and System Settings

Another useful usability testing method is performing quick usability checks that test the site using a variety of platforms, browsers, and different system settings. We each use

Navigation. The links, toolbar menus, search features, and other tools used to search and browse the site.

	Always	Sometimes	Never	Notes
All navigation systems and hyperlinks are functional and link to the appropriate pages.				
The navigation systems use descriptive labels.				
Navigation options are provided on all web pages.				

Consistency. The repeated use of elements and styles.

	Always	Sometimes	Never	Notes
All design elements and page layouts demonstrate consistent use of fonts, colors, spacing, and alignments.				
Navigation tools are placed in consistent locations.				
Interface layouts organize elements in consistent locations.				

Clarity. The level of clarity of the structure, textual content, and graphics.

	Always	Sometimes	Never	Notes
The site structure is easy to discern through the use of contextual clues, labels and/or site maps, and indices.				
Font faces, text styles, and colors use good contrast and are clear.				
Graphic content is clear and does not appear choppy or pixilated.				

Legibility. The ease at which textual content and graphics can be read or understood.

	Always	Sometimes	Never	Notes
Textual content is free of grammatical errors and conforms to appropriate style guidelines.				
Graphic content conforms to appropriate style guidelines and is easy to understand.				

■ **Figure 11–14** ■ Sample web-usability checklist

different versions of software and have our monitors set at different resolutions that may affect the display and function of certain pages and layouts. What you see on your screen may not match what other users see on theirs. Although it might not be possible in every case to design a site that works optimally for every possible system, you can make adjustments to reduce problems or at least to ensure every user has equal access to your site's content.

To test your site, create a working copy of your website on a CD or publish it on a web server. Test your site with at least two different browsers, preferably the most recent version of each. View the site using at least two different screen resolutions. Screen resolution is measured in *pixels,* or individual dots, that compose the picture you see on your monitor. The most commonly used screen resolution settings change with technology, but you should consider at least one lower setting (e.g., 640×480 or 800×600) and one higher setting (e.g., 1280×1024 or 1600×1200) for testing purposes. Check your screen resolution settings using your system's control panel (Windows) or system preferences (Mac). Color depth can also be checked in the display properties on your computer. Common settings are 16 colors, 256 colors, 16-bit (or 65,356) colors, and 32-bit (16.7 million) colors. Be sure to view and test your site on at least two color-depth settings, preferably 256 colors and 32-bit colors, to see if there are any problems with the clarity or readability of your site. You may also want to test your site on a laptop and a desktop computer, as well as with both standard and wide-screen monitors.

Once you have a list of the different system settings to test, create a quick-check worksheet to catalog each problem you find. When you test the site, examine the functionality of navigation, consistency of page layouts, and the readability of the textual and graphic content. Devise a system to rank the severity of the problem to help establish your redesign tasks later. For example, you might use the following system:

1. Severe—affects the accessibility of pages
2. Serious—affects the readability and legibility of pages
3. Concern—affects the clarity of pages

By identifying and ranking the severity level of problems, you can prioritize your work in editing and revision. After you complete your quick usability checks, be sure to devise solutions for each problem you identify and follow through on making changes to your site. You should consult with other team members in devising appropriate solutions.

Accessibility Guidelines

Another important aspect of web usability is checking the accessibility of your site's content. Whether your site's purpose is to solicit clients, sell products, or provide information, you want to make sure all users have equal access to the content.

Some users may have trouble accessing your content because of system limitations or settings; others may have trouble based on a specific disability, such as impaired vision. Many people with disabilities access the web on a daily basis using a variety of hardware and software products that can assist them. For example, users with visual impairments can use software that converts textual content into live audio that essentially reads the content aloud. Although you, as the designer, may not be able to control some of these issues, the best way to ensure that your site is accessible is to check its pages against at least one set of accessibility guidelines. Two sets most commonly used are the U.S. Government Section 508 Accessibility Guidelines and the World Wide Web Consortium Web Content Accessibility Guidelines.

The U.S. Government Section 508 Accessibility Guidelines, http://www.section508. gov, was enacted by the U.S. Congress in 1998 as an amendment to the Rehabilitation Act to ensure that all government public information websites and electronic materials be made accessible to all users, regardless of their disability. Although the law applies also to software programs, telecommunication products, and other electronic media, it provides a set of 13 guidelines for web-based materials, including websites. Figure 11–15 summarizes the U.S. Government Section 508 Accessibility Guidelines.

The World Wide Web Consortium's (W3C) Web Content Accessibility Guidelines 1.0, http://www.w3.org/TR/WCAG10, is another commonly used set of accessibility

1. Any websites designed for the United States government should meet the requirements of Section 508 of the Rehabilitation Act. Ideally, all websites should be accessible and compliant with Section 508.
2. Ensure that users using assistive technology can complete and submit online forms.
3. Ensure that all information conveyed with color is also available without color.
4. To aid those using assistive technologies, provide a means for users to skip repetitive navigation links.
5. Provide a text equivalent for every non-text element that conveys information.
6. To ensure accessibility, test any applets, plug-ins or other applications required to interpret page content to ensure that they can be used by assistive technologies.
7. When designing for accessibility, ensure that the information provided on pages that utilize scripting languages to display content or to create interface elements can be read by assistive technology.
8. Provide text-only pages with equivalent information and functionality if compliance with accessibility provisions cannot be accomplished in any other way.
9. To improve accessibility, provide client-side image maps instead of server-side image maps.
10. To ensure accessibility, provide equivalent alternatives for multimedia elements that are synchronized.
11. Organize documents so they are readable without requiring an associated style sheet.
12. To ensure accessibility, provide frame titles that facilitate frame identification and navigation.
13. Design web pages that do not cause the screen to flicker with a frequency greater than 2 Hz and lower than 55 Hz.

■ **Figure 11–15** ■
Summary of the U.S. Government Section 508 Accessibility Guidelines

Source: U. S. Government Section 508 Accessibility Guidelines, http://www.usability.gov/PDFs/Chapter3.pdf

guidelines that overlap with some of the Section 508 guidelines. The W3C is a group of institutions and individuals committed to creating accessible and consistent standards for web development. The W3C's set of guidelines is based mostly on recommendations about design proposed by the W3C, which are not mandated by any specific law.

Regardless of which set of accessibility guidelines you select, it is important to test your site to ensure you are providing equal access to your site to all users. You can print a set of these guidelines and use them to evaluate your site's level of accessibility.

Publishing Your Site

After completing all the tests and making appropriate revisions, the site should be ready for publication. To publish the site, you must to obtain dedicated storage space on a web server. If you have an Internet Service Provider (ISP) that you use to connect to the web or a company or school web server, most likely you already have space allocated. You must know the following:

- amount of storage space available
- process for logging in to the server
- address of the host server
- Uniform Resource Locator (URL) or web address

Normally, your ISP or system administrator assigns a URL for your site. If you want a custom URL, you can purchase one over the Internet by searching for companies that sell them. Another important issue to consider before publishing your site is its size. Check to ensure the total size of your website does not exceed the available storage space. In many cases, you can purchase or request more space if needed.

You can publish, or *upload,* your site to the web server in two ways: (1) upload it through your web-authoring software or by using a File Transfer Protocol (ftp) program; or (2) have a system administrator help you. Keep a protected copy of your site on a computer or disk after you publish it, in case you encounter problems with the uploaded version. This copy can also serve as a working copy for you to make changes and later upload the corrected files.

>>> Chapter Summary

This chapter provides an overview of the process and unique aspects of developing websites and web content. Your role may vary in developing professional websites. Typical roles include the project manager, programmer, graphic artists, writer/editor, content provider, and usability tester.

The five major phases of designing a website are planning, content development, structure, design, and usability.

1. *Planning* requires that you determine the audiences, purposes, and issues. User-centered design places the user at the center of the development process and focuses on how to design a product that is easy to use and understand.

2. *Content development* involves writing content chunks, adapting existing content, knowing scripting languages and software authoring tools, and converting documents to web-viewable formats.

3. *Structure* includes developing a site structure, labeling content chunks and pages, selecting the types of navigation, and using the schemes that govern the labeling.

4. *Design* involves developing a graphic identity, optimizing graphic content for the web, and creating effective interface designs and layouts.

5. *Usability* involves the testing and editing of your site to ensure it is readable and accessible by the broadest user base.

Publishing a website requires a web server and the following information: amount of available server space, process for logging in to the server, host address, and your URL or web address.

>>> Learning Portfolio

Communication Challenge "What Does Your Company Do, Anyway?"

In an effort to take a more aggressive marketing stance, the M-Global marketing department has decided to develop online marketing materials to complement the packages currently sent to potential clients. This case study provides background on M-Global's current website layout, the marketing team's plan for a redesign, and questions for discussion.

Background

Although M-Global already has a website, the site has less of a marketing emphasis and is not tailored to attract new business. Currently, the company website is organized into content sections that resemble the organization's major departments, such as foundation design, construction management, equipment development, environmental remediation, and training services (see Model 11–1). Each page can be accessed from a single navigation toolbar that uses department names as labels for links. Individual site pages provide each department's mission, supervisor, employees, list of past projects, and technical references on the Internet. This organization works well for current clients and employees, but it is less comprehensible to new and potential clients not familiar with M-Global's firm. The sales staff has noted many questions from customers inquiring on how to find information on specific services and past projects on the current website.

The Plan

To address these concerns, the marketing department has decided that a redesign of the current company website would be the best solution. The website must be redesigned and reorganized so it can serve both existing and potential customers and provide them with information on the range of M-Global's services. The marketing team identified the following goals in the site's redevelopment:

- Create a new labeling scheme that helps customers understand the range of services, projects, staff, and purpose of the firm.

- Implement multiple means of navigation to help users find information more easily.
- Develop a series of project description pages that provide the scope, work performed, services provided, and photos on specific projects completed by the firm.

Questions and Comments for Discussion

At this point, the marketing team has passed the project to a web-development team to determine how to best achieve the goals they have laid out. Although they identified rather high-level conceptual goals for the site redevelopment, they need a team to help them consider the following questions to determine specific approaches to meeting their goals:

1. Who should be involved on the web-development team and why?
2. What labeling schemes could be used to help achieve the team's goal of making the website easier to comprehend?
3. In addition to a navigation toolbar, what other navigation systems could be implemented to help users find information more easily?
4. What strategies would be most effective in designing the site to serve both customers and employees? Explain.
5. What other goals should the team consider in the site redevelopment project?

Write About It

Assume that you are the leader of the team that is redesigning the website. The team members have submitted the new website in Model 11–2 for your feedback. In a memo, evaluate the proposed website, discussing its strengths and weaknesses. Make any recommendations that you feel are necessary. As team leader, don't forget to thank your team for their hard work.

Collaboration at Work Usable Navigation

General Instructions

Each Collaboration at Work exercise applies strategies for working in teams to chapter topics. This exercise assumes you (1) have been divided into teams of about three to six

students, (2) will use team time inside or outside of class to complete the case, and (3) will produce an oral or written response. For guidelines about writing in teams, refer to pages 24–28 and to chapter 13.

Background for Assignment

One of the major hurdles in developing websites is developing a site that users can understand and navigate easily. Navigation systems should provide users with methods to search and browse content and help users understand how the site as a whole is organized through effective labeling. Good navigation systems outline the site's content and functions for users as well.

Although a design team may possess specialized knowledge of how navigation systems are developed, the end-users often express frustration with such systems. For example, they fail to grasp the function and organization of the tools or comprehend the site's content because of the inability to move around the site with ease. Therefore, design teams should anticipate problems potential users might face and design systems that are flexible, that are easy to use, and that help users understand the site's organization.

Team Assignment

Select a website that has multiple navigation systems on the home page and throughout the site, such as a news site or a school site, and complete the following steps in preparing a report: (1) Browse the site's content to determine the audience and purpose of the site. (2) Make a list of at least three types of navigation used in the site, such as navigation toolbars, drop-down menus, pull-down menus, search fields, and other hyperlinks. (3) Evaluate each navigation system used to determine its flexibility, function, and labeling scheme. Consider some or all of the following questions: Do the navigation systems provide users with enough flexibility to search and browse the site? Are there problems with regard to their functionality and use? Do labeling schemes used help users understand how the site is organized and functions? (4) Devise a strategy to improve the function, organization, or labeling schemes used in the navigation. (5) Redesign each navigation system, either on paper or as a single prototype web page.

Assignments

Each of the following assignments can be completed individually or with a small team.

1. Analysis of WWW Consortium's Home Page

Explore the World Wide Web Consortium's home page at http://www.w3.org. Perform an audience analysis on the site, identifying the primary, secondary, and tertiary audiences. Also, identify the primary and any secondary purposes. Finally, identify any contextual constraints, such as legal, ethical, social, cultural, technical, or other issues that govern its content. How effective is the site in targeting its audience(s), achieving its purpose(s), and dealing with contextual issues? List three suggestions that might improve the site and make it more tailored to these user specifications.

2. Analysis of Government Website

Select a government agency website (federal, state, or local) and make note of the types of content (e. g., descriptions, definitions, forms, pamphlets) that are used. Also, make note of the file format types used (e.g., HTML, doc, pdf). How effective is the organization of the content? Is it easy to discern the organization? Are multiple formats used for downloadable or printable documents? Are there clarity or readability issues with documents? List three to five suggestions to improve the organization, readability, or overall accessibility of the content.

3. Analysis of Business Website

Find a business website and browse it for 15 minutes to become familiar with its organization and purpose. Then, sketch a basic site architecture map of the site based on information you find in the navigation, use of titles, headers, visuals and other clues. For larger sites, sketch only the first three or four levels of the structure. Identify the type of structure it most resembles and then its breadth and depth. Based on your findings, make a list of three to five recommendations for optimizing the site structure and organization.

4. Analysis of Entertainment Website

Select and familiarize yourself with an entertainment website. Identify the elements used in the design of the site that make up its graphic identity or brand. Look for titles, logos, colors, graphics, and other visuals used consistently throughout the site. In a few sentences, describe the overall design theme or graphic identity used. What general impression do these elements convey? Are there elements that don't fit the design theme used?

5. Usability Check of Website

Perform a quick usability check on an e-commerce site, using at least two different screen resolution settings and two different color-depth settings. If possible, use two different browser types as well. Make note of any exceptions that impair the function, readability, or legibility of the site. For each exception you note, propose a possible solution for fixing the problem.

 ### 6. Ethics Assignment

Digital text and graphics generally are protected by the same copyright laws that cover printed matter. In consideration of this point, choose one or more of the following tasks, depending on the directions given by your instructor: (a) develop a list of specific ethical concerns that apply to the development of a website, (b) research current law as it applies to use of borrowed material on websites and provide a narrative summary of same, and (c) evaluate a particular website according to ethical concerns and guidelines with which you are familiar.

7. International Communication Assignment

Develop a list of general questions that should be asked in planning a website that will be written in English but that will be used primarily in cultures or countries other than your own. After locating an individual from another such culture or country, ask the questions you have developed and prepare a summary of the responses.

 ### 8. A.C.T.N.O.W. Assignment (Applying Communication To Nurture Our World)

Choose a local, regional, or national environmental or sustainability issue in which you have interest—for example, overdevelopment in your town, availability of clean water, contamination of trout waters, global warming, air quality, or energy use. Using the guidelines presented in this chapter, do preliminary work in planning a website that addresses the issue. Specifically, do the following: (a) define your audience, (b) develop some content of the website, (c) describe the web structure you plan to use, and (d) explain how graphics will be incorporated into the site. Then to gather some preliminary usability information, share your ideas with another person to get suggestions on developing the site.

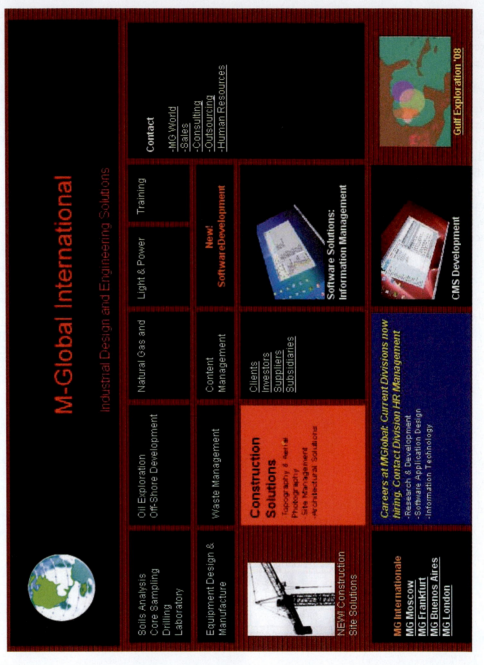

■ **Model 11–1** ■ Original M-Global website

M-Global International

Engineering Technologies

Today's Date

Customer Service

Careers with MGlobal

Employment

Safety

Contact HR

News

Welcome to MGlobal International! Our record speaks for itself. MGlobal provides innovative technologies, outstanding customer service and unparalleled expertise. We combine the latest technologies with 45 years' experience to offer unsurpassed services and equipment. because every exploration initiative is unique we customize our services and equipment to our customers' needs, creating cost effective solutions that are tailor made for various situations. Learn More ...

Departments

Marketing
Audit & Compliance
Public Information
Information
Systems
Accounting
Office of the
Controller
Equipment
Development
Publications
Development

Services

Soils Analysis
Oil Exploration
Site Management-Builders
Site Management-Waste
Light & Power
Heavy Equipment R&D
Software Development
CMS Design
Training

■ **Model 11-2** ■ Proposed M-Global website

Chapter | 12 | Graphics

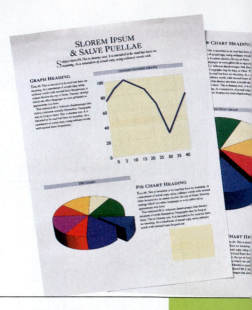

>>> Chapter Outline

Terms in Graphics   401

Reasons for Using Special Fonts, Color, and Graphics   402

Using Fonts   405
  Font Types   405
  General Guidelines   405

Using Color   408
  The Cost and Time of
    Using Color   408
  Developing a Color Style Sheet   409
  Color Terms   411
  Guidelines for Using Color   411

General Guidelines for Graphics   413

**Specific Guidelines for Seven
  Graphics**   416
  Tables   416

Pie Charts   420
Bar Charts   423
Line Charts   426
Flowcharts   428
Organization Charts   430
Technical Drawings   431

Misuse of Graphics   435
  Description of the Problem   435
  Examples of Distorted Graphics   435

Chapter Summary   440

Learning Portfolio   441
  Communication Challenge   441
  Collaboration at Work   443
  Assignments   443

Shon Williams, the Director of Public Information at M-Global, always contributes to the organization's annual report. Among the sections he has been asked to write is a description of M-Global's environmental work. This description will also be used by the Marketing Department in press releases as a way to interest industry magazines in profiling M-Global's environmental efforts.

Shon has gathered stacks of data about M-Global's environmental projects and internal eco-friendly initiatives from project reports and quarterly reports from each of the organization's branches and major departments. To make sense of the information, he has created a set of spreadsheets to track the various environmental activities at M-Global. Now his challenge is to put this information in a form that can be understood easily for stockholders and other readers. He decides to convert the data in the spreadsheets into line charts, bar charts, and pie charts to make the information easily accessible. Using his spreadsheet program, it will be easy to create graphics that can be published in print and electronic forms.

Technology has radically changed the world of graphics. Now, almost anyone with a computer and the right software can quickly produce illustrations that used to take hours to construct. As a result, today there are sophisticated graphics in every medium—newspapers, television, websites, and, of course, technical documents.

Because readers expect graphics to accompany text, you as a technical professional must respond to this need. Well-designed and well-placed graphics keep your documents clear and user-friendly. Your graphics do not have to be fancy, however. Nor is it true that adding graphics will necessarily improve a document. Readers are impressed by visuals only when they are well done and appropriate.

Because so many different types of graphics are available, technical professionals must first understand the basics of graphics before they apply sophisticated techniques. To emphasize these basics, this chapter (1) defines some common graphics terms, (2) explains the main reasons to use graphics, (3) covers the importance of font choice, (4) suggests guidelines for using color, (5) gives some general suggestions for incorporating graphics into text, (6) lists specific guidelines for seven common graphics, and (7) shows you how to avoid graphics misuse.

>>> Terms in Graphics

Terminology for graphics is not uniform in the professions, which can lead to some confusion. For the purposes of this chapter, however, some common definitions are adopted and listed here:

- **Graphics:** This generic term refers to any non-textual portion of documents or oral presentations. It can be used in two ways: (1) to designate the field (e.g., "Graphics is an area in which he showed great interest") or (2) to name individual graphical items ("She placed three graphics in her report"). Graphics can be divided into **tables** and **figures.**
- **Illustrations, visual aids:** Used synonymously with *graphics,* these terms also can refer to all non-textual parts of a document. The term *visual aids*, however, often is limited to the context of oral presentations.
- **Tables:** Illustrations that place numbers or words in columns and rows.

- **Figures:** All graphics other than tables. Examples include charts (pie, bar, line, flow, and organization), engineering drawings, maps, and photographs.
- **Charts, graphs:** A subset of *figures,* these synonymous terms refer to a type of graphic that displays data in visual form—as with bars, pie shapes, or lines on graphs. In this textbook, we use the term *chart* most often.
- **Technical drawing:** Another subset of *figures,* a technical drawing is a representation of a physical object. Such illustrations can be drawn from many perspectives and can include *exploded* or *cutaway* views.

Of course, you may see other graphics terms. For example, some technical companies use the word *plates* for figures. Be sure to know the terms your readers understand and the types of graphics they use.

>>> Reasons for Using Special Fonts, Color, and Graphics

Although the technology for producing graphics is constantly changing, the rationale for using them remains the same. Before exploring specific types of illustrations, this section covers some reasons why readers might choose special fonts, color, and graphics to accompany text.

>> Reason 1: Special Fonts, Color, and Graphics Simplify Ideas

Your readers usually know less about the subject than you. Font styles, color, and graphics can help them cut through technical details and grasp basic ideas. For example, if a *Danger* symbol styled with the boldface **Impact** font style is given a bright orange background, the message is much more powerful than a buried warning message that looks just like the text around it. Also, a simple illustration of a laboratory instrument, such as a Bunsen burner, can make the description of a lab procedure much easier to understand. In a more complex example, Figure 12–1 uses a group of four different charts to convey the one main point—that M-Global's Equipment Development group lags behind the company's other profit centers. A quick look at the charts tells the story of the group's difficulties much better than would several hundred words of text. Of course, the text itself must still focus the reader's attention on the relevant information.

>> Reason 2: Graphics Reinforce Ideas

When a point needs emphasis, you can use a graphic element like a special font style or color, or you can create a graphic. For example, if you need to get instant reader recognition of section or chapter breaks, you might use a distinctive font style or color to capture the reader's attention. Also, you might draw a map to show where computer terminals will be located within a building or use a pie chart to show how a budget will be spent. You might even include a drawing that indicates how to operate a scanner. Each of these solutions would reinforce points made in the accompanying text.

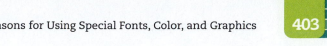

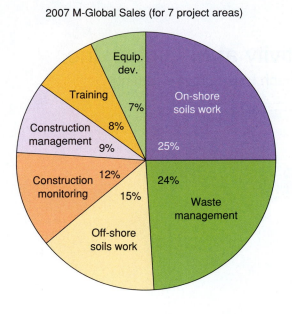

2007 M-Global Sales (for 7 project areas)

Graphics used to simplify ideas

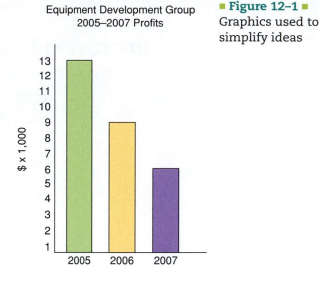

Equipment Development Group
2005–2007 Profits

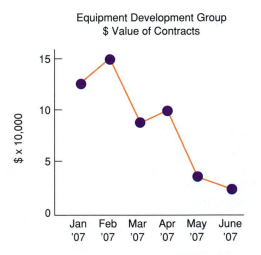

Equipment Development Group
$ Value of Contracts

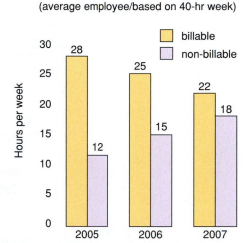

Equipment Development Group
Billable vs. Non-billable Time
(average employee/based on 40-hr week)

>> Reason 3: Graphics Create Interest

Tools such as font styles, color, and graphics are *attention grabbers* and can be used to engage readers' interest. If your customers have three reports on their desks and must quickly decide which one to read first, the one with a distinctive look that complements the text will probably be chosen first. It may be something as simple as (1) a map outline of the state, county, or city where you will be doing a project; (2) a picture of the product or service you are providing; or (3) a symbol of the purpose of your writing project.

Figure 12–2 shows how an outside consultant used a well-known Leonardo da Vinci drawing to attract attention to his M-Global proposal. The drawing helps (1) add a classical touch to the cover; (2) focus on the human side of employee testing; and (3) associate the innovation of Infinite Vision, Inc., with the creativity of da Vinci. Also, by using the

Improving Productivity at M-Global, Inc.
An Innovative Approach to Employee Testing

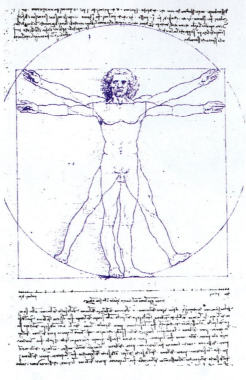

Prepared for James McDuff
President, M-Global, Inc.
by
James H. Stephens
Infinite Vision, Inc.

FEBRUARY 22, 2008

font style and color preference of the client, the writer allows the reader to "take owner-ship" of the proposal before ever reading the contents.

>> Reason 4: Graphics Are Universal

Some people wrongly associate the growing importance of graphics with today's reliance on television and other popular media—as if graphics pander to less-intellectual instincts. Although media such as television and the web obviously rely on visual effects, graphics have been humankind's universal language since cave drawings. Although those who sell or advertise products and services have long known the power of images, writers of technical documents have only recently started to merge the force of graphics with their text.

>>> Using Fonts

Chapter 4 gives an overview of two basic points about fonts to consider when designing documents: *fonts* and *font sizes*. This section provides more detail to help you meet the creative challenge of designing documents such as manuals, proposals, annual reports, and promotional materials.

Font Types

You already may have identified and grouped fonts into two categories—*sans serif* ("without feet or tails") and *serif* ("with feet or tails")—as you have explored the selection of fonts available to you on your desktop publishing software. Perhaps you have also noticed that there are many other font types, such as those shown next:

- *Freehand* and *Freestyle Script*—appear to be handwritten
- *Vivaldi* and 𝔚𝔢𝔡𝔡𝔦𝔫𝔤 𝔗𝔢𝔵𝔱—seem to have an austere and more formal appearance
- Parisian—reflects a high-fashion silhouette
- Kidprint—looks uneven or even awkward, as though it was printed by a child.

Your choice of fonts may determine whether your documents get read.

General Guidelines

How do you know which fonts to choose? Although there are no hard-and-fast rules to follow, keep the following five guidelines in mind to make the task easier:

> **Font Style Guidelines**
> **Consider:**
> 1. the reader's or company's preferences
> 2. the need for clarity
> 3. the space available
> 4. the purpose of your document
> 5. the tone you want to convey

>> Font Style Guideline 1: Consider the Reader's or Company's Preferences

Give font style the same consideration as you give your message. If your reader has clear preferences, by all means adhere to them. If, however, your audience is receptive to new ideas and images, you can become creative in selecting fonts.

>> Font Style Guideline 2: Consider the Need for Clarity

All technical writers recognize the need for their messages to be clear. As you select a font, ask yourself questions such as:

- Am I using this type font for captions or for long passages of text?

■ **Figure 12–3** ■ Font styles

SANS-SERIF FONT STYLES

Arial	Readability
Bauhaus Md	Readability
Basic Sans SF	Readability
Bernhard Fashion BT	Readability
Futura Md BT	**Readability**
Impact	**Readability**

Serif Font Styles

Caslon Bd BT	**Readability**
Americana XBdCn BT	**Readability**
Poster Bodni BT	**Readability**
GarmdITC Bk BT	**Readability**
Kuenstler script BT	*Readability*

■ Does the material I've written contain technical terms or formulae, or was it written for a general audience?

■ Will the font style enhance or detract from the readability of the material?

Readability of a font style partly depends on whether it uses sans-serif characters or serif characters. Sans-serif font styles (without feet or tails) and serif font styles (with feet or tails) are shown in 18 pt. in Figure 12–3 in the word *Readability*.

Sans-serif font styles are more effective when used for headers, numeric data, and transparencies for these reasons: characters appear cleaner and there is more white space between the individual characters. Serif font styles are more appropriate for normal text. The feet (or tails) of serif font styles seem to lead the readers' eyes across the text in a natural flow, making it easier to scan the page. The result is better readability.

Clarity of fonts is more than just a matter of size or serif. It also depends on how and where the document will be used. For example, the typeface **Clearview** was created to solve problems with legibility of highway signs, especially at night. The existing typeface, **Highway Gothic,** was inconsistent in the design of its letters, and the reflective letters often blurred at night. To solve these problems, Don Meeker and James Montalbano created a font with letters that were more open and easier to read, especially at night. (For a detailed discussion of the design of the **Clearview** typeface, visit http://www.clearviewhwy.com.)

>> Font Style Guideline 3: Consider the Space Available

Although all font styles are measured vertically using a scale of 72 points to an inch, they vary horizontally. If you are writing a long formal proposal, the space available might not be important. Adding or deleting one page might not matter. However, if you are writing the text for a software package, it may be necessary to select a font style that is clear but that—for this special assignment—fits into a $3'' \times 6''$ field.

Refer to Figure 12–3. What differences do you see? Wider letters? Thicker strokes? More or less white space between the characters (*kerning*)? More or less white space between rows of typed material (*leading*)? Each of these criteria varies from one font style to another.

In some cases, you may decide to use the same font style. Then you can make minor adjustments in line length by reducing the kerning between the characters to fit all your information on one page or to prevent the carryover of one word to the next page.

In other situations, you may want to adjust the leading. In this way, the lines are closed up by one point of space (of the 72 points per inch), and one more line can fit on the page. Unless your organization uses a mandatory style sheet, these choices are up to you.

>> Font Style Guideline 4: Consider the Purpose of the Document

Before making font-style decisions, evaluate how different font styles may reinforce your document's purpose. Ask yourself the following questions:

- Will the document be referred to frequently?
- Does the document present financial or statistical data that must be read and comprehended easily?
- Must it be eye-catching enough to make the audience eager to read it?
- Is it a routine document that needs to be read only once, handled, and filed?

Considering these questions will help you choose a font style that serves the purpose of your entire document.

>> Font Style Guideline 5: Consider the Tone You Want to Convey

Finally, give careful thought to which font style reinforces the tone of your message.

For example, assume you have worked diligently to develop an annual report reflecting serious growth problems for the company and for the company's industry. The document is formal in tone; however, it is also a no-nonsense business document. A font type such as **Lydian CSV BT** is formal; however, it does not even remotely present the tone required for this annual report. In this instance, you may want to use Arial or Avant Garde because of its crisp sans-serif image.

In another example, you have been asked to invite management and hourly employees to a retirement party given for a mid-level manager. The tone, you correctly assess, will be informal, warm, and hospitable. As you scroll down the list of font styles, you find several that seem appropriate—**Dom Casual BT**, **Zapfhumnst BT**, and **Ad Lib**—and you wonder which, if any, conveys the right image. The first one looks interesting and casual, but it appears too small. The second one looks too impersonal. The third, although a sans-serif font, appears casual, large, and powerful—like someone is shouting, *"Come on in!"*

In addition to the diversity of font styles available for your use, desktop publishing packages today offer you an opportunity to reconfigure your text into arcs, waves, slopes, and even circles. You can outline, shade, print vertically, and do much more whenever it is appropriate. That is the key: Your font styles must be *appropriate* for the tone you are trying to convey.

>>> Using Color

Just as people expect color in photographs, movies, advertisements, and TV, color has emerged as a necessity in technical communications. This tool adds excitement to and stimulates interest in your documents. Most writers have software that lets them simply click on the colors they think are appropriate for a document and, when the job is completed, send them to a color printer. However, their color choices—and even the decision to use color—may be inappropriate for a number of reasons.

Because color is so costly and requires a longer time to print, you must know more about its effective use. In this segment, we discuss what you must consider before you use color, including ways to use it effectively.

The Cost and Time of Using Color

The cost of using color in workplace communication is much like the cost of buying a luxury automobile. You receive many benefits, but you must be ready to pay for them. Before you spend your time and energy on using color, you must consider the answers to some important questions.

■ **Has a budget been established for the job? If so, what does it include?** Like it or not, budgets control our creative efforts. You won't want to invest massive amounts of time in thinking through your color strategy if you know up front that color is too costly. The seemingly high cost, however, may be justified by an anticipated return on investment.

Developing color documents is usually a less-significant cost factor than printing the materials. Most printer software gives you the option to print four-color documents. However, design your color graphics with clear, crisp images that retain their impact *even if they cannot be printed in color*.

■ **What is the distribution?** Ask yourself these questions about distribution: Will the document be sent to potential customers or key executives in your organization? How many other people will receive copies? Must all recipients have color copies?

The number of color copies required may be directly related to the hierarchy of receivers. If only five or six copies are needed and the pages are few, you may be able to print the documents on your desktop printer or to make copies on a color copier. If more copies are needed, it may be necessary to use a professional print shop to mass produce your documents. Four-color printing costs per document increase dramatically as volumes increase because the printer must make a separate *master* for each page and ink color used (Cyan, Magenta, Yellow, and Black, or *CMYK*) and the printer then runs each color for each page separately.

You may also want to consider electronic distribution of documents, especially documents used frequently for reference. These may be distributed as attachments to email, or they may be housed on a company's intranet or website. Electronic documents use RGB (Red, Green, and Blue) colors instead of the CMYK designations for ink colors, so if your organization has a color style sheet, it may be necessary to check the accuracy of a document's colors if you are converting between electronic and print formats.

■ **Does the company have printing/reproduction facilities?** Large companies frequently plan for high-volume reproduction of reports, advertising, training, and documentation. If they have the capability, you may receive more rapid turnaround times and benefit from lower costs.

■ **Are other options available?** If you have a tight budget and still want to use color to stimulate interest in your document, you may want to consider (1) using colored paper to distinguish one section from another, (2) printing one or two copies in color for the most important readers and printing the balance in black and white, and (3) using one color with black to add some interest (perhaps the color most readily identified with your company's image).

Developing a Color Style Sheet

Once you decide to use color in your documents, you'll notice a difference in how people react to your messages. Organizations often have color style sheets, such as the one shown in Figure 12–4, that provide guidelines for the use of color, using CMYK designations for print documents and RGB designations for electronic documents and web pages. If your company doesn't already have a color style sheet, develop your own so you can use color uniformly and consistently from one project to another.

OFFICIAL WESTERN COLORS

BLACK   PMS 123 COATED*

The official colors are black and yellow gold (referred to as Western gold). Due to the difficulty in reproducing the yellow gold on different papers, **two different inks should be used depending on the paper choice**.

How do I know which logo to use?
It depends upon your design, space restrictions and audience.
• If you have limited space use the Western logo with or without the Griffon. If you have very limited space use only the Griffon logo.
• In general if you are appealing to any external audience use the Discover Gold with Western logo.
• Always follow the standards for each logo on size restriction.

• *On coated paper (glossy, shiny paper) the Western gold is (PANTONE® Matching System) PMS 123C (as shown in this book).*
• *On uncoated paper (such as paper used in an office copier) the Western gold is PMS 109U.*
• When **four-color process** inks are used, Western gold can be produced by printing: *0C/25M/100Y/0K.*
• For **web publication or audio/visual usage**, the Western gold can be produced by: *R=254 G=194 B=10.*
If desired, it is acceptable to use a metallic gold, such as a foil or gold metallic ink:
• *On coated paper and uncoated paper the metallic gold is PMS 871.*

Western's Graphic Standards   3

■ **Figure 12–4** ■ Color style sheet

Source: Missouri Western Campus Printing and Design Services(2007). *Graphic Standards Manual.*

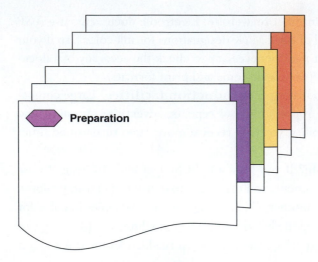

Avoid overusing color. Too many colors, like too much data, distort your message, confuse your audience, cost more, and take more time to produce.

To develop a style sheet, ask yourself the following four questions:

1. **How can I use color to help my audience read and retain the document's message?** You can define the levels of importance in a document by using a different color for different headings, and lighter or darker tones of a color to distinguish between subheadings or between major and minor concepts within a heading.

   For instance, in Figure 12–5, a report that contains five different sections uses colored icons that match section tabs. This technique adds color without giving up the readability of black font styles on white paper.

2. **How can I use color to attract attention to important data?** Use one color to highlight significant points or to provide a specific function. For example, if you consistently use color to frame tables or graphics, you create a visual cue that prompts your reader to look at the graphics. Alternatively, you may want to use one color to consistently emphasize key words, phrases, or specific actions. In so doing, you build continuity in your document and enhance readability.

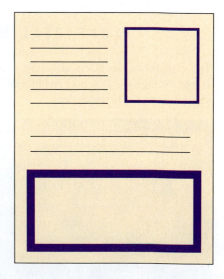

3. **Will my audience be able to see the differences in color?** About 10% of the male population has some degree of difficulty seeing differences in colors. Therefore, be sure that graphics use light and dark contrasts effectively. This also ensures that your graphics will still be clear, even if the document is copied in black and white.

4. **Will the document be distributed universally in color, or will one or more sections be printed in black and white?** If your document will be reproduced—either all or in part—in black and white, you must place greater emphasis on textural differences as well as use distinctive shading and tinting. Most important, do not use blue images; unless shaded considerably, they do not reproduce on a photocopier. You must do everything you can to sharpen the images you want your audience to read.

After you answer questions about the general use of color, you can begin the process of selecting and combining colors to enhance your message and your graphics.

Color Terms

Most of us are familiar with the spectrum of colors seen when light passes through a prism or when rainbows appear in the sky. In most instances, your desktop publishing software displays a full color array that resembles a color wheel. However, you may not be familiar with some common color-related terms such as *hues, primary, secondary, tertiary*, and *complementary*. Let us look at them briefly (see Figure 12–6):

- **Hues:** *Hues* are the intense true colors we see on a color wheel that displays the relationships and intensities of colors.
- **Primary Colors:** Red, blue, and yellow are considered *primary* colors. We learned ways to combine primary colors and make other colors that also appear on the color wheel through light, pigments, or inks.
- **Secondary Colors:** As children, we often learn that when we mix equal parts of red and blue pigments, the result is purple; in the same way, blue and yellow become green, and yellow and red become orange. Because purple, green, and orange are mixed from the primary colors, they are often referred to as *secondary colors*.
- **Tertiary Colors:** *Tertiary colors* are colors mixed by a combination of a primary color and a secondary color, and are shown in Figure 12–6. They are also considered hues because of their intensity.
- **Complementary Colors:** *Complementary colors* are those that appear immediately opposite one another on the color wheel. For example, in Figure 12–6, you can see that red and green are opposite each other—as are blue and orange, and yellow and violet. Because complementary colors are opposites, they are frequently used to give the best contrast between colors.

Guidelines for Using Color

This section includes four guidelines to help you select colors for your documents:

>> **Color Selection Guideline 1: Use Colors Your Audience Will Associate with Your Topic in a Positive Way**

Take time to research your reader's reactions to colors. In particular, if your document has an international audience, make certain you know what colors are considered appropriate or inappropriate in specific cultures. Take time to do your research before you add color.

■ **Figure 12-6** ■
Color classifications

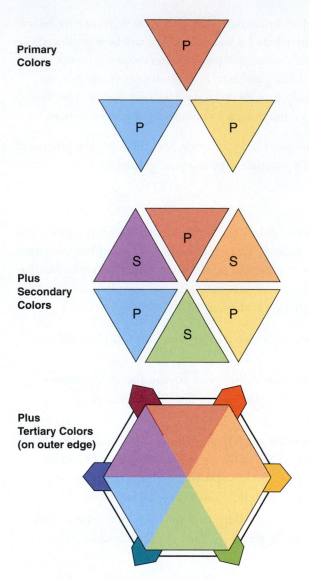

Primary Colors

Plus Secondary Colors

Plus Tertiary Colors (on outer edge)

>> Color Selection Guideline 2: Use Colors That Enhance Your Company's Logo

By using colors that complement and reinforce your organization's logo, you are more likely to get immediate acceptance of your ideas. Recently, organizations have become more aware of the importance of *branding*, careful use of a logo and colors to project a clear and easily identifiable image. The M-Global logo to the left includes colors that are used in a variety of ways. Color is an important part of establishing an organization's brand.

>> Color Selection Guideline 3: Use Dark or Textured Backgrounds Infrequently

For years, advertisers thought the most productive newspaper ads were those with solid black backgrounds and large, white letters. When writers learned more about the use of white space, the myth of the impact of these *reverse messages* was shattered. Although dark or textured backgrounds can be effective, like other design tools they can be overdone. Use them when you want to focus your reader's attention on a graphic by using darker tones as a frame.

If your documents will be printed in black, white, and gray, try using texture for only one or two elements of your graphic; it helps you avoid having a cluttered and busy appearance. If you decide to use shading to mask out—but not obliterate—important data, don't use shading any darker than a 10%–25% screen.

Finally, if you really want your message to be surrounded with a dark color, remember you must also use a larger and bolder font style and a lighter and brighter hue, or the information you want so much to be noticed will fade into the background.

>> Color Selection Guideline 4: Learn How Any Color Will Be Affected by the Medium It Will Be Printed on and Make Adjustments Accordingly

Textures and colors of paper stock can drastically alter the appearance of your document. It is best to test the final colors before you use them in a final draft. Ask your peers, your boss, and your public relations department for their opinions and suggestions. A little planning can help you strengthen your document.

>>> General Guidelines for Graphics

A few basic guidelines apply to all graphics. Keep the following fundamentals in mind as you move from one type of illustration to another.

>> Graphics Guideline 1: Determine the Purpose of the Graphic

Graphics, like text, should only be used if they serve a purpose. Ask yourself the following questions:

- What kind of information does your audience need to better understand the scope, problem, or solution?
- What type of graphic can be used to present the data in the most interesting and informative way?
- Are any special symbols, colors, or font styles needed to reinforce the data?

>> Graphics Guideline 2: Evaluate the Accuracy and Validity of the Data

Unless the information you plan to include in your document is accurate, you run the risk of presenting information that could damage your credibility as well as the credibility of the document. Remember, one false or inappropriate statistic tends to cast doubt on the balance of the data. Do the following:

- Check the accuracy of information
- Make sure the source is reputable
- Ensure that data are not distorted by flawed scales or images

>> Graphics Guideline 3: Refer to All Graphics in the Text

With a few exceptions—such as cover illustrations used to grab attention—graphics should be accompanied by clear references within your text. Specifically, you should follow these rules:

- Include the graphic number in arabic, not roman, when you are using more than one graphic
- Include the title, and sometimes the page number, if either is needed for clarity or emphasis
- Incorporate the reference smoothly into text wording
- Highlight significant information being communicated by the graphic

Following are two ways to phrase and position a graphics reference. In Example 1, there is the additional emphasis of the graphic's title, whereas in Example 2, the title is left out. Also, note that you can draw more attention to the graphic by placing the reference at the start of the sentence in a separate clause, or you can relegate the reference to a parenthetical expression at the end or middle of the passage. Choose the option that best suits your purposes.

- **Example 1:** In the past five years, 56 businesses in the county have started in-house recycling programs. The result has been a dramatic shift in the amount of property the county has bought for new waste sites, as shown in Figure 5 ("Landfill Purchases, 1985–1990").
- **Example 2:** As shown in Figure 5, the county has purchased much less land for landfills during the past five years. This dramatic reduction results from the fact that 56 businesses have started in-house recycling programs.

>> Graphics Guideline 4: Think About Where to Put Graphics

In most cases, locate a graphic close to the text in which it is mentioned. This immediate reinforcement of text by an illustration gives graphics their greatest strength. Variations of this option, as well as several other possibilities, include the following:

- **Same page as text reference:** A simple visual, such as an informal table, should go on the same page as the text reference if you think it too small for a separate page.
- **Page opposite text reference:** A complex graphic, such as a long table, that accompanies a specific page of text can go on the page opposite the text—that is, on the opposite page of a two-page spread. Usually this option is exercised *only* in documents that are printed on both sides of the paper throughout.
- **Page following first text reference:** Most text graphics appear on the page after the first reference. If the graphic is referred to throughout the text, it can be repeated at later points. (*Note:* Readers prefer to have graphics positioned exactly where they need them, rather than their having to refer to another part of the document.)
- **Attachments or appendices:** Graphics can go at the end of the document in two cases: first, if the text contains so many references to the graphic that placement in a

central location, such as an appendix, would make it more accessible; and second, if the graphic contains less important supporting material that would only interrupt the text.

>> Graphics Guideline 5: Position Graphics Vertically When Possible

Readers prefer graphics they can view without having to turn the document sideways. However, if the table or figure cannot fit vertically on a standard 8½- × 11-inch page, either use a foldout or place the graphic horizontally on the page. In the latter case, position the illustration so that the top is on the left margin. (In other words, the page must be turned clockwise to be viewed.)

>> Graphics Guideline 6: Avoid Clutter

Let simplicity be your guide. Readers go to graphics for relief from or reinforcement of the text. They do not want to be bombarded by visual clutter. Omit information that is not relevant to your purpose while still making the illustration clear and self-contained. Also, use enough white space so that the readers' eyes are drawn to the graphic. The final section of this chapter discusses graphics clutter in more detail.

>> Graphics Guideline 7: Provide Titles, Notes, Keys, and Source Data

Graphics should be as self-contained and self-explanatory as possible. Moreover, they must include any borrowed information. Follow these basic rules for format and acknowledgment of sources:

- **Title:** Follow the graphic number with a short, precise title—either on the line beneath the number or on the same line after a colon (e.g., "Figure 3: Salary Scales").
- **Tables:** The number and title go at the top. (As noted in Table Guideline 1 on page 416, one exception is informal tables, which have no table number or title.)
- **Figures:** The number and title usually go below the illustration. Center titles or place them flush with the left margin.
- **Notes for explanation:** When introductory information for the graphic is needed, place a note directly underneath the title or at the bottom of the graphic.
- **Keys or legends for simplicity:** If a graphic needs many labels, consider using a legend or key, which lists the labels and corresponding symbols on the graphic. For example, a pie chart might have the letters *A, B, C, D,* and *E* printed on the pie pieces and a legend at the top, bottom, or side of the figure listing what the letters represent.
- **Source information at the bottom:** You have an ethical, and sometimes legal, obligation to cite the person, organization, or publication from which you borrowed information for the figure. Either (1) precede the description with the word *Source* and a colon, or (2) if you borrowed just part of a graphic, introduce the citation with *Adapted from*.

As well as citing the source, it is sometimes necessary to request permission to use copyrighted or proprietary information, depending on how you use it and how much you are using. (A prominent exception is most information provided by the federal government;

most government publications are not copyrighted.) Consult a reference librarian for details about seeking permission.

>>> Specific Guidelines for Seven Graphics

Illustrations come in many forms. Almost any non-textual part of your document can be placed under the umbrella term *graphic*. Among the many types, the following are often used in technical communication: (1) tables, (2) pie charts, (3) bar charts, (4) line charts, (5) flowcharts, (6) organization charts, and (7) technical drawings. This section of the chapter highlights their different purposes and gives guidelines for using each type.

Tables

Tables present readers with raw data, usually in the form of numbers but sometimes in the form of words. Tables are classified as either *informal* or *formal*:

- **Informal tables:** limited data arranged in the form of either rows or columns
- **Formal tables:** data arranged in a grid, always with both horizontal rows and vertical columns

The following five guidelines help you design and position tables within the text of your documents.

>> Table Guideline 1: Use Informal Tables as Extensions of Text

Informal tables are usually merged with the text on a page, rather than isolated on a separate page or attachment. As Figure 12–7 shows, an informal table usually has (1) no table number or title, and (2) few if any headings for rows or columns. Also, it is not included in the list of illustrations in a formal document.

>> Table Guideline 2: Use Formal Tables for Complex Data Separated from Text

Formal tables may appear on the page of text that includes the table reference, on the page following the first text reference, or in an attachment or appendix. In every case, you should do the following:

1. Extract important data from the table and highlight them in the text
2. Make every formal table as clear and visually appealing as possible by doing the following:

   - Use color to designate positive or negative totals, increases, or decreases or very important points.
   - Use colored or dark gray borders to frame the information and call attention to the table.

FTC staff then posted sets of three of these newly-created email addresses consisting of an Unfiltered Address, an address at Filtered ISP 1, and an address at Filtered ISP 2 – on 50 Internet locations. The 50 Internet locations included websites controlled by the FTC[5] and several popular message boards, blogs, chat rooms, and USENET groups which had high hit/visit rates, according to ranking websites such as www.message-boards.com and Google popularity searches.[6] All of the 150 addresses were posted during a three day period in July 2005.

Graphic 1

Locations On Which Email Addresses Were Posted

Type	Number
FTC Website Pages	12
Message Boards	12
Blogs	12
Chat Rooms	12
USENET Groups	2

■ **Figure 12–7** ■
Informal table in a report
Source: Federal Trade Commission. (Nov. 2005). *Email Address Harvesting and the Effectiveness of Anti-Spam Filters: A Report by the Federal Trade Commission's Division of Marketing Practices,* p. 2.

- Use gray screens (no denser than 10%–25%) to subordinate less-important data that appear on the table.
- Avoid excessive use of heavy horizontal and vertical crossed lines that create a grid look; instead, consider using white lines on a light gray background or light gray lines on a white background to lead the eye across the page.

Figure 12–8 shows examples of three tables that use different designs to present the same information. The first version includes no shading; the second version uses shading to emphasize the age groups that are being reported on; and the third version, which is the original version, uses shading to emphasize the statistically significant changes.

>> Table Guideline 3: Use Plenty of White Space

Used around and within tables, white space guides the eye through a table much better than black lines. Avoid putting densely drawn black boxes around tables. Instead, leave 1 inch more of white space than you would normally leave around text and let it act as a frame.

Table 1B. Percentage of Students Ages 9,13, and 17, by Frequency of Reading for Fun: 1984,1999, and 2004

Age 9	Almost every day	Once or twice a week	Once or twice a month	A few times a year	Never or hardly ever
1984	53%	28%	7%	3%	9%
1999	54%	26%	6%	4%	10%
2004	54%	26%	7%	5%	8%
Age 13	Almost every day	Once or twice a week	Once or twice a month	A few times a year	Never or hardly ever
1984	35%	35%	14%	7%	8%
1999	28%	36%	17%	10%	9%
2004	30%	34%	15%	9%	13%
Age 17	Almost every day	Once or twice a week	Once or twice a month	A few times a year	Never or hardly ever
1984	31%	33%	17%	10%	9%
1999	25%	28%	19%	12%	16%
2004	22%	30%	15%	14%	19%

Source. U.S. Department of Education, National Center for Education Statistics

Table 1B. Percentage of Students Ages 9,13, and 17, by Frequency of Reading for Fun: 1984,1999, and 2004

Age 9	Almost every day	Once or twice a week	Once or twice a month	A few times a year	Never or hardly ever
1984	53%	28%	7%	3%	9%
1999	54%	26%	6%	4%	10%
2004	54%	26%	7%	5%	8%
Age 13	Almost every day	Once or twice a week	Once or twice a month	A few times a year	Never or hardly ever
1984	35%	35%	14%	7%	8%
1999	28%	36%	17%	10%	9%
2004	30%	34%	15%	9%	13%
Age 17	Almost every day	Once or twice a week	Once or twice a month	A few times a year	Never or hardly ever
1984	31%	33%	17%	10%	9%
1999	25%	28%	19%	12%	16%
2004	22%	30%	15%	14%	19%

Source. U.S. Department of Education, National Center for Education Statistics

Table 1B. Percentage of Students Ages 9,13, and 17, by Frequency of Reading for Fun: 1984,1999, and 2004

Age 9	Almost every day	Once or twice a week	Once or twice a month	A few times a year	Never or hardly ever
1984	53%	28%	7%	3%	9%
1999	54%	26%	6%	4%	10%
2004	54%	26%	7%	5%	8%
Age 13	Almost every day	Once or twice a week	Once or twice a month	A few times a year	Never or hardly ever
1984	35%	35%	14%	7%	8%
1999	28%	36%	17%	10%	9%
2004	30%	34%	15%	9%	13%
Age 17	Almost every day	Once or twice a week	Once or twice a month	A few times a year	Never or hardly ever
1984	31%	33%	17%	10%	9%
1999	25%	28%	19%	12%	16%
2004	22%	30%	15%	14%	19%

Source. U.S. Department of Education, National Center for Education Statistics

■ **Figure 12–8** ■ Use of shading and color highlights in 3 identical tables

Adapted from National Endowment for the Arts. (Nov. 2007). *To Read or Not to Read: A Question of National Consequence, p. 29.*

TABLE 22: M-Global's Employee Retirement Fund

Investment Type	Book Value	Market Value	% of Total Market Value
Temporary Securities	$ 434,084	434,084	5.9%
Bonds	3,679,081	3,842,056	52.4
Common Stocks	2,508,146	3,039,350	41.4
Mortgages	18,063	18,063	0.3
Real Estate	1,939	1,939	nil
Totals	$6,641,313	$7,335,492	100.0%

Note: This table contrasts the book value versus the market value of the M-Global Employee Retirement Fund, as of December 31, 2008.
Source: M-Global's accounting firm of Bumble and Bumble, Inc.

■ **Figure 12–9** ■

Example of formal table

>> **Table Guideline 4: Follow Usual Conventions for Dividing and Explaining Data**

Figure 12–9 shows a typical formal table. It satisfies the overriding goal of being clear and self-contained. To achieve that objective in your tables, use the following guidelines:

1. **Titles and numberings:** Give a title to each formal table, and place the title and number above the table. Number each table if the document contains two or more tables.

2. **Headings:** Create short, clear headings for all columns and rows.

3. **Abbreviations:** Include in the headings any necessary abbreviations or symbols, such as *lb* or %. Spell out abbreviations and define terms in a key or footnote if the reader may need such assistance.

4. **Numbers:** For ease of reading, round off numbers when possible. Also, align multi-digit numbers on the right edge, or at the decimal when shown.

5. **Notes:** Place any necessary explanatory headnotes either between the title and the table (if the notes are short) or at the bottom of the table.

6. **Footnotes:** Place any necessary footnotes below the table.

7. **Sources:** Place any necessary source references beneath the footnotes.

8. **Caps:** Use uppercase and lowercase letters rather than all caps.

>> **Table Guideline 5: Pay Special Attention to Cost Data**

Most readers prefer to have complicated financial information placed in tabular form. Given the importance of such data, edit cost tables with great care. Devote extra attention to the following two issues:

■ Placement of decimals in costs
■ Correct totals of figures

■ **Figure 12–10** ■
Pie chart with as few pieces as possible. (Chart shows M-Global workforce breakdown for an offshore Atlantic project. M-Global can draw most project workers from its East Coast offices.)

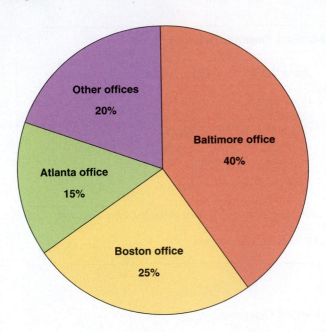

Documents like proposals can be considered contracts in some courts of law, so there is no room for error in relating costs.

Pie Charts

Familiar to most readers, *pie charts* show approximate relationships between the parts and the whole. Their simple circles with clear labels can provide comforting simplicity within even the most complicated report. Yet the simple form keeps them from being useful when you must reveal detailed information or changes over time. Following are specific guidelines for constructing pie charts.

>> Pie Chart Guideline 1: Use Pie Charts Especially for Percentages and Money

Pie charts catch the readers' eyes best when they represent items divisible by 100—as with percentages and dollars. Figure 12–10 shows percentages; Figure 12–11 shows cents. Using the pie chart for money breakdowns is made even more appropriate by the coinlike shape of the chart. *In every case, make sure your percentages or cents add up to 100.*

>> Pie Chart Guideline 2: Use No More Than Six or Seven Divisions

To make pie charts work well, limit the number of pie pieces to no more than six or seven. In fact, the fewer segments the better. This approach lets the reader grasp major relationships without having to wade through the clutter of tiny divisions that are difficult to label and read. In Figure 12–10, for example, M-Global's client can readily see that most of the project staff will come from the three M-Global offices closest to the project site.

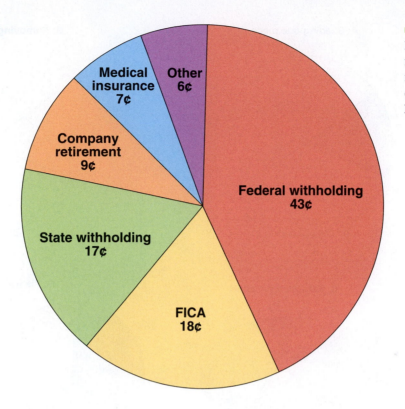

>> Pie Chart Guideline 3: Move Clockwise from 12:00, from Largest to Smallest Wedge

Readers prefer pie charts oriented like a clock, with the first wedge starting at 12:00. Move from the largest to the smallest wedge to provide a convenient organizing principle.

Make exceptions to this design only for good reason. In Figure 12–10, for example, the last wedge represents a greater percentage than the previous wedge. In this way, it does not break up the sequence the writer wants to establish by grouping the three M-Global offices with the three largest percentages of project workers.

Another exception over which you have no control is your use of software packages that do not permit you to begin your pie at 12:00.

>> Pie Chart Guideline 4: Be Creative, But Stay Simple

Figure 12–12 shows a number of options for designing pie charts, including

1. Shading a wedge
2. Removing a wedge from the main pie

Be aware that some desktop publishing programs automatically format your pie charts using complex backgrounds and shading. However, these are often difficult to read and may distort the pieces of the pie. Remember to keep your pie charts simple and clean looking.

■ **Figure 12–12** ■
Techniques for
emphasis in pie
charts

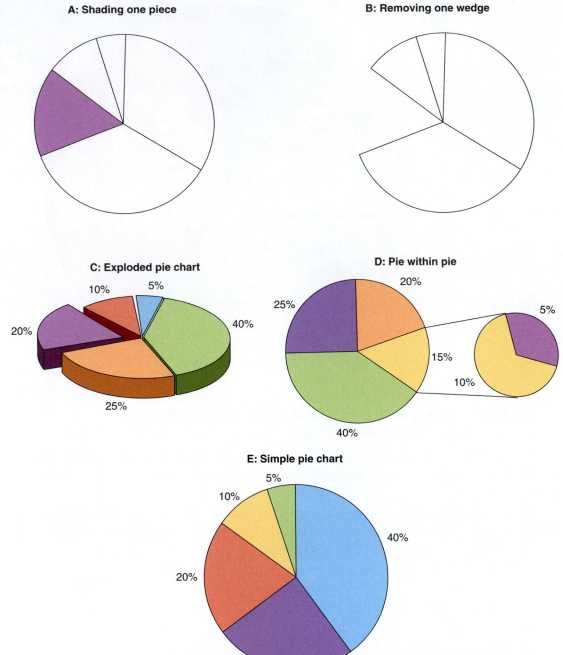

A: Shading one piece

B: Removing one wedge

C: Exploded pie chart
10% 5%
40%
20%
25%

D: Pie within pie
20%
25%
15%
10%
5%
40%

E: Simple pie chart
5%
10%
40%
20%
25%

>> Pie Chart Guideline 5: Draw and Label Carefully

The most common pie chart errors are (1) wedge sizes that do not correspond correctly to percentages or money amounts and (2) pie sizes that are too small to accommodate the information placed in them. Following are some suggestions for avoiding these mistakes:

■ **Pie size:** Make sure the chart occupies enough of the page. On a standard $8\frac{1}{2} \times 11$-inch sheet with only one pie chart, your circle should be from 3 to 6 inches

in diameter—large enough not to be dwarfed by labels and small enough to leave sufficient white space in the margins.

■ **Labels:** Place the wedge labels either inside the pie or outside, depending on the number of wedges, the number of wedge labels, or the length of the labels. Choose the option that produces the cleanest-looking chart.

Remember, however, that a pie chart does not reveal fine distinctions very well; it is best
used for showing larger differences.

Bar Charts

Like pie charts, bar charts are easily recognized, because they are seen every day in newspapers and magazines. Unlike pie charts, however, bar charts can accommodate a good deal of
technical detail. Comparisons are provided by means of two or more bars running horizontally or vertically on the page. Use the following five guidelines to create effective bar charts.

>> Bar Chart Guideline 1: Use a Limited Number of Bars

Although bar charts can show more information than pie charts, both illustrations have
their limits. Bar charts begin to break down when there are so many bars that information
is not easily grasped. The maximum bar number can vary according to chart size, of
course. Figure 12–13 shows two multi-bar charts. The impact of the charts is enhanced by
the limited number of bars.

>> Bar Chart Guideline 2: Show Comparisons Clearly

Bar lengths should be varied enough to show comparisons quickly and clearly. Avoid using
bars that are too close in length, because then readers must study the chart before understanding it. Such a chart lacks immediate visual impact.

Also, avoid the opposite tendency of using bar charts to show data that are much different in magnitude. To relate such differences, some writers resort to the dubious technique

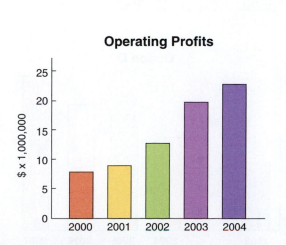

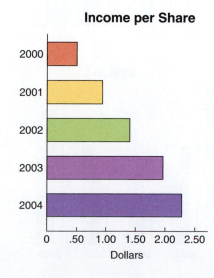

■ **Figure 12–13** ■
Bar charts

■ **Figure 12–14** ■
Hash marks on bar
charts—a technique
that can lead to
misunderstanding

Source: Toyota Motor
Corporation. (May 2003).
Toyota Hybrid System,
p. 18.

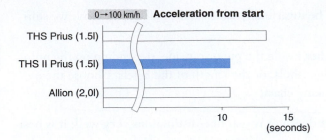

of inserting *break lines* (two parallel lines) on an axis to reflect breaks in scale (Figure 12–14).
Although this approach at least reminds readers of the breaks, it is still deceptive. The reader
must think about these differences before making sense out of the chart. In other words,
the use of hash marks runs counter to a main goal of graphics—creating an immediate and
accurate visual impact.

>> **Bar Chart Guideline 3: Keep Bar Widths Equal
and Adjust Space Between Bars Carefully**

Although bar length varies, bar width must remain constant. As for distance between the
bars, following are four options (along with examples in Figure 12–15):

■ **Option A: Use no space** when there are close comparisons or many bars, so that
differences are easier to grasp.

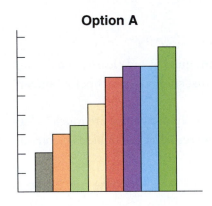

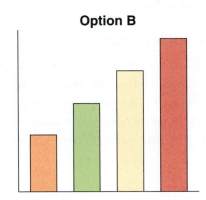

■ **Figure 12–15** ■
Bar chart variations

Adapted from William S.
Pfeiffer. (1989). *Proposal
Writing.* Columbus,
OH: Merrill, p. 148.

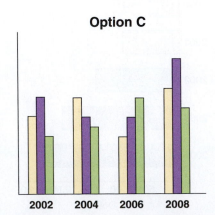

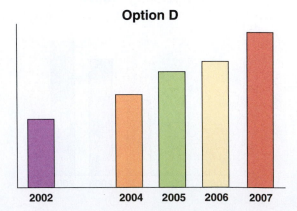

- **Option B: Use equal space, but less than bar width** when bar height differences are great enough to be seen in spite of the distance between bars.
- **Option C: Group related bars** to emphasize related data.
- **Option D: Use variable space** when gaps between some bars are needed to reflect gaps in the data.

>> **Bar Chart Guideline 4:** Carefully Arrange the Order of Bars

The arrangement of bars is what reveals meaning to readers. Following are two common approaches:

- **Sequential:** used when the progress of the bars shows a trend—for example, M-Global's increasing number of environmental projects in the past five years
- **Ascending or descending order:** used when you want to make a point by the rising or falling of the bars—for example, the 2008 profits of M-Global's seven international offices, from lowest to highest

>> **Bar Chart Guideline 5:** Be Creative

Figure 12–16 shows two bar chart variations that help display multiple trends. The *segmented bars* in Option A produce four types of information: the total percentage each of

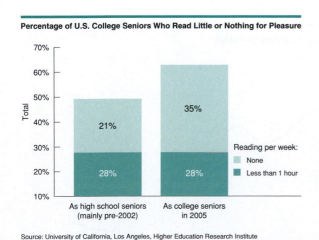

Source: University of California, Los Angeles, Higher Education Research Institute

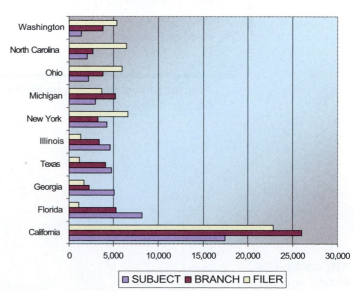

- **Figure 12–16** - Bar chart variations for multiple trends

Source (Figure 12–16a): Office of Regulatory Analysis. (Nov. 2006). *Mortgage Loan Fraud: An Industry Assessment Based upon Suspicious Activity Report Analysis*, p. 10.

Source (Figure 12–16b): National Endowment for the Arts. (Nov. 2007). *To Read or Not to Read: A Question of National Consequence*, p. 9.

high school seniors and college freshmen who spend little time reading each week, and the percentage of high school seniors and college freshman who do not spend any time reading. The *grouped bars* in Option B show the top ten states for reported mortgage fraud, and compare the source of the reports in each state.

Line Charts

Line charts are a common graphic. Almost every newspaper contains a few charts covering topics such as stock trends, car prices, or weather. More than other graphics, line charts telegraph complex trends immediately.

They work by using vertical and horizontal axes to reflect quantities of two different variables. The vertical (or *y*) axis usually plots the dependent variable; the horizontal (or *x*) axis usually plots the independent variable. (The dependent variable is affected by changes in the independent variable.) Lines then connect points that have been plotted on the chart. When drawing line charts, use the following five main guidelines.

>> Line Chart Guideline 1: Use Line Charts for Trends

Readers are affected by the direction and angle of the chart's line(s), so take advantage of this persuasive potential. In Figure 12–17, for example, the writer wants to show the decline in spending on books. Including a line chart in the study gives immediate emphasis to the recent downward trend in book purchases.

>> Line Chart Guideline 2: Locate Line Charts with Care

Given their strong impact, line charts can be especially useful as attention-grabbers. Consider placing them (1) on cover pages (to engage reader interest in the document), (2) at the beginning of sections that describe trends, and (3) in conclusions (to reinforce a major point of your document).

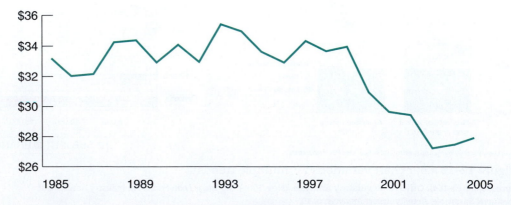

Average Annual Spending on Books, by Consumer Unit
Adjusted for Inflation

■ **Figure 12–17** ■
Basic line chart

Source: National Endowment for the Arts. (Nov. 2007). *To Read or Not to Read: A Question of National Consequence,* p. 11.

The Consumer Price Index, 1982–1984 (less food and energy), was used to adjust for inflation.
Source: U.S. Department of Labor, Bureau of Labor Statistics

>> Line Chart Guideline 3: Strive for Accuracy and Clarity

Like bar charts, line charts can be misused or just poorly constructed. Be sure that the line or lines on the graph truly reflect the data from which you have drawn. Also, select a scale that does not mislead readers with visual gimmicks. Following are some specific suggestions to keep your line charts accurate and clear:

■ Start all scales from zero to eliminate the possible confusion of breaks in amounts (see Bar Chart Guideline 3).

■ Select a vertical-to-horizontal ratio for axis lengths that is pleasing to the eye (three vertical to four horizontal is common).

■ Make chart lines as thick as, or thicker than, the axis lines.

>> Line Chart Guideline 4: Do Not Place Numbers on the Chart Itself

Line charts derive their main effect from the simplicity of lines that show trends. Avoid cluttering the chart with a lot of numbers that only detract from the visual impact.

>> Line Chart Guideline 5: Use Multiple Lines with Care

Like bar charts, line charts can show multiple trends. Simply add another line or two. To help readers quickly distinguish between lines, assign a differently shaped data point to each line. If you place too many lines on one chart, however, you run the risk of confusing the reader with too much data. Use no more than four or five lines on a single chart (Figure 12–18).

Figure A9-3: Industrial Sector Gas Prices in the United States, OECD Europe, Japan, and Taiwan, 1994-2002, in 2003 Dollars

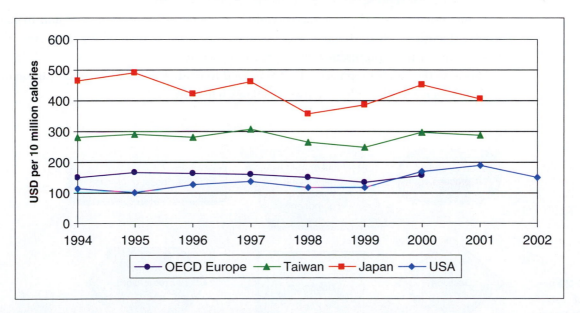

■ **Figure 12–18** ■ Line chart using multiple lines

Source: The Economic Future of Nuclear Power: A Study Conducted at The University of Chicago (August 2004), p. A9–A13.

■ **Figure 12–19** ■
Flowchart for basic
M-Global project

Adapted from William
S. Pfeiffer. (1989). *Proposal
Writing.* Columbus, OH:
Merrill, p. 155.

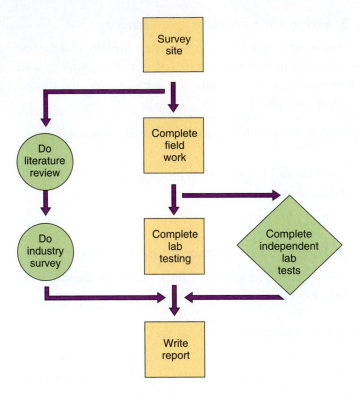

Flowcharts

Flowcharts tell a story about a process, usually by stringing together a series of boxes and other shapes that represent separate activities (Figure 12–19).

Some flowcharts use standardized symbols to represent steps in the decision-making process (Figure 12–20). Although these symbols were originally used for programming, they are now used to represent a wide range of processes.

Because they have a reputation for being hard to read, you must take extra care in designing flowcharts. The following five guidelines will help.

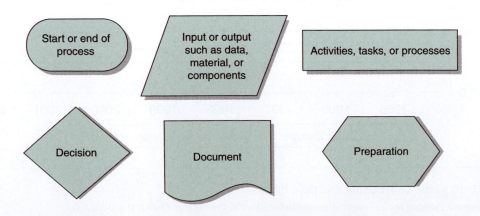

■ **Figure 12–20** ■
Selection of
standard flowchart
symbols

>> Flowchart Guideline 1: Present Only Overviews

Readers usually want flowcharts to give them only a capsule version of the process, not all the details. Reserve your list of particulars for the text or the appendices, where readers expect it.

>> Flowchart Guideline 2: Limit the Number of Shapes

Flowcharts rely on rectangles and other shapes to relate a process—in effect, to tell a story. Different shapes represent different types of activities. Some flowcharts, like the one in Figure 12–21, use icons and images to present information. This variety helps in describing a complex process, but it can also produce confusion. For the sake of clarity and simplicity, limit the number of different shapes in your flowcharts.

>> Flowchart Guideline 3: Provide a Legend When Necessary

Simple flowcharts often need no legend. The few shapes on the chart may already be labeled by their specific steps. When charts get more complex, however, include a legend that identifies the meaning of each shape used.

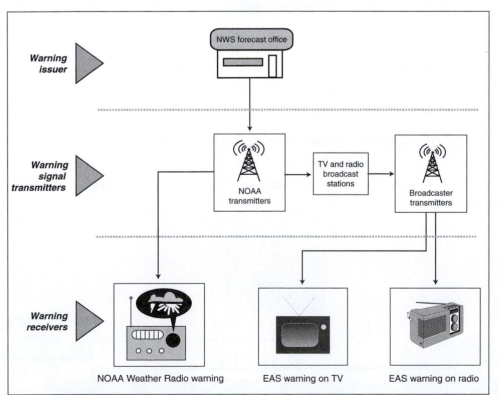

Figure 8: Tsunami Warning Signal Transmission for EAS and NOAA Weather Radio

Source: GAO analysis and Art Explosion.

■ **Figure 12–21** ■
Flowchart using icons and images

Source: Government Accounting Office. (June 2006). *U.S. Tsunami Preparedness: Federal and State Partners Collaborate to Help Communities Reduce Potential Impacts, but Significant Challenges Remain*, p. 27.

>> **Flowchart Guideline 4: Run the Sequence from Top to Bottom or from Left to Right**

Long flowcharts may cover the page with several columns or rows; however, they should always show some degree of uniformity by assuming either a basically vertical or horizontal direction.

>> **Flowchart Guideline 5: Label All Shapes Clearly**

Besides a legend that defines meanings of different shapes, the chart usually includes a label for each individual shape or step. Follow one of these approaches:

- Place the label inside the shape.
- Place the label immediately outside the shape.
- Put a number in each shape and place a legend for all numbers in another location (preferably on the same page).

Organization Charts

Organization charts reveal the structure of a company or other organization—the people, positions, or work units. The challenge in producing this graphic is to make sure that the arrangement of information accurately reflects the organization.

>> **Organization Chart Guideline 1: Use the Linear Boxes Approach to Emphasize High-Level Positions**

This traditional format uses rectangles connected by lines to represent some or all of the positions in an organization (Figure 12–22). Because high-level positions usually appear at the top of the chart, where the attention of most readers is focused, this design tends to emphasize upper management.

>> **Organization Chart Guideline 2: Connect Boxes with Solid or Dotted Lines**

Solid lines show direct reporting relationships; dotted lines show indirect or staff relationships (see Figure 12–22).

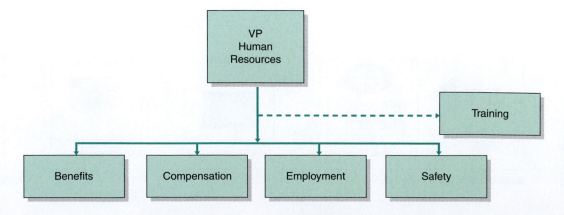

■ **Figure 12–22** ■
Basic organization chart, M-Global Human Resources Division

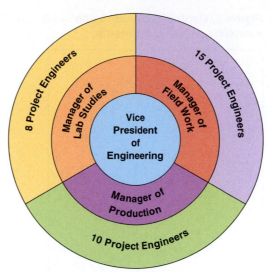

■ **Figure 12–23** ■
Concentric
organization chart
Adapted from William
S. Pfeiffer. (1989).
Proposal Writing.
Columbus, OH: Merrill,
p. 159.

>> Organization Chart Guideline 3: Use a Circular Design to Emphasize Mid- and Low-Level Positions

An arrangement of concentric circles gives more visibility to workers outside upper management. These are often the technical workers most deeply involved in the details of a project. For example, Figure 12–23 draws attention to the project engineers perched on the chart's outer ring.

>> Organization Chart Guideline 4: Use Varied Shapes Carefully

Like flowcharts, organization charts can use different shapes to indicate different levels or types of jobs. However, beware of introducing more complexity than you need. Use more than one shape only if you are convinced this approach is needed to convey meaning to the reader.

>> Organization Chart Guideline 5: Be Creative

When standard forms will not work, create new ones. For example, the organization charts in Figure 12–24 use circles and lines to compare responsibilities and reporting structures in two different kinds of organizations.

Technical Drawings

Technical drawings are important tools of companies that produce or use technical products. These drawings can accompany documents such as instructions, reports, sales orders, proposals, brochures, and posters. They are preferred over photographs when specific views are more important than photographic detail. Whereas all drawings used to be produced mainly by hand, now they are usually created by computer-assisted design (CAD) systems. Use the following guidelines for producing and using technical drawings that complement your text.

■ **Figure 12–24** ■
Creative
organization charts
to show work
distribution
systems

Source: JoAnn T. Hackos.
(Jan. 2008). *"Information
Development in a Flat
World." Intercom*
(January 2008), p. 25.

Figure 2. Centralized organization with a satellite set of departments.

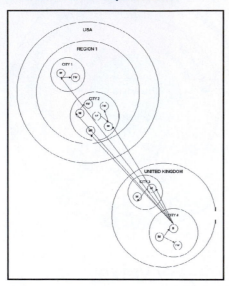

Figure 3. Two core organizations with several lone writers.

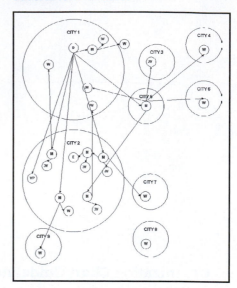

>> Drawing Guideline 1: Choose the Right Amount of Detail

Keep drawings as simple as possible. Use only the level of detail that serves the purpose of your document and satisfies your readers' needs. For example, Figure 12–25 uses an exploded view to show a gear and its placement in a hybrid engine.

>> Drawing Guideline 2: Label Parts Well

A common complaint of drawings is that parts included in the illustration are not carefully or clearly labeled. Place labels on every part you want your reader to see. (Conversely, you can also choose not to label those parts that are irrelevant to your purpose.)

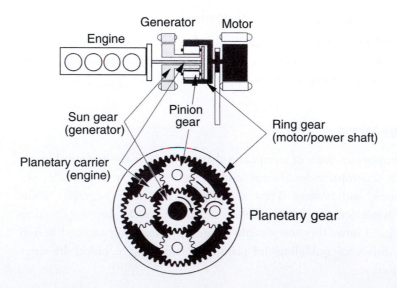

■ **Figure 12–25** ■
Technical drawing
(exploded view)

Source: Toyota Motor
Corporation. (May 2003).
Toyota Hybrid System,
p. 10

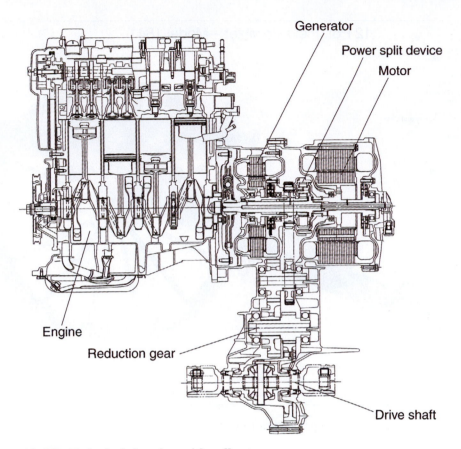

■ **Figure 12–26** ■ Technical drawing with callouts

Source: Toyota Motor Corporation. (May 2003). *Toyota Hybrid System*, p. 21.

When you label parts, use a typeface large enough for easy reading. Also, arrange labels so that (1) they are as easy as possible for your reader to locate and (2) they do not detract from the importance of the drawing itself. The simple labeling in Figure 12–26 fulfills these objectives.

>> Drawing Guideline 3: Choose the Most Appropriate View

As noted previously, illustrations—unlike photographs—permit you to choose the level of detail needed. In addition, drawings offer you a number of options for perspective or view:

■ **Exterior view** shows surface features with either a two- or three-dimensional appearance—see Figure 12–27.

■ **Cross-section view** shows a "slice" of the object so that interiors can be viewed—Figure 12–28.

■ **Cutaway view** is similar to a cross section view, but only part of the exterior is removed to show the inner workings of the object.

■ **Exploded view** shows relationship of parts to each other by "exploding" the mechanism—see Figure 12–25.

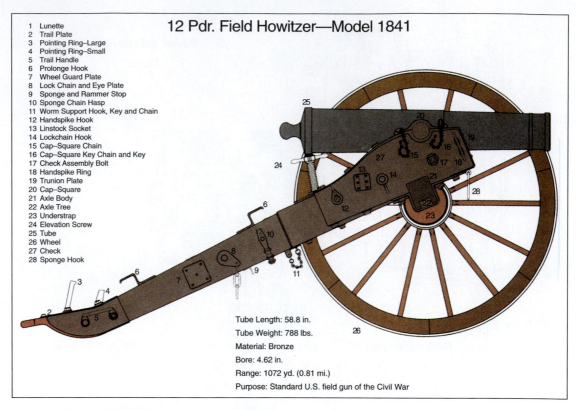

1 Lunette
2 Trail Plate
3 Pointing Ring–Large
4 Pointing Ring–Small
5 Trail Handle
6 Prolonge Hook
7 Wheel Guard Plate
8 Lock Chain and Eye Plate
9 Sponge and Rammer Stop
10 Sponge Chain Hasp
11 Worm Support Hook, Key and Chain
12 Handspike Hook
13 Linstock Socket
14 Lockchain Hook
15 Cap–Square Chain
16 Cap–Square Key Chain and Key
17 Check Assembly Bolt
18 Handspike Ring
19 Trunion Plate
20 Cap–Square
21 Axle Body
22 Axle Tree
23 Understrap
24 Elevation Screw
25 Tube
26 Wheel
27 Check
28 Sponge Hook

12 Pdr. Field Howitzer—Model 1841

Tube Length: 58.8 in.

Tube Weight: 788 lbs.

Material: Bronze

Bore: 4.62 in.

Range: 1072 yd. (0.81 mi.)

Purpose: Standard U.S. field gun of the Civil War

■ **Figure 12–27** ■ Line drawing of exterior view

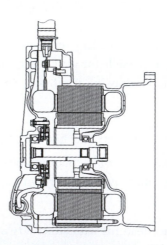

■ **Figure 12–28** ■ Cross section view of hybrid engine

Source: Toyota Motor Corporation. (May 2003). *Toyota Hybrid System*, p. 6.

>> Drawing Guideline 4: Use Legends When There Are Many Parts

In complex drawings, avoid cluttering the illustration with many labels. Figure 12–27, for example, places all labels in one easy-to-find spot, rather than leaving them on the drawing.

>>> Misuse of Graphics

Technology has revolutionized the world of graphics by placing sophisticated tools in the hands of many writers. Yet this largely positive event has its dark side; many graphics—in spite of their slickness—distort data and misinform the reader. The previous sections of this chapter establish principles and guidelines to help writers avoid such distortion and misinformation. This final section shows what can happen to graphics when sound design principles are not applied.

Description of the Problem

Through clutter or distortion, graphics can oversimplify data, be confusing, or be misleading. Writers can miss problems with graphics as easily as they can miss problems in spelling or punctuation, when they are facing deadlines or they have become so familiar with their documents that they don't see the errors. To avoid misleading or confusing graphics, it is important to proofread and edit them carefully. If you are using charts or tables, it may be useful to ask someone else to look at them carefully to see if they interpret the graphics in the way that you intended.

Edward R. Tufte analyzes graphics errors in detail in his excellent work, *The Visual Display of Quantitative Information*. In setting forth his main principles, Tufte notes that "graphical excellence is the well designed presentation of interesting data—a matter of *substance* of *statistics*, and of *design*." He further contends that graphics must "give to the viewer the greatest number of ideas in the shortest time with the least ink in the smallest space."[1]

One of Tufte's main criticisms is that charts are often disproportional to the actual differences in the data represented. The next subsection shows some specific ways that this error has worked its way into contemporary graphics.

Examples of Distorted Graphics

There are probably as many ways to distort graphics as there are graphics types. This section gets at the problem of misrepresentation by showing several examples and describing the errors involved. None of the examples commits major errors, yet each one fails to represent the data accurately.

>> Example 1: Confusing Bar Charts

Figure 12–29 accompanied a newspaper article about changes in mailing costs and service. The problem here is that the chart's decoration—the mailboxes—inhibits rather than promotes clear communication. Although the writer intends to use mailbox symbolism in lieu of precise bars, the height of the mailboxes does not correspond to the actual increase in second-class postage rates.

A revised graph should include either (1) mailboxes that correctly approximate the actual differences in second-class rates or (2) a traditional bar chart without the mailboxes.

Another problem with bar charts is that they become busy or confusing, especially when the options for creating charts in spreadsheet programs are not used carefully.

[1] Tufte, Edward R. (1983). *The Visual Display of Quantitative Information.* Cheshire, CT: Graphics Press, p. 51.

■ **Figure 12–29** ■
Faulty comparisons
on modified bar
chart

*Adapted from
Atlanta Constitution,
30 Nov. 1989, p. H–1. Used
by permission.*

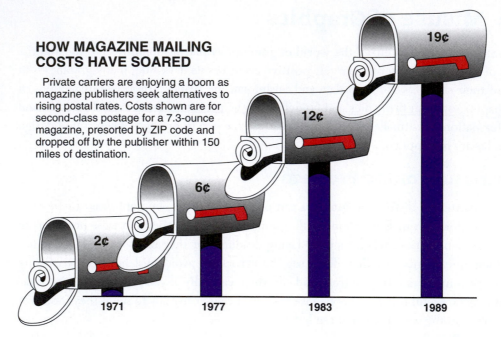

**HOW MAGAZINE MAILING
COSTS HAVE SOARED**

Private carriers are enjoying a boom as
magazine publishers seek alternatives to
rising postal rates. Costs shown are for
second-class postage for a 7.3-ounce
magazine, presorted by ZIP code and
dropped off by the publisher within 150
miles of destination.

2¢   6¢   12¢   19¢

1971   1977   1983   1989

Source: U.S. Postal Service

Figure 12–30 uses several stacked bars, which make it hard to compare the data clearly. It also uses too many colors and patterns, as well as colors that are similar, and that would not reproduce well on a black-and-white copier. Figure 12–31 tries to avoid the problem caused by reproduction on a copier by using patterns instead of colors, but the patterns are so busy and similar that it is difficult to distinguish among them. Figure 12–32 uses a three-dimensional chart that tries to represent perspective in such a way that it is difficult to understand the data. Some of the bars are also hidden behind others, making it difficult to read the data that they represent.

>> Example 2: *Chartjunk* That Confuses the Reader

Figure 12–33 concludes a report from a county government to its citizens. Whereas the dollar backdrop is meant to reinforce the topic—that is, the use to which tax funds are put—in

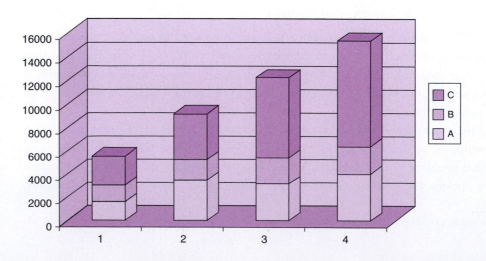

■ **Figure 12–30** ■
Confusing bar
chart

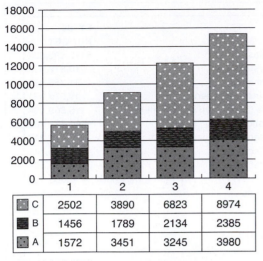

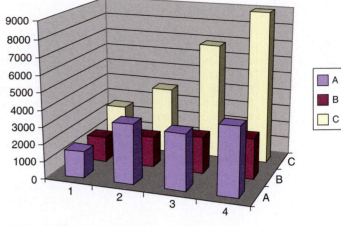

■ **Figure 12–31** ■ Confusing bar chart   ■ **Figure 12–32** ■ Confusing bar chart

fact, it impedes communication. Readers cannot quickly see comparisons. Instead, they must read the entire list below the illustration, mentally rearranging the items into some order.

At the very least, the expenditures should have been placed in sequence, from least to greatest percentage or vice versa. Even with this order, however, one could argue that the dollar bill is a piece of what Edward Tufte calls *chartjunk,* which fails to display the data effectively.

>> **Example 3: Confusing Pie Charts**

The pie chart in Figure 12–34 (1) omits percentages that should be attached to each of the budgetary expenditures; (2) fails to move in a largest-to-smallest clockwise sequence; (3) includes too many divisions, many of which are about the same size and thus difficult to distinguish; and (4) introduces a third dimension that adds no value to the graphic.

Figure 12–35 attempts the visual strategy of alternating shades, but it succeeds only in overloading the chart with too many small percentage divisions in uncertain order. Moreover, the reader cannot easily see how the pie slices should be grouped under the four headings listed beneath the chart. A grouped bar chart would better serve the purpose, with *Southeast, New England, Mid-Atlantic,* and *Other* providing the groupings.

■ **Figure 12–33** ■ *Chartjunk* that confuses the reader

	FUND NAME	DESCRIPTION	FY '89 BUDGET
A	General Fund	Basic government activities	$112,895,822
B	Transit Fund	Implementation of bus system	10,812,522
C	Fire District Fund	Operation of Fire Department	21,253,523
D	Bond Funds	General obligation bond issue proceeds	11,073,371
E	Road Sales Tax Fund	1% special purpose sales tax for road improvements	116,869,904
F	Water & Pollution Control Fund	Daily water system operation	60,572,506
G	Debt Service Fund	Principal & interest payments for general obligation bonds	8,240,313
H	Water RE&I Fund	Maintenance of existing facilities	27,365,744
I	Solid Waste RE&I Fund	Maintenance of existing facilities	1,111,237
J	Solid Waste Disposal Facilities	Landfill operations	6,003,367
K	Water Construction Fund	Construction of new facilities	110,884,507
L	Other Uses*		18,384,364
		SUB-TOTAL	$505,467,180
		LESS INTERFUND ACTIVITY	− 23,393,042
		TOTAL EXPENDITURES	**$482,074,138**

*Other Uses includes: Community Service Block Grants, Law Library, Claims Fund, Capital Projects, Senior Services, Community Development Block Grant, Grant Fund

In addition to the General Fund, the county budgets a number of other specialized funds. These include the Fire District Fund, Transit Fund, Road Sales Tax Fund, and enterprise funds such as Water and Pollution Control, and Solid Waste Disposal Facilities.

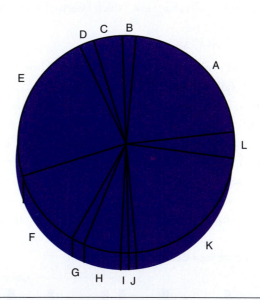

■ **Figure 12–34** ■ Confusing pie chart

Source: Adapted from Cobb County 1988–89 Annual Report (Cobb County, GA), 14. Used by permission.

Figure 12–36 negates the value of the pie chart by assuming an oblong shape, rather than a circle. This distortion can make it difficult for the reader to distinguish among sections that are similar in size, such as sections 1 and 2 in Figure 12–36. The pie chart should be a perfect circle, should have percentages on the circle, and should move in large-to-small sequence from the 12:00 position.

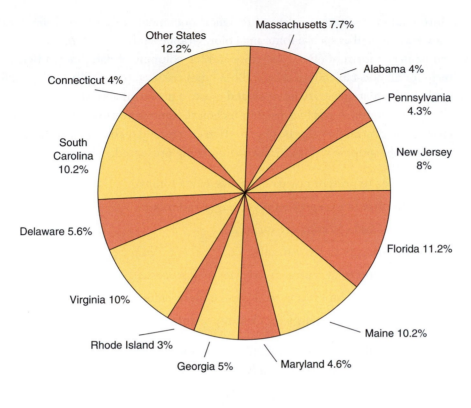

Location of All Ding-Dong Convenience Stores

Southeast	30.4
New England	24.9
Mid-Atlantic	32.5
Other	12.2
	100%

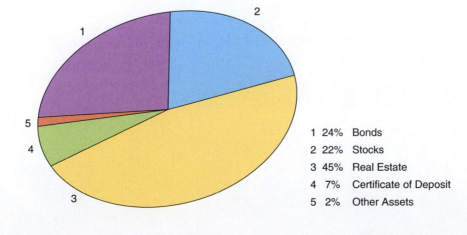

ASSETS OF JONES RETIREMENT FUND

1	24%	Bonds
2	22%	Stocks
3	45%	Real Estate
4	7%	Certificate of Deposit
5	2%	Other Assets

■ **Figure 12–36** ■
Confusing
pie chart

>>> **Chapter Summary**

More than ever before, readers of technical documents expect good graphics to accompany text, as well as special fonts and color. Graphics (also called *illustrations* or *visual aids*) can be in the form of (1) tables (rows and/or columns of data) or (2) figures (a catchall term for all non-table illustrations). Both types are used to simplify ideas, reinforce points made in the text, generate interest, and create a universal appeal.

Seven common graphics used in technical communication are tables, pie charts, bar charts, line charts, flowcharts, organization charts, and technical drawings. It is important to match the graphic with the information being presented:

- Use **tables** to present raw data in rows and columns
- Use **pie charts** to represent percentages and money.
- Use **bar charts** to compare quantities.
- Use **line charts** to show trends.
- Use **flowcharts** to represent overviews of processes.
- Use **organization charts** to represent hierarchies.
- Use **technical drawings** for clear representations of objects.

As detailed in this chapter, you should follow specific guidelines in constructing each type. The following basic guidelines apply to all graphics:

1. Determine the purpose of the graphic.
2. Evaluate the accuracy and validity of the data.
3. Refer to all graphics in the text.
4. Think about where to put graphics.
5. Position graphics vertically when possible.
6. Avoid clutter.
7. Provide titles, notes, keys, and source data.

>>> Learning Portfolio

Communication Challenge "Massaging M-Global's Annual Report"

Just about everyone considered M-Global's 2007 annual report a boring piece of work. It contained pages of text with only a few tables for visual relief. Wanting to spice up the next annual report, the company hired a graphics firm to create a more appealing design and graphics. GeeWhiz Graphics is completing the 2008 annual report draft for review by M-Global's corporate staff. What follows are highlights of the graphics challenge, some questions and comments for discussion, and an assignment for a written response to the Challenge.

Your Part of the Project

When GeeWhiz got the M-Global job, you and Rick Ford were assigned the account. The two of you were to take text and data provided by M-Global's PR department and create a graphically interesting format for readers of the annual report—who are mostly stockholders, M-Global employees, or clients. You and Rick have been asked to make data as appealing as possible, especially in light of criticism the company received for its previous annual report. You are creating graphics for four pieces of information that M-Global wants emphasized in the report:

1. **International Sales:** M-Global has become a more international firm. You have been asked to create some graphics that reflect this shift. In 2008, the international offices accounted for 35% of the total $96 million sales. Of that 35%, the Tokyo office was highest with $12 million in sales, Munich was next with $10 million, and the other four international offices shared the rest.
2. **Total Sales:** Over the past 10 years, total sales have gone up steadily. Figures for 1998 through 2008 are, in millions: $65, $70, $73, $74, $80, $83.5, $87, $90, $90.5, $96.
3. **Number of Employees:** Except for one year, when there was a minor layoff to reduce costs, the number of employees has risen over the past five years, as follows: 1,800 (in 2004), 1,950 (in 2005), 1,925 (in 2006), 2,200 (in 2007), 2,500 (in 2008).
4. **Corporate Overhead:** Of the six service areas covered by the corporate office in Baltimore (see pages 51–54 in chapter 2), the corporate budget spends 40% on human resources, 20% on research, 15% on computer operations, 10% on training, 10% on project management, and 5% on marketing. The company wants to emphasize that the employee-related portion of overhead grew since last year—for example, training went from 5% to 10%, and human resources went from 35% to 40%.

Rick's Part of the Project

Your colleague at GeeWhiz Graphics, Rick Ford, was given a similar assignment—that is, to create interesting graphical representations of data about the company. He produced three graphics in his initial work on the report:

1. **Bar Chart:** Rick drew a bar chart to reflect the growth of the international offices in the past five years.

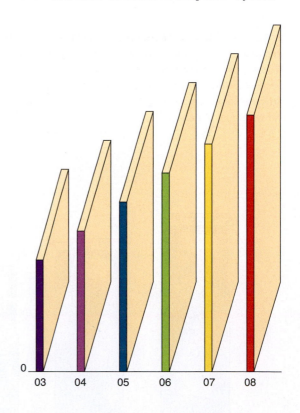

2. **Pie Chart:** To show the investment portfolio of M-Global's retirement plan, Rick produced a pie chart.

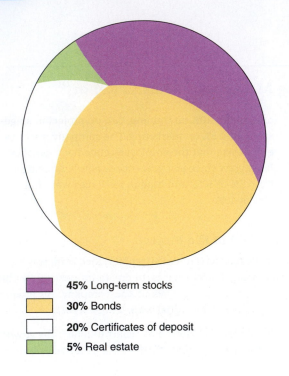

45% Long-term stocks

30% Bonds

20% Certificates of deposit

5% Real estate

3. **Bar Chart without Bars:** In the past five years, M-Global has made a concerted effort to emphasize preventive medical care and wellness programs among its employees. A variation on a bar chart was meant to show this progress in a more visually appealing manner than a conventional bar chart.

Questions and Comments for Discussion

1. Separate into groups and discuss the kinds of graphics that would be most effective, appropriate, and clear for the four assignments given to you. Report the results of your discussion to the entire class.
2. Are there any ethical implications that you should consider in producing your graphics? If so, what are they?
3. Discuss any ethical or clarity issues suggested by the graphics produced by Rick Ford. Should these graphics be included in M-Global's annual report? If so, why? If not, why not, and what would you change? If useful, refer to the Equal Consideration of Interests (ECI) principle discussed on page 43.

Write About It

Assume that Rick sent you an email with his graphics attached. Rick is a new employee at Gee Whiz, so he has asked you to comment on the graphics and make suggestions for improving them. Write the email message that you would send to him. Remember that you will be working with him on a daily basis, and that the two of you are working on this project together. Then create your own graphics, using the guidelines listed in this chapter, to send to Rick as examples. Turn in the text of your email message and your graphics.

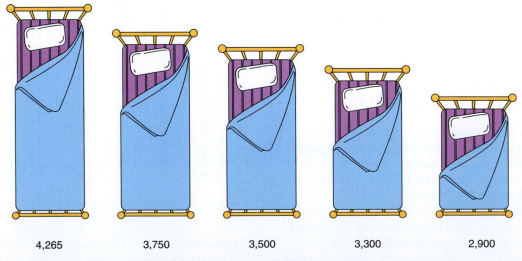

4,265   3,750   3,500   3,300   2,900

Sick Days Taken Last 5 Years: M-Global, Inc.

Collaboration at Work Critiquing an Annual Report

General Instructions

Each Collaboration at Work exercise applies strategies for working in teams to chapter topics. The exercise assumes you (1) have been divided into teams of about three to six students, (2) will use team time inside or outside of class to complete the case, and (3) will produce an oral or written response. For guidelines about writing in teams, refer to pages 24–28.

Background for Assignment

While planning and writing, you make two main decisions about the use of graphics—first, when they should be used; and second, what types to select. This chapter helps you make such decisions. Yet you already possess the quality that is most useful in your study of graphics: common sense. Whether consciously or subconsciously, most of us tend to seek answers to basic questions like the following when we read a document:

1. Is there an appropriate mix of text and graphics?
2. Are the graphics really useful or are they just visual "fluff"?
3. Can information in the graphic be understood right away?
4. Was the correct type of graphic selected for the context?
5. Do any of the graphics include errors, such as in proportion?

Your answers to these questions often determine whether you continue reading a document—or at least whether you enjoy the experience.

Team Assignment

Choose one document that includes a variety of graphics—newspaper, magazine, report, textbook, catalog, website, and so on. Using the questions previously listed, work with your team to evaluate the use of graphics in all or part of the document. Whether you think a graphic is successful or not, give specific reasons to support your analysis.

Assignments

Your instructor may want you to practice graphics in the context of some of the writing assignments in this textbook, especially in chapters 8, 9, and 10. Following are a few additional exercises that can be completed either as individual or collaborative projects, depending on directions from your instructor.

1. Pie, Bar, and Line Charts

Figure 12–37 shows employment by industry from 2000 through 2006, while also breaking down the 2006 data into four categories by race. Use those data to complete the following charts:

- A pie chart that shows the groupings of race in 2006.
- A bar chart that shows the trend in total employment during 2000, 2004, 2005, and 2006. Indicate the gap in data.
- A segmented bar chart that compares employment in the production of durable goods to employment in the production of nondurable goods within the manufacturing sector for 2004, 2005, and 2006.
- A single-line chart showing employment in agriculture and related industries for 2000 through 2006.

- A multiple-line chart that contrasts employment in retail trade, professional and business services, and leisure and hospitality for 2004 through 2006.

2. Flowcharts

Identify the main activities involved in enrolling in classes on your campus. Then draw two flowcharts that outline the main activities involved in this process. In the first chart, use the standard flowchart symbols shown in Figure 12–20 on page 428. In the second flowchart, use images and symbols creatively to explain the process.

3. Organization Chart

Select an organization with which you are familiar or one about which you can find information. Then construct a linear flowchart that helps an outsider understand the management structure of all or part of the organization.

4. Technical Drawing

Drawing freehand, using the draw function in your word processor, or using a computer illustration or design program available to you, produce a simple technical drawing

Table 602.　**Employment by Industry: 2000 to 2006**

[In thousands (136,891 represents 136,891,000), except percent. See Table 584 regarding coverage and headnote Table 587 regarding industries]

Industry	2000	2004 [1]	2005 [1]	2006 [1]	2006, percent [1] Female	Black [2]	Asian [2]	His-panic [3]
Total employed	**136,891**	**139,252**	**141,730**	**144,427**	**46.3**	**10.9**	**4.5**	**13.6**
Agriculture and related industries.	2,464	2,232	2,197	2,206	24.6	2.7	1.2	19.4
Mining	475	539	624	687	13.0	4.9	0.7	13.6
Construction	9,931	10,768	11,197	11,749	9.6	5.5	1.4	25.1
Manufacturing	19,644	16,484	16,253	16,377	29.5	9.5	5.2	14.7
Durable goods	12,519	10,329	10,333	10,499	25.8	8.5	5.8	12.4
Nondurable goods	7,125	6,155	5,919	5,877	36.1	11.4	4.2	18.7
Wholesale trade	4,216	4,600	4,579	4,561	29.0	6.5	4.1	13.5
Retail trade	15,763	16,269	16,825	16,767	48.9	10.1	4.2	12.7
Transportation and utilities	7,380	7,013	7,360	7,455	24.2	16.5	3.6	12.7
Transportation and warehousing.	6,096	5,844	6,184	6,269	24.7	17.6	3.8	13.5
Utilities	1,284	1,168	1,176	1,186	21.9	10.9	2.5	8.2
Information	4,059	3,463	3,402	3,573	44.4	11.7	5.2	9.4
Financial activities.	9,374	9,969	10,203	10,490	55.5	10.2	5.1	10.0
Finance and insurance	6,641	6,940	7,035	7,254	58.2	10.5	5.6	8.5
Real estate and rental and leasing . . .	2,734	3,029	3,168	3,237	49.4	9.5	4.1	13.4
Professional and business services . . .	13,649	14,108	14,294	14,868	42.5	9.8	5.7	13.0
Professional and technical services. . .	8,266	8,386	8,584	8,776	44.4	6.4	7.6	6.2
Management, administrative, and waste services	5,383	5,722	5,709	6,092	39.8	14.8	3.0	22.9
Education and health services.	26,188	28,719	29,174	29,938	74.9	14.2	4.7	9.1
Educational services	11,255	12,058	12,264	12,522	68.9	10.8	3.6	8.5
Health care and social assistance. . . .	14,933	16,661	16,910	17,416	79.1	16.7	5.4	9.5
Hospitals	5,202	5,700	5,719	5,712	76.6	16.4	7.0	7.6
Health services, except hospitals . .	7,009	8,118	8,332	8,639	78.6	15.3	5.3	9.5
Social assistance	2,722	2,844	2,860	3,065	85.4	21.2	2.9	12.9
Leisure and hospitality	11,186	11,820	12,071	12,145	51.3	10.5	5.9	19.4
Arts, entertainment, and recreation . . .	2,539	2,690	2,765	2,671	45.2	8.3	3.6	11.9
Accommodation and food services . . .	8,647	9,131	9,306	9,474	53.0	11.2	6.5	21.6
Other services	6,450	6,903	7,020	7,088	51.7	9.8	5.8	15.5
Other services, except private households.	5,731	6,124	6,208	6,285	46.5	9.6	6.2	13.3
Private households	718	779	812	803	92.5	11.1	2.5	32.8
Government workers	6,113	6,365	6,530	6,524	45.4	16.2	3.5	8.6

[1] See footnote 2, Table 569.　[2] Persons in this race group only. See footnote 3, Table 570.　[3] Persons of Hispanic or Latino origin may be of any race.

Source: U.S. Bureau of Labor Statistics, *Employment and Earnings,* monthly, January 2007 issue. See Internet site <http://www.bls.gov/cps/home.htm>.

■ **Figure 12–37** ■ Reference for Assignment 1

Source: U.S. Census Bureau. (2008). *The 2008 Statistical Abstract.* http://www.census.gov/compendia/statab/tables/08s0602.pdf

of an object with which you are familiar through work, school, or home use.

5. Table

Using the bar charts in Figure 12–38, create a table that shows home ownership by region and total home ownership for the United States for each quarter of 2007.

6. Misuse of Graphics

Find three deficient graphics in newspapers, magazines, reports, or other technical documents. Submit copies of the graphics along with a written critique that (1) describes in detail the deficiencies of the graphics and (2) offers suggestions for improving them.

7. Misuse of Graphics

Analyze the graphics in Figures 12–39, 12–40, and 12–41. Describe any deficiencies and offer suggestions for improvement.

? 8. Ethics Assignment

Develop a list of practical guidelines that helps writers like you avoid ethical errors in creating and using graphics on the job. To create this list, complete the following steps:

1. Review parts of this chapter that deal with ethical issues, especially the "Misuse of Graphics" and "Communication Challenge" sections.
2. Interview someone who creates or uses graphics frequently, such as a member of the public relations or admissions departments at the college or university that you attend.
3. Review a variety of graphics in diverse media such as popular magazines, textbooks, and the Internet.

Remember to focus on concrete guidelines that could be used during the process of producing a document.

First Quarter 2007: Graph of Homeownership Rates

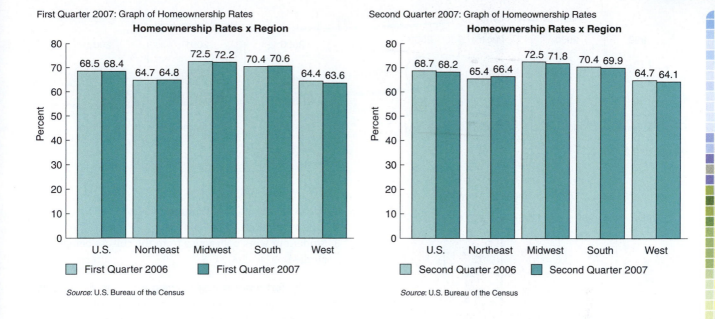

Second Quarter 2007: Graph of Homeownership Rates

Third Quarter 2007: Graph of Homeownership Rates

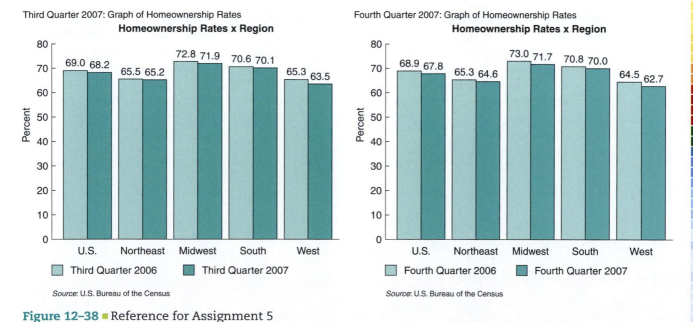

Fourth Quarter 2007: Graph of Homeownership Rates

Figure 12–38 ■ Reference for Assignment 5

Source: U.S. Census Bureau. (2008). *Housing Vacancies and Homeownership*. http://www.census.gov/hhes/
www/housing/hvs/hvs.html

Average KWH Usage

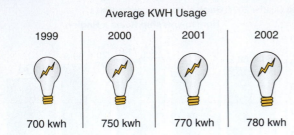

1999	2000	2001	2002
700 kwh	750 kwh	770 kwh	780 kwh

■ **Figure 12–39**

Hamburgers Sold

2000	2001	2002
1 million	3 million	5 million

■ **Figure 12–41**

Credit Union Membership Growth

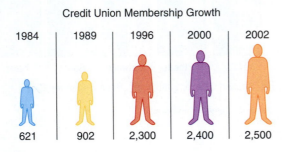

1984	1989	1996	2000	2002
621	902	2,300	2,400	2,500

■ **Figure 12–40**

1. Books on international communication
2. Internet sites on global cultures
3. Student or work colleagues from the culture being studied

You may be fortunate enough to gain access to documents either written in the culture being studied or written by culturally knowledgeable people for use in the culture. If so, you often can draw useful conclusions from reading such primary sources.

 ### 9. International Communication Assignment

Examine and report on the graphics preferences of a country other than the United States. For example, you could choose one of the countries in which M-Global has an international office—Venezuela, England, Germany, Kenya, Saudi Arabia, Russia, or Japan. Sources of information might include one or more of the following:

 ### 10. A.C.T. N.O.W. Assignment (Applying Communication To Nurture Our World)

Choose a campus, local, regional, national, or international issue about which you have fairly strong views and about which you can locate supporting data. Produce and submit one or two graphics that reinforce your views clearly and well, along with a brief narrative summary of your view.

Chapter | 13 | Collaboration and Writing

>>> Chapter Outline

Approaches to Collaboration   449

Collaboration and the Writing Process   449

The Writing Team   450

Planning   451

Budgeting Time and Money   451

Communication   454

Modular Writing   455

Teamwork   456

Running Effective Meetings   456

Writers and Subject Matter Experts   459

Chapter Summary   461

Learning Portfolio   462

Communication Challenge   462

Collaboration at Work   463

Assignments   463

Models for Good Writing   466

Model 13–1 Example of M-Global modular writing   466

Model 13–2 Meeting agenda   467

Model 13–3 Meeting minutes   468

As the leader of an engineering team in M-Global's Equipment Development department, Scott Montgomery guided the design of a new sensor that generates quicker and more accurate results from on-site tests for soil contamination. Now he is preparing to write a *white paper* about the new sensor. In some industries, white papers are a common way to share information about new developments as well as to publicize the organization. Scott is beginning to write such a paper.[1] He begins by gathering documents that have already been written about the sensor. The Equipment Development team has kept meticulous records during the development and testing of the sensor, and now that M-Global is preparing to offer the new sensor to its clients, the marketing team has created additional documents that he has access to through the company intranet. Scott gathers the written materials, creates a framework, and decides how the existing text will fit into his plan. He writes some new sections himself, connecting the existing text, creating a framework, and revising for unified voice and purpose. Scott then asks two of his team members, engineers who have written much of the documentation for the sensor, to serve as co-authors. The three of them review and revise drafts until they are satisfied and are ready to pass the paper along to the vice president for research and training and to the legal department to make sure that the paper does not make any unsubstantiated claims.

By the time it is presented at a conference and published on the M-Global website, the white paper will include the work of dozens of people, even though only three will be listed as authors. This is common in organizations, as most publications are considered a collaborative product of the entire organization.

In the workplace, correspondence and some short documents may be written by a single author, but most documents are the result of some kind of collaboration between writers. In one study, technical communication managers listed collaboration with subject matter experts and collaboration with co-workers as the two most important competencies for technical communicators.[2] This collaboration may be as simple as asking a co-worker to read through a report before turning it in to a supervisor, or you may become part of a standing team that creates documents. You may collaborate with others in the development and delivery of products or services, in the marketing of those products or services, or in creating documentation to support those products or services. *Collaborative writing* (also called *team writing*) can be defined as follows:

Collaborative writing: the creation of a document by two or more people. Documents are created collaboratively to meet the common purposes and goals of a community of writers, editors, and readers.

This chapter focuses on collaboration strategies as they are used in the writing process, but many of these strategies can contribute to the success of any team project.

---

[1]This scenario draws on a case study published by Dorothy A. Winsor. (1989), "An Engineer's Writing and the Corporate Construction of Knowledge." *Written Communication* 6, pp. 270–285.

[2]Rainey, Kenneth T., Roy K. Turner, and David Dayton. (2005). "Do Curricula Correspond to Managerial Expectations? Core Competencies for Technical Communicators." *Technical Communication* 52 (3), pp. 323–352.

>>> Approaches to Collaboration

The scope of the writing project, the setting in which it is written, and the number of people involved can all influence the form that collaborative writing takes. There are five common approaches to writing collaboratively.

■ **Divide and conquer:** When the writing project is large and has clearly defined sections, it may be helpful to assign individual sections of the document to specific writers. Later in the process the parts of the document are brought together and combined. Many documents today are produced using a version of this approach that depends on modular writing, discussed later in this chapter.

■ **Specialization:** Often referred to as *writing in cross-functional teams,* this is a version of divide and conquer in which the parts of the project are assigned to team members based on their specialty. For example, on a proposal writing team, an engineer might write the technical descriptions and specifications; an accountant might write up the budget projections; someone from marketing who is familiar with the potential client might write the final, persuasive sections of the proposal; and a technical communicator might provide the overall plan for the document, assemble the parts, and provide the document design and final editing work.

■ **Sequence:** In this approach, several people are involved in creating a document, but instead of working on it at the same time, they pass it from one person to the next. An engineer may write a description of a new product, then pass that along to a documentation specialist, who revises the description for readers who don't have the engineer's expertise. The documentation specialist may then pass the document along to a marketing communication writer, who uses it to create a description of the product for the company's website.

■ **Dialog:** When two writers are working together on a project, they may work best by sending drafts back and forth to each other, commenting and revising until they are both pleased with the final draft. This is a common practice in settings where supervisors comment on the documents that their employees write, or when a writer is collaborating with an editor. When writing in this back-and-forth dialog, it is important to keep versions of each draft separate, in case the writers decide that an earlier version was more appropriate for the document's purpose.

■ **Synthesis:** This approach to team writing works best with two or three writers, and with shorter documents. The team sits together and writes together, adding ideas and commenting on the work as it progresses. This is the most seamlessly collaborative approach to writing, and is most successful when the members of the team have worked together long enough to know each other well.

>>> Collaboration and the Writing Process

Writing collaboratively uses the same steps in the writing process as those introduced in chapter 1. The team must identify the purpose of the document and the needs of its audience. It must collect information, plan the document, draft, and revise. And the team must do this task together, creating a cohesive and useful document. Let's quickly review the guidelines for team writing that were introduced in chapter 1:

- Get to know your team
- Set clear goals and ground rules
- Use brainstorming techniques for planning
- Use storyboarding techniques for drafting
- Agree on a thorough revision process
- Use computers to communicate

The Writing Team

Some organizations have standing teams for common types of projects such as proposal writing, or for ongoing projects such as compliance with regulations. Teams may also be temporary, coming together for one project, and then separating, each member moving on to another project. Whether the team is a permanent (or standing) team or one that has been brought together for a single project, it is important to be aware of the roles that team members play. Begin by identifying the skills that each member can contribute to the project and assign tasks based on those skills. Don't just assume that skills are limited to the team members' job titles. In the example at the beginning of this chapter, Scott is an engineer, but he has moved into a supervisory position in Research and Development because of his creativity and his communication skills. Effective teams include the following roles:

- **The team leader** is the central contact person for team members, but is also the contact for people who aren't on the team. This person may also be working as the manager for the project.
- **The planning coordinator** is responsible for managing communication among team members, for keeping track of benchmarks and deadlines, and for preparing for meetings. On small teams, the team leader may serve as the planning coordinator.

- **The archivist** keeps minutes of meetings, copies of all written communication, and copies of all written material related to the project. At the end of the project, the archivist creates the material that is stored in the organization's library or archives.
- **Devil's advocate** is a role that often occurs spontaneously, as one member of a team raises concerns or points out problems. This is an important role, and helps avoid *groupthink*, when members of a group begin to echo each other and stop looking critically at the work they are doing. Some teams formally assign this role, rotating it from meeting to meeting. If you find

yourself raising concerns about a project during a meeting, it is helpful to announce it—"I'm just playing devil's advocate here, but …" —as a way of keeping the focus on the project and avoiding the temptation to make disagreements personal.

Planning

As with any writing project, team projects must be planned carefully. The Planning Form in the back of this book can be used for team writing in the same way that you have been using it for individual writing projects. Begin by identifying your audience. Who will be reading this document? What do they expect to learn from it? You should also identify the stakeholders in your team project. Obviously, the team members themselves have a stake in the success of the project, but there are others who will be interested in its success as well. This may include members of management, employees in other departments, and the organization as a whole. Clients are important stakeholders, especially if they have hired your organization for the project that your team is working on. If a client hired your organization for a project such as creating a website or training materials, you should work closely with the client and consider the client's representative a member of your team.

As part of the planning process, you must state clearly the desired outcome of the project. How will you know if you have completed it successfully? Your team's goal should be more than simply producing the required document. You should decide what information makes that document successful, where to find the information, and how best to organize the information. Then identify the tasks that must be accomplished to achieve the project's goals, and assign the tasks to team members.

Budgeting Time and Money

Once you have identified the tasks to be accomplished, you should identify *benchmarks*—the deadlines for specific tasks that keep the project focused and on schedule. These benchmarks vary from project to project, but common benchmarks for writing projects include the following:

- Completion of preliminary research
- Organization of collected information
- Planning of graphics
- Completion of first draft
- Editing of late draft
- Document design
- Publication of document

After identifying the benchmarks, your team can plan the calendar for the project. It is rare for a team to be able to set their own deadline. Team projects usually have a deadline that has been imposed from outside, so it is helpful to *backplan* the schedule for the project. Backplanning begins with the due date, and works backwards. For example, if a

■ **Figure 13–1** ■

Gantt and milestone schedule charts

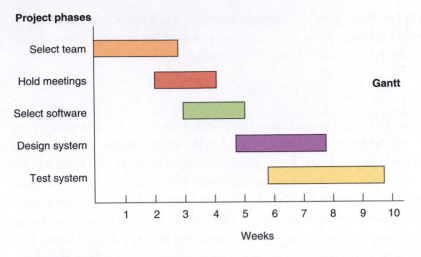

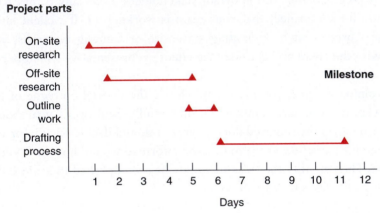

project is due July 1, the project coordinator may ask how long it will take to complete the final edit on your document. If it will take two days, then the benchmark to have the draft ready for final editing is two days before the due date. Working backward through the benchmarks that the team has identified, the project manager plans the rest of the schedule.

Using Schedule Charts

Schedule charts provide a graphic representation of a project plan. Many documents, especially proposals and feasibility studies include schedule charts to shows readers when specific activities will be accomplished. Often called a milestone or Gantt chart (after Henry Laurence Gantt, 1861–1919), it usually includes these parts (Figure 13–1):

■ **Vertical axis,** which lists the various parts of the project, in sequential order

■ **Horizontal axis,** which registers the appropriate time units

■ **Horizontal bar lines** (Gantt) or separate markers (milestone), which show the starting and ending times for each task

Follow these basic guidelines for constructing effective schedule charts for your projects.

>> Schedule Chart Guideline 1: Include Only Main Activities

Keep readers focused on no more than 10 or 15 main activities. If more detail is needed, construct a series of schedule charts linked to the main "overview" chart.

>> Schedule Chart Guideline 2: List Activities in Sequence, Starting at the Top of the Chart

As shown in Figure 13–1, the convention is to list activities from the top to the bottom of the vertical axis. Thus the readers' eyes move from the top left to the bottom right of the page, the most natural flow for most readers of English.

>> Schedule Chart Guideline 3: Create New Formats When Needed

Figure 13–1 shows only two common types of schedule charts; you should devise your own hybrid form when it suits your purposes. Your goal is to find the simplest format for helping team members know when a task will be completed, when a product will be delivered, and so forth. Figure 13–2 includes one such variation.

>> Schedule Chart Guideline 4: Be Realistic About the Schedule

Schedule charts can come back to haunt you if you do not include feasible deadlines. As you set dates for activities, be realistic about the likely time something can be accomplished. Your managers and clients understand delays caused by weather, equipment breakdowns, and other unforeseen events. However, they will be less charitable about schedule errors that result from sloppy planning.

You can use project management software to help you plan your project, create schedule charts, and keep track of the progress toward the benchmarks that your team has set.

Managing Finances

Your team may also have to work within a monetary budget. From the beginning of the project, you should know how much you have to spend on your project. If you must travel to collect information, you must make sure that you have the funds you need for expenses. You should also be familiar with your organization's policies concerning copying expenses, meals, and so on. Make sure that all of the team members know these policies. You may also have a budget for publication of the final document, especially if it is a document you are creating for a client. As you are beginning to plan the project, you should be aware of printing

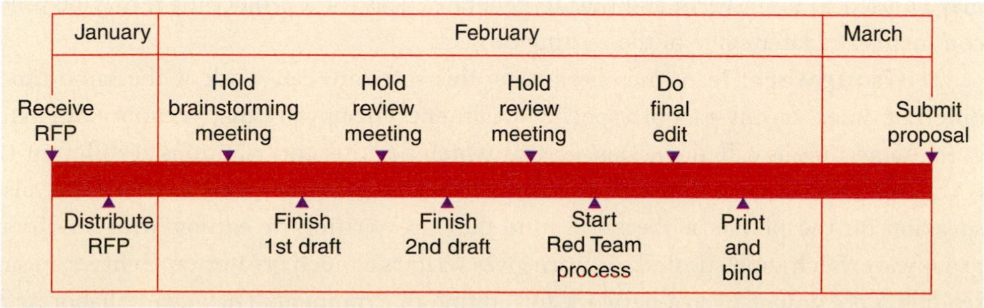

■ Figure 13–2 ■
Schedule chart variation

costs and plan the document's format and use of color to stay within your budget. (Read more about the use of color in chapter 12.)

Communication

Face-to-face meetings are the best way to keep a team running smoothly. Today, however, many teams are spread across different branches and even different countries, so this isn't always possible. However, it is beneficial if teams can meet in person at least once at the beginning of a project and once near the end of a project.

Computers can be used to overcome many obstacles for writers and editors in different locations. Indeed, electronic communication can help accomplish all the guidelines in this chapter. Specifically, (1) email can be used by group members to get to know each other; (2) email or a computer conference can be used to establish goals and ground rules; (3) synchronous, or real-time, groupware can help a team brainstorm about approaches to the project (and may, in fact, encourage more openness than a face-to-face brainstorming session); (4) computer conferences combined with groupware can approximate the storyboard process; and (5) either synchronous or asynchronous groupware can be used to approximate the editing process.

■ **Electronic mail (email):** Team members can send and receive messages from their office computers or from remote locations. They can also attach documents in a variety of forms. When attaching a document to an email, you should identify by file name and type of document (e.g., as a PDF) in the email.

■ **Computer conference:** Members of a team can make their own comments and respond to comments of others on a specific topic or project. Computer conferences may be open to all interested users or open only to a particular group. For the purposes of collaborative writing, the conference probably would be open only to members of the writing team. A leader may be chosen to monitor the contributions and keep the discussion focused. Contributions may be made over a long period, as opposed to a conventional face-to-face meeting, wherein all team members are present at the same time. Accumulated comments in the conference can be organized or indexed by topic. The conference may be used to brainstorm and thus to generate ideas for a project, or it may be used for comments at a later stage of the writing project.

■ **Groupware:** Team members using this software can work at the same time, or different times, on any part of a specific document. Groupware that permits contributions at the same time is called *synchronous*; that which permits contributions at different times is called *asynchronous*. Because team members are at different locations, they may also be speaking on the phone at the same time they are writing or editing with synchronous groupware. Such sophisticated software gives writers a much greater capability than simply sending a document over a network for editing or comment. They can collaborate with team members on a document at the same time, almost as if they were in the same room. With several windows on the screen, they can view the document itself on one screen and make comments and changes on another screen.

Granted, such techniques lack the body language used in face-to-face meetings. Yet when personal meetings are not possible, computerized communication can provide a substitute that allows writers in different locations to work together to meet their deadline.

Of course, computers can create problems during a group writing project, if you are not careful. When different parts of a document have been written and stored by different writers, your group must be vigilant during the final editing and proofreading stages. Before submitting the document, review it for consistency and correctness.

Modular Writing

In the past, team members of a collaborative writing project could assume that before the final version of the document was released, they would have a chance to review the entire document. Today, however, the writing process in organizations is changing. Documents are broken into small sections, with different people responsible for each section. Variations of this practice go by many names—single sourcing, structured authoring, or content management. In this book we refer to the general process as *modular writing*.

Modular writing: A process in which large documents are broken down into smaller elements, and different people are given responsibility for each element. These smaller elements are usually stored electronically so that they can be retrieved and edited or assembled into larger documents, help files, or web pages as they are needed.

For example, in a company that produces a number of owner's manuals for maintenance equipment, several people may share responsibility for all of the documents at once. One writer may be responsible for technical descriptions and another for instructions. An engineer may be responsible for technical specifications and a graphic artist may be responsible for schematics and illustrations. Each person saves his or her work on a server where it can be accessed by anyone who uses it in a document. Someone writing a proposal to sell the equipment to a client may use the technical description and specifications. A user's manual can be assembled from the elements that are specific to the equipment and to the user's needs. If the company sells its products overseas, translators in other countries can begin working on sections of a user's manual as soon as the individual sections are saved to the server, instead of waiting to receive the whole document before translation begins. Model 13–1 is an example of modular writing at M-Global. The information about the organization's history is used in sales brochures, proposals, annual reports, and on the company website.

Modular writing requires careful planning. The writing team must identify all the elements needed in the final project, and assign those elements to different writers. Individual writers may never see a draft of the complete document. In order to ensure consistency throughout all documents created from the separate elements, the writing team must create a thorough style guide and adhere to it, even if the team includes an editor whose job is to check all documents for consistency.

Although organizations that begin using modular writing face many challenges, it has benefits that make the effort worthwhile. If a product is improved, the elements that are affected by the change can be updated easily. Then, any documentation about the product includes accurate information automatically. In the M-Global history in Model 13–1, new information about the organization's accomplishments can be added to the source element, and then all documents that use this history are automatically updated before they are printed or published. Because all the updates are kept in one file, there is no problem with someone missing an important update.

>>> Teamwork

Whether you are an engineer creating a document with other engineers, a technical communicator assigned to a company branch, or a documentation specialist on a *cross-functional team* (a team that includes people from different departments, each contributing his or her own expertise to the project), you should understand and stay focused the on the project goals.

Running Effective Meetings

Like formal presentations, meetings are a form of spoken communication that go hand in hand with written work. Important reports and proposals—even many routine

ones—often are followed or preceded by a meeting. For example, you may meet with your colleagues to prepare a team-written report, with your clients to discuss a proposal, or with your department staff to outline recommendations to appear in a yearly report to management. This section will make you a first-class meeting leader by (1) highlighting some common problems with meetings, along with their associated costs to organizations; and (2) describing 10 guidelines for overcoming these problems.

Common Problems with Meetings

Following are six major complaints about meetings held in all types of organizations:

1. They start and end too late
2. Their purpose is unclear
3. Not everyone in the meeting really needs to be there

4. Conversations get off track

5. Some people dominate while others do not contribute at all

6. Meetings end with no sense of accomplishment

As a result of these frustrations, career professionals waste much of their time in poorly run meetings.

Because they waste participants' time, bad meetings also waste a lot of money. To find out what meetings cost an organization, do this rough calculation. Use information about an organization for which you work or for which a friend or family member works.

1. Take the average weekly number of meetings in an office.

2. Multiply that number by the average length of each meeting, in hours.

3. Multiply the result of step 2 by the average number of participants in each meeting.

4. Multiply the result of step 3 by the average hourly salary or billable amount of the participants.

The result, which may surprise you, is the average weekly cost of meetings in the office that you investigated. With these heavy costs in mind, the next section presents some simple guidelines for running good meetings.

Guidelines for Good Meetings

When you choose (or are chosen) to run a meeting, your professional reputation is at stake—let alone the costs just mentioned. Therefore, it is in your own best interests to make sure meetings run well. When you are a meeting participant, you also have an obligation to speak up and help accomplish the goals of the meeting.

The guidelines that follow help create successful meetings. They fall into three main stages:

Stage 1: Before the meeting (Guidelines 1–4)

Stage 2: During the meeting (Guidelines 5–9)

Stage 3: After the meeting (Guideline 10)

These 10 guidelines apply to *working* meetings—that is, those in which participants use their talents to accomplish specific objectives. Such meetings usually involve a lot of conversation. The guidelines do not apply as well to *informational* meetings, wherein a large number of people are assembled only to listen to announcements.

>> Meeting Guideline 1: Involve Only Necessary People

Necessary means those people who, because of their position or knowledge, can contribute to the meeting. Your goal should be a small working group—four to six people is ideal. If others must know what occurs, send them a copy of the minutes after the meeting.

>> Meeting Guideline 2: Distribute an Agenda Before the Meeting

A meeting agenda should identify the objectives of the session (Model 13–2) clearly and should include a report by each team member, sharing what he or she has accomplished since the last meeting. The agenda also gives you, as leader, a way to keep the meeting on schedule. If you are worried about having time to cover the agenda items, consider attaching time limits to each item. This technique helps the meeting leader keep the discussion moving.

>> Meeting Guideline 3: Distribute Readings Before the Meeting

Jealously guard time at a meeting, making sure to use it for productive discussion. If any member has reading materials that committee members should review as a basis for these discussions, such readings should be handed out ahead of time. Do not use meeting time for reading. Even worse, do not refer to handouts that all members have not had the opportunity to go over.

>> Meeting Guideline 4: Have Only One Meeting Leader

To prevent confusion, one person should always be in charge. The meeting leader should be able to perform the following tasks:

- **Listen carefully** so that all views get a fair hearing
- **Generalize accurately** so that earlier points can be brought back into the discussion when appropriate
- **Give credit to participants** so that they receive reinforcement for their efforts
- **Move toward consensus** so that the meeting does not involve endless discussion

>> Meeting Guideline 5: Start and End on Time

Nothing deadens a meeting more than a late start, particularly when it is caused by people arriving late. Tardy participants are given no incentive to arrive on time when a meeting leader waits for them. Even worse, prompt members become demoralized by such delays. Latecomers will mend their ways if you make a practice of starting right on time.

It is also important to set an ending time for meetings so that members have a clear view of the time available. Most people do their best work in the first hour of a meeting. After that, productive discussion reaches a point of diminishing returns. If working meetings must last longer than an hour, make sure to build in short breaks and stay on the agenda.

>> Meeting Guideline 6: Keep Meetings on Track

By far, the biggest challenge for a meeting leader is to encourage open discussion while still moving toward resolution of agenda items. As a leader, you must be assertive yet tactful in your efforts to discourage the following three main time-wasters:

- Long-winded digressions by the entire committee
- Domination by one or two outspoken participants
- Interruptions from outside the meeting

>> Meeting Guideline 7: Strive for Consensus

Consensus means agreement by all those present. Your goal should be to orchestrate a meeting such that all members, after a bit of compromise, feel comfortable with a decision.

Such a compromise, when it flows from healthy discussion, is far preferable to a decision generated by voting on alternatives. After all, you are trying to reach a conclusion that everyone helps produce, rather than one only part of the committee embraces.

>> Meeting Guideline 8: Use Visuals

Graphics help make points more vivid at a meeting. They are especially useful for recording ideas that are being generated rapidly during a discussion. Toward this end, you may want someone from outside the discussion to write important points on a flip chart, a whiteboard, or an overhead transparency.

>> Meeting Guideline 9: End with a Summary

Before the meeting adjourns, take a few minutes to summarize what items have been discussed and agreed to. Review the team's progress toward benchmarks and project outcomes. This wrap-up gives everyone the opportunity to clarify any point brought up during the meeting. Also, identify clearly what each team member is responsible for accomplishing before the next meeting.

>> Meeting Guideline 10: Distribute Minutes Soon

Write and send out minutes within 48 hours of the meeting (Model 13–3). Even for routine meetings, it is important that there be a record of the meeting's accomplishments. Meeting minutes should include the date and location of the meetings, list attendees and absences, and summarize discussions and decisions. If any discussion items are particularly controversial, consider having committee members approve minutes with their signature and return them to you before final distribution.

Writers and Subject Matter Experts

Many articles have been written about the importance of collaboration between technical communicators and the engineers, programmers, scientists, and other specialists that they work with. These *subject matter experts* (*SMEs,* often pronounced "Smees") often contribute the technical content of documents, whereas technical communicators contribute their expertise in document design, writing, and editing. Good communication is important from the beginning of any project where technical communicators and SMEs are collaborating. The SME's misunderstanding of what technical communicators contribute to a project is one common cause of frustration for documentation specialists. However, the lack of technical knowledge on part of technical communicators can be source of frustration for SMEs. By keeping a few important guidelines in mind, technical communicators and SMEs can collaborate more effectively.

Guidelines for Collaborating with SMEs

>> Technical Communicator Guideline 1: Use the SME's Time Wisely

Do your background research before contacting the SME. Don't waste the specialist's time with questions that can be answered through other sources.

>> **Technical Communicator Guideline 2:** Put Questions in Writing When Possible

Make sure that email questions are clear. You won't get useful answers if your questions are ambiguous or confusing.

>> **Technical Communicator Guideline 3:** Prepare for Interviews and Meetings

Have clear goals. If you want to ask for feedback on documentation, send it to the SME beforehand and bring a copy with you.

>> **Technical Communicator Guideline 4:** Treat the SME with Respect

When you are making changes in text that has been supplied by a technical specialist, remember that you are reading a draft, not a polished document. Never make negative comments to other employees (including fellow writers), about the writing ability of SMEs.

Guidelines for Being a Collaborative SME

>> **SME Guideline 1:** Keep Technical Communicators Informed

Provide technical communicators with the information they need, even if they don't ask for it. This includes keeping them informed of changes or updates of products or projects that they are documenting.

>> **SME Guideline 2:** Respond to Emails and Phone Calls Promptly

If you aren't sure what is being requested, ask for clarification. If you are being asked to a meeting or an interview, make time for it. Delays in providing necessary information to a documentation specialist can delay and entire project.

>> **SME Guideline 3:** Prepare for Interviews and Meetings

Find out ahead of time what you are going to be asked to explain or provide. Have all appropriate prototypes, samples, or products on hand, if possible. If something comes up that you can't answer right away, make a note of it and respond as soon as possible.

>> **SME Guideline 4:** Treat the technical communicator with respect

If the technical communicator has revised text that you provided, this is not a criticism of your writing ability. The changes were probably made to shift the focus of the text to the users' needs. Clearly written documentation is an important part of a well-run organization, as well as being important to the products or services that your organization provides to its clients.

>>> Chapter Summary

Most organizations rely on people collaborating throughout the writing process to produce documents. The success of writing projects depends on information and skills contributed by varied employees. Collaborative writing can be done by pairs of people working side by side or by several people each contributing their own expertise. Writing teams can have standing tasks such as proposal writing or they can be put together for a single project. In some organizations, most print and electronic documents, even web sites, are created collaboratively through modular writing. Organizations depend on skilled individuals working together to create the final product.

In collaborative writing, the whole is greater than the sum of the parts. In other words, benefits go beyond the collective specialties and experience of individual group members. Participants create new knowledge as they plan, draft, and edit their work together. They become better contributors and faster learners simply by being a part of the social process of a team. Discussion with fellow participants moves them toward new ways of thinking and inspires them to contribute their best. This collaborative effort yields ideas, writing strategies, and editorial decisions that result from the mixing of many perspectives.

Of course, team writing does have drawbacks. Most notably, the group must make decisions without falling into time traps that slow down the process. There must be procedures for getting everyone's ideas on the table and for reaching decisions on time. A leader with good interpersonal skills helps the group reach its potential, whereas an indecisive or autocratic leader is an obstacle to progress. Good leadership rests at the core of every effective writing team.

In addition to good leadership, shared decision making is at the heart of every successful writing team. Members of the writing team must communicate information and expectations clearly. Participants in a team must work together during the planning, drafting, and revising stages of writing. They must respect each other, and remember that successful writing projects contribute to the success of their organization.

>>> Learning Portfolio

Communication Challenge   "A Field Guide: Planning a User's Manual"

The Research and Training division of M-Global is responsible for all in-house documentation, especially documentation of new equipment that has been created by the Equipment Development teams. This documentation is usually written by cross-functional teams that are brought together for a specific project. (A *cross-functional team* includes people from different departments, each contributing his or her own expertise to the project.) A documentation team has been assembled to create a user's manual for a new Lab-in-A-Box (LAB). What follows is background about the equipment, information about the team members, information from the team's first meeting, some questions and comments for discussion, and an assignment for a written response to the Challenge.

Background on Soils LABs

The new Lab-in-A-Box (LAB) improves field testing for contaminated soils. The Soils LAB includes equipment for collecting soil samples and analyzing them chemically, a new sensor for measuring for volatile organic compounds (VOCs) at and below the surface (see page 448 at the beginning of this chapter), and a notebook computer equipped with GPS and satellite communication capabilities so that the data gathered in the field can be analyzed and reports can be easily sent to M-Global labs, government agencies, and clients.

Documentation Team

The documentation team consists of team leader Rob McCulley, a documentation specialist; Mike Sealy, an engineer who helped develop the new sensor and who is the co-author of a white paper about the sensor; Shauna Hill, an M-Global chemist; and Joe Freeman, an editor from the Publications Development department.

The First Meeting

After everyone has introduced themselves at the first meeting, they begin brainstorming about what should be included in the manual. Mike wants to include detailed descriptions of the equipment in the Soils LAB. He argues that the people using it in the field must have a thorough understanding of the new sensor, including how it was developed and its improvements over older equipment. With this information, Mike argues, the users will understand how to take care of the equipment, how to use it, and how to fix it if something

goes wrong. Shauna argues that because the Soils LAB will be used in the field, a complete user's manual isn't necessary. All that is needed is a sort of quick reference to remind people what steps to take to make sure that the tests are accurate and results are reliable. After all, she argues, the technicians using the LAB will have been trained on it, and with the notebook computer, it will be easy to contact a specialist for troubleshooting help. Joe agrees with Shauna that a large user's manual doesn't make much sense, and he suggests that they begin by deciding what format they will use for the information in the manual. He suggests a booklet, or maybe a quick reference sheet attached to the lid of the box. He adds that a help file can be installed on the notebook computer for more complete information, such as maintenance and troubleshooting. Mike likes the help file idea, because it doesn't have the size limitations of a printed manual.

After an hour of discussion, Rob hands out assignments for the next meeting. Joe will create a framework for the help files, Mike will draft information about the sensor, and Shauna will draft instructions for the gathering and analysis of data. After a little more discussion, the group decides that it would be helpful to watch the Soils LAB used on site and to be able to use it themselves. Rob knows that he is on a limited budget for this project, but he might be able to argue that one or two people should be sent to a brownfield or other site with known soil contamination to try out the Soils LAB themselves.

Questions and Comments for Discussion

As Rob sits down to read through his notes and write up the minutes of the first meeting, he thinks about the discussion during the meeting. Although the members of the team have very different backgrounds (and priorities), they seem to get along well, and they respect each other. They even built on each other's ideas during the brainstorming. However, he wonders if he should have started the meeting by discussing the contents of the manual. As he prepares his minutes and the agenda for the next meeting, he begins asking himself questions such as the following:

1. What should the team focus on first as they are planning this project? The users? The equipment? The physical context in which the Soils LAB will be used? The format that the manual will take? The timeline and budget? Why do you think this should be the first step in planning the user's manual?

2. What should the team know about the people who will be using the Soils LAB, and where they will be using it, in order to decide what should go in the user's guide?

3. Look at what each member of the team is advocating for. What does this reveal about each person's interests in the project? As team leader, how would you deal with the competing interests?

4. Identify all of the stakeholders in this project. Think about everyone who will have an interest in the successful use of the Soils LABs. Which of these stakeholders is most important to the documentation team? Why?

5. Think about user's guides that you have seen for portable equipment. How and where were they meant to be used? How did this affect the content and design of the guide? How might lessons from the guides you have seen be applied to the M-Global situation?

Write About It

Assume the role of Rob McCulley. Write memo to Janet Remington, director of the Publications Development department, proposing that someone be sent to see how the Soils LAB will be used in the field. You must make a case for funding the travel (the nearest suitable site is about a three-hour drive away). You must also identify who will be making the trip, and why this choice is necessary for successful completion of your documentation project.

Collaboration at Work Advice About Advising

General Instructions

Each Collaboration at Work exercise applies strategies for working in teams to chapter topics. The exercise assumes you (1) have been divided into teams of about three to six students, (2) will use time inside or outside of class to complete the case, and (3) will produce an oral or written response. For guidelines about writing in teams, refer to pages 24–28 in chapter 1.

Background for Assignment

Academic advising can be one of the most important as well as the most confusing activities for college students, especially for students who are going through the process for the first time. Students depend on advice from other students, from seminars and workshops, and from teachers. This advice may not always fit the student's situation, or steps for advising and enrollment may change. This assignment asks you to create a document or web page to help your fellow students get the most out of advising.

Team Assignment

In your groups, brainstorm the questions that you have had about advising and enrollment. What advice would you give to fellow students? Identify the steps in the process, where information is currently available, and other sources of information (such as faculty members, the Registrar's Office, or Student Services). Your instructor may assign a team leader, or ask each team to choose its own leader. Decide what information you must gather, how you will gather it, and how you are going to make it available to students. Your team should also decide what approach it will take to gathering and writing the information—divide and conquer, writing in sequence, or working at the same time (see page 449).

Your instructor may decide to make this an assignment in modular writing. If so, the class can brainstorm about the content and sources of information, and then teams will be assigned specific tasks. One team will be responsible for creating style guidelines and a document template to ensure a uniform voice and appearance throughout the document. This team, or another team (depending on how large the class is), will also have responsibility for final editing on the project. Other teams will be assigned to gather information from various sources and to write specific sections of the document. Your instructor will help you decide how your project will be made available to students.

Assignments

1. Survey—Your Experience with Teams

Answer following questions about your experience collaborating on projects, either in school or at work. In teams of five or more, compile and present information in a meaningful way. Discuss the responses.

A. Briefly describe your experiences with the following:

- Divide and conquer: The team planned the project together and randomly assigned tasks to each member
- Specialization: The team planned the project together and assigned tasks according to each person's expertise

- Sequence: One person drafted the project, passed it along to the next person who revised the project, who passed it along to the next person, and so on.
- Dialog: Two people worked on the project; one drafted it and gave it to the other, who revised it and returned it for more revisions, until both partners were happy with the result (or until the deadline)
- Synthesis: Two or three people created the project together, working side by side. Every responsibility in the project was shared completely.

B. What makes a good member of a project team?

C. What problems have you encountered in collaborative projects?

2. Schedule Charts

Using any options discussed in this chapter, draw a schedule chart that reflects your work on one of the following:

- A project at work
- A laboratory course at school
- A lengthy project in a course such as this one

All of the following assignments should be completed in teams of four or five students.

3. Short Report

In teams, write a brief evaluation of the teaching effectiveness of either the room in which your class is held or some other room or building of your instructor's choice. In following the tasks listed in this chapter, the team must establish criteria for evaluation, apply these criteria, and report on the results.

Your brief report should have three parts: (1) a one-paragraph summary of the room's effectiveness, (2) a list of the criteria used for evaluation, and (3) details of how the room met or did not meet the criteria you established.

Besides preparing the written report, be prepared to discuss the relative effectiveness with which the team followed this chapter's guidelines for collaborative writing. What problems were encountered? How did you overcome them? How would you do things differently next time?

4. Collabration—M-Global Context

Using one of the seven project sheets included at the end of chapter 2, your team will write a generic abstract for a report of the completed project. Follow the guidelines for abstracts in chapter 15.

5. Computer Communication 1

As in Assignment 4, for this assignment you will (a) work in teams established by your instructor, (b) write a generic abstract for a report on one of the seven project sheets included at the end of chapter 2, and (c) follow the abstract guidelines included in chapter 15. In addition, you are to conduct at least part of your team business by email. The degree to which your team uses email depends on the technical resources of team members and the campus. At a minimum, you should plan for each member to send a message to every other member concerning, for example, the drafting or editing process. At a maximum, and if computer resources permit, you may develop on-screen windows whereby you conduct a conversation with each fellow member in one window and make changes in text in another window. The point of this assignment, in other words, is that team members can use email in a substantive way to communicate with each other in the completion of team projects.

6. Computerized Communication 2

If your campus computer facilities permit, set up a groupware folder with members of a writing team to which you have been assigned by your instructor. Decide on a topic on which you and your team members will write. Each team member should post one short document to the folder, and each team member should contribute to the other documents in the folder. Print the contents of the team's folder and submit it to your instructor. Depending on the instructions you have been given, this assignment may be independent or it may be related to a larger collaborative writing assignment.

7. Research and Presentation

Using the working groups your instructor has established, collect information on collaborative learning and then make a brief oral presentation on your findings to the entire class. Your sources may involve print media or computer sources such as the Internet.

8. Ethics and Collaboration

Create an evaluation sheet that could be used for any collaborative projects that your instructor assigns. Decide if the whole group should sign one document or if individuals should write their own. Explain your decision in a cover memo to your instructor.

9. International Communication Assignment

This chapter offers guidelines on team writing because collaborative communication is essential for success in most careers. However, world cultures differ in the degree to which they use and require collaboration on the job. For this assignment, interview someone who is from a culture different from your own. Using information supplied by this informant, write a brief essay in which you (1) describe the importance of collaboration in the individual's home culture and workplace, (2) give specific examples of how and when collaborative strategies would be used, and (3) modify or expand this chapter's "Guidelines for Group Writing" to suit the culture you are describing, on the basis of suggestions provided by the person you interviewed.

10. A.C.T. N.O.W. Assignment (Applying Communication To Nurture Our World)

Find out what resources your campus offers to help students find opportunities for community service. Does the student employment office or another office on campus keep a list of organizations that are looking for help? Is there an office in student government that helps campus and community organizations connect? If so, working with the campus office, develop materials that help publicize their services. If the office hosts a campus-wide initiative, such as a weekend of cleaning up area parks, create materials to help publicize the event. If your campus does not have such a resource, create a proposal for such an office, using the guidelines in chapter 10.

History

M-Global, Inc. was founded in 1963 as McDuff, Inc. by Rob McDuff, as a firm that specialized in soils analysis. From its founding in 1963 until about 1967, the company worked mostly for construction firms in the Baltimore area. By the late 1960s, the firm enjoyed a first-rate reputation. It had offices in Baltimore and Boston and about 80 employees.

McDuff, Inc., kept growing steadily, with a large spurt in the mid-1970s and another in the 1980s. The first growth period was tied to increased oil exploration in all parts of the world. Oil firms needed experts to test soils, especially in offshore areas. The results of these projects were used to position oil rigs at locations where they could withstand rough seas. The second growth period was tied to environmental work required by the federal government, state agencies, and private firms. McDuff became a major player in the waste-management business, consulting with clients about ways to store or clean up hazardous waste. The third growth period has moved the firm into diverse service industries, such as security systems, hotel management, and landscaping.

In 2008, Rob McDuff announced his retirement and turned the company over to his son Jim. With the change in management, McDuff announced a name change to reflect their more diversified and global scope, becoming M-Global. Although engineering and environmental services still remain important to the company, it has expanded its activities in equipment development, and business services. Today, after 50 years of business, M-Global, Inc., has about 2,500 employees. There are nine offices in the United States and six overseas, as well as a corporate headquarters in Baltimore. M-Global performs a wide variety of work. What started as a technical consulting engineering firm has expanded into a firm that does both technical and non-technical work for a variety of customers.

■ **Model 13–1** ■ Example of M-Global modular writing

Slide Presentation Software Training Team
Meeting Agenda

To Attend: Sally Harkin, Bill Samuelson, Jody Simmons

From: Jody Simmons

Meeting Date: May 8, 2008

Time: 2:00 p.m.

Place: Conference room 2

Objective: Compile results of research

Reports: Sally—results of library research
                  Bill—examples of current M-Global PowerPoint presentations
                  Jody—results of survey

Action: Begin preparing material for slide presentation workshops

■ **Model 13–2** ■ Meeting agenda

M-Global Slide Presentation Software Training Team
Meeting Minutes
Thursday, May 8, 2008
2:00–3:00 Conference Room 2

Attendees:	Sally Harkin, Bill Samuelson, Jody Simmons
	(All members of the M-Global Training Group)
Absentees:	—
Objective:	Compile results of research on effective slide presentations to begin planning of workshops

Sally reported that several articles have been published that make suggestions for effective computer slide presentations. She handed out a summary of the main findings and a bibliography of the articles that she read. She recommended that M-Global use the sentence/graphic design described by Michael Alley and Kathryn A. Neeley in their article, "Rethinking the Design of Presentation Slides: A Case for Sentence Headlines and Visual Evidence" in *Technical Communication*'s Nov. 2005 issue.

Bill shared 10 examples of past and current slide presentations given by a variety of M-Global divisions, branches, and departments. Four were chosen to use as good examples to build on during the training workshops. It was decided that it would be more productive to focus on good examples instead of singling out bad examples.

Bill recommended that the team create an M-Global slide presentation template that would be placed on the company server and made available to all employees.

Jody shared the results of a survey of M-Global managers. The survey revealed that slide presentation software was used most commonly in North America and in Europe. It was used for internal meetings about 70% of the time, and for presentations to potential or existing clients about 30% of the time. Most external presentations were made at meetings of fewer than 20 people. About 15% of internal presentations were made to large groups of 50 or more.

The majority of managers agreed that presentation slides were commonly bulleted lists and seemed to be used more for the reference of the speaker than to provide information for the audience.

Post-meeting actions:

Sally will begin planning activities for the workshops

Bill will develop the presentation software template

Jody will begin drafting an introduction to the workshops that explains why effective presentation slides are important to M-Global

Next meeting:

Wednesday, May 14, 2008, 2:00 P.M. in Conference Room 2

■ **Model 13–3** ■ Meeting minutes

Chapter | **14** | Oral Communication

>>> Chapter Outline

Presentations and Your Career   470

Guidelines for Preparation and Delivery   471

Guidelines for Presentation Graphics   478

Overcoming Nervousness   481

Why Do We Fear Presentations?   482

A Strategy for Staying Calm   482

An Example of an M-Global Oral Presentation   485

Chapter Summary   485

Learning Portfolio   487

Communication Challenge   487

Collaboration at Work   488

Assignments   488

Model for Good Writing   490

Model 14–1: Text and graphics of sample M-Global presentation   490

As the leader of the "Commute Group" in the Boston Branch, Larry Beeman was responsible for preparing the report on the branch's experiment with telecommuting. (See the chapter 3 Communication Challenge, pages 90–92.) Because of the interested generated by the project, Larry was invited to present information about the project and answer questions about it at a meeting of all of the managers of M-Global's domestic branches. At the corporate headquarters in Baltimore, Larry enters the elevator and pushes the button for the executive floor. Just as the door is about to close, a woman carrying a briefcase squeezes into the elevator. She notices that the button for her floor has already been pushed, and asks him, "Are you here for the Branch Managers' meeting?"

Larry responds, "Yes, as a matter of fact, I'm going to be presenting."

"What are you going to be talking about?" she asks.

"I'm from the Boston Branch. I'm going to be talking about how the lessons we learned when we started our telecommuting project can be useful to other branches that are interested in trying it."

"I'm looking forward to hearing about that," she says as the elevator door opens on their floor.

As the woman walks down the hall, and is greeted with, "Good Morning, Ms. McDuff! Good Morning Jeannie!" Larry realizes that he's been in the elevator with the vice president of domestic operations, Jeannie McDuff, granddaughter of the company's founder. He's glad that he practiced an *elevator speech,* a summary of his talk that is short enough to be delivered during an elevator ride.

Your career will present you with many opportunities for oral presentations, both formal and informal. At the time they arise, however, you may not consider them to be "opportunities." They may seem to loom on the horizon as stressful obstacles. That response is normal. The purpose of this chapter is to provide the tools that help oral presentations contribute to your self-esteem and career success. There are guidelines for preparation and delivery, techniques for dealing with anxiety, and an example of a technical presentation. The chapter also addresses the related topic of running effective meetings.

The entire chapter is based on one simple principle: *Almost anyone can become an excellent speaker.* Put aside the myth that competent speakers are born with the talent—that "Either they have it or they don't." Certainly some people have more natural talent at thinking on their feet or have a more resonant voice, but success at speaking can come to all speakers, whatever their talent, if they follow the 3 *Ps:*

Step 1: **P**repare carefully

Step 2: **P**ractice often

Step 3: **P**erform with enthusiasm

These steps form the foundation for all specific guidelines that follow. Before presenting these guidelines, this chapter examines specific ways that formal and informal presentations become part of your professional life.

>>> Presentations and Your Career

Some oral presentations you will choose to give; others will be "command performances" thrust upon you. Using M-Global, Inc., as a backdrop, the following examples present some realistic situations in which the ability to speak well can lead to success for you and your organization:

- **Getting hired:** As a job applicant with a business degree, you are asked to present several M-Global managers with a 10-minute summary of your education, previous experience, and career goals.

- **Getting customers:** As coordinator of an M-Global proposal team, you have just been informed that M-Global made the "shortlist" of companies bidding on a contract to manage a large construction project. You and your three team members must deliver a 20-minute oral presentation that highlights the written proposal. Given in five days at the client's office in Grand Rapids, Michigan, the presentation begins and ends with comments by you in your role as coordinator. Your three colleagues each contribute a five-minute talk.

- **Keeping customers:** As a field engineer at M-Global's St. Louis office, you recently submitted a report on your evaluation of a 50-year-old dam in the Ozarks. Now your clients, the commissioners of the county that owns the dam, have asked you to attend their monthly meeting to present an overview of your findings and respond to questions.

- **Contributing to your profession:** As a laboratory supervisor for M-Global, you belong to a professional society that meets yearly to discuss issues in your field. This year you have been asked to deliver a 15-minute presentation on new procedures for testing toxic-waste samples in the laboratory.

- **Contributing to your community:** As an environmental scientist at M-Global's San Francisco office, you have been asked to speak at the quarterly meeting of OceanSave, an activist environmental organization. The group suggests that you speak for half an hour on environmental threats to aquatic life. You accept the invitation because you know that M-Global management encourages such community service.

- **Getting promoted:** As an employee about to be considered for promotion, you are evaluated on your ability to present information orally. Supervisors will discuss whether they themselves have heard—or heard from others—about your effective presentations to colleagues, clients, or community representatives.

As you can see from this list, oral presentations are defined quite broadly. Usually they can be classified according to criteria such as the following:

1. **Format:** from informal question/answer sessions to formal speeches
2. **Length:** from several-minute overviews to long sessions of an hour or more
3. **Number of presenters:** from solo performances to group presentations
4. **Content:** from a few highlights to detailed coverage

Throughout your career, you will speak to different-size groups, on diverse topics, and in varied formats. The next two sections provide some common guidelines on preparation, delivery, and graphics.

>>> Guidelines for Preparation and Delivery

The goal of most oral presentations is quite simple: You must present a few basic points, in a fairly brief time, to an interested but usually impatient audience. Simplicity, brevity, and interest are the keys to success. If you deliver what *you* expect when *you* hear a speech, then you will give good presentations yourself.

Although the guidelines here apply to any presentation, they relate best to those that precede or follow a written report, proposal, memo, or letter. Few career presentations are isolated from written work. With this connection in mind, note that there are many similarities between the guidelines for good speaking and those for good writing covered in earlier chapters—especially the importance of analyzing the needs of the audience.

>> Presentation Guideline 1: Know Your Listeners

The following features are common to most listeners:

- They cannot "rewind the tape" of your presentation, as opposed to the way they can skip back and forth through the text of a report.
- They are impatient after the first few minutes, particularly if they do not know where a speech is going.
- They will daydream and often must have their attention brought back to the matter at hand (expect a 30-second attention span).
- They have heard so many disappointing presentations that they might not have high expectations for yours.

To respond to these realities, you must learn as much as possible about your listeners. For example, you can (1) consider what you already know about your audience, (2) talk with colleagues who have spoken to the same group, and (3) find out which listeners make the decisions.

Most important, make sure not to talk over anyone's head. If there are several levels of technical expertise represented by the group, find the lowest common denominator and decrease the technical level of your presentation accordingly. Remember—decision makers are often the ones without current technical experience. They may want only highlights; later, they can review written documents for details or solicit more technical information during the question-and-answer session after you speak.

>> Presentation Guideline 2: Use the Preacher's Maxim

The well-known preacher's maxim goes like this:

> First you tell 'em what you're gonna tell 'em, then you tell 'em, and then you tell 'em what you told 'em.

Why should most speakers follow this plan? Because it gives the speech a simple three-part structure that most listeners can grasp easily. Following is how your speech should be organized (note that it corresponds to the ABC format used throughout this text for writing):

1. **Abstract (beginning of presentation):** Right at the outset, you should (1) get the listeners' interest (with an anecdote, a statistic, or other technique), (2) state the exact purpose of the speech, and (3) list the main points you will cover. Do not try the patience of your audience with an extended introduction—use no more than a minute.

Example: "Last year, Jones Engineering had 56% more field accidents than the year before. This morning, I'll examine a proposed safety plan that aims to solve this problem. My presentation will focus on three main benefits of the new plan: lower insurance premiums, less lost time from accidents, and better morale among the employees."

2. **Body (middle of presentation):** Here you discuss the points mentioned briefly in the introduction, in the same order that they were mentioned. Provide the kinds of obvious transitions that help your listeners stay on track.

Example: "The final benefit of the new safety plan will be improved morale among the field-workers at all our job sites."

3. **Conclusion (end of presentation):** In the conclusion, review the main ideas covered in the body of the speech and specify actions you want to occur as a result of your presentation.

Example: "Jones Engineering can benefit from this new safety plan in three main ways. . . . If Jones implements the new plan next month, I believe you will see a dramatic reduction in on-site accidents during the second half of the year."

This simple three-part plan for all presentations gives listeners the handle they need to understand your speech. First, there is a clear *road map* in the introduction so that they know what lies ahead in the rest of the speech. Second, there is an organized pattern in the body, with clear transitions between points. And third, there is a strong finish that brings the audience back full circle to the main thrust of the presentation.

>> Presentation Guideline 3: Stick to a Few Main Points

Our short-term memory holds limited items. It follows that listeners are most attentive to speeches organized around a few major points. In fact, a good argument can be made for organizing information in groups of threes whenever possible. For reasons that are not totally understood, listeners seem to remember groups of three items more than they do any other size groupings—perhaps because

- The number is simple
- It parallels the overall three-part structure of most speeches and documents (beginning, middle, end)
- Many good speakers have used triads (Winston Churchill's "Blood, sweat, and tears," Caesar's "I came, I saw, I conquered," etc.)

Whatever the reason, groupings of three make your speech more memorable to the audience.

>> Presentation Guideline 4: Put Your Outline on Cards or Paper

The best presentations are *extemporaneous,* meaning the speaker shows great familiarity with the material but uses notes for occasional reference. Avoid extremes of (1) reading a speech verbatim, which many listeners consider the ultimate insult, or (2) memorizing a speech, which can make your presentation seem somewhat wooden and artificial.

Ironically, you appear more natural if you refer to notes during a presentation. Such extemporaneous speaking allows you to make last-minute changes in phrasing and emphasis that may improve delivery, rather than locking you into specific phrasing that is memorized or written out word for word.

Depending on your personal preference, you may choose to write speech notes on (1) index cards, (2) a sheet or two of paper, or (3) the "Notes View" available with most presentation software. The main advantages and disadvantages of each are presented in the list that follows.

1. *Notes on Cards (3" × 5" or 4" × 6")*

   Advantages

   ◆ Are easy to carry in a shirt pocket, coat, or purse
   ◆ Provide a way to organize points, through ordering of cards
   ◆ Can lead to smooth delivery in that each card contains only one or two points
   ◆ Can be held in one hand, allowing you to move away from lectern while speaking

   Disadvantages

   ◆ Keep you from viewing outline of entire speech
   ◆ Require that you flip through cards repeatedly in speech
   ◆ Can limit use of gestures with hands
   ◆ Can cause confusion if they are not in correct order

2. *Notes on Sheets of Papers*

   Advantages

   ◆ Help you quickly view outline of entire speech
   ◆ Leave your hands free to use gestures
   ◆ Are less obvious than note cards, for no flipping is needed

   Disadvantages

   ◆ Tend to tie you to lectern, where the sheets lie
   ◆ May cause slipups in delivery if you lose your place on the page

3. *Printout of "Notes View" pages from presentation software*

   Advantages

   ◆ Reminds you of text to accompany each slide
   ◆ Can lead to smooth delivery in that each page contains only a few points
   ◆ Can include special notes and reminders about information to be highlighted
   ◆ Can be used as record of the presentation

   Disadvantages

   ◆ Requires you to flip pages with each new slide
   ◆ Tends to tie you to lectern, where the sheets lie
   ◆ Can cause confusion if they are not in correct order

>> Presentation Guideline 5: Practice, Practice, Practice

Many speakers prepare a well-organized speech but then fail to add the essential ingredient: practice. Constant practice distinguishes superior presentations from mediocre ones. It also helps eliminate the nervousness that most speakers feel at one time or another.

In practicing your presentation, make use of four main techniques, listed here from least to most effective:

- **Practice before a mirror:** This old-fashioned approach allows you to hear and see yourself in action. The drawback, of course, is that it is difficult to evaluate your own performance while you are speaking. Nevertheless, such run-throughs definitely make you more comfortable with the material.

- **Use of an audio recording:** Most presenters have access to an audio recorder, so this approach is quite practical. The portability of electronics allows you to practice almost anywhere. Although recording a presentation does not improve gestures, it helps you discover and eliminate verbal distractions such as *filler words* (e.g., *uhhhh, um, ya know*).

- **Use of live audience:** Groups of your colleagues, friends, or family—simulating a real audience—can provide the kinds of responses that approximate those of a real audience. In setting up this type of practice session, however, make certain that observers understand the criteria for a good presentation and are prepared to give an honest and forthright critique.

- **Use of video recording:** This practice technique allows you to see and hear yourself as others do. Your careful review of the recording, particularly when done with another qualified observer, can help you identify and eliminate problems with posture, eye contact, vocal patterns, and gestures. At first it can be a chilling experience, but you soon get over the awkwardness of seeing yourself on the screen.

>> Presentation Guideline 6: Speak Vigorously and Deliberately

Vigorously means with enthusiasm; *deliberately* means with care, attention, and appropriate emphasis on words and phrases. The importance of this guideline becomes clear when you think back to how you felt during the last speech you heard. At the very least, you expected the speaker to show interest in the subject and demonstrate enthusiasm. Good information is not enough—you must arouse the interest of the listeners.

You may wonder, "How much enthusiasm is enough?" The best way to answer this question is to hear or (preferably) watch yourself on tape. Your delivery should incorporate just enough enthusiasm so that it sounds and looks a bit unnatural to you. Few if any listeners ever complain about a speech being too enthusiastic or a speaker being too energetic, but many people complain about dull speakers who fail to show that they themselves are excited about the topic. Remember—every presentation is, in a sense, *showtime*.

>> Presentation Guideline 7: Avoid Filler Words

Avoiding filler words presents a tremendous challenge to most speakers. When they think about what comes next or encounter a break in the speech, they may tend to fill the gap with filler words and phrases such as these:

uhhhhh . . .

ya know . . .

okay . . .

well . . . uh . . .

like . . .

I mean . . .

umm . . .

These gap-fillers are a bit like spelling errors in written work: Once your listeners find a few, they start looking for more and are distracted from your presentation. To eliminate such distractions, follow these three steps:

Step 1: **Use pauses to your advantage.** Short gaps or pauses inform the listener that you are shifting from one point to another. In signaling a transition, a pause serves to draw attention to the point you make right after the pause. Note how listeners look at you when you pause. Do *not* fill these strategic pauses with filler words.

Step 2: **Practice with a recorder.** A recording is brutally honest: When you play it back, you become instantly aware of fillers that occur more than once or twice. Keep a tally sheet of the fillers you use and their frequency. Your goal is to reduce this frequency with every practice session.

Step 3: **Ask for help from others.** After working with audio recorders in step 2, give your speech to an individual who has been instructed to stop you after each filler. This technique gives immediate reinforcement.

>> Presentation Guideline 8: Use Rhetorical Questions

Enthusiasm, of course, is your best delivery technique for capturing the attention of the audience. Another technique is the use of rhetorical questions at pivotal points in your presentation.

Rhetorical questions are those you ask to get listeners thinking about a topic, not those that you would expect them to answer out loud. They prod listeners to think about your point and set up an expectation that important information follows. Also, they break the monotony of standard declarative sentence patterns. One example of a rhetorical question could be used by a computer salesperson in proposing a purchase by one of M-Global's small offices:

Example: I've discussed the three main advantages that a centralized copy center would provide your office staff. But is this an approach that you can afford at this point in the company's growth?

Next the speaker could follow the question with remarks supporting the position that the system is affordable.

"What if" scenarios provide another way to introduce rhetorical questions. They gain listeners' attention by having them envision a situation that might occur. For example, a safety engineer could use this kind of rhetorical question in proposing M-Global's asbestos-removal services to a regional bank:

Example: What if you repossessed a building that contained dangerous levels of asbestos? Do you think that your bank would then be liable for removing all the asbestos?

Again, the question pattern heightens listener interest.

Rhetorical questions do not come naturally. You must make a conscious effort to insert them at points when it is most important to gain or redirect the attention of the audience. Three particularly effective uses follow:

1. **As a grabber at the beginning of a speech:** "Have you ever wondered how you might improve the productivity of your clerical staff?"

2. **As a transition between major points:** "We've seen that a centralized copy center can improve the efficiency of report production, but will it simplify report production for your staff?"

3. **As an attention-getter right before your conclusion:** "Now that we've examined the features of a centralized copy center, what's the next step you should make at M-Global?"

>> Presentation Guideline 9: Maintain Eye Contact

Your main goal—always—is to keep listeners interested in what you are saying. This goal requires that you maintain control, using whatever techniques possible to direct the attention of the audience. Frequent eye contact is one good strategy.

The simple truth is that listeners pay closer attention to what you are saying when you look at them. Think how you react when a speaker makes constant eye contact with you. If you are like most people, you feel as if the speaker is speaking to you personally—even if there are 100 people in the audience. Also, you tend to feel more obligated to listen when you know that the speaker's eyes will be meeting yours throughout the presentation. Following are some ways you can make eye contact a natural part of your own strategy for effective oral presentations:

■ **With audiences of about 30 or less:** Make regular eye contact with everyone in the room. Be particularly careful not to ignore members of the audience who are seated to your far right and far left (Figure 14–1). Many speakers tend to focus on the listeners in section B. Instead, make wide sweeps so that listeners in sections A and C get equal attention.

■ **With large audiences:** There may be too many people or a room too large for you to make individual eye contact with all listeners. In this case, focus on just a few people in all three sections of the audience, as noted in Figure 14–1. This approach gives the appearance that you are making eye contact with the entire audience.

■ **Figure 14–1** ■
Audience sections

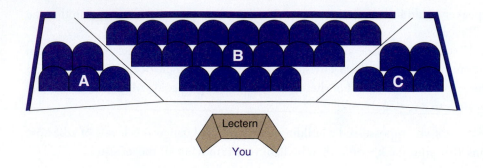

- ■ **With any size audience:** Occasionally, look away from the audience—either to your notes or toward a part of the room where there are no faces looking back. In this way, you avoid the appearance of staring too intensely at your audience. Also, these breaks give you the chance to collect your thoughts or check your notes.

>> **Presentation Guideline 10:** Use Appropriate Gestures and Posture

Speaking is only one part of giving a speech; another is adopting appropriate posture and using gestures that reinforce what you are saying. Note that good speakers are much more than "talking heads" before a lectern. Instead, they

1. Use their hands and fingers to emphasize major points
2. Stand straight, without leaning on or gripping the lectern
3. Step out from behind the lectern on occasion, to decrease the distance between them and the audience
4. Point toward visuals on screens or charts, without losing eye contact with the audience

The audience judges you by what you say and what they see, a fact that again makes video recording a crucial part of your preparation. With work on this facet of your presentation, you can avoid problems like keeping your hands constantly in your pockets, rustling change (remove pocket change and keys beforehand), tapping a pencil, scratching nervously, slouching over a lectern, and shifting from foot to foot.

>>> Guidelines for Presentation Graphics

More than ever before, listeners expect good graphics during oral presentations. Much like gestures, graphics transform the words of your presentation into true communication with the audience. When you display graphics and text during a presentation, they should illustrate and clarify your speech. Therefore, we include displayed text in our discussion of graphics in this section.

>> Graphics Guideline 1: Discover Listener Preferences

Some professionals prefer simple speech graphics, such as flip charts or transparencies. Others prefer more sophisticated presentations, such as animations, audio, or video. Some audiences prefer simple outlines, some may prefer charts and graphs, and still others may prefer full-color images.

Your listeners are usually willing to indicate their preferences when you call on them. Contact the audience ahead of time and make some inquiries. Also ask for information about the room in which you will be speaking. If possible, request a setting that allows you to make best use of your graphics choice. If you have no control over the setting, then choose graphics that best fit the constraints. Details about lighting, wall space, and chair configuration can greatly influence your selection.

>> Graphics Guideline 2: Match the Graphics to the Content

Plan graphics while you prepare the text so that the final presentation seems fluid. Remember that everything you project on a screen or present on a flip chart should support and enhance your presentation. Choose images and words that help your reader focus on what you are saying.

>> Graphics Guideline 3: Keep the Message Simple

When Edward Tufte critiqued PowerPoint slides in *The Cognitive Style of PowerPoint*, one of the problems he pointed to was the use of too many graphic elements on each slide, the equivalent of the *chartjunk* that he had argued against in his earlier studies of graphics[1] (see chapter 12). Some basic design guidelines apply, whether you are using posters, overhead transparencies, or computer-aided graphics such as PowerPoint.

- Use few words, emphasizing just one idea on each frame.
  Note: A common PowerPoint mistake is the use of too much text, which then gets read to the audience by the speaker.

- Use more white space, perhaps as much as 60%–70% per frame.

- Use landscape format more often than portrait, especially because it is the preferred default setting for most presentation software.

- Use sans-serif large print, from 14 pt. to 18 pt. minimum for text to 48 pt. for titles.

Your goal should be to create graphics that are seen easily from anywhere in the room and that complement—but do not overpower—your presentation.

You should also use audio and video elements sparingly. Most presentation software programs include sound effects to accompany slide changes or the appearance of text or images. These are distracting and annoying, and should be avoided. You should also use video carefully. For example, during a presentation on automobile safety, the presenter included an animation of a crash that ran in a constant loop throughout her presentation. This made it difficult for the audience to focus on the findings of the presenter's study.

---

[1]Tufte, E. R. (2003). *The Cognitive Style of PowerPoint*. Cheshire, CT: Graphics Press.

However, if the presenter had allowed the video to run once, pointing out the important aspects of the video, and then stopped the action (or even blanked it from the screen), the audience would have focused on the information she was presenting.

>> Graphics Guideline 4: Consider Alternatives to Bulleted Lists

Recently, there has been a move away from the default slide layouts in most presentation software. One recommendation is to use full-sentence headings on slides to help the audience understand and remember the information being presented; another is to combine text with graphics on slides when appropriate.[2]

>> Graphics Guideline 5: Use Colors Carefully

Colors can add flair to visuals. Use the following simple guidelines to make colors work for you:

- Have a good reason for using color (such as the need to highlight three different bars on a graph with three distinct colors).
- Be sure that a color contrasts with its background (e.g., yellow on white does not work well).
- Use no more than three or four colors in each graphic (to avoid a confused effect).

>> Graphics Guideline 6: Leave Graphics Up Long Enough

Because graphics reinforce text, they should be shown only while you address the particular point at hand. For example, reveal a graph just as you are saying, "As you can see from the graph, the projected revenue reaches a peak in 2007." Then pause and leave the graph up a bit longer for the audience to absorb your point.

How long is too long? A graphic outlives its usefulness when it remains in sight after you have moved on to another topic. Listeners continue to study it and ignore what you are now saying. If you use a graphic once and plan to return to it, take it down after its first use and show it again later.

>> Graphics Guideline 7: Avoid Handouts

Because timing is so important in your use of speech graphics, handouts are usually a bad idea. Readers move through a handout at their own pace, rather than at the pace the speaker might prefer. Thus handouts cause you to lose the attention of your audience. Use them only if (1) no other visual will do, (2) your listener has requested them, or (3) you distribute them as reference material after you have finished talking.

>> Graphics Guideline 8: Maintain Eye Contact While Using Graphics

Do not stare at your visuals while you speak. Maintain control of listeners' responses by looking back and forth from the visual to faces in the audience. To point to the graphic aid,

---

[2]Alley, M., and Neeley, K. A. (2005). "Rethinking the Design of Presentations Slides: A Case for Sentences Headlines and Visual Evidence." *Technical Communication* 52 (4): 417–426.

use the hand closest to the visual. Using the opposite hand causes you to cross over your torso, forcing you to turn your neck and head away from the audience.

>> Graphics Guideline 9: Include All Graphics in Your Practice Sessions

Dry runs before the actual presentation should include every graphic you plan to use, in its final form. This is a good reason to prepare graphics as you prepare text, rather than as an afterthought. If you are going to be projecting images from a transparency or computer program, the projected image may appear different than the original image. Colors may look slightly different, and text and images that are clear on your computer screen or transparency may seem out of focus or too small for an audience looking at a screen. By previewing your graphics, you are able to fix them before your presentation. The goal is to use graphics you can be proud of. Never put yourself in the position of having to apologize for the quality of your graphic material. If an illustration is not up to the quality your audience would expect, do not use it.

You should also practice timing your graphics with your speech. Running through a final practice without graphics would be much like doing a dress rehearsal for a play without costumes and props—you would be leaving out parts that require the greatest degree of timing and orchestration. Practicing with graphics helps you improve transitions.

>> Graphics Guideline 10: Plan for Technology to Fail

Murphy's Law always seems to apply when you use another person's audiovisual equipment: Whatever can go wrong, will, and at the worst possible moment. For example, a new bulb burns out and there is no extra bulb in the equipment drawer, an extension cord is too short, the screen does not stay down, the client's computer doesn't read your file—many speakers have experienced these problems and more. Even if the equipment works, it often operates differently from what you are used to. The only sure way to put the odds in your favor is to carry your own equipment and set it up in advance.

However, most of us must rely on someone else's equipment, at least some of the time. Following are a few ways to ward off disaster:

- Find out exactly who is responsible for providing the equipment and contact that person in advance.
- Have some easy-to-carry backup supplies in your car—an extension cord, an overhead projector bulb, felt-tip markers, and chalk, for example.
- Bring handout versions of your visuals to use as a last resort.
- In short, you want to avoid putting yourself in the position of having to apologize. Plan well.

>>> Overcoming Nervousness

The problem of nervousness deserves special mention because it is so common. Virtually everyone who gives speeches feels some degree of nervousness before "the event." An instinctive "fight or flight" response kicks in for the many people who have an absolute dread

of presentations. In fact, surveys have determined that most of us rate public speaking at the top of our list of fears, even above sickness and death! Given this common response, we next consider the problem and offer suggestions for overcoming it.

Why Do We Fear Presentations?

Most of us feel comfortable with informal conversations, when we can voice our views to friends and indulge in impromptu exchanges. We are used to this type of casual presenting of our ideas. Formal presentations, however, put us into a more structured, more awkward, and thus more tense environment. Despite the fact that we may know the audience is friendly and interested in our success, the formal context triggers nervousness that is sometimes difficult to control.

This nervous response is normal and, to some degree, useful. It gets you "up" for the speech. That adrenaline pumping through your body can generate a degree of enthusiasm that propels the presentation forward and creates a lively performance. Just as veteran actors admit to some nervousness helping improve their performance, excellent speakers usually can benefit from the same effect.

The problem occurs when nervousness felt before or during a speech becomes so overwhelming that it affects the quality of the presentation. Because sympathy is the last feeling a speaker wants the audience to have, it is worth considering some techniques to combat nervousness.

A Strategy for Staying Calm

As the cliché goes, do not try to eliminate butterflies before a presentation—just get them to fly in formation. In other words, it is best to acknowledge that a certain degree of nervousness will always remain; then go about the business of getting it to work for you. Following are a few suggestions:

>> No Nerves Guideline 1: Know Your Speech

The most obvious suggestion is also the most important one. If you prepare your speech well, your command of the material helps conquer any queasiness you feel—particularly at the beginning of the speech, when nervousness is usually at its peak. Be so sure of the material that your listeners overlook any initial discomfort you may feel.

>> No Nerves Guideline 2: Prepare Yourself Physically

Your physical well-being before the speech can have a direct bearing on anxiety. More than ever before, most cultures understand the essential connection between mental and physical well-being. This connection suggests you should take the following precautions before your presentation:

- **Avoid caffeine or alcohol for at least several hours before you speak.** You do not need the additional jitters brought on by caffeine or the false sense of ease brought on by alcohol.

- **Eat a light and well-balanced meal within a few hours of speaking.** However, do not overdo it—particularly if a meal comes right before your speech. If you are convinced that any eating will increase your anxiety, wait to eat until after speaking.

- **Practice deep-breathing exercises before you speak.** Inhale and exhale slowly, making your body slow down to a pace you can control. If you can control your breathing, you can probably keep the butterflies flying in formation.

- **Exercise normally the same day of the presentation.** A good walk helps invigorate you and reduces nervousness; however, do not wear yourself out by exercising more than you would normally.

>> No Nerves Guideline 3: Picture Yourself Giving a Great Presentation

Many speakers become nervous because their imaginations are working overtime. They envision the kinds of failure that almost never occur. Instead, speakers should be constantly bombarding their psyches with images of success. Mentally take yourself through the following steps of the presentation:

- Arriving at the room
- Feeling comfortable at your chair
- Getting encouraging looks from your audience
- Giving an attention-getting introduction
- Presenting your supporting points with clarity and smoothness
- Ending with an effective wrap-up
- Fielding questions with confidence

Sometimes called *imaging,* this technique helps program success into your thinking and control negative feelings that pass through the minds of even the best speakers.

>> No Nerves Guideline 4: Arrange the Room as You Want

To control your anxiety, assert some control over the physical environment as well. You need everything going for you if you are to feel at ease. Make sure that chairs are arranged to your satisfaction, that the lectern is positioned to your taste, that the lighting is adequate, and so on. These features of the setting can almost always be adjusted if you make the effort to ask. Again, it is a matter of your asserting control to increase your overall confidence.

>> **No Nerves Guideline 5:** Have a Glass of Water Nearby

Extreme thirst and a dry throat are physical symptoms of nervousness that can affect delivery. There is nothing to worry about as long as you have water available. Think about this need ahead of time so that you do not have to interrupt your presentation to pour a glass of water.

>> **No Nerves Guideline 6:** Engage in Casual Banter Before the Speech

If you have the opportunity, chat with members of the audience before the speech. This ice-breaking technique reduces your nervousness and helps start your relationship with the audience.

>> **No Nerves Guideline 7:** Remember That You Are the Expert

As a final psyching up exercise before you speak, remind yourself that you have been invited or hired to speak on a topic about which you have useful knowledge. Your listeners want to hear what you have to say and are eager for you to provide useful information to them. So tell yourself, "I'm the expert here!"

>> **No Nerves Guideline 8:** Do Not Admit Nervousness to the Audience

No matter how anxious you may feel, never admit it to others. First of all, you do not want listeners to feel sorry for you—that is not an emotion that leads to a positive critique of your speech. Second, nervousness is almost never apparent to the audience. Your heart may be pounding, your knees may be shaking, and your throat may be dry, but few if any members of the audience can see these symptoms. Why draw attention to the problem by admitting to it? Third, you can best defeat initial anxiety by simply pushing right on through.

>> **No Nerves Guideline 9:** Slow Down

Some speakers who feel nervous tend to speed through their presentations. If you have prepared well and practiced the speech on an audio recording, you are not likely to let this happen. Having heard yourself on tape, you will be better able to sense that the pace is too quick. As you speak, constantly remind yourself to maintain an appropriate pace. If you have had this problem before, you might even write "Slow down!" in the margin of your notes.

>> **No Nerves Guideline 10:** Join a Speaking Organization

The previous nine guidelines will help reduce your anxiety about a particular speech. To help solve the problem over the long term, however, consider joining an organization like Toastmasters International, which promotes the speaking skills of all its members. Like some other speech organizations, Toastmasters has chapters that meet at many companies and campuses. These meetings provide an excellent and supportive environment in which all members can refine their speaking skills.

>>> An Example of an M-Global Oral Presentation

Model 14–1 presents the text and visuals of a short presentation given by Kim Mason, an environmental expert for M-Global's Atlanta office. She has been invited to speak at the monthly lunch meeting of an organization of building owners in the Atlanta region. The agreed-on topic is the problem of asbestos contamination.

The members of Kim's audience have an obvious interest in the problem: They own buildings at risk. Yet they know little more about asbestos than that it is a health issue they must consider when they renovate. Kim's job is to inform them and heighten their awareness. She must cover only the highlights, however, because the presentation will be followed by a detailed question-and-answer session. Although some of these owners have been and will be clients of M-Global, she has an ethical obligation to avoid promoting M-Global during her presentation.

>>> Chapter Summary

The main theme of this chapter is that anyone can become a good speaker by preparing well, practicing often, and giving an energetic performance. This effort pays off richly by helping you deal effectively with employers, customers, and professional colleagues.

Ten guidelines for preparation and delivery lead to first-class presentations:

1. Know your listeners.
2. Use the preacher's maxim.
3. Stick to a few main points.
4. Put your outline on cards or paper.
5. Practice, practice, practice.
6. Speak vigorously and deliberately.
7. Avoid filler words.
8. Use rhetorical questions.
9. Maintain eye contact.
10. Use appropriate gestures and posture.

You should also strive to incorporate illustrations into your speeches by following these 10 guidelines:

1. Discover listener preferences.
2. Match the graphics to the content.
3. Keep the message simple.
4. Consider alternatives to bulleted lists.
5. Use colors carefully.

6. Leave graphics up long enough.

7. Avoid handouts.

8. Maintain eye contact while using graphics.

9. Include all graphics in your practice sessions.

10. Plan for technology to fail.

A major problem for many speakers is fear of giving presentations. You can control this fear by following these guidelines:

1. Know your speech!

2. Prepare yourself physically.

3. Picture yourself giving a great presentation.

4. Arrange the room as you want.

5. Have a glass of water nearby.

6. Engage in casual banter before the speech.

7. Remember that you are the expert.

8. Do not admit nervousness to the audience.

9. Slow down.

10. Join a speaking organization.

>>> Learning Portfolio

Communication Challenge "Ethics and the Technical Presentation"

Carlos Santiago had been looking forward to attending his first international conference. The Association of Medical Technology invited him to speak at its annual meeting. Now, however, some of his research presents him with an ethical dilemma. The following sections present background on his invitation and an assignment for a written response to the Challenge questions and comments for discussion.

An Opportunity in Caracas

Carlos Santiago works as an equipment development specialist in M-Global's San Francisco office. For two years he has been helping a large medical equipment firm, MedExcel, Inc., improve its C-2000 electrocardiograph. An electrocardiograph records electrical changes that occur during the human heartbeat. One of the most important pieces of equipment in diagnosing heart problems, it is made up of four main parts:

- Electrodes attached to the patient's body
- Electronics that convert and amplify the signal traveling from the electrodes to the computer
- A computer that interprets information provided by the electrodes
- An output screen and printer that provide information to the medical professional

In his work on the C-2000, Carlos (1) developed new electrodes that take extremely accurate readings, (2) designed a panel that is easier to read, and (3) created electronics that convert and amplify signals better than any model he has seen produced by any firm. These major refinements, as well as a host of smaller ones, created what he and MedExcel believe is the most user-friendly and accurate electrocardiograph on the market.

As a result, Carlos' work has been recognized in the industry. Next week he will fly from San Francisco to Caracas, Venezuela, to speak at the annual meeting of the Association of Medical Technology (AMT). MedExel, which conveyed the invitation to him from the AMT, will be paying all his expenses for the eight-day trip. He has been asked to speak on "Advances in the Electrocardiograph." As you might expect, MedExcel expects that Carlos will showcase the C-2000 as an example of the most current technological advances.

A Late-Breaking Surprise

Everything has come together for Carlos. His client is happy, his branch manager is overjoyed that M-Global's small equipment development team is gaining recognition in a major industry, and a high-visibility professional association is about to showcase his work. The public relations and marketing potential is significant, both for M-Global and MedExcel.

Now the other shoe has dropped. Yesterday, just a week before the conference, Carlos was shocked to learn that Worldwide Medical, the main competitor of MedExcel, just came out with an updated version of its electrocardiograph, the HeartCart 300. From the article he reviewed in a weekly health-care newspaper, the machine is competitive with the C-2000 in accuracy of electrode readings and in electronics. In addition, the panel pictured in the article appears to be a significant improvement over the previous HeartCart model. After reading the article, Carlos went to his computer to find the home page for Worldwide Medical on the Internet. From the additional information he found there, it appears the HeartCart 300 is competitive with the C-2000.

Carlos has mixed emotions about this finding. On the one hand, he is glad to see that another firm is investing research dollars in this technology that is crucial to health care throughout the world. On the other hand, he is understandably concerned that his two years of work may become upstaged by the work of a competing company. For the sake of the investment his client has made in this research, he was hoping that the C-2000 would enjoy sales that would justify the research investment.

Questions and Comments for Discussion

More to the immediate point, Carlos is wondering what effect, if any, this new information should have on his presentation—a presentation that is being underwritten by a $5,000 grant from MedExcel. Following are some questions for you to consider:

1. Do you think Carlos is obligated to revamp his presentation to include information about the HeartCart 300? If your answer is yes, explain exactly how he should proceed. If your answer is no, give your rationale.
2. Should his association with MedExcel have any bearing on his decision?
3. Should the manner by which Carlos received this new information have any effect on the answer to question 1?
4. Give what you think would be the opinion of the AMT in response to question 1.
5. Explain how your answer to question 1 relates to the Equal Consideration of Interests (ECI) ethical principle discussed in chapter 2 (see page 43).

6. Generally, do you see any larger concern here in the relationship between professional associations and private enterprise? Explain your answer.

Write About It

Assume the role of Carlos. You feel that you must contact MedExcel about the HeartCart 300. Decide what you will say.

Will you ask questions? Make suggestions? Propose changes in your presentation? Write the text of an email that you will send to Robert Morgan, the president of MedExcel. Remember that this is professional correspondence, so use appropriate tone. Provide the context, and include all of the information that Robert needs to give you an appropriate response.

Collaboration at Work Speeches You Have Heard

General Instructions

Each Collaboration at Work exercise applies strategies for working in teams to chapter topics. The exercise assumes you (1) have been divided into teams of about three to six students, (2) will use team time inside or outside of class to complete the case, and (3) will produce an oral or written response. For guidelines about writing in teams, refer to pages 24–28.

Background for Assignment

Even if you have little experience as a public speaker, and even if you have not read this chapter, you already know a lot about what makes a good or bad speech because you have listened to so many presentations in your life, from informal lectures in a classroom to famous speeches by national figures. Every day you see or hear snippets of presentations in the media, so your exposure has been high.

Considering all the visual input, you probably have developed preferences for certain features in presentations.

Team Assignment

In this exercise, you and your fellow team members will do the following:

1. Share anecdotes about good and bad presentations you have heard, focusing on criteria such as content, organization, delivery, graphics, and gestures.
2. Assemble a first list that includes features of speeches and characteristics of speakers that you consider worthy of modeling.
3. Assemble a second list that includes features of speeches and characteristics of speakers that you think should be avoided.

Be able to provide reasons why these features are positive or negative.

Assignments

1. A 2–3 Minute Presentation Based on M-Global Projects

Select one of the projects described at the end of chapter 2. Use information from the project and, if you wish, additional details you invent that could relate to the project. For this presentation, assume you are an M-Global marketing specialist talking to a group of potential clients in a meeting. They may hire M-Global for a similar project, and thus want a summary of the job described in the project sheet.

2. A 2–3 Minute Presentation Based on Your Academic Major

Give a presentation in which you discuss (a) your major field, (b) reasons for your interest in this major, and (c) specific career paths you may pursue. Assume your audience is

a group of students, with undecided majors, who may want to select your major.

3. A 5–6 Minute Presentation Based on Short Report

Select any of the short written assignments in chapters that you have already completed. Prepare a presentation based on the report you have chosen. Assume that your main objective is to present the audience with the major highlights of the written report, which they have all read. Use at least one visual aid.

4. A 5–6 Minute Presentation Based on Proposal

Prepare a presentation based on the proposal assignment at the end of chapter 10. Assume that your audience wants highlights of your written proposal, which they have read.

5. A 10–12 Minute Presentation Based on Formal Report

Prepare a presentation based on any of the long-report assignments at the end of chapter 9. Assume that your audience has read or skimmed the report. Your main objective is to present highlights along with some important supporting details. Use at least three visual aids.

6. Team Presentation

Prepare a team presentation in the size teams indicated by your instructor. It may be related to a collaborative writing assignment in an earlier chapter, or it may be done as a separate project. Review the chapter 13 guidelines for collaborative work. Although related to writing, some of these suggestions apply to any team work.

Your instructor will set time limits for the entire presentation and perhaps for individual presentations. Make sure that your team's members move smoothly from one speech to the next; the individual presentations should work together for a unified effect.

7. Ethics Assignment

Suzanne Anthony, a prominent ecologist with M-Global's Atlanta office, has been asked to make a 30-minute speech to a public workshop on environmentalism, sponsored by SprawlStopper, a regional environmental action organization. She agrees to give the talk—for which she will receive an "honorarium" of $500—on her area of expertise: the effects of unplanned growth on biological diversity of plant and animal species. Suzanne views the talk as a public service and has no knowledge of the sponsoring organization.

A few weeks before the speech, Suzanne's boss, Paul Finn, gets heartburn over his morning coffee as he reads an announcement about Suzanne's speech in the "Community Events" column of the local paper. Just yesterday, a large local builder, Action Homes, accepted his proposal for M-Global to complete environmental site assessments on all of Action Homes' construction sites for the next three years. Paul is aware of the fact that Action Home's has an ongoing court battle with SprawlStopper concerning Action Homes' desire to develop a large site adjacent to a Civil War national park north of Atlanta. Although M-Global is not now involved in the suit, Paul is worried that if Action

Homes sees the name of an M-Global scientist associated with an event sponsored by SprawlStopper, Action Homes may have second thoughts about having chosen M-Global for its site work.

Do you think Paul should say something to Suzanne? If so, what should he say, and why should he say it? If not, why not? If you were in Suzanne's place, how would you respond to a suggestion by Paul that her speech might be inappropriate? Are there any similarities between the situation described here and the one characterized in this chapter's "Communication Challenge"?

8. International Communication Assignment

Prepare a team presentation that results from research your team does on the Internet concerning speech communication in a country outside the United States.

Option A: Retrieve information about one or more businesses or careers in a particular country. Once you have split up the team's initial tasks, conduct some of your business by email, and then present the results of your investigation in a panel presentation to the class. For example, your topic could be the computer software industry in England, the tourist industry in Costa Rica, or the textile industry in Malaysia.

Option B: Retrieve information on subjects related to this chapter—for example, features of public speaking, business presentations, presentation graphics, and meeting management.

9. A.C.T.N.O.W. Assignment (Applying Communication To Nurture Our World)

For this assignment, use the same context as that described in the A.C.T.N.O.W. exercise in chapter 12 (see page 446). Prepare the text of a short speech on the topic—with at least one presentation graphic—as if the speech is to be delivered to an audience with a strong interest in the topic. Assume the speech narrative is the basis for notes you develop for the speech. Alternatively, your instructor may ask that you prepare the speech notes and deliver the presentation either to the class or to another appropriate group.

Good evening. My name is Kim Mason, and I work for the asbestos-abatement division of M-Global, Inc., in Atlanta. I've been asked to give a short presentation on the problem of asbestos and then to respond to your questions about the importance of removing it from buildings. I'll focus on three main reasons why you, as building owners, should be concerned about the asbestos problem: (NEXT SLIDE)

How to avoid asbestos contamination

Kim Mason, environmental engineer M-Global, Inc., Atlanta

krmason@mglobal.com

1. To prevent future health problems of your tenants
2. To satisfy regulatory requirements of the government
3. To give yourself peace of mind for the future

Again, my comments will provide just an overview, serving as a basis for the question session that follows in a few minutes. (NEXT SLIDE)

By avoiding asbestos contamination, you can

1. Prevent health problems
2. Satisfy regulatory requirements
3. Give yourself peace of mind

Three main reasons why building owners should be concerned about asbestors:

1. To prevent future health problems of your tenants.
2. To satisfy regulatory requirements of the government.
3. To give yourself peace of mind for the future.

■ **Model 14–1** ■ Text and graphics of sample M-Global presentation

Question: What is the most important reason you need to be concerned about asbestos?

Answer: The long-term health of the tenants, workers, and other people in buildings that contain asbestos. Research has clearly linked asbestos with a variety of diseases, including lung cancer, colon cancer, and asbestosis (a debilitating lung disease). Although this connection was first documented in the 1920s, it has only been taken seriously in the last few decades. Unfortunately, by that time asbestos had already been commonly used in many building materials that are part of many structures today. (NEXT SLIDE)

What are some of the most common building products containing asbestos?

Asbestos was used in materials as varied as floor tiles, pipe wrap, roof felt, and insulation around heating systems. (NEXT SLIDE)

Prevent future health problems of your tenants.

Most important reason be concerned?

Long term health of your tenants.
Asbestos linked to:
Lung cancer
Colon cancer
Asbestosis (debilitating lung disease)

Connection first documented in 1920s, but only taken seriously in last few decades.

By this time, asbestos common in buildings.

Asbestos is found in

• Paper products

• Plastic products

• Insulating products

Where will you find asbestos in buildings? [wait for response—discuss response]

[reveal list]
Plastic products: floor tile, coatings, sealants
Paper products: roof felt, gaskete, pipe wrap
Insulating products: sprayed coating preformed pipe wrap, insulation board, boiler insulation

An abundant and naturally occurring mineral, asbestos was fashioned into construction materials through processes such as packing, weaving, and spraying. Its property of heat resistance, as well as its availability, was the main reason for such widespread use.

While still embedded in material, asbestos causes no real problems. However, when it deteriorates or is damaged, fibers may become airborne. In this state, they can enter the lungs and cause the health problems mentioned a minute ago. This risk prompted the Environmental Protection Agency in the mid-1970s to ban the use of certain asbestos products in most new construction. But today the decay and renovation of many asbestos-containing building materials may put many of our citizens at risk for years to come. (NEXT SLIDE)

After your concern for occupants' health, what's the next best reason to learn more about asbestos? It's the *law*.

Both the Occupational Safety and Health Administration (OSHA) and the Georgia Department of Natural Resources (DNR) require that you follow certain procedures when structures you own could endanger tenants and asbestos-removal workers with contamination. For example, when a structure undergoes renovation that will involve any asbestos-containing material (ACM), the ACM must be removed by following approved engineering procedures. Also, the contaminated refuse must be disposed of in approved landfills. Considering the well-documented potential for health problems related to airborne asbestos, this legislative focus on asbestos contamination makes good sense. (NEXT SLIDE)

Abundant, naturally occuring mineral

Fashioned into construction materials by packing, wearing, and spraying

Heal resistance, and availability, main reason for widespread used

While embedded, asbestos causes no real problems.

When deteriorates/is damaged, fiber become airbome—enter lungs, cause health problems.

Mid-1970s—EP A banned use of certain asbesto products in most new construction.

Satisfy regulatory requirements of the government.

Occupational Safety and Health Administration (OSHA)

Georgia Department of Natural Resources (DNR)

Next reason?

It's the law

OSHA and Georgia DNR require follow certain procedures w/structures you own—don't endanger tenants & asbestos removal workers with contamination.

When structure undergoes renovation involving ACM—follow engineering procedures.

Contaminated refuse—disposed on in approval landfills.

Both OSHA & DNR require removal of asbestos by licensed contractors.

Contractors assume liability only for what they have been told to remove.

By the way, both OSHA and DNR regulations require removal of asbestos by licensed contractors. These contractors, however, will assume liability only for what they have been told to remove. They may or may not have credentials and training in health and safety. Therefore, building owners should hire a firm with a professional who will (1) survey the building and present a professional report on the degree of asbestos contamination and (2) monitor the work of the contractor in removing the asbestos. By taking this approach, you as an owner stand a good chance of eliminating all problems with *your* asbestos. (NEXT SLIDE)

> ## Find a qualified asbestos removal professional.
>
> 1. The professional should survey the building and present a professional report on the degree of asbestos contamination.
> 2. The professional should monitor the work of the contractor.

By taking this approach, you stand a good chance of eliminating all problems with your asbestos.

Yes, it is *your* asbestos. As owner of a building, you also legally own the asbestos associated with that building—*forever*. For example, if a tenant claims to have been exposed to asbestos because of your abatement activity and then brings a lawsuit, you must have documentation showing that you contracted to have the work performed in a "state-of-the-art" manner. If, as recommended, you have hired a qualified monitoring firm and a reputable contracting firm, liability will be focused on the contractor and the monitoring firm— *not* on you. (NEXT SLIDE)

> ## It's *your* asbestos!

Yes, it is your asbestos!

As owner, you legally own the asbestos forever.

A tenant claims to have been exposed to asbestos because of abatement, brings lawsuit?

Have documentation showing you contracted to have work performed in "state-of-the-art" manner.

Liability will be on contractor & monitoring fine, not you

■ **Model 14–1** ■ continued

Which brings me to the last reason for concerning yourself with any potential asbestos problem: *peace of mind*. If you examine and then effectively deal with any asbestos contamination that exists in your buildings, you will sleep better at night. (NEXT SLIDE)

Give yourself peace of mind for the future.

Last reason—Peace of mind

If you effectively deal w/asbestos contamination— you will sleep better at night

For one thing, you will have done your level best to preserve the health of your tenants. For another, as previously noted, you will have shifted any potential liability from you to the professionals you hired to solve the problem—assuming you hired professionals. Your monitoring firm will have continuously documented the contractor's operations and will have provided you with reports to keep in your files, in the event of later questions by lawyers or regulatory agencies. (NEXT SLIDE)

You benefit from safe asbestos removal because
- You have preserved the health of your tenants.
- You are no longer legally liable.
- You have helped your community.
- You have helped the environment.

Assuming you have hired professionals ....

Your monitoring firm will have continuously document contractor's operations; provided reports to keet in your files

In just these few minutes, I have given only highlights about asbestos. It poses a considerable challenge for all of us who own buildings or work in the abatement business. Yet the current diagnostic and cleanup methods are sophisticated enough to suggest that this problem, over time, *will* be solved. Now I would be glad to answer questions.

Questions?

These are highlights about asbestos.

Poses challenge for all who own buildings or work in abatement business.

Current diagnostic & cleanup methods sophisticated enough to suggest this problem will be solved over time.

Questions???

■ **Model 14–1** ■ continued

Chapter | **15** | Technical
Research

>>> Chapter Outline

Getting Started   496

Searching Online Catalogs   498

Author or Title Search   499

Subject Search   499

Keyword Search   499

Advanced Search Techniques   500

Searching in the Library   505

Library Services   505

Library Resources   506

Searching the Web   514

Fundamentals of Web Searching   514

Web Search Options   516

Using Questionnaires and Interviews   520

Questionnaires   520

Interviews   525

Using Borrowed Information Correctly   527

Avoiding Plagiarism   527

Following the Research Process   528

Selecting and Following a Documentation System   532

Writing Research Abstracts   538

Types of Abstracts   538

Guidelines for Writing Research Abstracts   539

Chapter Summary   541

Learning Portfolio   542

Communication Challenge   542

Collaboration at Work   543

Assignments   543

Model for Good Writing   546

Model 15–1: Memo report citing research—APA style   546

Research does not end with the last college term paper. In fact, your career often will require you to gather technical information from libraries, interviews, questionnaires, the Internet, and other sources. Such on-the-job research produces documents as diverse as reports, proposals, conference presentations, published papers, newspaper articles, websites, or essays in company magazines. Your professional reputation may depend on your ability to locate information, evaluate it, and use it effectively in everyday research tasks.

This chapter takes you through the research process used in college courses and practiced on the job. Specifically, the chapter has eight main sections:

1. Getting started
2. Searching online catalogs
3. Searching in the library
4. Searching the Web
5. Using questionnaires and interviews
6. Using borrowed information correctly
7. Selecting and following a documentation system
8. Writing research abstracts

A common thread throughout the chapter is an M-Global case study. Tanya Grant, who works in marketing at M-Global's Atlanta office, is searching for information to complete a project for Jim McDuff, the president of the firm. We'll observe Tanya as she gathers research material.

Like Tanya, later in your career you will have to apply the research process to a specific technical-writing task in college or on the job. It is one thing to read about doing research; it is quite another to dive into your project and work directly with the books, periodicals, electronic databases, websites, and other resources in the library and on the Internet. You should seek firsthand research experience as soon as you can.

Finally, remember that the best research writing smoothly merges the writer's ideas with supporting data. Such writing should (1) impress the reader with its clarity and simplicity and (2) avoid sounding like a strung-together series of quotations. These two goals present a challenge in research writing.

>>> Getting Started

In chapter 1, you learned about the three phases of any writing project: planning, drafting, and revising. Research can occur in the planning stage, right before you complete an outline, but it can also occur again and again throughout the project. Before starting your research, ask yourself questions like the following to give direction for your work:

■ What questions must be answered during the research phase?

■ What print, multimedia, and electronic sources are most useful?

■ What is the nature and extent of information that is needed? Should it be scholarly or popular? Current or historical?

■ What are the best strategies and research tools for locating information?

■ To whom can you talk who might have expertise on the topic?

■ What are the best criteria for critically evaluating information for reliability, validity, accuracy, timeliness, or point of view or bias?

■ What format must be used to document material borrowed from sources, and what copyright permissions must be acquired to use the information?

Let's take a look at the way Tanya Grant answers these questions during her project.

Tanya has just been given an important task. The company president, Jim McDuff, wants her to examine the feasibility of the company switching to hybrid electric-powered cars in their American offices. McDuff believes the company's success in using these vehicles in their Asian offices is so significant that the company should consider using them in the U.S. offices. Success with hybrid electric cars could help address serious air-quality issues and offset escalating petroleum prices. Also, Jim hopes his firm becomes a major player in refining the technology and fostering its use. In short, this move could serve as a public relations effort, a budgetary control, and a marketing tool for M-Global products.

Jim told Tanya she should write a report investigating the success and failures of hybrid electric cars. At an upcoming meeting, upper-level management will review the report. She should study the impending tax and other legislation affecting fuel-efficient vehicles. In addition, she should investigate potential of hybrid vehicles. Some of her report will look at start-up costs for switching the vehicle fleet, employee needs, and other in-house matters. The following outline shows how Tanya would probably answer the questions previously noted as she begins her research:

1. Main question: Should M-Global switch to hybrid electric vehicles?

2. Main types of information needed:

   - What are the advantages and disadvantages of this technology?

   - What is the research saying about the outlook of hybrid electric vehicles?

   - What tax or other legislation is pending at both state and federal levels?

   - Who are the current and potential consumers, and what do they think about hybrid electric cars?

   - What are the real costs in maintaining a fleet?

   - What could M-Global gain by the switch? What would they lose?

   - What impact would such a switch have on the competition, potential clients, or employees?

3. Possible sources:

   - Memos, reports, and other M-Global documents related to the use of hybrid cars in the organization's Asian offices

   - Directories (of periodicals, newsletters, newspapers, electronic journals, organizations)

   - Journal and newspaper articles found in indexes, abstracts, and electronic databases

   - Bibliographies and literature reviews

   - Government documents

   - Books

   - Websites

   - Questionnaires

   - Interviews

4. Format for documentation: Tanya submits a short report to Jim McDuff, documenting her research using the *Publication Manual of the American Psychological Association's* (APA) system for citing borrowed information (the same format used by M-Global engineers and scientists in their research reports).

With this basic plan in mind, Tanya can begin her work. Her first decision is where to start her research. Choices include the following:

- The public library
- Her corporate library
- The university library
- The World Wide Web

She eliminates the public library as too general, and the M-Global corporate library contains only a copy of the report about the Asian offices' use of hybrid vehicles. Jim McDuff has already given her a copy of this report. The remaining options are the web and the local university library. Tanya must use both to do a thorough job of research. She knows about the university's extensive print collections in science and technology, their web-based online catalog, and their extensive collection of electronic databases. In addition, the staff in the reference department and interlibrary loan office will help her identify and track down sources. She also uses the web regularly for finding business information, news, entertainment, and discussion on just about anything. The two, the web and the library, work amazingly well together—they intersect, overlap, and complement one another, and each contributes unique sources, provided you know the basic research techniques and tools. A good researcher uses both the web and the library to take advantage of the strengths and overcome the weaknesses of both.

A planning trip to the library can save you hours of time searching the web in unfamiliar subject areas. However, a few hours' preparation online—for example, searching a library's web-based catalog—can make your trip to the library more productive.

Tanya schedules two days this week and one next week to work in the library and begins her research from her computer in the office. The next sections introduce some of the strategies, tools, and basic concepts you need for searching library online catalogs, searching in the library, and searching on the web.

>>> Searching Online Catalogs

Both the library and the Internet can seem like intimidating places when you first start a project. Once you learn a few basics, however, you will become comfortable and even confident about using these resources. This section includes information on using online library catalogs to locate books, journals, and other resources. We look at the basics as well as more advanced techniques you need for searching not only library catalogs, but also most web search engines and electronic databases.

Books and other printed sources provide well-supported and tested information about a topic, but, by definition, the information is often dated. Even a book just published has information that is one to two years old, given the time it takes to put a book-length manuscript into print. Keep this limitation in mind as you search.

The library's catalog is a road map to its collection of books, periodicals, and other material; it is an alphabetical list by author, title, and subject. These days, the majority of college and university libraries offer sophisticated online catalogs that can be searched at the library or remotely from home or office. The online catalog often has additional features such as keyword and Boolean searching and information about whether a book is available or checked out.

The current trend in university and college libraries is to provide access to the library's catalog through the World Wide Web. Now you can search not only your library's catalog for the best books and other sources, but also the catalogs of many other libraries around the world. Your library's website often provides additional helpful information about services and hours, electronic databases, access to other library catalogs, and links to important websites. One of the easiest ways to locate the address for your library's website is to telephone the library staff and ask.

The rules for searching online catalogs vary depending on the computer program used by the library. The online catalog's Help screen is the best guide to search techniques. Following are some general strategies for effective searching.

Author or Title Search

If you know specific authors or titles of potentially useful books, conduct an author or title search to locate the call numbers. Study the catalog entry, especially the subject heading; similar books can be found if you search by subject using these terms. Often online catalogs feature automatic links to these terms.

Subject Search

Once you know the subject headings assigned to books or resources on your topic, searching by subject can be very efficient. Libraries select the subject headings from *The Library of Congress Subject Headings*. Unlike the web, the subject terms in a library's catalog are controlled and very specific in order to bring all the material together. Knowing exactly which subject words to use can be a matter of trial and error, but once found, these headings can serve as powerful tools to gather information on your topic.

Keyword Search

This strategy is probably your best choice, because it allows you to scan through all the fields in a book's library record—author field, title, subject headings, dates, notes, publishers, etc.—to locate books that match your request. Searching by keyword permits use of natural language instead of rigid subject headings and is more like a web search. Pay close attention to the catalog's rules for keyword searches; you can often improve your results by limiting searches to particular fields. Figure 15–1 shows a typical keyword search.

> **TIP:** One of the easiest ways to find material on your topic is to switch between keyword and subject searching. For example, begin with a keyword search, selecting the books that match your request and the subject headings used to describe those books. Then do a subject search.

Advanced Search Techniques

A library catalog may also include advance strategies such as Boolean searching, positional operators, and truncation. Many times you may not be aware that you are using these tools because they are built into the catalog's search functions. However, learning to use these techniques is important, because they are used in library catalogs, most periodical databases, and web search engines. Look for the Help screens in your catalog that describe the advanced search options and practice using them whenever possible; they can save you

Georgia Tech Library Catalog (GTEC) - Basic Search Page
For database status and news, see GTEC News

Need help? Check our Search Tips page. Please logout when you are done. [Logout]

For more options, go to our Advanced Search page.

Search Tip: Do not use stop words in a search. Stop words are A, AN, BY, FOR, IN, OF, ON, THE, TO.
To search a title such as *"IBM Journal of Research and Development"* enter "ibm journal research development."

| hybrid electric cars | Search Everything ⇕ |

[Start Search] [Clear Form]

No. of Records to display per page: [20 ⇕]

View Search History: [⇕]

| about us | architecture library | ask a librarian | contact us | site map | help |

■ **Figure 15–1** ■ Results of a typical keyword search
Source: GIL Across Georgia Union Catalog. Used by permission.

Georgia Tech Library Catalog (GTEC)
Search Results for:

Please logout when you are done. [Logout]

S1:
6 Record(s) found. (This page: 1 ~ 6)

[search] [prev. group] [next group]

Results are sorted by Year, newest-oldest. *Want to change sort order?* [Year ▼] [Descending ▼] Sort by in order [Re-Sort]	[Format for printing]

[Select marked records]

1 ☐	**Title:**	Forward drive : the race to build "clean" cars for the future / Jim Motavalli.
	Author:	Motavalli, Jim.
	Call Number:	TL221.15 .M68 2000.
	Year:	2000.
2 ☐	**Title:**	Electric vehicles : socio-economic prospects and technological challenges / edited by Robin Cowan, Staffan Hulten.
	Author:	
	Call Number:	TL220 .E4485X 2000.
	Year:	2000.
3 ☐	**Title:**	History of the electric automobile : hybrid electric vehicles / Ernest Henry Wakefield.
	Author:	Wakefield, Ernest Henry, 1915-.
	Call Number:	TL220 .W343 1998.
	Year:	1998.
4 ☐	**Title:**	Evaluation of a Toyota prius hybrid system (THS) [[microform] /] Karl H. Hellman, Maria R. Peralta, Gregory K. Piotrowski.
	Author:	Hellman, Karl H.
	Call Number:	EP6.2:P93.
	Year:	1998.
5 ☐	**Title:**	The household market for **electric** vehicles : testing the **hybrid** household hypothesis-- reflexively designed survey of new-**car**-buying, multip
	Author:	Turrentine, Thomas.
	Call Number:	TL220 .T88X 1995.
	Year:	1995.
6 ☐	**Title:**	**Electric and hybrid** vehicles : selected papers through 1980 / prepared under the auspices of the **Electric** Vehicle Committee, Passenger **Car** Act
	Author:	
	Call Number:	TL220 .E37 1981.
	Year:	1981.

■ **Figure 15–1** ■ continued

Georgia Tech Library Catalog (GTEC)

Please logout when you are done.   [**Logout**]

Search Results for:

S2:
Record 3 out of 6

| [search] | [result list] | prev. group | next group | [first record] | [last record] |

| [prev. record] | [next record] |

[**Request Document Delivery**]

Call Number: **TL220 .W343 1998.**
Main Author: <u>Wakefield, Ernest Henry, 1915-.</u>
Main Title: **History of the electric automobile : hybrid electric vehicles / Ernest Henry Wakefield.**
Pub Info: Warrendale, Pa. : Society of Automotive Engineers, c1998.
Physical Desc: xxii, 332 p. : ill. ; 26 cm.
Notes: Includes bibliographical references and index.
Subject: <u>Automobiles, Electric--History.</u>
Subject: <u>Hybrid electric cars--History.</u>
Record Type: MON.
Language: ENG.
ISSN/ISBN: 0768001250.
Voyager Number: 461227.
OCLC: 38566039.
LCCN: 8003420.

Location and availability:

1) TL220 .W343 1998
Location 1: Currently checked out--to recall, ask at Circulation desk, 1st floor West

---

| [search] | [result list] | prev. group | next group | [first record] | [last record] |

| [prev. record] | [next record] |

[Logout] Please logout when you are done.

■ **Figure 15–1** ■ continued

time and produce excellent results. Following is a brief description of some of the most common search options:

■ A **Boolean search** outlines the relationship of words and phrases using simple AND, OR, NOT statements (Figure 15–2).

■ **Positional operators** stipulate the relative location of each term within the record. For example, you can often specify that terms must be adjacent or within a certain number of words.

BOOLEAN SEARCHING

AND: Example: periodicals AND directories.
        Locates only those records where both terms are present.
        Use this to narrow your search and reduce the number of matches.

OR   Example: periodicals OR journals OR zines.
        Locates records in which any one of these terms can appear.
        Use this to broaden or enlarge your search.

NOT: Example: periodicals NOT magazines.
        Eliminates records containing the excluded term.
        Use this sparingly to narrow your search.

■ **Truncation** allows for variant spelling or plurals. For example, in some catalogs entering *wom*n* retrieves records with either the words *woman* or *women*.

Searching other library catalogs can sometimes be as simple as selecting a link from your library's website to a library consortium or union catalog of university and college libraries within your region. Searching libraries close to home has the advantage of easier access to their collection, whether you visit in person or gain access through your library's interlibrary loan service. If you want to see "what's out there" in larger or more specialized libraries, try searching for library catalogs on the web or ask the reference librarian if your library provides access to OCLC's *Worldcat*. Two web directories of library catalogs that have been around for some time are *Libcat: A Guide to Library Resources on the Internet—Libraries in the United States* (http://www.librarysites.info) and *Libweb: Library Servers via WWW* (http://sunsite.berkeley.edu/libweb). (Keep in mind, however, that the nature of the web is such that they may have disappeared since this book was written.)

> **TIP:** Cite your sources as you go. Keep close track of what you find and where you find it so that you don't waste your time searching for books on the shelves of your library when they are actually located elsewhere. Consult the reference department of your library to learn about your library's interlibrary loan service or borrowing privileges at other libraries.

When you use the web for searching catalogs, remember to evaluate what you find as you search. Begin thinking critically as soon as you start, and work to keep this perspective throughout your research. For example, when looking at a reference to a book, a journal, an article, or a website, ask yourself the following questions:

■ What are the author's academic qualifications?

■ Who is the publisher, and what is its reputation?

■ What are the scope and content of the work?

■ How does this information fit in with what you know about this topic?

■ What are the trends in information on this topic, and how does this book, article, journal, or website fit in?

■ How current is this information?

Finally, use your library's online catalog to find out what electronic databases are available to you. You may be able to search important research databases with access to periodicals, newspapers, encyclopedias, dictionaries, directories, statistics, and other reference sources (Figure 15–3). Many of these databases can be searched remotely from your home or office, but some are restricted to in-library searching only. Policies governing who can search, from where, passwords, and whether searching is fee-based or free vary widely depending on the contracts between the library and the database vendor. The next section of this chapter covers electronic databases in more depth; for now, keep in mind that the quality of information you retrieve from research databases is usually superior to the material you may locate searching the vast World Wide Web. In addition, online catalogs may provide links to recommended high-quality websites you might not otherwise locate. Explore your online options and discuss your needs with the reference staff at your library.

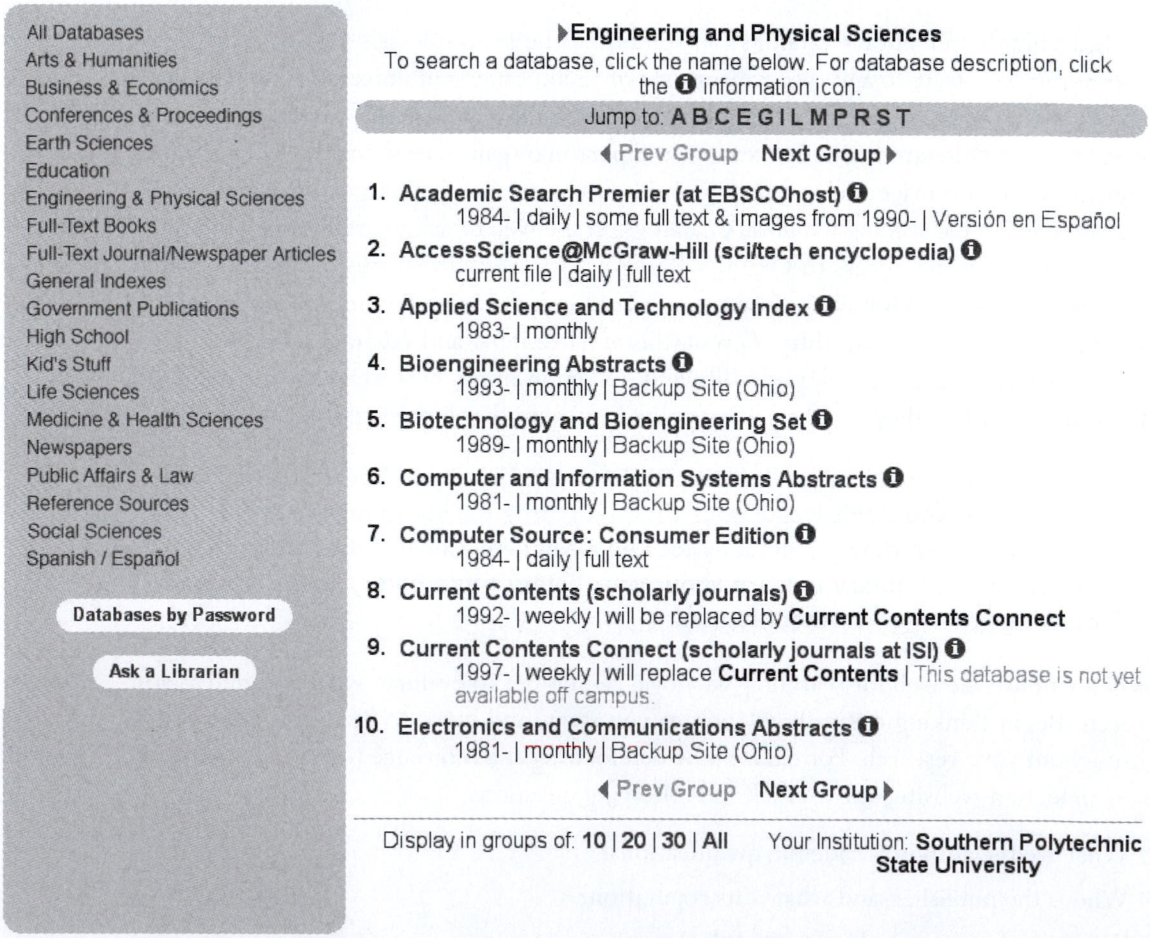

All Databases
Arts & Humanities
Business & Economics
Conferences & Proceedings
Earth Sciences
Education
Engineering & Physical Sciences
Full-Text Books
Full-Text Journal/Newspaper Articles
General Indexes
Government Publications
High School
Kid's Stuff
Life Sciences
Medicine & Health Sciences
Newspapers
Public Affairs & Law
Reference Sources
Social Sciences
Spanish / Español

Databases by Password

Ask a Librarian

▶**Engineering and Physical Sciences**
To search a database, click the name below. For database description, click the ❶ information icon.
Jump to: **A B C E G I L M P R S T**
◀ Prev Group   Next Group ▶

1. **Academic Search Premier (at EBSCOhost)** ❶
   1984- | daily | some full text & images from 1990- | Versión en Español
2. **AccessScience@McGraw-Hill (sci/tech encyclopedia)** ❶
   current file | daily | full text
3. **Applied Science and Technology Index** ❶
   1983- | monthly
4. **Bioengineering Abstracts** ❶
   1993- | monthly | Backup Site (Ohio)
5. **Biotechnology and Bioengineering Set** ❶
   1989- | monthly | Backup Site (Ohio)
6. **Computer and Information Systems Abstracts** ❶
   1981- | monthly | Backup Site (Ohio)
7. **Computer Source: Consumer Edition** ❶
   1984- | daily | full text
8. **Current Contents (scholarly journals)** ❶
   1992- | weekly | will be replaced by **Current Contents Connect**
9. **Current Contents Connect (scholarly journals at ISI)** ❶
   1997- | weekly | will replace **Current Contents** | This database is not yet available off campus.
10. **Electronics and Communications Abstracts** ❶
    1981- | monthly | Backup Site (Ohio)

◀ Prev Group   Next Group ▶

Display in groups of: **10** | **20** | **30** | **All**   Your Institution: **Southern Polytechnic State University**

■ **Figure 15–3** ■ Research databases available on one library's website

Source: Galileo—Georgia Library Learning Online. Courtesy of the Board of Regents of the University System of Georgia.

Tanya Grant's M-Global Project

Tanya Grant, for example, rightly thinks that books will not be her main source of information about hybrid electric cars, because the topic has developed relatively recently, but she at least wants to see what range of sources the catalog offers. Tanya does a keyword search and determines that the correct subject heading is "hybrid electric vehicles." Her own library's holdings are somewhat limited, but one item is worth reviewing. She decides to search the online catalog from another local university with an automotive engineering school, where she finds a better selection of books, and because her library card gives her borrowing privileges at all state university system libraries, she decides to take a trip to this library. Her final search is in OCLC's Webcat, a comprehensive database that leads her to a few more noteworthy titles that she will borrow through her library interlibrary loan department.

>>> Searching in the Library

At some point during your search for secondary resources, you must visit your academic library. The library's services and collections of books, journals, electronic databases, microforms, and reference materials, although complex, supports your research and helps you locate information. Fortunately, academic and research libraries are organized along similar principles, and the skills you gain from using one library can generally be transferred to other libraries. This section highlights some of the services and resources you can expect to find as you conduct your research in the library.

Library Services

The services that your library can provide during the research process are many. The three most important ones are (1) reference and information, (2) interlibrary loan, and (3) circulation.

Reference and Information

Discussing your research topic with a reference librarian can be a very productive first step. The librarian can recommend reference books; provide instruction on how to search indexes, abstracts, and electronic databases; and guide you to collections such as government documents, microform sets, and noteworthy websites that you may not have found on your own. Make the most of your time with the reference staff by asking specific questions and returning whenever you need more help in locating or using sources.

Interlibrary Loan

In the course of research, you will identify excellent sources not owned by your library. Fortunately, in most libraries you can arrange to borrow or photocopy these sources through the interlibrary loan service. Be aware that the lending process can take a number of weeks and may require payment for the service. Interlibrary loan and other document delivery services have become increasingly popular as electronic databases have expanded.

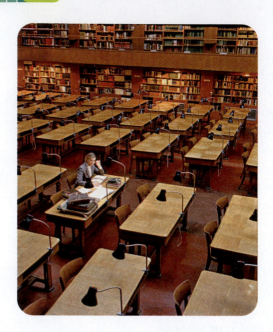

Circulation

The circulation department is responsible for the lending and returning of library materials. In addition, the department may register library users, provide access to reserve materials, recall checked-out books, renew materials to extend the loan, and search for material not on the shelf. Circulation services are generally highly automated with a growing trend toward self-service checkout and account maintenance.

Library Resources

This section includes information on the following resources: books; periodicals; newspapers; company directories; and dictionaries, encyclopedias, and other general references.

>> Resource 1: Books

As previously discussed, the library catalogs these days are generally automated, and very few traditional wooden card catalogs remain. Once you locate the exact book for which you are searching, browse through the books located beside this title. You will likely find other useful and related material. Ask for assistance at the reference or circulation desk if you cannot locate the books on your topic.

>> Resource 2: Periodicals

Periodicals are publications that are issued on a regular basis, usually weekly, monthly, or quarterly. The term encompasses

- Popular magazines that take commercial advertising, such as *Time, Science,* and *National Geographic*
- Professional and scholarly journals such as *IEEE Transactions on Professional Communication*

Most library visitors are familiar with the section that houses current periodicals, either in alphabetical order or by call number. However, they are less familiar with the part of the library containing back issues. Libraries keep back issues of the periodicals considered most important to its users. They may be in the form of bound volumes, microfilm or microfiche, or full-text electronic versions of periodicals.

Your key tool to locating information within periodicals is an electronic database, a periodical index, or an abstract. By looking up your subject in the index, you can find articles that provide the information you need. Some databases like *Academic Search Premier* include popular periodicals. Others, like the *Engineering Index,* deal with a broad range of technical information. Still others, like *Mechanical Engineering Abstracts,* focus on periodicals, books, websites, and papers in specialized technical fields. The periodicals covered in the index or abstract are listed in the volumes or in the online information screen, along with the inclusive dates of the issues indexed.

Sometimes an abstract or a summary of the article is all that is available to you, providing a brief description of articles so that you can decide whether the entire article is worth finding. Abstracts are especially useful when the article being summarized is not available in your library. The abstract can help you decide whether to (1) visit another library, (2) order an article through the interlibrary loan service, or (3) disregard an article altogether.

Most periodicals are available as electronic databases that can be searched quickly and thoroughly once you know the basics. Libraries typically purchase subscriptions to electronic databases and link to the website from the library's website or online catalog.

Increasingly, electronic databases provide full-text copies of the periodical articles. Some libraries permit you to search these databases from your home or office, whereas other libraries, because of license requirements with the database vendors, permit searching within the library only. Still other libraries provide professional search services where, for a fee, the research staff conducts the search for you.

The rules for searching electronic databases vary widely. Each database has unique features and searching requirements. You must invest time and energy to learn these rules to take full advantage of the information the database offers. Start your search by reading the Help screens and the instructional materials about the database or any support materials that the library provides. You will save yourself time and improve your search results if you understand the basic search strategies and have a grasp of the scope of the database. At a minimum, make sure that you know the rules for printing, emailing, or saving to disc the results of your search before you get too far into your research.

Most of the electronic databases have search strategies similar to what you may have encountered when searching the online catalog for books, and likely include subject, keyword searching, advanced search techniques using Boolean and positional operators, truncation options, and language- and date-limiting options. Also common are options to limit searches to scholarly or peer-reviewed journals only, or to full-text journals only. The more you practice, the better your searching and the more precise your results.

TIP: Emailing results from a search in an electronic database is an efficient and accurate way to collect the information you need to document your research and build your works-cited page.

Once you have evaluated and narrowed the list of articles on your topic that you wish to read, your next step is to determine which are owned by the library, which must be requested through interlibrary loan, and which are owned by other nearby libraries. Libraries are often able to flag their holdings in the online database. Also, database vendors sometimes provide online full-text copies of articles that can be printed immediately or emailed to your home or office. Keep in mind, however, that many excellent periodical articles are not available electronically. Try to avoid the trap of arbitrarily limiting your research only to those periodicals with readily available full text. Be aware that the process of getting copies of the article can be tedious and time-consuming; be sure to reserve sufficient time for this important step.

There are hundreds of electronic databases and print indexes or abstracts available. Many libraries provide guides to these resources. Ask the reference staff to help you locate the most appropriate ones for your topic. Following is a list of a few of the well-known titles available in print or electronically:

- *Academic Search Premier*
- *Applied Science and Technology Abstracts* (print title: *Applied Science and Technology Index*)
- *ABI/Inform Complete at ProQuest*
- *BIOSIS: Biological Abstracts*
- *CSA: Cambridge Scientific Abstracts*
- *CAS: Chemical Abstracts*
- *Computer Abstracts International Database*
- *Current Contents*
- *EI: Engineering Information*
- *General Science Abstracts* (print index: *General Science Index*)
- *GPO Monthly Catalog* (index to government documents)
- *INSPEC*
- *LEXIS-NEXIS Academic Universe*
- *Psychological Abstracts* (electronic title: *PsycINFO*)
- *Science Citation Index, Social Science Citation Index, Arts & Humanities Citation Index* (online through the web of Science)

Tanya Grant's M-Global Project

Tanya decided to consult a few of the electronic databases recommended by the reference librarian.

1. She conducted a search using *Environmental Engineering Abstracts,* which the library subscribed to electronically through Cambridge Scientific Abstracts. The scope and content of the abstract was exactly what she wanted because it targeted the technological and engineering aspect of hybrid electric vehicles. Because the database was new to her, she spent time learning how to conduct a search and save her results. She limited her search to articles from the last few years. The search not only retrieved useful articles, but also provided links to six high-quality websites. Scanning the results, she selected the most promising articles and website and emailed a copy to herself and printed a copy of the list to use for locating the periodicals in the library. Figure 15–4 shows Tanya's primary search.

2. Next she consulted *ABI/Inform Global,* an online database that covers business and management trade journals produced by ProQuest. She was interested in looking at business viewpoint on hybrid vehicles. Her search produced 72 items published since 1999, many of which had full-text copies of the article available for her to read immediately. After sampling a few articles, she flagged those she wanted and emailed them to herself. Tanya decided to redo her search and narrow it to peer-reviewed articles only. The nine articles she retrieved in her

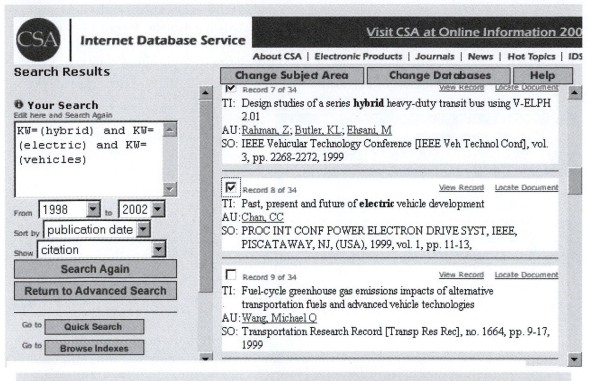

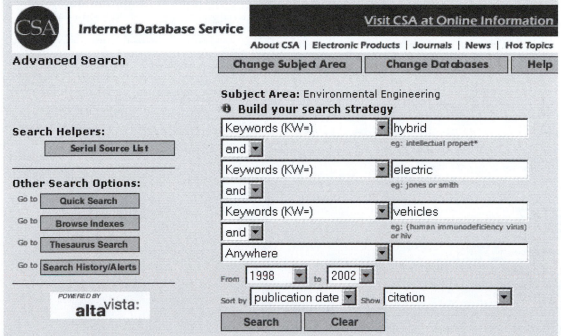

■ **Figure 15–4** ■ Results of a search conducted in Cambridge Scientific Abstracts (CSA) through a library's website

Source: Cambridge Scientific Abstracts. Used by permission.

second search have undergone review and evaluation by experts in the field prior to publishing. These articles will be particularly noteworthy. Figure 15–5 shows Tanya's peer-reviewed search.

3. Finally, Tanya consulted *Academic Search Premier,* a comprehensive, general-purpose database. It covers almost 4,000 periodicals, 2,300 of which are scholarly.

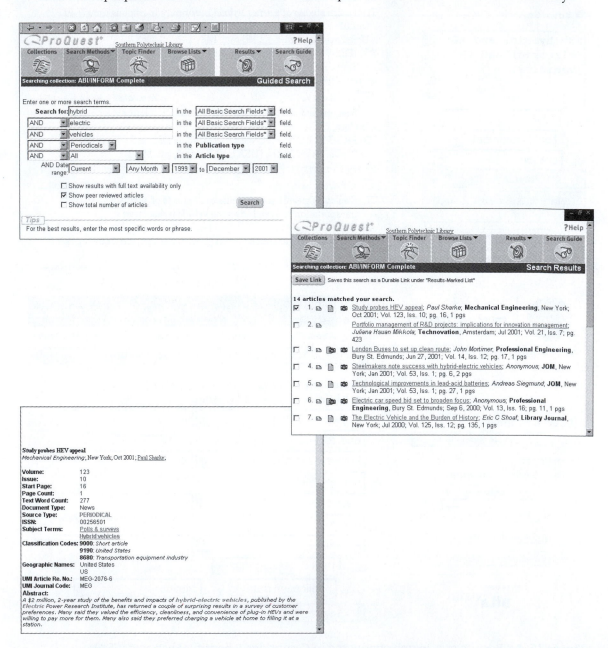

■ **Figure 15–5** ■ Results of a search conducted in *ABI/Inform* database through a library's website

Again, she was able to narrow her search to peer-reviewed articles and located some very current and useful information. One of the full-text articles referred to an organization she wanted to investigate further, the Partnership for a New Generation of Vehicles.

>> Resource 3: Newspapers

If your research topic demands the most current information, newspapers provide an excellent source. One disadvantage is that newspaper information has not "stood the test of time" to the same extent as information in journals and books. Despite this drawback, newspaper articles can give you insight, facts, and opinion on many contemporary issues. Two particularly noteworthy newspapers are *The New York Times* and *The Wall Street Journal*. These well-respected newspapers have a long tradition of high-quality journalism. Both titles are thoroughly indexed, and many libraries either provide access through an electronic database or keep print indexes and back issues in microfilm or microfiche.

Many other regional, national, and international newspapers have established elaborate websites in which you can frequently locate the archives or find additional information not available in the print version. Two significant electronic databases for newspapers include *Lexis-Nexis' Academic Universe,* a full-text index to some 5,000 publications including newspapers, wire services, legal news, and government publications and *ProQuest Newspapers,* an index to five major newspapers. Check in your library's online catalog to see if it provides additional links to some of the web-based news services.

Tanya Grant's M-Global Project

Tanya decided to see what kind of newspaper coverage hybrid electric cars were receiving and try to uncover some of the tax legislation being proposed by each state. Her first search in *ProQuest Newspapers* located 327 newspaper articles—some written just the previous week. Reading through a few full-text articles convinced her that this would be her best approach to get the consumer perspective she needed. With her second search, Tanya added the concept "tax" and uncovered 64 articles from major newspapers from around the country describing various tax legislation efforts under way. These articles served as a starting point for studying the complex tax legislation being proposed by various state legislatures. Figure 15–6 shows Tanya's search results.

>> Resource 4: Company Directories

Often your research needs may require that you find detailed information about specific firms. For example, you could be completing research about a company that may hire you, or you may seek information about companies that compete with your own. Today, you can find many databases of company information on line. In addition, most companies now produce sophisticated websites about their services and products. Although not without bias, these can be an excellent source of information. The following is a small sample of some useful directories that are available; ask the reference

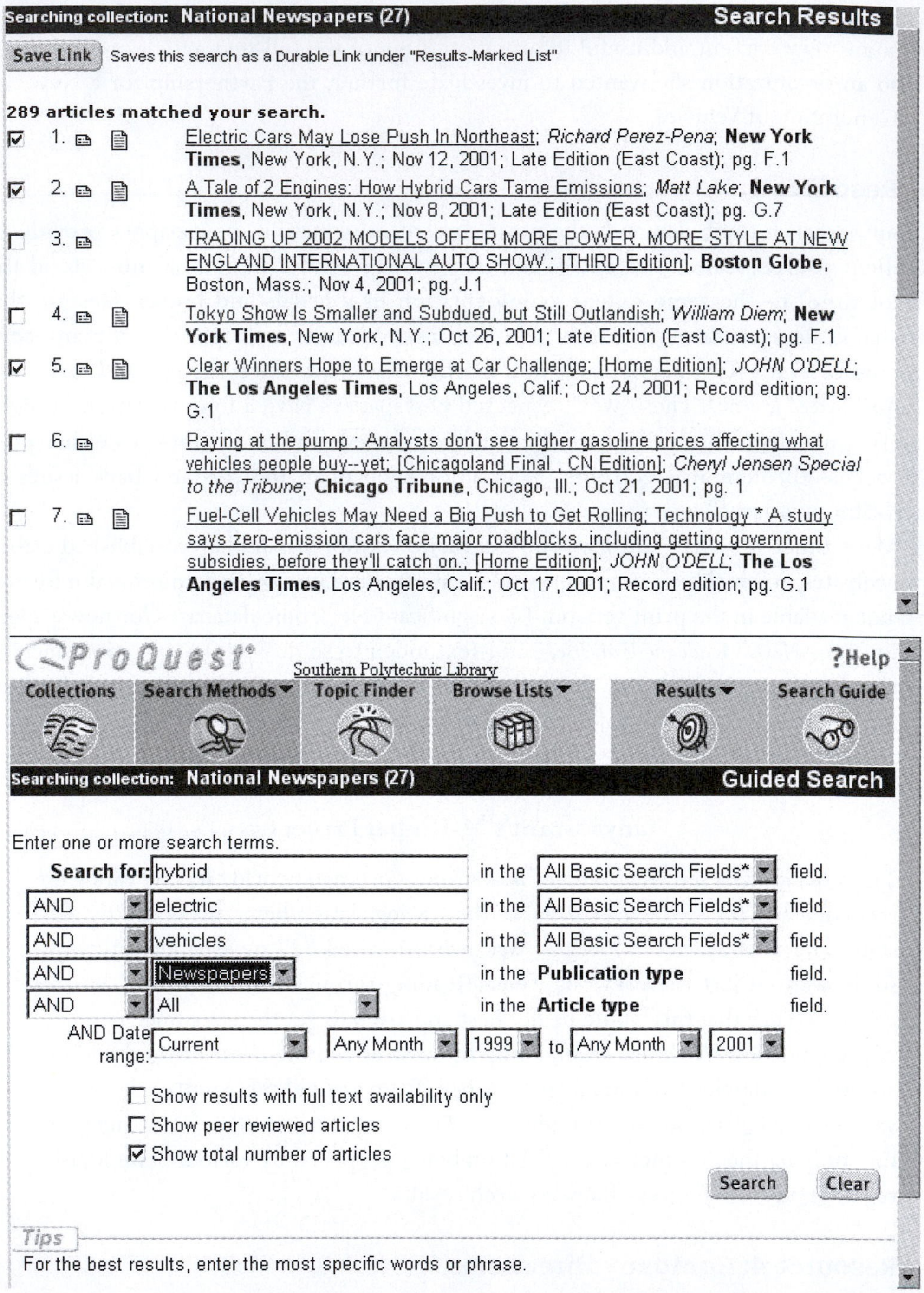

■ **Figure 15–6** ■ Results of a search conducted in the ProQuest Newspapers through a library's website

Source: Image published with permission of ProQuest Information and Learning Company. Further reproduction is prohibited without permission. Image produced by ProQuest Information and Learning Company. Inquiries may be made to: ProQuest Information and Learning Company, 300 North Zeeb Road, Ann Arbor, MI 48106–1346 USA. Telephone (734) 761–7400; E-mail: info@il.proquest.com; webpage: www.il.proquest.com.

librarian to recommend others and for assistance in using the online versions of these and other directories.

Compact D/SEC

Corp Tech Directory of Technology Companies

D & B Million Dollar Directory

Mergent Online

Standard & Poor's Register of Corporations, Directors, and Executives

Ward's Business Directory of U.S. Private and Public Companies

Who's Who in Science and Engineering

>> Resource 5: Dictionaries, Encyclopedias, and Other General References

Sometimes you may need some general information to help you get started on a research project. In this case, you may wish to consult specialized dictionaries, handbooks, or encyclopedias. Most general encyclopedias are available in some electronic format, generally as web-based products, such as the *Encyclopedia Britannica* online. There are, however, advantages with using a specialized subject-based encyclopedia or dictionary over a general one in that the articles target a more scholarly audience, assume greater subject expertise, and reference more scholarly materials in their bibliographies. Following is a list of a few specialized dictionaries, handbooks, and encyclopedias you may find in the reference collection:

Blackwell Encyclopedia of Management

CRC Handbook of Chemistry and Physics

Encyclopedia of Associations

Encyclopedia of Business Information Sources

Handbook of Industrial Engineering

Handbook of Technology and Operations

International Business Information

McGraw-Hill Encyclopedia of Science and Technology

Van Nostrand's Scientific Encyclopedia

Tanya Grant's M-Global Project

At this point, Tanya had spent many hours examining the library's online catalog, searching in online periodicals' indexes and abstracts and in newspaper indexes. She located books and articles in scholarly and technical periodicals as well as articles in popular magazines and newspapers. She requested a few promising items not locally owned through interlibrary loans. In the meantime, she has plenty to read and begin creating notes. She has a couple of leads to reliable websites from the library's online catalog and an organization she wants to research. She has a good start, and plenty of work ahead.

>>> Searching the Web

Throughout this chapter you have seen references to the World Wide Web. This vast global computer network has changed the way we communicate, market services and products, and collect and distribute information. The web is the largest and fastest growing portion of the Internet, with its appealing graphic interface that incorporates text, images, and sound, and its ability to move from one web page to another through hyperlinks. We next highlight some of the terms and concepts, challenges, and strategies associated with using the web as a research tool and information source.

Fundamentals of Web Searching

Mining the web for useful resources is always challenging and frequently frustrating, but it can yield terrific results. Why is searching such a challenge?

- The web is huge; it contains tens of millions of documents and is growing at an astounding rate.
- The web is constantly changing—sites are added, altered, moved, and disappear without warning.
- Search engines and subject directories don't work very well—they retrieve too much, they don't cover the entire web, the relevancy ranking defies logic, and no two search engines work alike.
- The content of the web is unregulated; anyone can add anything—fact, fiction, or fiction that looks like fact.
- There is no central index to the web and few rules for describing web pages.
- The process of searching, sifting through results, downloading pages, and evaluating each web page critically is time-consuming.
- The web is full of distractions that make it difficult to stay focused.

Despite these challenges, the web offers access to extraordinary resources that often have no print counterpart. Because of the web's sheer size, a search usually finds something on any topic—possibly something of value or perhaps something useless. Some studies have estimated that scholarly sites represent only 10%–20% of the web, but this number is still significant. Most people agree that the web's strength lies in its information on current events, business and industry, popular culture, the government, computing, and technology, but all disciplines are represented in some way. Some resources you can find on the web are

1. Directories of people, business, and organizations
2. Advertising, marketing materials, and product catalogs
3. Government documents
4. Periodicals, newspapers, and magazines
5. Books

6. Conference proceeding and reports

7. Reference tools like guides, indexes to periodicals, dictionaries

8. An increasing number of "by subscription only" information sources

9. Sound and video clips

10. Images

>> Using Your Evaluation Skills

When you search the web, be prepared to invest time and effort in evaluating critically what you find. Unlike books and articles that undergo a rigorous editing and review process, any website can be loaded directly onto the Internet. You will encounter misinformation, grossly biased content, and poor text and graphic design. Evaluate web sources using the criteria discussed earlier in this chapter. However, because web sources do not generally follow standard publishing practices, be prepared to invest your valuable research time determining the authority, timeliness, reliability, accuracy, point-of-view, and validity of the source. Once you develop a systematic approach to evaluating sources, you will quickly recognize both the high- and low-quality web sources. Be particularly alert to the following:

- Obscured authorship: Often a web designer is credited as the author when in fact an organization or a corporation is the real source.

- Out-of-date information: The web is littered with abandoned and unmaintained websites. A high-quality website displays the date prominently.

- Subtle and obvious bias: Many websites are elaborate advertisements promoting products, services, causes, or points-of-view. Data manipulation, false arguments, and unsubstantiated opinions are common.

- Poor-quality links: Links from a high-quality website usually lead you to other valuable sites; links from a poor-quality site usually lead you to other poor-quality sites. Spending time examining the links helps you determine the quality of the site.

- Flawed style and design: Well-organized and accessible websites support the research process. Although there are many cases of good research in poorly designed sites, be aware that extracting the information from overly complex sites drains away your research time.

Spend time evaluating the source up front before you spend time reading the document. If you cannot determine the scope, authority, or date of the website, don't use it.

>> Learning the Basics

The web is made up of millions of *web pages,* each uniquely identified by an address or *Uniform Resource Locator (URL).* This *address* often contains important clues to the website's authorship, country of origin or domain, or the type of organization sponsoring the site. Figure 15–7 shows a list of common domains and examples.

Web browsers, such as Microsoft Internet Explorer and Mozilla Firefox, are software applications for viewing web documents and navigating the web. Both share similar features: a line

■ **Figure 15–7** ■
Common Internet
domain extensions

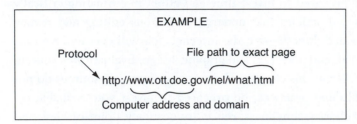

com	Commercial organization (for profit)
edu	Educational institution
gov	Government organization (non-military)
int	International non-profit organizations
mil	Military organization (US)
net	Networking organization
org	Non-profit organization
ca	Canada (country of origin)
uk	United Kingdom (country of origin)

EXAMPLE

Protocol   File path to exact page

http://www.ott.doe.gov/hel/what.html

Computer address and domain

for entering URLs; options for creating *bookmark* (or *favorite*) sites; basic navigational features for moving forward and backward, and stopping; and options for setting preferences to customize the browser.

Web Search Options

Searching the web has become second nature for many of us. More than ever, we turn to the web for basic news and information, to conduct business, and for entertainment. Invitations to "Visit our website" are everywhere, and companies have invested heavily to guarantee that their websites rank high in the results of a web search. Developing effective web research skill is critical and requires continuous updating as new techniques and search tools emerge. Options for searching include the following:

■ Searching by a specific address or URL

■ Searching by keyword in an index-type *search engine* or *meta-search engine*

■ Drilling through subject category using a *subject directory*

■ Using guides to reviewed and recommended websites

Figure 15–8 lists some of the most popular search tools. Keep in mind that web address changes or improved applications may have appeared since this list was created.

TIP: **Competition among search engines is high, and new features and applications appear regularly. To keep up with search engine development and testing, try** *Search Engine Watch* **at** *http://searchenginewatch.com*

>> Searching by URL: Uniform Resource Locator

Searching by a specific web address or URL is a very effective strategy, provided you have complete information and the web page still exists. References to URLs are regularly

■ **Figure 15–8** ■
Popular search
engines and
subject guides

AllTheWeb	http://ww/alltheweb.com	Keyword
AltaVista	http://www.altavista.com	Keyword, directory, owned by Yahoo
Ask.com	http://www.ask.com/	Keyword & subject prompting using natural language
Dogpile	http://www.dogpile.com	Meta-search engine
Excite	http:www.excite.com	Keyword & subject directory
Google	http://www.google.com/	Keyword results based on links
HotBot	http://www.hotbot.com	Keyword & subject directory
Internet Public Library	http://www.ipl.org	Subject guide plus
Librarians' Internet Index	http://lii.org	Subject guide plus
Lycos	http://www.lycos.com	Keyword & subject directory
MSN Search	http://search.msn.com	Keyword & subject
WebCrawler	http://www.webcrawler.com	Metasearch engine Keyword & subject directory
Yahoo!	http://www.yahoo.com	Keyword & subject directory

included in books, journals, television and radio broadcasts, and marketing and advertising literature. One good website can lead you to other well-written and maintained websites.

>> Searching by Keywords Using Search Engines and Meta-Search Engines

Hundreds of search engine companies on the web have created massive databases of websites and provide keyword searching. Keep in mind that these search engine companies are actually in the business of selling advertising, leasing keywords, and attracting potential customers for the companies that pay to advertise.

Their databases are usually built without human intervention using computer programs called *robots* or *spiders* that move throughout the web. No single search engine indexes the entire web, and there is fierce competition among companies for the distinction of having the largest, most current, or most useful database.

The value of a search engine from a research viewpoint depends on the relevancy ranking of the results, speed, quantity, and currency of information it retrieves. The best search engines provide simple and clear instructions that allow you to refine the results. Pay particular attention to the advanced search features.

TIP: **Master the features of one search engine before moving on to the next.**

Meta-search engines simultaneously use the databases of a number of search engines to respond to a request. The keyword search is forwarded to a variety of search engines; then the database results are collected and displayed. You can save time using a meta-search engine, particularly in narrow, well-defined topics, but you often lose the ability to refine a search using the features of the individual search engines.

There are hundreds of search engines, and a new and better one is always on the way. Second-generation search engines feature *intelligent agents* designed to help refine your question by providing suggestions and alternate lines of inquiry. Other second-generation search engines provide continual updating services using *push technology* that store your search profile, run searches, and report results automatically.

Keeping up with developments in search engines is a challenge. Search for new ones periodically or ask colleagues to recommend one. Keep trying different ones until you find a few that meet your needs.

>> Searching with Subject Directories

Searching the web with a subject directory is similar to browsing in a library; it can be very useful when you are not exactly sure what you are looking for or just want to get a feel for where sources are located. *Yahoo!* is the best-known subject directory. The distinction between a subject directory and a true keyword search engine is disappearing as companies rush to add both features to attract potential customers and advertisers.

The strength of a subject directory lies in how selective it is and how well it classifies websites into subject categories. Subject directories are generally compiled by people who review and index sites, although robots are used to retrieve potential additions. Subject directories, unlike robot-driven search engines, do not attempt to be a comprehensive index to the web. Instead, they try to capture a segment of the web that appeals to their customers and advertisers. The addition of the keyword search feature in subject directories has improved access and eliminated burrowing through multiple layers of information.

>> Using Reviewed and Recommended Subject Guides

Finding reviewed and recommended guides to websites is the most efficient route to high-quality information. One list of recommended sources compiled and maintained by an expert can be all you need to do a thorough job of research. Naturally, the reviewer's reliability and authority, the scope and criteria for what is included, and the timeliness of the list determine its value. The best guides publish a clear statement of their criteria, background information on their reviewers, and a "Last Updated" statement. Many lists have been compiled by librarians, scholars, and other subject experts associated with professional organizations or government agencies. You may discover your library's online catalog includes links to such lists. Figure 15–9 shows an example of a guide to recommended websites found in an online catalog.

Tanya Grant's Web Search

Tanya began her web search using a URL one of the M-Global engineers gave her, which led her to a website maintained by the U.S. Environmental Protection Agency. This comprehensive site helped her organize the issues, policies,

To browse recommended Web sites for freely available resources, click on a selection from the categories on the left.

View recent additions to Internet Resources.

To access databases purchased or licensed for use by GALILEO participants, click the GALILEO Home Page button below, then choose "Databases."

Search our collection of recommended Internet resources:

◉ All Terms ○ Any Term

[   ] [Search] [Search Tips]

Exploring the Internet

Internet Directories
Browse Internet sites by subject category (including one for kids).

Internet Search Engines
Search the Internet for key words or phrases.

About the Internet

Information about the Internet
User Guides, Definitions, E-mail Addresses.

How to cite Internet Resources
Guidelines for citing Internet sites & articles from GALILEO databases (using APA, MLA formats, etc.).

GALILEO Home Page   Leave a Comment

Academia & Libraries
Area Studies & Ethnic Studies
Arts & Humanities
Business & Economics
Career & Job Information
Consumer Resources
Demographics & Census Data
Education
Electronic Journals & Texts
General & Reference Sources
Government & Politics
News, Media, & Publishing
Sciences & Medicine
Social Sciences
Student Resources, K-12
Technology
Weather & Climate

Special thanks to Eric Griffith and the University of Georgia Libraries, for kindly allowing us to use their collection of web links.

■ **Figure 15–9** ■ Recommended websites found in one library's online catalog

Source: Galileo—Georgia Library Learning Online. Courtesy of the Board of Regents of the University System of Georgia.

and research trends, as well as locate articles, reports, and other information sources on the subject. Next, she followed up on a reference to an organization she found mentioned in a journal article. Using the advanced search feature in *Google*, she entered the organization's name as a phrase and located the website immediately. It was here that she located a number of useful Canadian and international documents. She spent 3–4 hours reviewing the sites and following up the links.

Tanya spent half an hour searching for a guide to recommended and reviewed websites on the topic. First she checked the library's online catalog subject guide to the World Wide Web. Although she did find a few useful guides on more general topics, there was nothing exactly on topic. She next checked *BUBL Information Service* and *Argus Clearinghouse* for prepared guides, but neither had anything on target. Finally, she tried *Ask.com* and had better results. From here she was guided to the websites for a number of government agencies, private companies, and universities conducting studies on hybrid vehicle. She came across a website for a local research center at a nearby university. She bookmarked the page and made a note to contact the center later that day.

Tanya needed current and specific information on hybrid vehicles and tax incentives. This narrow search worked well using the advanced search features of *Google*, *AltaVista*, and *Ask.com*, which allowed multiple domain limits such as *.gov*,

.edu, and *.org*. Just to double-check, Tanya tried *Dogpile*, a meta-search engine; the results were mixed and she found she was spending too much time trying to evaluate sources. Ultimately, because Tanya needed to be confident she had located the most accurate and current information, she returned to her library's online catalogs for a guide to government documents on the web and was referred to USA.gov (*www.usa.gov*). Her search here was uncluttered and produced information that she could use confidently.

Tanya's final search used the popular subject directory *Yahoo!* After a few false starts using the subject tree, she used the keyword feature to locate appropriate sites. Although she found a few new sites, most of them were familiar. This is a sure sign that she had completed her web research and should move on.

>>> Using Questionnaires and Interviews

Sometimes, your research project may require collecting firsthand information yourself, such as questionnaires or personal interviews. This section covers both questionnaires and interviews. You learn how to (1) prepare, send out, and report the results of a questionnaire; and (2) prepare, conduct, and summarize a personal interview.

Tanya Grant's M-Global Project

Recall that Tanya, who works in Marketing at M-Global, has been asked by the company president to write a report that examines the successes and failures of hybrid electric cars. This report will look at start-up costs for switching the vehicle fleet, tax and other incentives, and the potential of the technology.

Now, before reporting her findings to Jim McDuff, she wants to find out what corporate users of the technology think of its potential. She believes her best approach is to (1) send a questionnaire to companies that have hybrid vehicle fleets, and (2) personally interview three or four respondents, including employees at M-Global's Asian offices, who will help management decide on the company's direction.

Questionnaires

Tanya Grant has the same challenge you would face in developing a questionnaire. Like you, she receives many questionnaires herself. Most of them she tosses in the recycle bin because they don't warrant her time, are too long, or seem confusing. Now that the shoe is on the other foot, she wants to design a questionnaire that attracts the attention of readers and entices them to complete it. To accomplish this feat, she goes through the following three-stage process:

>> Step 1: Preparing the Questionnaire

Obviously, your questionnaire is useful only if readers complete and return it. You must focus just as much on your readers' needs as you do on your own objectives. Before readers complete a form, they must perceive that (1) it benefits them personally or professionally and (2) it is easy to fill out and return. Keep these two points in mind as you design the form and the cover letter. Following are some specific guidelines for preparing a reader-focused document:

1. **Write a precise purpose statement.** As in other documents, a one-sentence statement of purpose provides a good lead-in for your cover letter that accompanies the survey (see next section). For example, Tanya prepared the following purpose statement for her survey concerning hybrid electric vehicles: "The purpose of this survey is to find out ways in which your experience with hybrid cars can benefit others." As obvious as that statement sounds, it helps busy readers who don't have time to wade through long rationales.

2. **Limit the number of questions.** Every question must serve to draw out information that relates to your purpose statement. For example, Tanya knows her questions must focus on the reader's experience with hybrid electric vehicles. She must resist the temptation to clutter the questionnaire with irrelevant questions on other alternative fuel vehicles such as natural gas or electric cars.

3. **Ask mostly objective questions.** You must design your form so that (1) questions are easy to answer and (2) responses are easy to compile. Although open-ended questions yield more detailed information, the answers take time to write and are difficult to analyze. Instead, your goal is breadth, not depth, of response. With the exception of one or two open-ended questions at the end of your questionnaire, reserve long-answer responses for personal interviews you conduct with a select audience. For example, Tanya decided to include an optional open-ended question at the end of her questionnaire, where she asks hybrid users to recommend design improvements for hybrid electric vehicles.

Objective questions come in several forms. Four common types are described next, along with examples of each.

- **Either/Or Questions:** Such questions give the reader a choice between two options, such as "yes" or "no." They are useful only when your questions present clear, obvious choices.

  Example: "Do you believe your hybrid vehicles accelerate well in all driving situations?" (followed by "yes" and "no" blocks), or "The hybrid accelerates well in all driving situations."

- **Multiple-Choice Questions:** These questions expand the range of possibilities for the reader to three or more, requiring a longer response time.

  Example: "If you answered 'yes' to the preceding question [a question asking if the hybrid vehicle accelerates well], what is your typical driving terrain? (a) Flat; (b) Hilly; (c) Combination of flat and hilly; (d) Mountainous"

- **Graded-Scale Questions:** By permitting degrees of response, these questions help gauge the relative strength of the reader's opinion.

  Example: "Using a hybrid vehicle has met our day-to-day driving needs. (a) Strongly agree; (b) Agree; (c) Disagree; (d) Strongly disagree; (e) Have no opinion"

- **Short-Answer Questions:** Use these questions when the possible short answers are too numerous to list on your form.

  Example: "List the makes of vehicles that your company has purchased in the last five years."

4. **Provide clear questions that are easy to answer.** Like other forms of technical writing, questionnaires can frustrate readers when individual questions are unclear. Four common problems are (1) bias in phrasing, (2) use of undefined terms, (3) use of more than one variable, and (4) questions that require too much homework. Following are some examples of right and wrong ways to phrase questions, along with a brief comment on each problem:

Biased Question:

Original question: "Are the federal and state government's excessive tax credits for purchasing alternative fueled vehicles affecting your purchasing decision?" (Words like *excessive* reflect a bias in the question, pushing a point of view and thus skewing the response.)

Revised question: "Do you believe that the federal and state tax credits affected your purchasing decision?"

Undefined Technical Terms:

Original question: "Are you familiar with the work of the PNGV on AFVs?" (Your reader may not know that PNGV is short for Partnership for a New Generation of Vehicles, or that AFV stands for Alternative Fuel Vehicle. Thus some "no" answers may be generated by confusion about terminology.)

Revised question: "Are you familiar with the work of the Partnership for a New Generation of Vehicles on alternative fuel vehicles?"

Mixed Variables:

Original question: "Were the dealer's maintenance technicians prompt and thorough in their work?" (There are two questions here, one dealing with promptness and the other with thoroughness.)

Revised question: (two separate questions): "Were the dealer's maintenance technicians prompt?" "Were the dealer's maintenance technicians thorough?"

Question That Requires Too Much Homework:

Original question: "What other alternative fuel vehicles has your company researched, tested, or purchased in the last ten years?" (This question asks the reader to conduct research for an accurate answer. If they do not have the time for that research, they may leave the answer blank or provide an inaccurate guess. In either case, you are not getting valid information.)

Revised question: "Has your company tried other alternative fuel vehicles?"

5. **Include precise and concise instructions at the top of the form.** Your instructions can be in the form of an easy-to-read list of points that start with action verbs, such as the following list:

- Answer Questions 1–20 by checking the correct box.
- Answer Questions 21–30 by completing the sentences in the blanks provided.
- Return the completed form in the envelope provided by October 15, 2009.

Or, if instructions are brief, they can be in the form of a short, action-centered paragraph, such as: "After completing this form, please return it in the enclosed stamped envelope by October 15, 2009."

6. **Apply principles of document design.** Although you must strive for economy of space when designing a questionnaire, use adequate white space and other design principles to make the document attractive to the eye.

7. **Test the questionnaire on a sample audience.** Some sort of "user test" is a must for every questionnaire. For example, after completing her survey, Tanya decided to test it on three people:

- A fellow marketing colleague at M-Global who has conducted several questionnaires for the firm

- A psychologist Tanya knows through a local professional association

- A vehicle fleet manager whom she knows well enough to ask for constructive criticism on the form

Thus her user test will solicit views from people with three quite different perspectives.

>> Step 2: Conducting the Project

After you have designed a good form, the next task is to distribute it. Following are guidelines for selecting a good sampling of potential respondents, introducing the questionnaire to your audience, and encouraging a quick response from a high percentage of readers.

1. **Choose an appropriate audience.** Selecting your audience depends on the purpose of your questionnaire. If you manage a 100-employee engineering firm and want to gauge customer satisfaction with recent construction jobs, you might send your questionnaire to all 156 clients you have served in the past two years. Restricting the mailing list would be unnecessary, because you have a small sample.

However, if you are in Tanya's position at M-Global, with a mailing list totaling about 3,200 corporations that have purchased hybrid vehicles in 2007 and 2008, you must select a random sample. Tanya's research suggests that she will receive about a 25% rate of return on her questionnaires. (Actually, this rate would be quite good for an anonymous questionnaire.) Given that she wants about 200 returned forms, she must send out about 800 questionnaires in expectation of the 25% return rate.

With a client list of 3,200, she simply selects every fourth name from the alphabetized list to achieve a random list of 800 names. Note that the selection of client names from an alphabetized list preserves what is essential—that is, the random nature of the process.

Of course, you can create more sophisticated sampling techniques if necessary. For example, let's assume Tanya wants an equal sampling of companies that purchased in each of the two years—2007 (with 1,200 names) and 2008 (with 2,000 names). In other words, she wants to send an equal number of forms to each year's hybrid owners, even though the number of corporate hybrid owners varies from year to year. In this case, first she would select 400 names—or every third name—from the 1,200 alphabetized names for 2005. Then she would select the other 400 names—or every fifth name—from the 2,000 alphabetized names for 2008. As a result, she has done all she can do to equalize the return rate for two years.

This strategy helps you choose the audience for simple questionnaire projects. You may want to consult a specialist in statistics if you face a sophisticated problem in developing an appropriate sampling.

2. **Introduce the questionnaire with a clear and concise cover letter.** In 15 or 20 seconds, your letter of transmittal must persuade readers that the questionnaire is worth their time. Toward this end, it should include three main sections (which correspond to the letter pattern presented in chapter 7):

- **Opening paragraph:** State precisely the purpose of the questionnaire and perhaps indicate why this reader was selected.

- **Middle paragraph(s):** State the importance of the project and strive to emphasize ways that it may benefit the reader.

- **Concluding paragraph:** Specify when the questionnaire should be returned, even though this information will be included in the directions on the questionnaire itself.

3. **Encourage a quick response.** If your questionnaire is not anonymous, you may need to offer an incentive for respondents to submit the form by the due date. For example, you can offer to send them a report of survey results, a complimentary pamphlet or article related to their field, or even something more obviously commercial, when appropriate. Obviously, any incentive must be fitting for the context. Keep in mind also that some experts believe an incentive of any kind introduces a bias to the sample.

If the questionnaire is anonymous or if complimentary gifts are inappropriate or impractical, then you must encourage a quick response simply by making the form as easy as possible to complete. Clear instructions, frequent use of white space, a limited number of questions, and other design features mentioned earlier must be your selling points.

>> Step 3: Reporting the Results

After you tabulate results of the survey, you must return to the needs of your original audience—the persons who asked you to complete the questionnaire. They expect you to report the results of your work. Described next are the major features of such a report.

First, you must show your audience that you did a competent job of preparing, distributing, and collecting the questionnaire; thus, the body of your report should give details about your procedures. Appendices may include a sample form, a list of respondents, your schedule, extensive tabulated data, and other supporting information.

Second, you must reveal the results of the survey. This is where you must be especially careful. Present only those conclusions that flow clearly from data. Choose a tone that is more one of suggesting than declaring. In this way, you give readers the chance to draw their own conclusions and to feel more involved in final decision making. Graphs are an especially useful way to present statistical information (see chapter 12, "Graphics").

Finally, remember that your report and the completed questionnaires may remain on file for later reference by employees who know nothing about your project. Be sure that your document is self-contained. Later readers who uncover the "time capsule" of your project should be able to understand its procedures and significance from the report you have written.

Interviews

Besides questionnaires, interviews are another common way to gather primary research. Often they are conducted after a questionnaire has been completed, as a follow-up activity with selected respondents. Interviews also may be done as primary research independent of questionnaires. In either case, you must follow some common guidelines to achieve success in the interview. Following are a few basic pointers for preparing, conducting, and recording the results of your interviews:

>> Step 1: Preparing for the Interview

Put at least as much effort into planning the interview as you do into conducting it. Good planning puts you at ease and shows interviewees that you value their time. Specifically, follow these guidelines:

- **Develop a list of specific objectives for the interview.** Know exactly what you want to accomplish so that you can convey this significance to the person you interview.

- **Make clear your main objectives when you make contact for the interview.** This conversation should (1) stress the uniqueness of the person's contribution, (2) put him or her at ease with your goals and the general content of the proposed discussion, and (3) set a starting time and approximate length for the interview. If handled well, this preliminary conversation will serve as a prelude to the interview, giving direction to the next meeting.

- **Prepare an interview outline.** People you interview understand your need for written reference during the interview. Indeed, they expect it of any well-prepared interviewer. A written outline should include (1) a sequential list of topics and subtopics you want to cover and (2) specific questions you plan to ask.

- **Show that you value your interviewee's time.** You can do this first by showing up a few minutes early so that the interview can begin on time. You also show this courtesy by staying on track and ending on time. Never go beyond your promised time limit unless it is absolutely clear that the person being interviewed wants to extend the conversation further than planned.

>> Step 2: Conducting the Interview

Your interview will be successful if you stay in control of it. Maintaining control has little to do with force of personality, so don't worry if you are not an especially assertive person. Instead, keep control by sticking to your outline and not letting time get away from you. If you find your speakers straying, for example, gently bring them back to the point with another question from your list. Following are additional pointers for conducting the interview:

- **Ask mostly open-ended questions.** Open-ended questions require your respondent to say something other than "yes," "no," or other short answers. They are useful to the speaker because they offer an opportunity to clarify an opinion or a fact. They are useful to you because you get the chance to listen to the speaker, digest information, and prepare for the next question.

  For example M-Global's Tanya Grant may ask questions such as: "Could you describe two or three ways in which your expectations for hybrid vehicles have been met? For what purposes is your company currently using its fleet of hybrid vehicles?" or "I've been told that your company has a high commitment to environmental issues in the Denver area. How has purchasing and using hybrid vehicles been part of that commitment?"

- **Ask close-ended questions when you need to nail down an answer.** For example, Tanya may ask persons she interviews, "Would you be willing to meet with our fleet supervisor to discuss your experience with maintaining hybrid vehicles?" A "yes" or "perhaps" answer will give her an opening for calling this person several months later. A close-ended question works when commitment is needed.

- **Use summaries throughout the interview.** Brief and frequent summaries serve as important resting points during the conversation. They give you the chance to make sure you understand the answers that have been given, and they give your counterpart the chance to amplify or correct previous comments. For example, Tanya may comment to her speaker, "So, in other words, you are saying that hybrid vehicles make most sense right now for in-city driving where only one or two people share the vehicle." This summary elicits either a "yes" or a clarification, either of which helps Tanya record the interview accurately.

>> Step 3: Recording the Results

You should take notes throughout the interview. The actual mechanics of this process may influence the accuracy of your note taking. Following are three possible approaches:

- **Option 1: Number reference:** Using this approach, you begin the interview with a list of numbered questions on your outline page; then, when you take notes, simply list the number of the question, followed by your notes. This approach gives you as much space as you want to write questions, but it does require that you move back and forth between your numbered question list and note page.

- **Option 2: Combined question-and-answer page:** For this approach, place a major question or two on each page, leaving the rest of the page to record answers to these and related questions that may be discussed. Although this strategy requires considerably more paper and separates your prepared list of questions, it does help you focus quickly on each specific question and answer.

- **Option 3: Split page:** Some interviewers prefer to split each page lengthwise, writing questions in the left column and corresponding answers in the right column. Some questions may have been prepared ahead of time, as in option 2; others may be written as they are asked. In either case, you have a clear visual break between questions on one side and answers on the other. The advantage over option 2 is that you have a visual map that shows you your progress during the conversation. Questions and answers are woven together into the fabric of your interview.

>>> Using Borrowed Information Correctly

In some workplace writing, issues of citation can become complicated, especially in collaborative projects that use documents published by the writer's organization, and that will be published under the organization's name. (See chapter 13 for more on collaborative writing.) However, whenever you are using material that has been published in a book, periodical, or on another organization's website, you should cite your sources.

Most errors in research papers occur in transferring borrowed information. This section has three goals: (1) to explain why you must acknowledge sources you have used, (2) to outline a research process from the point at which you identify sources of information, and (3) to provide sample documentation styles from three well-known style manuals.

Avoiding Plagiarism

One basic rule underlies the mechanical steps described in the rest of this chapter:

> With the exception of *common knowledge,* you should cite sources for all borrowed information used in your final document. This includes quotations, paraphrases, and summaries.

Common knowledge is information generally available from basic sources in the field. In the case of Tanya's research project, *common knowledge* is a definition of hybrid electric vehicles. When you are uncertain whether a piece of borrowed information is common knowledge, go ahead and cite the source. It is better to err on the side of excessive documentation than to leave out a citation and risk a charge of *plagiarism* (the intentional or unintentional use of the ideas of others as your own). Following are three main reasons for documenting sources thoroughly and accurately:

1. **Courtesy:** You owe readers the *courtesy* of citing sources where they can seek additional information on the subject. Presumably, sources for quotations, paraphrases, and summaries provide such a reference point.

2. **Ethics:** You have an *ethical* obligation to show your reader where your ideas stop and those of another person begin; otherwise, you are parading the ideas of others as your own.

3. **Law:** You have a *legal* obligation to acknowledge information borrowed from a copyrighted source. In fact, you should seek written permission for the use of borrowed information that is copyrighted when you plan to publish your document or when

you are using your document to bring in profit to your firm (as in a proposal or report). If you need more specific information about copyright laws or about the legalities of documentation, see a research librarian.

Certainly some plagiarism occurs when unscrupulous writers intentionally copy the writing of others without acknowledging sources. However, most plagiarism results from sloppy work during the research and writing process. Described next are two common types of unintentional plagiarism. Although the errors are unintentional—that is, the writer did not intend to cheat—both result in the inappropriate use of another person's work. That's plagiarism.

Mike Pierson, a supervisor at M-Global's Cleveland office, has been asked to deliver a presentation at an upcoming conference on hybrid electric vehicles. In his last-minute rush to complete the presentation—which will be published in a collection of papers from the meeting—Mike is taking notes from a source in the company library. He hurriedly writes a note card from a source but fails to indicate the source. Later, when he is writing the paper draft, he finds the card and does not know whether it contains information that was borrowed from a source or ideas that came to him during the research process. If he incorporates the passage into his paper without a source, he will have committed plagiarism.

In our second case, Mike transfers a direct quotation from a source into a computer document file but forgets to include quotation marks. If he were to incorporate the quotation into his presentation later *with* the source citation but *without* quotation marks, he would have plagiarized. Why? Because he would be presenting the exact words of another writer as his own paraphrase. The passage would give the appearance of being his own words that are supported by the ideas of another, when in fact the passage would be a direct quote. Again, remember that the test for plagiarism is not one's intent; it is the result.

The next section shows you how to avoid plagiarism by completing the research process carefully. In particular, it focuses on a methodical process that involves (1) bibliography notes, (2) a rough outline, (3) notes of three main kinds, (4) a final outline, and (5) drafts.

Following the Research Process

The research process is no different from most other technical tasks. One of the early steps is to determine to which style the project must adhere. For her research project, Tanya decided to use the *Publication Manual of the American Psychological Association*, 5th ed., 2001, which is used widely in her company. We discuss more about selecting and following a document style system shortly. It is very important to select one at the outset of note taking. Once the style is determined, carefully attend the five-step procedure that follows and you will avoid most documentation problems.

>> Step 1: Begin a Research File

Plan one location for all of the information you gather during your research. This could be a set of file folders in your desk or on your computer, or it could be a loose-leaf notebook. Divide your file folders or notebook into three subfolders or sections: Prewriting, Working Bibliography, and Notes. Keep your working research question, brainstorming, versions

■ **Figure 15–10** ■
Sample bibliography entry

GALILEO
Academic Search Premier

Henry E., and C. McGrath 2000.
Coast to coast on fewer fill-ups. Kiplinger's Personal Finance Magazine, 548 (August), 160+.

Mentions the Insight Cult and "fudoxeia nervosa"

of your outline, and other prewriting in the first section. You may also find it useful to create a calendar for your research and put it at the beginning of this folder or notebook section. Your working bibliography includes your bibliographic entries for your sources as well as notes about leads on other sources. For example, if you see an authority referred to in several sources, make a note to look up articles and books published by that person. You should keep all of your notes in the final section of your file.

>> **Step 2: Record Complete Bibliographic Information**

Record complete bibliographic information of each source you consult in your working bibliography. These records become the foundation for the rest of your research. (Figure 15–10 shows a bibliography record that Tanya might have collected in her research on hybrid vehicles.)

Errors made at this stage—in transferring information from sources to your file— can easily work their way into a final document. In recording information, therefore, be sure to take these precautions:

1. **Include all information needed for the final-copy citation in your paper.** Use the exact wording for titles and publication information. Common errors are to leave off articles (*a*, *an*, *the*) and to abbreviate words in titles, with the writer thinking there will be time later to double-check the original source. In fact, that final check often does not occur, leading to errors in the final citation. Also, for web sources from electronic databases, include the name of the database (e.g., EBSCOhost Galileo), the date retrieved, and the URL (e.g., *http://www.GALILEO.usg.edu*).

2. **Save space at the bottom of the entry for a reminder to yourself about the usefulness of the source.** For example, a notation such as "includes excellent chapter on water resources" may help you later as you begin your research. Although they are mainly to record source information, bibliography records can also provide some guidance in the next stage of note taking.

3. **Include information in the exact format as it will appear in your final bibliography, down to the indenting, punctuation, and capitalization.** Tanya's record in Figure 15–10, for example, follows the capitalization guidelines

described for the APA system. Again, do not assume that you will have time later to transcribe every card into another format. That time will not be there. Also, using the same format ensures that you make sure to take down all information needed later for the source page.

>> Step 3: Develop a Rough Outline

Record this in your prewriting subfolder. This outline is not the one you will use to write the first draft. Instead, it is essentially a list of topics in the approximate sequence they will be covered in the paper. It serves to direct your writing of notes during the next step.

>> Step 4: Take Careful Notes

Most plagiarism results from sloppy note taking. This important stage requires that you attend to detail and follow a rigorous procedure. The procedure suggested here divides note cards into three types: summary, paraphrase, and quotation. Figure 15–11 gives examples of all three types.

■ **Summaries** are written in your words and reduce a good deal of borrowed information to a few sentences. They are best written by reading a section of source material,

Henry & McGrath:

The Insight has modest power but can't carry much weight.

Summary, p. 161

Henry & McGrath:

The Insight won't win any races, but it will get you safely on the freeway. Its load weight (365 pounds) is poor, however. Honda says that is wouldn't damage the car to overload it, but that performance and mileage would suffer.

paraphrase, p. 161

Henry & McGrath:

"Although the car [Insight] will never lead the pack at Indy, you won't fear for your life getting on the freeway. But not this. The Insight's 365-pound payload capacity means a couple of hefty passengers can overload [sic] the car. Honda says that wouldn't damage the vehicle, but would impede performance and mileage."

p. 161

■ **Figure 15–11** ■
Sample notes on same passage

looking away from the source, and summarizing the passage in your own words. In this way, you can later use any of this information with confidence in your paper, without worrying about the absence of quotation marks. It is useful to write a summary of every source that you read. This provides a context for your other notes and helps you understand the main points in the source.

■ **Paraphrases** include a close rephrasing of material from your sources. Unlike summaries, which condense a considerable amount of information, paraphrase notes usually include more of the original text. Thus they demand even more attention than summary notes to avoid the problem of plagiarism. Like summaries, they are best written by looking away from the source for a moment and then rewriting the passage in your own words. You can use a few key words from the passage, but do not duplicate exact phrasing or sentence structure. Using the "look away" technique helps you avoid creating a paraphrase that too closely resembles the original.

■ **Quotations** include only words taken directly from the source. Your main concern should be the care with which you transfer sentences from source to your notes—with absolute accuracy. Even include grammatical or spelling errors that the original source may contain (and, in so doing, use the word *sic*—see Handbook). Also, you should use *ellipses* (spaced dots) when you leave out words that you deem unnecessary. When using ellipses, however, be sure not to alter the meaning of the passage you are quoting.

>> Step 5: Organize Research in an Outline

With your notes at hand, you are now ready to render order from chaos—to create an outline that flows from the technical research related in your notes. You may have already been developing a working outline in your prewriting folder. If you haven't, review your notes and make a list of the important topics that you want to include in your document, and then decide on the best way to organize those ideas. (See information about organization in chapter 3.) Some writers find it useful to use the Outline function in their word-processing programs. These outlines can later be converted into documents. After you create your outline, copy and paste from your notes file (or copy from your notebook). Make sure that you include at parenthetical citation for each note that you copy into your outline to avoid accidental plagiarism.

>> Step 6: Write the Draft

This step poses the greatest challenge because it is here that you must incorporate borrowed information with your own ideas to create fluid prose. Your goal should be to demonstrate (a) a smooth transition between your ideas and those you have borrowed and (b) absolute clarity about when borrowed ideas and quotations start and end.

>>> Selecting and Following a Documentation System

Documentation refers to the mechanical system you use to cite sources from which you borrow information. This section briefly compares documentation styles from three important style manuals—the previously mentioned APA, the Modern Language Association (MLA), and the Council of Science Editors (CSE)—and provides examples for the most common citations. For complete details about a particular documentation system you are using, consult one of the manuals in the list that follows or consult the website of the organization that publishes the manual. Pay special attention to new guidelines these manuals may provide for documenting information from online databases and the Internet.

There are almost as many systems as there are professional organizations, but all have the same goal of showing readers the sources from which you gathered information. One of your early steps in research is to determine which style manual to use. Often your instructors select a discipline-specific style manual. Style manuals guide the writer through the editorial rules governing everything from use of headers and pagination and graphic and text layout, to managing data display and, of course, the rule for documenting sources. Style manuals are regularly revised by the organizations that publish them. One of the areas of greatest changes is the rules for citing electronic resources. As the variety and use of electronic materials continue to evolve, so too do the style manuals. Be sure to check your edition of the style manual you are using to make sure it is the latest available.

Following are just a few documentation manuals commonly used in business, industry, and the professions. You can often locate useful tips and examples at the websites maintained by each of these organizations in addition to the purchasing information for the style manual itself.

American Psychological Association (APA)

Publication Manual of the American Psychological Association, 5th ed. 2001.
website: *http://www.apastyle.apa.org*

Also useful from APA:
Electronic Reference Formats Recommended by the American Psychological Association,
website: *http://www.apastyle.org/elecref.html*

Council of Science Editors (CSE)

Scientific Style and Format: The CSE Manual for Authors, Editors, and Publishers,
7th ed. 2006.
website: *http://www.councilscienceeditors.org/publications/style.cfm*

Modern Language Association (MLA)

MLA Handbook for Writers of Research Papers, 6th ed. 2003.

University of Chicago Press.

Chicago Manual of Style, 15th ed. 2003.
website: *http: // www.chicagomanualofstyle.org / tools.html*
A Manual for Writers of Research Papers, Theses, and Dissertations, 7th ed., 2007
Also noteworthy:

University of Wisconsin's Writing Center

Writer's Handbook
website: *http: // www.wisc.edu / writing / Handbook*

Purdue Online Writing Lab

website: *http: // owl.english.purdue.edu /*

Locate helpful hints by searching the web by the name of the documentation style. Many libraries, writing centers, and universities have posted reliable and useful sample style sheets. Remember—you must evaluate the quality, currency, and authorship of the website carefully before using it.

We focus briefly on APA, MLA, and CSE and compare documentation styles for citing works. The three systems share some common characteristics. Each uses parenthetical references in the body of the report that lead the reader to a separate works cited or reference page. Each system cites the author's name and either the publication year (APA and CSE) or, if using MLA, the relevant page number where the fact, quote, or observation can be located. Frequently, the content of the parenthetical references is blended into the text with perhaps only the date or page in parentheses. The works cited or reference page is arranged alphabetically by the author's last name for APA, MLA, and CSE.

CSE offers a three style choices: the *name–year* system, the *citation–sequence* system, and the *citation–name* system. The name–year system is similar to APA style, using a parenthetical reference to the date. In the citation–sequence system and the citation–name system, the parenthetical citation refers to a numbered list of citations at the end of the document. For the citation–sequence system, the sources in the bibliography are numbered sequentially in the order in which they appear in the document. In the citation–name system, the sources in the bibliography are alphabetized by the authors' last names, and then numbered in that order. When using CSE, you must determine which system is preferred—the name–year system, the citation–sequence system, or the citation–author system. Check with your instructor or editor.

There are significant and subtle variations in parenthetical entries and works cited listings when the style manuals are closely compared. The examples in Figure 15–12 outline a few basic examples. Writers must consult the style manual itself for a thorough discussion.

See Model 15–1 at the end of this chapter for Tanya Grant's complete memo report that cites research.

■ **Figure 15–12** ■
Sample list of
references

APA Parenthetical References and Work Cited Examples

The American Psychological Association (APA) uses the parenthetical author and year system in the text that leads the reader to an alphabetically arranged works cited page at the end of the text. Anonymous works are cited using the first few title words and year.

In-text Reference	Entry in Works Cited List
1.1 Book, one author (Wakefield, 1998, p. 138)	Wakefield, E. (1998). *History of the electric automobile: Hybrid electric vehicles.* Warrendale, PA: Society of Automotive Engineers.
1.2 Book, two or more authors (Hodkinson & Fenton, 2001, p. 93)	Hodkinson, R., & Fenton, J. (2001). *Lightweight electric/hybrid vehicle design.* Woburn, MA: Butterworth-Heinemann.
1.3 Edited book (Johnson, 1993, p. 123)	Johnson, A. E. (Ed.). (1993). *Future of vehicle transportation.* London: Sage.
1.4 Work in an Anthology (Seal, 1993, p. 330)	Seal, M. (1993). Feasibility studies of solar electric hybrids. In A. Johnson (Ed.), *Transportation* (pp. 321–332). London: Sage.
1.5 Encyclopedia article (unsigned) (Alternative, 2003)	Alternative automobiles. (2003) In K. Kramer (Ed.), *Encyclopedia of automotive history.* (3rd ed., Vol. 2, pp. 235–239). New York: Harper.
1.6 Newspaper (Kiley & Healey, 2004)	Kiley, D., & Healey, J. R. (2004, May 14). Hybrid SUV getting big response; Ford says 30,000 want to buy one. *USA Today*, p. 1A.
1.7 Magazine article (Wouk, 1997, p. 71)	Wouk, V. (1997, October). Hybrid electric vehicles. *Scientific American*, 277, 70–74.
1.8 Journal article (Kim, Jung, & Nam, 2004, p. 312)	Kim, J., Jung, J., & Nam, K. (2004). Dual-inverter control strategy for high-speed operation of EV induction motors. *IEEE Transactions on Industrial Electronics*, 51(2), 312–321.
1.9 Article from an electronic database (Einstein, 1999, para. 1)	Einstein, P. (1999). The Benefits of Insight. *Professional Engineering, 12(19),* 23. Retrieved August 3, 2001, from Academic Search Premier database at EBSCOhost.
1.10 WWW site (no author) (Toyota Prius, 2001)	*Toyota Prius.* (2001). Cartalk.com. Retrieved August 19, 2001, from http://cartalk.cars.com/Info/Testdrive/Reviews/toyota-prius-2001.html

■ **Figure 15–12** ■
continued

MLA Parenthetical Reference and Work Cited Examples

The Modern Language Association (MLA) uses an author and page reference system. The author and specific page number where the fact, quote, or reference can be located are referenced in the text, either in parentheses or as part of the text. The works cited page is arranged alphabetically by author. Titles may be underlined or italicized with title words generally capitalized.

In-text Reference	Entry in Works Cited List
1.1 Book, one author (Wakefield 125)	Wakefield, Earl. *History of the Electric Automobile: Hybrid Electric Vehicles.* Warrendale, PA: Society of Automotive Engineers, 1998.
1.2 Book, two or more authors Hodkinson and Fenton in their 2001 book (311)	Hodkinson, Richard, and John Fenton. *Lightweight Electric/Hybrid Vehicle Design.* Woburn, MA: Butterworth-Heinemann, 2001.
1.3 Edited book Johnson argues (225)	Johnson, Arthur, ed. *Future of Vehicle Transportation.* London: Sage, 1993.
1.4 Work in an anthology (Seal 321)	Seal, Martin. "Feasibility Studies of Solar Electric Hybrids." *Future of Vehicle Transportation.* Ed. Arthur Johnson. London: Sage, 1993. 321–32.
1.5 Encyclopedia article (lesser known encyclopedia) (Murray 325)	Murray, Kim L. "Alternative Automobiles." *Encyclopedia of Automotive History.* Eds. David Jones and Jack Smith. 3rd ed. 2 vol. New York: Harper, 2003.
1.6 Newspaper (Kiley and Healey 1A)	Kiley, Dan, and John Healey. "Hybrid SUV Getting Big Response Ford Says 30,000 Want to Buy One." *USA Today*, 14 May 2004, 1: A.
1.7 Magazine article (Wouk 71)	Wouk, V. "Hybrid Electric Vehicles." *Scientific American* Oct. 1997: 70–74.
1.8 Journal article (Kim and Nam 314–15)	Kim, J., J. Jung, and K. Nam. "Dual-inverter Control Strategy for High-Speed Operation of EV Induction Motors." *IEEE Transactions on Industrial Electronics* 51 (2004): 312–21.
1.9 Article from an electronic database (Einstein)	Einstein, Paul. "The Benefits of Insight." *Professional Engineering.* 12.19 (1999): 23. *Academic Search Premier* EBSCOhost. GALILEO. 3 Aug. 2001 <http://www.galileo.usg.edu.>.
1.10 WWW site (Toyota Prius)	"Toyota Prius." *Cartalk.com.* 2001. 19 Aug. 2001 <http://cartalk.cars.com/Info/Testdrive/Reviews/toyota-prius-2001.html>.

■ **Figure 15–12** ■
continued

CSE References and Work Cited Examples

The *CSE Manual for Authors, Editors, and Publishers* (7th edition) prepared by the Council of Science Editors Style Manual Committee, offers a choice in citation systems—by name (for the name–year system) or by reference number (for the citation–sequence system or citation-name system). You must decide early on which CSE system to use. Ask your instructor or the editor. Each system is demonstrated here.

In the citation-sequence system, sources are listed in the order they appear in the paper. The sources in the list are then numbered, and the numbers are used in the references in the text.

CSE's Citation–Sequence and Citation–Name Examples

In-text Reference

1.1 Book, one author
As Wakefield[1] claims

1.2 Book, two or more authors
Hodkinson and Fenton[2]

1.3 Edited book
As described by Johnson[3]

1.4 Work in an anthology
Early experiments demonstrated[4]

1.5 Encyclopedia article (unsigned) Alternatives[5]

1.6 Newspaper
Kiley and Healey[6]

1.7 Magazine article
Wouk[7]

1.8 Journal article
The breakthrough in the power system[8]

1.9 Article from an electronic database Einstein[9]

1.10 WWW site (no author)
The performance of the Toyota Prius[10]

Entry in Works Cited List

1. Wakefield E. History of the electric automobile: hybrid electric vehicles. Warrendale, PA: Society of Automotive Engineers; 1998.

2. Hodkinson R, Fenton J. Lightweight electric/hybrid vehicle design. Woburn, MA: Butterworth-Heinemann; 2001.

3. Johnson AE, editor. Future of vehicle transportation. London: Sage; 1993.

4. Seal M. Feasibility studies of solar electric hybrids. In: Johnson A, editor. Transportation. London: Sage; 1993. p. 321–332.

5. Alternative automobiles. Encyclopedia of automotive history. 3rd ed. vol. 1. New York: Harper; 2003; 2: 235–239.

6. Kiley D, Healey JR. Hybrid SUV getting big response Ford says 30,000 want to buy one. USA Today. 2004 May 14; Section A:1.

7. Wouk V. Hybrid electric vehicles. Scientific American. 1997 Oct; 277: 70–74.

8. Kim J. Jung J, Nam, K. Dual-inverter control strategy for high-speed operation of EV induction motors. IEEE Transactions on Industrial Electronics. 2004; 51(2): 312–321.

9. Einstein P. The benefits of Insight. Professional Engineering, 1999 Oct; 12: 23(19). Academic Search Premier EBSCOhost [database on the Internet]. [cited 2008 Apr 30]. Available from: http://web.ebscohost.com/ehost/pdf?vid=5&hid=106&sid=7fd7ffb0-7cda-4694-a7fe392d429d476%40sessionmgr107; Accession nr: 2479269.

10. Toyota Prius. Cartalk [Internet]. c2001 [cited 2001 Aug 19]. Available from: http://cartalk.cars.com/Info/Testdrive/Reviews/toyota-prius-2001.html

■ **Figure 15–12** ■
continued

In the name-year system, sources are listed alphabetically by the author's last name. Parenthetical citations are used in the paper. Note that in the name-year system, the date of publication is placed after the last author's name, in Works Cited entries.

CSE's Name–Year System

Parenthetical Reference	Entry in Works Cited List
1.1 Book, one author As demonstrated by Wakefield (1998)	Wakefield E. 1998. *History of the electric automobile: Hybrid electric vehicles.* Warrendale, PA: Society of Automotive Engineers.
1.2 Book, two or more authors (Hodkinson and Fenton 2001)	Hodkinson R, Fenton J. 2001. *Lightweight electric/hybrid vehicle design.* Woburn, MA: Butterworth-Heinemann.
1.3 Edited book (Johnson 1993)	Johnson AE, editor. 1993. *Future of vehicle transportation.* London: Sage.
1.4 Work in an anthology Seal argues (1993)	Seal M. 1993. Feasibility studies of solar electric hybrids. In Johnson. A, editor. Transportation. London: Sage. p. 321–332.
1.5 Encyclopedia article (unsigned) HEV are defined as (Alternative 2003)	Alternative automobiles. (2003) In Kramer K. editor *Encyclopedia of automotive history.* 3rd ed. vol. 1. New York: Harper; p. 235–239.
1.6 Newspaper (Kiley and Healey 2004)	Kiley D, Healey JR. 2004, May 14. Hybrid SUV getting big response Ford says 30,000 want to buy one. *USA Today* 1A.
1.7 Magazine article (Wouk 1997)	Wouk V. 1997 Oct. Hybrid electric vehicles. *Scientific American* 277: 70–74.
1.8 Journal article (Kim and others 2004)	Kim J, Jung J, Nam, K. 2004. Dual-inverter control strategy for high-speed operation of EV induction motors. *IEEE Transactions on Industrial Electronics* 51(2): 312–321.
1.9 Article from an electronic database (Einstein 1999)	Einstein P. The benefits of Insight. Professional Engineering 1999 Oct; 12:23. Academic Search Premier EBSCOhost [Internet]. [cited 2008 Apr 30]. Available from: http://web.ebscohost.com/ehost/pdf?vid=5&hid=106&sid=7fd7ffb0-7cda-4694-a7fe392d429d476%40sessionmgr107; Accession nr: 2479269.
1.10 WWW site (no author) (Toyota Prius, 2001)	Toyota Prius. Cartalk [Internet]. c2001 [cited 2001 Aug 19]. Available from: http://cartalk.cars.com/Info/Testdrive/Reviews/toyota-prius-2001.html

>>> **Writing Research Abstracts**

The term *abstract* has been used throughout this book to describe the summary component of any technical document. As the first part of the ABC pattern, it gives decision makers the most important information they need. However, here we use *abstract* for a narrower purpose—it is a stand-alone summary that provides readers with a capsule version of a piece of research, such as an article or a book. This section (1) describes the two main types of research abstracts, with examples of each, and (2) gives five guidelines for writing research abstracts.

Types of Abstracts

There are two types of abstracts: informational and descriptive. As the following definitions indicate, informational abstracts include more detail than descriptive abstracts:

Informational Abstract.

- **Format:** This type of abstract includes the major points from the original document.
- **Purpose:** Given their level of detail, informational abstracts give readers enough information to grasp the main findings, conclusions, and recommendations of the original document.
- **Length:** Although longer than descriptive abstracts, informational abstracts are still best kept to one to three paragraphs.
- **Example:** A sentence from such an abstract might read, "The article notes that functional resumes should include a career objective, academic experience, and a list of the applicant's skills." (See corresponding example in definition of a descriptive abstract.)

Descriptive Abstract.

- **Format:** This type of abstract gives only main topics of the document, without supplying supporting details such as findings, conclusions, or recommendations.
- **Purpose:** Given their lack of detail, descriptive abstracts can only help readers decide whether they want to read the original document.
- **Length:** Their lack of detail usually ensures that descriptive abstracts are no more than one paragraph.
- **Example:** A sentence from such an abstract might read, "The article lists the main parts of the functional resume." (See corresponding example in definition of an informational abstract.)

You may wonder when you'll need to write abstracts during your career. First, your boss may ask you to summarize some research, perhaps because he or she lacks your technical background. Second, you may want to collect abstracts as part of your own research project. In either case, you must write abstracts that reflect the tone and content of the original document accurately.

Assume, for example, that your M-Global supervisor asked you to read some current research on strategies for negotiating. Later, your boss plans to use your abstracts to get an overview of the field and to decide which, if any, of the original full-length documents should be read in full. The examples that follow show both informational and descriptive abstracts of the section of chapter 16 that covers negotiating. Note that the informational abstract actually lists the guidelines contained in the chapter, whereas the descriptive abstract notes only that the article includes the guidelines.

Informational Abstract: "Guidelines for Negotiating"

This article suggests that modern negotiations should replace "I win, you lose" thinking with a "We can both win" attitude. To achieve this change, these six main guidelines are prescribed: (1) think long term, (2) explore many options, (3) find the shared interests, (4) listen carefully, (5) be patient, and (6) do look back. Although this strategy applies to all types of negotiation, this article focuses on a business context. It includes an extended example that involves establishing an appropriate entry salary for a job applicant in computer systems engineering.

Descriptive Abstract: "Guidelines for Negotiating"

This article describes six main guidelines that apply to all types of negotiations. The emphasis is on strategies to be used in the context of business. All the suggestions in the article support the need for a "We both win" attitude in negotiating, rather than an "I win, you lose" approach.

Guidelines for Writing Research Abstracts

The following guidelines help you (1) locate the important information in a document written by you or someone else and (2) present it with clarity and precision in an abstract. In every case, you must present a capsule version of the document in language the reader can understand. The ultimate goal is to save the reader's time.

>> Abstracting Guideline 1: Highlight the Main Points

This guideline applies whether you are abstracting a document written by you or someone else. To extract information to be used in your abstract, follow these steps:

1. Find a purpose statement in the first few paragraphs.
2. Skim the entire piece quickly, getting a sense of its organization.
3. Read the piece more carefully, underlining main points and placing comments in margins.
4. Pay special attention to information gained from headings, first sentences of paragraphs, listings, graphics, and beginning and ending sections.

>> Abstracting Guideline 2: Sketch an Outline

From the notes and marginal comments gathered in Abstracting Guideline 1, write a brief outline that contains the main points of the piece. If you are dealing with a well-organized piece of writing, it is an easy task; if not, it is a challenge. Following is an outline for the negotiation section of chapter 16, as abstracted in the previous examples.

Outline for "Guidelines for Negotiating." Purpose: to provide rules that help readers adopt a "We both win" strategy

I. Think long-term
  A. Focus on building mutual trust
  B. Project the long-term attitude into every part of the negotiation process

II. Explore many options
  A. Get away from thinking there are only two choices
  B. Put diverse options on the table early in the negotiation

III. Find the shared interests
  A. Stress points of agreement, rather than conflict
  B. Use mutual concerns to defuse contentious issues

IV. Listen carefully
  A. Ask questions and listen, rather than talk
  B. Use probing questions to move discussion along in salary discussion
    1. Break out of attack/counterattack cycle
    2. Uncover motivations
    3. Expose careless logic and unsupported demands
    4. Move both sides closer to objective standards

V. Be patient
  A. Avoid the mistakes that come from hasty decisions
  B. Avoid the bad feeling that results when people feel pressured

VI. Do look back
  A. Keep a journal in which you reflect on your negotiations
  B. Analyze the degree to which you followed the previous five guidelines

>> Abstracting Guideline 3: Begin with a Short Purpose Statement

Both descriptive and informational abstracts should start with a concise overview sentence. This sentence acquaints the reader with the document's main purpose. Stylistically, it should include an action verb and a clear subject. Following are three options that can be adapted to any abstract:

- The article "Recycle Now!" states that Georgia must intensify its effort to recycle all types of waste.

- In "Recycle Now!" Laurie Hellman claims that Georgia must intensify its effort to recycle all types of waste.

- According to "Recycle Now!" Georgians must intensify their efforts to recycle all types of waste.

>> Abstracting Guideline 4: Maintain a Fluid Style

One potential hazard of the abstracting process is that you may produce disjointed and awkward paragraphs. You can reduce the possibility of this stylistic flaw by following these steps:

- Write in complete sentences, without deleting articles (*a, an, the*)

- Use transitional words and phrases between sentences

- Follow the natural logic and flow of the original document itself

>> Abstracting Guideline 5: Avoid Technical Terms Readers May Not Know

Another potential hazard is that the abstract writer, in pursuit of brevity, will use terms unfamiliar to the readers of the abstract. This flaw is especially bothersome to readers who do not have access to the original document. As a general rule, use no technical terms that may be unclear to your intended audience. If a term or two are needed, provide a brief definition in the abstract itself.

Note, also, that abstracts that might become separated from the original document should include a bibliographic citation (see the previous examples).

>>> Chapter Summary

This chapter highlights the process of conducting technical research and writing about the results. Much of the information is presented within the context of a research project conducted at M-Global.

Before starting your search for information, you must decide what main question you are trying to answer. Also, think about the types of information you need, the types of sources that would be useful, and the format required for the final document. Once in the library, you have many sources available to you: books, periodicals, newspapers, company directories, general references such as dictionaries and encyclopedias, abstracts—both online and in print. You can also use the Internet, World Wide Web, questionnaires, and interviews.

As you begin to locate sources, follow this six-step research process: (1) begin a research file, (2) record complete bibliographic information, (3) develop a rough outline, (4) take careful notes, (5) organize research in an outline, and (6) write the draft. For the final paper, choose a documentation system appropriate for your field or organization.

Another research skill is writing research abstracts (summaries) of articles, books, or other sources of information. Abstracts can be either descriptive (quite brief) or informational (somewhat more detailed).

>>> Learning Portfolio

Communication Challenge "To Cite or Not to Cite"

Dan Gibbs works as a benefits and finance specialist at M-Global's corporate office in Baltimore. As the number of M-Global employees has grown, he has received many inquiries about ways to save for retirement. He recently wrote and distributed a four-page flyer on the topic using materials from print and online sources. The response was so positive that his boss wants to send the flyer to clients as a "freebie"—both to help clients' employees and to create good will in marketing. This use of the flyer has made Dan rethink how he developed the piece. The following sections present Dan's research process, his results, some questions for discussion, and an assignment for a written response to the Challenge.

Background of Retirement Booklet

Unlike Tanya Grant in the hybrid vehicle project described in this chapter, Dan didn't have time or interest in pursuing a full-scale library search about retirement strategies. Besides, he has personnel magazines in the office with data that support his points. In addition, he is connected to the Internet, a series of loosely connected databases that might provide information directly to him at his office computer.

Like many companies, M-Global has a retirement plan largely in the form of what is called a 401k program. It allows employees to contribute a percentage of their salaries into a tax-deferred retirement account, a portion of which is matched by the employer. Even though M-Global has a generous matching arrangement, many employees do not take full advantage of the program. Therefore, Dan wrote the retirement flyer to remind them that it is never too early to plan for retirement. As it happens, he learned that many U.S. workers are failing to put away enough money for their retirement years.

The Research Process

After outlining his goals for the booklet, Dan began surfing through related information on the Internet. He made use of four pieces of information from an *Atlanta-Journal Constitution*[1] article found on an Internet database called "Human Resources Retrieval." Following are three of the themes he stressed, along with related information he used from the newspaper article.

1. **Theme 1:** *We're living longer past retirement.* The following changes occurred in years of life expected after age 65: for men, 12.8 in 1960, 13.1 in 1970, 14.1 in 1980, and 15.1 in 1990; for women, 15.8 in 1960, 17.0 in 1970, 18.3 in 1980, and 18.9 in 1990.

2. **Theme 2:** *We cannot depend exclusively on Social Security.* As many more people retire from the "Baby Boom" generation born between 1946 and 1964, fewer workers paying Social Security are supporting each person getting it. The following numbers are actual and projected number of workers supporting each retiree: 41.9 in 1945, 8.6 in 1955, 4.0 in 1965, 3.2 in 1975, 3.3 in 1985, 3.3 in 1995, 3.1 in 2005, 2.7 in 2015, 2.2 in 2025, and 2.0 in 2035.

3. **Theme 3:** *We should begin saving when we're young.* If you start saving $3,000 yearly in a tax-deferred account at age 25, with a 9% annual return, you'll accumulate $166,971 by age 45, $457,685 by age 55, and $1,170,330 by age 65. If you start at age 35, you'll have $48,378 by 45, $166,971 by 55, and $457,685 by 65. If you start at age 45, you'll have $48,378 by 55 and $166,971 by 65.

Questions and Comments for Discussion

1. If you were presenting the previously mentioned data in a research report, what format would you choose? Why? (See chapter 12.)

2. Considering the four data sets Dan took from the newspaper article/Internet sources, which ones need documentation and which, if any, do not? Explain your answer.

3. Does the fact that the flyer will be sent to clients have any effect on your answer to question 2?

4. Do a works cited reference for the data in theme 1 previously stated. See the footnote to this Communication Challenge for actual source information. Assume that Dan found the piece on the Internet on March 4, 2008.

Write About It

Using your library periodical databases, find an article that explains 401k retirement plans. Write an informational abstract of the article; and include a full citation of the article in APA style.

---

[1]The database is a fictional one created just for this case. However, the data are real and included in "A Special Report: The Rocky Road to Retirement," by Hank Ezell, *Atlanta Journal-Constitution*, September 25, 1995, pp. E1, E4–6. The article cites the following sources for data listed: The Rand Corporation, National Vital Statistics System, and Ibbotson Associates for item 1; the Social Security Administration for item 2; and Hewitt Associates for item 3.

Collaboration at Work Surfing the Turf

General Instructions

Each Collaboration at Work exercise applies strategies for working in teams to chapter topics. The exercise assumes you (1) have been divided into teams of about three to six students, (2) will use team time inside or outside of class to complete the case, and (3) will produce an oral or written response. For guidelines about writing in teams, refer to pages 24–28 in chapter 1 and refer to chapter 13.

Background for Assignment

The Internet has greatly expanded the range of information you can find on a subject and the speed with which it can be found. Yet the Internet also has introduced new challenges. While conducting this type of research, you must do the following:

- Stay focused so that the inevitable distractions of new links and fascinating data do not detract from your main purpose.

- Evaluate the reliability of sources that are quite different from traditional hardcopy sources found in a library.
- Determine the mix of Internet and traditional sources that provide the best support for your topic.
- Keep good notes so that later you can properly document information that has been secured from the Internet.

To start you thinking about the process of Internet research, this exercise asks you to work with your team on a short project.

Team Assignment

First, agree as a team on a topic about which you want to find information on the Internet. Second, work individually to locate three to five sources of information on the topic (the information itself—not just a list of sources). Third, come back together as a team and discuss the relative value of the sources of the information you found.

Assignments

If your instructor considers it appropriate, use a copy of the Planning Form at the end of the book for completing these assignments.

1. General Research Paper

Using a topic approved by your instructor, follow the procedure suggested in this chapter for writing a paper that results from some technical research. Be sure that your topic (1) relates to a technical field in which you have an interest, by virtue of your career or academic experience; and (2) is in a field about which you can find information in nearby libraries.

2. Research Paper—Your Major Field

Write a research paper on your major field. Consider some of these questions in arriving at your thesis for the paper: What is the history of your major? What types of jobs do majors in your discipline pursue? Do their job responsibilities change after 5, 10, or 15 years in the field? What kinds of professional organizations exist to support your field?

3. Research Paper—M-Global

As an M-Global engineer or scientist, you have been asked to write a research paper for M-Global's upper management. Choose your topic from one of the technical fields listed

below. Assume that your readers are gathering information about the topic because they may want to conduct consulting work for companies or government agencies involved in these fields. Focus on advantages and disadvantages associated with the particular technology you choose. Follow the procedure outlined in this chapter.

- Artificial intelligence
- Chemical hazards in the home
- Fiber optics
- Forestry management
- Geothermal energy
- Human-powered vehicles
- Lignite-coal mining
- Organic farming
- Satellite surveying
- Solar power
- Wind power

4. Abstract—One Article or Several Articles

Option A Visit your college library and find a magazine or journal in a technical area, perhaps your major field. Then photocopy a short article (about five pages) that does not already contain a separate abstract or summary at the beginning of the article. Using the guidelines in this chapter, write both an informative and a descriptive abstract for a nontechnical audience. Submit the two abstracts, along with the copy of the article.

Option B   Follow the instructions in option A, but use a short article that has been selected or provided by your instructor.

Option C   Read three to five current articles in your major field. Write an abstract that summarizes all of them on one page.

5. Questionnaire—Analysis

Using the guidelines in this chapter for questionnaires, point out problems posed by the following questions:

A. Is the poor economy affecting your opinion about the current Congress?

B. Do you think the company's severe morale problem is being caused by excessive layoffs?

C. Was the response of our salespeople both courteous and efficient?

D. Of all the computer consultants you have used in the previous 15 years, which category most accurately reflects your ranking of our firm: (a) the top 5%, (b) the top 10%, (c) the top 25%, (d) the top 50%, or (e) the bottom 50%?

E. In choosing your next writing consultant, would you consider seeking the advice of a professional association such as the STC or the CPTSC?

F. Besides the position just filled, how many job openings at your firm have been handled by Dowry Personnel Services?

6. Questionnaire—Writing

Design a brief questionnaire to be completed by students on your campus. Select a topic of general interest, such as the special needs of evening students or the level of satisfaction with certain college facilities or services. Administer the questionnaire to at least 20 individuals (in classes, at the student union, in dormitories, etc.). After you analyze the results, write a brief report that summarizes your findings. *Note:* Before completing this exercise, make sure that you gain any necessary approvals of college officials, if required.

7. Interviews—Simulation and Analysis

Divide into teams of three or four students, as your instructor directs. Two members of the team will take part in a simulated interview between a placement specialist at your school and a personnel representative from an area company. Assume that the firm might have a number of openings for your school's graduates in the next few years.

As a team, create some questions that would be useful during the simulated interview. Then have the two members perform the role-playing exercise for 15 to 20 minutes. Finally, as a team, critique the interview according to the suggestions in this chapter and share your findings with the entire class.

8. Interview—Real-World

Select a simple research project that would benefit from information gained from an interview. (Your project may or may not be associated with a written assignment in this course.) Using the suggestions in this chapter, conduct the interview with the appropriate official.

(?) 9. Ethics Assignment

This assignment is best completed as a team exercise.

Assume your team has been chosen to develop a web-based course in technical communication. Team members are assembling materials on a website that can be used by students like you—materials such as (1) guidelines and examples from this book, (2) scholarly articles on communication, (3) newspaper articles and graphics from print and online sources, and (4) examples of technical writing that have been borrowed from various engineering firms.

Your team has been told that generally speaking, the "fair use" provision of the Copyright Act permits use of limited amounts of photocopied material from copyrighted sources without the need to seek permission from, or provide payment to, the authors—as long as use is related to a not-for-profit organization, such as a college. Your tasks are as follows:

A. Research the Copyright Act to make sure you understand its application to conventional classroom use. If possible, also locate any guidelines that relate to the Internet.

B. Develop a list of some specific borrowed materials your team would want to include on the site for the technical communication course. These materials may fall inside or outside the four general groupings noted previously.

C. Discuss how the medium of the Internet may influence the degree to which the fair use provision would be applicable to your web course. Be specific about the various potential uses of the material.

D. Consult an actual web-based college course in any field and evaluate the degree to which you think it follows legal and ethical guidelines for usage.

E. Prepare a report on your findings (written or oral, depending on the directions you have been given by your instructor).

10. International Communication Assignment

Using interviews, books, periodicals, or the Internet, investigate the degree to which writers in one or more cultures besides your own acknowledge borrowed information in research documents. For example, you may want to seek answers to one or more of the following questions: Do you believe acknowledging the assistance of others is a matter of absolute ethics, or should such issues be considered relative and therefore influenced by the culture in which they arise? For example, would a culture that highly values teamwork and group consensus take a more lenient attitude toward acknowledging the work of others? These are not simple questions. Think them through carefully.

 ## 11. A.C.T.N.O.W. Assignment (Applying Communication To Nurture Our World)

Interview two or three students to find out why they do or do not participate in student elections on campus. On the basis on information you gather from the interviews, develop a survey form by which you solicit information on the topic in a systematic manner from a wider audience. Administer the survey to at least 10 students and include the results in an oral or written report, depending on the instructions you are given.

MEMORANDUM

DATE: May 7, 2008
TO: Jim McDuff, President
FROM: Tanya Grant, Technical Writer
SUBJECT: Hybrid Vehicle Research and Recommendations

INTRODUCTORY SUMMARY

When you heard that our Tokyo office had purchased some Toyota vehicles powered by a combination gasoline–electric engine, you asked me to research the hybrid vehicles and to recommend whether M-Global should use them in our offices in the United States. After making several phone calls, checking useful websites, and reviewing several magazine and newspaper articles, I recommend that M-Global replace our conventional company vehicles with the more fuel-economic and energy-efficient hybrid vehicles.

ADVANTAGES

Hybrid vehicles offer several advantages, especially in cost and impact on the environment.

• **Excellent Fuel Economy**

Hybrid vehicles are able to double the fuel economy of many of today's conventional cars. The midsize Toyota Prius is rated by the U.S. Environmental Protection Agency at 48 miles per gallon (mpg) city driving, and at 45 mpg on the highway. According to the Environmental Protection Agency (EPA), because of regenerative braking, electric motor drive/assist, and automatic start/shutoff, hybrid vehicles are able to utilize stop-and-go traffic to their advantage by conserving and regenerating energy.

• **Low Emissions**

In a hybrid vehicle, the electric motor releases no emissions, and the small gasoline engine recharges the battery and has to work less often, so its emissions are much lower. On average, hybrid cars produce about 80% less harmful pollutants and greenhouse gases than comparable gasoline cars, according to HybridCar.org.

As hybrid technology advances over the next few years, additional hybrid vehicles, such as heavy-duty trucks and plug-in hybrids, could also be added to the fleet in order to help reduce the amount of greenhouse gases released into our atmosphere. A recent study shows that conventional vehicles average 452 grams of CO_2 emissions per mile, whereas plug-in hybrid vehicles average only 294 grams of CO_2 emissions per mile (Richard, 2008). And according to a study performed by the Northeast Advanced Vehicle Consortium (NAVC), heavy-duty hybrid vehicles produce about 50% to 70% less particulate matter emissions, resulting in a lesser amount of hazardous airborne particles, like carbon.

• **Long Range**

Because the gasoline engine automatically recharges the electrical motor, hybrid vehicles can go hundreds of miles before refueling because of their excellent fuel economy. With a fuel tank of 11.9 gallons, the Toyota Prius can easily go more than 500 miles between gas stations.

■ **Model 15–1** ■ Memo report citing research—APA style

- **Conventional Fueling**

  Hybrid vehicles run on gasoline, so fuel is no different than for a standard automobile.

- **Adequate Power**

  Some power is lost between the conventional and hybrid versions, yet both have adequate power. The conventional 2008 Chevy Malibu has a V6 engine that holds 252 horsepower, and the hybrid Malibu holds 164 horsepower. Both can exceed 100 mph, but the hybrid Malibu averages about 7 more miles per gallon.

DISADVANTAGES

Hybrid technology is new, so there are still questions and concerns about it. Two disadvantages are especially important.

- **Electromagnetic Field Risk Still Uncertain**

  Although electromagnetic fields (EMFs) are all around us in cell phones, microwaves, televisions, and utility lines, the risk of exposure in hybrids is still uncertain. The kind of EMF most likely to be found in a hybrid is considered to be a lower frequency—which means that it dissipates at a very short distance from the source (HybridCars.com, 2006). In a company statement, Toyota said: "The measured electromagnetic fields inside and outside of Toyota hybrid vehicles in the 50 to 60 hertz range are at the same low levels as conventional gasoline vehicles. Therefore there are no additional health risks to drivers, passengers or bystanders" (Motivalli, 2008).

- **Complexity**

  With two powertrains, hybrids are far more complex than conventional cars, which could mean more down time and higher maintenance costs. However, with the recent rise in production of hybrid vehicles, there has also been a rise in the number of certified technicians that can perform the maintenance on them.

TAX AND LEGISLATIVE ADVANTAGES

Although there may be changes in tax incentives for hybrids, we should take advantage of them as long as they are available.

- **Tax Credits**

  State

  Right now, about 35 states offer tax incentives (ranging from $12 to $3,000) for the purchase of hybrid vehicles. Our Boston office could benefit from several incentives according to Senate Bill 1380, most notably a $2,000 income tax deduction. Other states where our offices are located, such as Missouri, New York, and Ohio, offer $1,500 to $3,000 in tax incentives for the purchase of hybrid vehicles.

  Federal

  According to the Internal Revenue Service (IRS), hybrid vehicles purchased or placed into service after December 31, 2005, and before December 31, 2010, may be eligible for a federal income tax credit of up to $3,400, depending on the make and model of the vehicle.

■ **Model 15–1** ■ continued

COSTS

Initial costs for a fleet of hybrids may seem high, but there are ways of reducing the cost of maintenance.

- **Purchase**

  Hybrid vehicles are more expensive than conventional gasoline vehicles. The current base model Chevy Malibu starts at $20,295 and the hybrid Malibu starts at $24,290. The base model Tahoe starts at $35,530, whereas the hybrid Tahoe retails at $50,490 (Chevrolet, 2008). Two of the least expensive hybrid sedans are the Toyota Prius ($21,100) and the Honda Civic ($22,600), and the least expensive hybrid SUV is the Ford Escape ($26,600) (Yahoo, 2008). Official prices for the upcoming hybrid trucks have not yet been released.

- **Maintenance**

  Most of the hybrid vehicles on the market come with standard 3-year/36,000-mile warranties and 8-year/100,000-mile warranties on the batteries. Routine maintenance should probably be performed at a dealer, especially if it is under warranty. Another option could be to provide certification to the M-Global fleet mechanics so that they can perform maintenance on hybrid vehicles themselves. Because the market for hybrid vehicles has become so popular, technician certification programs are popping up all over the country in order to keep pace with the demand.

THE HYBRID MARKET

The hybrid market is expanding. Each year, car companies are releasing new hybrid models and hybrid alternatives to existing models. The Toyota Prius remains as the top-selling hybrid vehicle (MSN Autos, 2008). When hybrids were first released in 2000, there were two—the Honda Insight and the Toyota Prius; now there are 25. According to JD Power and Associates, by 2012, hybrids will account for 3.5 percent of the new-car market, with 44 different hybrids for sale (The Automotive Editors, 2006).

THE OUTLOOK FOR HYBRID VEHICLES

Hybrid vehicles have a promising future, even as research continues on other technology.

- **Near Future**

  With the rising gas prices and growing concern for the environment, hybrid vehicles are an excellent way to save money and contribute to the reduction of greenhouse gases. Although most experts agree that hybrid vehicles are only a temporary alternative to conventional internal combustion automobiles, the next technology (fuel cell) is still many years from making itself into the mass market—particularly because fueling stations will have to become readily available to the general public.

- **2015 Predictions**

  Average gas prices are estimated to be around $4.50/gallon. There may be higher gas prices between now and then, but the investment to find additional reserves will be justified by the oil companies and prices could fluctuate (Belzowski, 2006). This rise in gas prices could affect the way that Americans purchase their vehicles, and it is possible that hybrids could become the predominant vehicle on the road.

■ **Model 15–1** ■ continued

• **2020 Predictions**

Average gas prices are estimated to be around $6.00/gallon. It is estimated that replacement energy will start and gas prices will drop (Belzowski, 2006). Many experts believe that this replacement energy will be in the form of hydrogen fuel cells, but others disagree because of the challenges that fuel cell technology must overcome. The biggest challenges are a low cost source of hydrogen and a hydrogen infrastructure (Belzowski, 2006). The possibility remains that hybrid vehicles could still be the predominant vehicle on the road, even after 2020.

CONCLUSION

Although the purchase of hybrid vehicles would be more expensive, we have a good chance of making up the initial expense in saved fuel costs and tax savings.

Because the market for hybrid vehicles now offers a sedan, SUV, and truck, we should be able to replace every vehicle in our fleet with a hybrid version in the same class. Furthermore, we can consider buying our vehicles using our discount arrangement with General Motors, as they now offer hybrid versions of the Malibu, Tahoe, and Silverado.

M-Global is well known for its environmental services, and having part or all of our fleet go green could help further emphasize this positive image in the minds of our customers.

REFERENCES

Belzowski, B. M. (2006, October). *Powertrain strategies for the 21st century.* University of Michigan Transportation Research Institute. Retrieved April 29, 2008, from http://www.osat.umich.edu/research/powertrain/NAPowertrainReportFinal1.pdf

Bickerstaffe, S. (2007). Cutting the cost of hybrids. *Automotive Engineer,* 32 (5), 34

Bullis, K. (2007). Electric cars 2.0. *Technology Review,* 110 (5), 100–101

Chevrolet. (n.d.). *2008 Chevy Showroom.* Retrieved April 28, 2008, from http://www.chevrolet.com/lineup

Global business briefs. (2007, June 10). *Wall Street Journal,* p. 11A

HybridCars.com. (2006, April 3). *Electromagnetic fields in hybrids.* Retrieved April 29, 2008, from http://www.hybridcars.com/safety/electromagnetic-fields-in-hybrids.html

HybridCars.com. (2008, April 27). *Hybrid EMF risk still uncertain.* Retrieved April 29, 2008, from http://www.hybridcars.com/safety/hybrid-emf-risk-still-uncertain.html

HybridCar.org. (n.d.). *Hybrid car emissions.* Retrieved April 28, 2008, from http://www.hybrid-car.org/hybrid-car-emissions.html

■ **Model 15–1** ■ continued

In search of the perfect battery. (2008, March 8). *The Economist,* 386, 22–24

Internal Revenue Service. (2007, November 8). *Summary of the credit for qualified hybrid vehicles.* Retrieved April 29, 2008, from http://www.irs.gov/newsroom/article/0,,id=157557,00.html

Motivalli, J. (2008, April 27). Fear, but few facts, on hybrid risk. *New York Times.* Retrieved April 29, 2008, from http://www.nytimes.com/2008/04/27/automobiles/27EMF.html?_r=1&oref=slogin

MSN Autos Editors. (n.d.) *Most popular hybrids on MSN.* Retrieved April 28, 2008, from http://editorial.autos.msn.com/article.aspx?cp-documentid=435590

Northeast Advanced Vehicle Consortium. (n.d.) *Heavy duty hybrid vehicle testing: Particulate matter (PM) emissions.* Retrieved April 29, 2008, from http://www.navc.org/HDPM.html

Richard, M. G. (2008, April 15). *Plug-in hybrid cars: Chart of CO_2 emissions ranked by power source.* Retrieved April 28, 2008, from http://www.treehugger.com/files/2008/04/plug-in-hybrid-cars-co2-emissions-electricity-energy.php

The Automotive Editors. (2006, February). Comparison—hybrids at the crossroads. *Popular Mechanics, February 2006.* Retrieved April 28, 2008, from http://www.popularmechanics.com/automotive/new_cars/2154662.html?page=5

U.S. Environmental Protection Agency. (n.d.). *How hybrids work.* Retrieved April 29, 2008, from http://www.fueleconomy.gov/feg/hybridtech.shtml

U.S. Environmental Protection Agency. (2008, January 17). *New energy tax credits for hybrids.* Retrieved April 28, 2008, from http://www.fueleconomy.gov/feg/tax_hybrid.shtml

U.S. Environmental Protection Agency. (2008, January 16). *2008 Fuel economy guide.* Retrieved April 29, 2008, from http://www.fueleconomy.gov/feg/FEG2008.pdf

Weisenfelder, J. (n.d.). *Top 10 2008 hybrids.* Retrieved April 28, 2008, from http://autos.yahoo.com/articles/autos_content_landing_pages/469/top-10-2

■ **Model 15–1** ■ continued

Chapter | 16 | The Job Search

>>> Chapter Outline

Researching Occupations and Companies   552

Job Correspondence   555

Job Letters   556

Resumes   558

Job Interviews   562

Preparation   562

Performance   565

Follow-Up Letters   566

Negotiating   567

Chapter Summary   571

Learning Portfolio   572

Communication Challenge   572

Collaboration at Work   573

Assignments   574

Models for Good Writing   577

Model 16–1: Job letter (modified block) and chronological resume   577

Model 16–2: Job letter (block style) and chronological resume   579

Model 16–3: Job letter (modified block) and functional resume   581

Model 16–4: Job letter (modified block) and functional resume   583

Model 16–5: Combined resume   585

Model 16–6: Combined resume formatted for submission online   586

Model 16–7: Resume with graphics—not effective for computer scanning   587

In applying for a job, you are selling yourself. It is no time for either self-delusion or false modesty. You must first assess your abilities, then find an appropriate job match, and finally persuade a potential employer that you are the right one for the job.

This chapter offers suggestions for landing a job in your profession. You'll find information on these main activities:

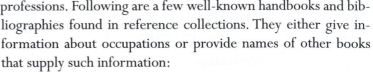

- Researching occupations and companies
- Writing job letters and resumes
- Succeeding in job interviews
- Negotiating with current and potential employers

>>> Researching Occupations and Companies

Before writing a job letter and resume, you may need information about (1) career fields that interest you (if you have not already chosen one), (2) specific companies that hire graduates in your field, and, obviously, (3) specific jobs that are available. Following are some pointers for finding such information—both from traditional sources and from your computer.

>> Do Basic Research in Your College Library or Placement Office

Libraries and placement centers offer one starting point for getting information about professions. Following are a few well-known handbooks and bibliographies found in reference collections. They either give information about occupations or provide names of other books that supply such information:

Career Choices Encyclopedia: Guide to Entry-Level Jobs

Dictionary of Occupational Titles

Directory of Career Training and Development Programs

Encyclopedia of Business Information Sources

Encyclopedia of Careers and Vocational Guidance

High-Technology Careers

Occupational Outlook Handbook

Professional Careers Sourcebook: An Information Guide for Career Planning

You can also check online sources such as *careeroverview.com*.

>> Interview Someone in Your Field of Interest

To get the most current information, arrange an interview with someone working in an occupation that interests you. This abundant source of information often goes untapped by college students, who mistakenly think such interviews are difficult to arrange. In fact, you can usually locate people to interview through (1) your college placement office, (2) your college alumni association, or (3) your own network of family and friends. Another possibility is to call a reputable firm in the field and explain that you wish to interview someone in a certain occupation. Make it clear, however, that you are not looking for a job—only information about a profession.

Once you set up the interview, prepare well by listing your questions in a notebook or on a clipboard that you take with you to the interview. This preparation keeps you on track and shows persons being interviewed that you value their time and information. Following are some questions to ask:

- How did you prepare for the career or position you now have?
- What college course work or other training was most useful?
- What types of activities fill your typical working day?
- What features of your career do you like the most? The least?
- What personality characteristics are most useful to someone in your career?
- How would you describe the long-term outlook of your field?
- How do you expect your career to develop in the next 5 years, 10 years, or 15 years?
- Do you know any books, periodicals, or online sources that might help me find out more about your field?
- Do you know any individuals who, like you, might permit themselves to be interviewed about their choice of a profession?

Although this interview may lead to a discussion about job openings in the interviewer's organization, the main purpose of the conversation is to retrieve information about an occupation.

>> Find Information on Companies in Your Field

With your focus on a profession, you can begin screening companies that employ people in the field chosen. First, determine the types of information you want to find; examples include location, net worth, number of employees, number of workers in your specific field, number of divisions, types of products or services, financial rating, and names and titles of company officers. The following are some sources that might include such information. They can be found in the reference sections of libraries and in college placement centers.

The Career Guide gives overviews of many American companies and includes information such as types of employees hired, training opportunities, and fringe benefits.

Corporate Technology Directory profiles high-tech firms and covers topics such as sales figures, number of employees, locations, and names of executives.

Facts on File Directory of Major Public Corporations gives essential information on 5,700 of the largest U.S. companies listed on major stock exchanges.

Job Choices in Science and Engineering, an annual magazine published by the College Placement Council, includes helpful articles and information about hundreds of companies that hire technical graduates.

Peterson's Business and Management Jobs provides background on employers of business, management, and liberal-arts graduates.

Peterson's Engineering, Science, and Computer Jobs provides background information on employers of technical graduates.

Standard & Poor's Register of Corporations, Directors, and Executives lists names and titles of officials at 55,000 public and private U.S. corporations.

>> Do Intensive Research on a Selected List of Potential Employers

The previous steps help get you started finding information on occupations and firms. Ultimately, you will develop a selected list of firms that interest you. Your research may have led you to these companies, or your college placement office may have told you that there are job openings there. Now you must conduct an intensive search to learn as much as you can about the firms. Following are a few sources of information, along with the kinds of questions each source helps answer.

- **Annual reports** (often available at company web sites and in your library or placement office): How does the firm describe its year's activities to stockholders? What are its products or services?

- **Websites or media kits** (available online or from public relations offices): How does the firm portray itself to the public? Is this firm "Internet-savvy"? What can you infer about the firm's corporate culture?

- **Personnel manuals and other policy guidelines:** What are features of the firm's *corporate culture*? How committed is the firm to training? What are the benefits and retirement programs? Where are its branches? What are its customary career paths?

- **Graduates of your college or university now working for the firm:** What sort of reputation does your school have among decision makers at the firm?

- **Company newsletters and in-house magazines:** How open and informative is the firm's internal communication?

- **Business sections of newspapers and magazines:** What kind of news gets generated about the firm?

- **Professional organizations or associations:** Is the firm active within its profession?

- **Stock reports:** Is the firm making money? How has it done in the past five years?

- **Accrediting agencies or organizations:** How has the firm fared during peer evaluations?

- **Former employees of the company:** Why have people left the firm?

- **Current employees of the company:** What do employees like, or dislike, about the company? Why do they stay?

Other good sources include the Better Business Bureau, Chamber of Commerce, and local newspapers. In other words, you should thoroughly examine an organization from the outside. The information you gather helps you decide where to apply and, if you later receive a job offer, where to begin or continue your career.

>> Use Your Computer to Gather Data

Today, many applicants go directly to the computer to find information about professions, organizations, graduate schools, and job openings. No doubt between the time this book is written and then published, the names and number of online resources will change dramatically. Generally, some of the information available include the following:

- College and university catalogs
- Websites for companies, organizations, and schools
- Employment listings from local and national sources
- Online discussion forums involving recent graduates of colleges and universities

There are a number of job search sites available on the Internet. The following list is current, as of this writing.

America's Career InfoNet www.acinet.org

CareerBuilder www.careerbuilder.com

College Grad Job Hunter www.collegegrad.com

Jammin' Jobs www.jamminjobs.com

Monster www.monster.com

NationJob Network www.nationjob.com

At many of these sites, you can

1. Get tips on writing your resume and searching for jobs
2. Post your resume at the site, in multiple forms for varied employers
3. Peruse lists of job openings
4. Receive advice from professionals who are in careers you wish to pursue
5. Read in-depth reports on particular companies
6. Share comments with other job-seekers

In other words, the Internet helps you locate a variety of information during your job search. Moreover, you can use your computer to search for openings and respond to job ads, as mentioned in the next section.

>>> Job Correspondence

Job letters and resumes must grab the attention of busy readers, who may spend only 60 seconds deciding whether to consider you further. This section gives you the tools to write a successful letter and resume. *Successful,* of course, means a letter and resume that get you an interview. After that, your interpersonal skills help you land the job. The letter and resume only aim to get you to the next step—the personal interview.

Most job letters and resumes still get sent through the mail. However, a growing number of applicants use the Internet to apply for jobs.

For example, online services can place resumes into a bank used by hundreds and perhaps thousands of companies. The resumes are scanned with software that searches for key words that reflect abilities needed for specific jobs. The program then sends selected resumes to companies. If you use this kind of service, remember one point: When you send credentials into cyberspace, you cannot be sure where they will land. Do not expect the level of confidentiality and security that you have with personal mail.

Whether you use online techniques like email and resume services or stick with the traditional approach, the same basic writing guidelines apply. Your letter, no longer than one page, should be specific about the job you seek and your main selling points. Then the resume—one page or two at most—should simply, specifically, and neatly highlight your background.

Job Letters

A job letter is just another type of sales letter—except that you are selling yourself, not a product or service. In preparing to write one, take the point of view of the persons to whom you are writing. What criteria do they use to evaluate your credentials? How much or how little do they want in the letter? What main points are they hunting for as they scan your resume? Accordingly, this section first examines the needs of these readers and then gives guidelines for you, the writer. Models 16–1 through 16–4, on pages 577–584 include sample job letters and resumes.

The Readers' Needs

You probably will not know personally the readers of your job letter, so you must think hard about what they may want. Your task is complicated by the fact that often there are several readers of your letter and resume who may have quite different backgrounds.

One possible scenario follows:

Step 1: The letter may go first to the personnel office, where a staff member specializing in employment selects letters and resumes that meet the criteria stated in the position announcement. (In some large employers, letters and resumes may even be stored in a computer, where they are scanned for key words that relate to specific jobs.)

Step 2: Applications that pass this screening are sent to the department manager who supervises the employee that is hired. The manager may then select a group to be interviewed. This manager interviews applicants and ultimately hires the employee.

One variation of this process has the personnel department doing an interview as well as screening letters and resumes—before the department manager even hears about any applications. Another variation, as noted earlier, has the employer relying on an online resume service for the initial screening.

Yet sooner or later, a supervisor or manager reads your letter and resume. And most readers, whatever their professional background, have the following five characteristics in common:

>> Feature 1: They Read Job Letters in Stacks

Most search-and-screen processes are such that letters get filed until there are many to evaluate. Your reader faces this intimidating pile of paper, from which you want your letter to emerge as the victor.

>> Feature 2: They Are Tired

Some employment specialists may save job letters for their fresher moments, but many people who do the hiring get to job letters at the end of a busy day or at home in the evening, so they have even less patience than usual for flowery wording or hard-to-read print.

>> Feature 3: They Are Impatient

Your readers expect major points to jump right out at them. In most cases, they will not dig for information that cannot be found quickly.

>> Feature 4: They Become Picky Grammarians

Readers of all professional and academic backgrounds expect good writing when they read job letters. There is an unspoken assumption that a letter asking for a chance at a career should reflect solid use of the language. Furthermore, it should have no typographical errors. If the letter does contain a typo or grammar error, the reader may wonder about the quality of writing you will produce on the job.

>> Feature 5: They Want Attention Grabbers but Not Slickness

You want your letter and resume to stand out without the use of gimmicks. Most readers prefer a tasteful and reserved format that does not draw too much attention to itself. For example, white or off-white stationary is still the standard, along with traditional fonts with lots of white space for easy reading. If you want to attract attention in a professional manner, consider attaching a business card to your letter so that the reader has your name and number handy. Students can design and print business cards with software available in many college computer labs.

Of course, likes and dislikes vary. An advertising director, who works all day with graphics, may want a bolder format design than an engineering manager, who works with documents that are less flashy. If you cannot decide, it is best to use a conservative format and style.

The Letter's Organization

The job-letter guidelines that follow relate to the features mentioned about readers. Your one and only goal is to tantalize the reader enough to want to interview you—that is all.

ABC Format: Job Letters

- **ABSTRACT:** Apply for a specific job
  - Refer to ad, mutual friend, or other source of information about the job
  - Briefly state how you can meet the main need of your potential employer
- **BODY:** Specify your understanding of the reader's main needs
  - Provide main qualifications that satisfy these needs (but only highlight points from resume—do not simply repeat all resume information)
  - Address specific qualifications mentioned in a job announcement
  - Avoid mentioning weak points or deficiencies
  - Keep body paragraphs to six or fewer lines
  - Use a bulleted or numbered list if it helps draw attention to three or four main points
  - Maintain the "you" attitude throughout
- **CONCLUSION:** Tie the letter together with one main theme or selling point, as you would a sales letter
  - Refer to your resume
  - Explain how and when the reader can contact you for an interview

With that goal and the reader's needs in mind, your job letter should follow the ABC format on the left.

This pattern gives you a starting point, but it is not the whole story. There is one feature of application letters that cannot be placed easily in a formula—style. Work hard with your draft to develop a unity and flow that, by itself, sets you apart from the crowd. Your attention grabber engages interest, but the clarity of your prose keeps readers attentive and persuades them that you are an applicant who should be interviewed.

Resumes

Resumes usually accompany application letters. Three points make writing resumes a challenge:

1. **Emphasis:** You should select just a *few major points of emphasis* from your personal and professional life. Avoid the tendency to include college and employment details best left for the interview.

2. **Length:** You usually should use only *one page*. For individuals with extensive experience, a two-page resume is acceptable—if it is arranged evenly over both pages.

3. **Arrangement:** You should arrange information so that it is *pleasing to the eye and easy to scan*. Prospective employers spend less than a minute assessing your application. They may even use computers to scan resumes, taking even less time.

Computers pose a special challenge to you as a resume writer, because they fail to appreciate some of the elegant variations and innovations sometimes used to get the human reader's attention. If you are writing a resume that may be read by a computer, you may want to (1) use white or very light-colored paper, (2) focus on key words—especially job skills—that might be picked up by the computer scan, and (3) avoid design features that might present obstacles to the scan such as italics, fancy typefaces, and graphics like those in Model 16–7. Actually, you may find that your placement office uses a computer resume program that requires a particular format, effectively removing decisions about style from your consideration.

You may also need to format a resume that can be copied and pasted into forms for online

applications. Such resumes should be saved as text files (.txt). It is a good idea to create this kind of resume in Notepad or similar program on your computer. Avoid using tabs, italics, bolding, bullets, or other special characters, as these do not translate to the text file. (For an example of a resume formatted to be submitted online, see Model 16–6.)

This section distills the best qualities of many formats into three basic patterns:

1. The chronological resume, which emphasizes employment history
2. The functional resume, which emphasizes the skills you have developed
3. The combined resume, which merges features of both the chronological and functional formats. (See the "Experience" section that follows to learn when to use each format. Choose the pattern that best demonstrates your strengths.)

The following paragraphs describe the main parts of the resume. The "Experience" section explains the differences between chronological, functional, and combined resumes. Refer to the models on pages 577–587 for resume examples.

Objective

Personnel directors, other people in the employment cycle, and even computers may sort resumes by the Objective statement. Writing a good objective is hard work, especially for new graduates, who often just want a chance to start working at a firm at any level. Despite this eagerness to please, do not make the mistake of writing an all-encompassing statement such as: "Seeking challenging position in innovative firm in civil-engineering field." Your reader will find such a general statement of little use in sorting your application. It gives the impression that you have not set clear professional goals.

Most objectives should be short, preferably one sentence. Also, they should be detailed enough to show that you have prepared for, and are interested in, a specific career, yet open-ended enough to reflect a degree of flexibility. If you have several quite different career options, you might want to design a different resume for each job description, rather than trying to write a job objective that takes in too much territory.

Note: Some employers prefer that you not include an objective. For example, you may be applying for an entry-level job for which an objective would be inappropriate. As always, consider your reader's needs as you make decisions about objectives.

Education

Whether you follow the objective with the "Education" or "Experience" section depends on the answer to one question: Which topic is most important to the reader? Most recent college graduates lead off with "Education," particularly if the completion of the degree prompted the job search.

This section seems simple at the outset. Obligatory information includes your school, school location, degree, and date of graduation. It is what you include beyond the bare details, however, that most interests employers. Following are some possibilities:

■ **Grade point average:** Include it if you are proud of it; do not if it fails to help your case.

- **Honors:** List anything that sets you apart from the crowd—such as dean's list or individual awards in your major department. If you have many, include a separate "Recognitions" heading toward the end of the resume.

- **Minors:** Highlight any minors or degree options, whether they are inside or outside your major field. Employers place value on this specialized training, even if (and sometimes especially if) it is outside your major field.

- **Key courses:** When there is room, provide a short list of courses you consider most appropriate for the kind of position you are seeking. Because the employer probably will not look at your transcripts until a later stage of the hiring process, use this brief listing as an attention grabber.

Experience

This section poses a problem for many applicants just graduating from college. Students often comment that experience is what they are looking for, not what they have yet. Depending on the amount of work experience you have gained, consider three options for completing this section of the resume: (1) emphasize specific positions you have held (chronological resume), (2) emphasize specific skills you have developed in your experience (functional resume), or (3) emphasize both experience and skills (combined resume).

Option 1: Chronological Format

This option works best if your job experience has led logically toward the job you now seek. Models 16–1 and 16–2 include examples of chronological resumes that meet these guidelines:

- List relevant full-time or part-time experience, including internships, in reverse chronological order.

- Be specific about your job responsibilities while still being brief.

- Be selective if you have had more jobs than can fit on a one-page resume.

- Include nonprofessional tasks (such as working on the campus custodial staff) if it helps your case (e.g., the employer might want to know that you worked your way through college).

- Remember that if you leave out some jobs, the interview will give you the chance to elaborate on your work experience.

- Select a readable format with appropriate white space.

- Use action verbs and lists to emphasize what you did or what you learned at jobs—for example, "Provided telephone support to users of System/23." Use parallel form in each list.

Option 2: Functional Format

This approach works best if (1) you wish to emphasize the skills and strengths you have developed in your career rather than specific jobs you have had, or (2) you have had "gaps" in your work history that would be obvious if you used the chronological format. Although it is sometimes used by those whose job experience is not a selling point, this is not always the case. Sometimes your skills built up over time may be the best argument for your being considered for a position, even if your job experience also is strong. For example, you may have five years' experience in responsible positions at four different retailers. You then decide to write a functional resume focusing on the three skill areas you developed: sales, inventory control, and management. Models 16–3 and 16–4 include functional resumes.

If you write a functional resume that stresses skills, you may still want to follow this section with a brief employment history (see option 3). Most potential employers want to know where and when you worked, even though this issue is not a high priority. *Note:* If you decide to leave out the history, bring it with you to the interview on a separate sheet.

Option 3: Combined Format

The combined format uses features of both chronological and functional formats. This format works best when you want to emphasize the skills you have developed while still giving limited information on the chronology of your employment.

Models 16–5 and 16–6 on pages 585–586 show two variations of the combined format. Model 16–5 integrates chronological information into the skills section. The positions held may not be prestigious, but together they show that the applicant has considerable experience developing the two sets of skills listed: Editing/Writing and Teaching/Research. In Model 16–6, the experience section looks exactly as it would in a functional resume, with subheadings giving the names of skills. However, the writer adds a brief skeleton work history near the end of the page; she believes the reader will want some sort of chronological work history, even if it is not the writer's strength.

Activities, Recognitions, Interests

Most resumes use one or two of these headings to provide the reader with additional background information. The choice of which, if any, to use depends on what you think best support your job objective. Following are some possibilities:

- **Activities:** selected items that show your involvement in your college or your community or both.
- **Recognitions:** awards and other specific honors that set you apart from other applicants. (Do not include awards that might appear obscure, meaningless, or dated to the reader, such as most high-school honors.)
- **Interests:** hobbies or other interests that give the reader a brief look at the "other" you.

However you handle these sections, they should be fairly brief and should not detract from the longer and more significant sections described previously.

References

Your resume opens the door to the job interview and later stages of the job process, when references will be called. There are two main approaches to the reference section of the resume:

1. Writing "Available upon request" at the end of the page
2. Listing names, addresses, and phone numbers at the end of the resume or on a separate page

The first approach assumes that the reader prefers the intermediate step of contacting you before references are sent or solicited. The second approach assumes that the reader prefers to call or write references directly, without having to contact you first. Use the format most commonly used in your field or, most important, the one most likely to meet the needs of a particular employer. As always, be ready to tailor your letter and resume each time you put it in the mail.

Your goal is to write an honest resume that emphasizes your good points and minimizes your deficiencies. To repeat a point made at the outset, you want your resume and job letter to open the door for later stages of the application process. Look on this writing task as your greatest persuasive challenge. Indeed, it is the ultimate sales letter, for what you are selling is the potential you offer to change an organization and, perhaps, the world as well. Considering such heady possibilities, make sure to spend the time necessary to produce first-rate results.

>>> Job Interviews

Your job letter and resume have only one purpose: to secure a personal interview by the personnel director or other official who screens applicants for a position.

Much has been written about job interviews. Fortunately, most of the good advice about interviewing goes back to just plain common sense about dealing with people. Following are some suggestions to show you how to prepare for a job interview, perform at your best, and send a follow-up letter.

Preparation

>> Do Your Homework on the Organization

You have learned how to locate data about specific companies. Once you have been selected for an interview, review whatever information you have already gathered about the employer, and then go one step further by searching for the most current information you can find. Your last source may be someone you know at the organization, or a friend of a friend.

When you don't have personal contacts, use your research skills again. For large firms, locate recent periodical or newspaper articles by consulting general indexes—such as the *Business Periodicals Index, Wall Street Journal Index, Readers' Guide to Periodicals, New York Times Index*, or the index for any newspaper in a large metropolitan area. For smaller firms, consult recent issues of local newspapers for announcements about the company. Being aware of current company issues demonstrates your initiative and shows your interest in the firm.

As noted earlier, the company's website can also be a good source of current information about an organization.

>> Write Out Answers to the Questions You Consider Likely

You probably would not take written answers with you to the interview, but writing them out will give you a level of confidence unmatched by candidates who only ponder possible questions that might come their way. This technique resembles the manner in which some people prepare for oral presentations: First they write out a speech, then they commit it to notes, and finally they give an extemporaneous presentation that reflects confidence in themselves and knowledge of the material. This degree of preparation places you ahead of the competition.

There are few, if any, original questions asked in job interviews. Most interviewers simply select from some standard questions to help them find out more about you and your background. Following are some typical questions, along with tips for responses:

1. **Tell me a little about yourself.** Keep your answer brief and relate it to the position and company—do not wander off into unrelated issues, like hobbies, unless asked to do so.

2. **Why did you choose your college or university?** Be sure your main reasons relate to academics—for example, the academic standing of the department, the reputation of the faculty, or the job placement statistics in your field.

3. **What are your strengths?** Focus on two or three qualities that would directly or indirectly lead to success in the position for which you are applying.

4. **What are your weaknesses?** Choose weaknesses that if viewed from another perspective, could be considered strengths—for example, your perfectionism or overattention to detail.

5. **Why do you think you would fit into this company?** Using your research on the firm, cite several points about the company that correspond to your own professional interests—for example, the firm may offer services in three fields that relate to your academic or work experience.

6. **What jobs have you held?** Use this question as a way to show that each previous position, no matter how modest, has helped prepare you for this position—for example, part-time employment in a fast-food restaurant developed teamwork and interpersonal skills.

7. **What are your long-term goals?** Be ready to give a 5- or 10-year plan that, preferably, fits within the corporate goals and structure of the firm to which you are

applying—for example, you may want to move from the position of technical field engineer into the role of a project manager, to develop your management skills.

8. **What salary range are you considering?** Avoid discussing salary if you can. Instead, note that you are most interested in criteria such as job satisfaction and professional growth. If pushed, give a salary range that is in line with the research you did on the career field in general and this company in particular; see the section in this chapter regarding negotiating.

9. **Do you like working in teams or prefer working alone?** Most employers want to know that you have interest and experience in teamwork—whether in college courses or previous jobs, but they also admire and reward individual accomplishment. In deciding what part of your background to emphasize, consider the corporate culture of the organization interviewing you.

10. **Do you have any questions of me?** Always be ready with questions that reinforce your interest in the organization and your knowledge of the position—for example, "Given the recent opening of your Tucson warehouse, do you plan other expansions in the Southwest?" or "What types of in-house or off-site training do you offer new engineers who are moving toward project management?" Other questions can concern issues such as (a) benefits, (b) promotions, (c) type of computer network, and (d) travel requirements.

>> Do Mock Interviews

You can improve your chances considerably by practicing for job interviews. One of the easiest and best techniques is role-playing. Ask a friend to serve as the interviewer and give him or her a list of questions from which to choose. Also, inform that person about the company so that he or she can improvise during the session. This way, you are prepared for the real thing.

You can get additional information about your interviewing abilities by videotaping your role-playing session. Reviewing the videotape helps you highlight (1) questions that pose special problems for you and for which you need further preparation and (2) mannerisms that need correction. This preparation technique is especially useful if you are one of the growing number of applicants who take part in a video interview with a recruiter.

>> Be Physically Prepared for the Interview

Like oral presentations, job interviews work best when you are physically at your best; therefore, all the old standbys apply:

- Get a good night's rest before the interview.
- Avoid caffeine or other stimulants.
- Eat about an hour beforehand so that you are not distracted by hunger pangs during the session.
- Take a brisk walk to dispel nervous energy.

Performance

Good planning is your best assurance of a successful interview. Of course, there are always surprises that may catch you. Remember, however, that most interviewers are seriously interested in your application and want you to succeed. Help them by selling yourself and thus giving them a reason to hire you. Following are some guidelines for the interview:

>> Dress Appropriately

Much has been written on the topic of appropriate attire for interviews. Some practical suggestions that are often emphasized include:

- Dress conservatively and thus avoid drawing attention to your clothing—for example, do not use the interview as an opportunity to break in a garment in the newest style.
- Consider the organization—for example, a brokerage-firm interview may require a dark suit for a man and a tailored suit for a woman, whereas an interview at a construction firm may require less formal attire.
- Avoid excessive jewelry.
- Pay attention to the fine points—for example, wear shined shoes and carry a tasteful briefcase or notebook.

>> Take an Assertive Approach

Either directly or indirectly, use everything you say to make the case for your hiring. Be positive, direct, and unflappable. Use every question as a springboard to show your capabilities and interest, rather than waiting for point-blank questions about your qualifications. To be sure, the degree to which you assert yourself partly depends on your interpretations of the interviewer's preference and style. Although you do not want to appear pushy, you should take the right opportunities to sell yourself and your abilities.

>> Use the First Few Minutes to Set the Tone

What you have heard about first impressions is true: Interviewers draw conclusions quickly. Having given many interviews, they are looking for an applicant who injects vitality into the interview and makes their job easier. Within a minute or two, establish the themes and the tone that will be reinforced throughout the conversation—that is, your relevant background, your promising future, and your eagerness (not pushiness). In this sense, the interview subscribes to the Preacher's Maxim mentioned in chapter 14: "First you tell 'em what you're gonna tell 'em, then you tell 'em, and then you tell 'em what you told 'em."

>> Maintain Eye Contact While You Speak

Although you may want to look away occasionally, much of the time your eyes should remain fixed on the person interviewing you. This way, you show interest in what she or he is saying.

If you are being interviewed by several people, make eye contact with all of them throughout the interview. No one should feel ignored. You are never quite certain exactly who may be the decision maker in your case.

>> Be Specific in the Body of the Interview

In every question, you should see the opportunity to say something specific about you and your background. For example, rather than simply stating that your degree program in computer science prepared you for the open position, cite three specific courses and briefly summarize their relevance to the job.

>> Do Not Hesitate

A job interview is no time to hesitate, unless you are convinced the job is not for you. If the interviewer notes that the position involves travel 40% of the time, quickly respond that the prospect of working around the country excites you. The question is this: Do you want the job or not? If you do, then accept the requirements of the position and show excitement about the possibilities. You can always turn down the job if you receive an offer and decide later that some restrictions, like travel, are too demanding.

>> Reinforce Main Points

The interviewer has no text for the session other than your resume; therefore, you should drive home main points by injecting short summaries into the conversation. After a 5-minute discussion of your recent work experience, take 15 seconds to present a capsule version of relevant employment. Similarly, orchestrate the end of the interview so that you have the chance to summarize your interest in the position and your qualifications. Here is your chance to follow through on the "Tell 'em what you told 'em" part of the Preacher's Maxim.

Follow-Up Letters

Follow every personal contact with a letter or email—whatever is most appropriate—to the person with whom you spoke. Send it within 24 hours of the interview or meeting so that it immediately reinforces the person's recollection of you. This simple strategy gives you a powerful tool for showing interest in a job.

Follow-up letters abide by the same basic letter pattern discussed in chapter 7. In particular, follow these guidelines:

- Write no more than one page.
- Use a short first paragraph to express appreciation for the interview.
- Use the middle paragraph(s) to (a) reinforce a few reasons why you would be the right choice for the position or (b) express interest in something specific about the organization.
- Use a short last paragraph to restate your interest in the job and to provide a hopeful closing.

See chapter 5 for the various formats appropriate for all types of business letters. A sample thank-you letter follows:

Dear Ms. Ferguson:

I enjoyed meeting with you yesterday about the career possibilities at Klub Kola's district headquarters. The growth that you are experiencing makes Klub an especially exciting company to join.

As I mentioned, my marketing background at Seville College has prepared me for the challenge of working in your new Business Development Department. Several courses last semester focused specifically on sales strategies for consumer goods. In addition, an internship this semester has given me the chance to try out marketing strategies in the context of a local firm.

Again, thank you for the chance to learn about your firm's current success and promising future. I remain very interested in joining the Klub Kola team.

Sincerely,

Marcia B. Mahoney

Marcia B. Mahoney

When your audience might appreciate a less formal response, consider writing your interviewer a personal note instead of a typed letter. This sort of note is most appropriate when you plan a short message.

>>> Negotiating

All of us negotiate every day of our lives. Both on the job and in our personal lives, we constantly find ourselves in give-and-take discussions to negotiate issues as diverse as those that follow:

- Major and minor purchases
- Relationships with spouses and friends
- Performance evaluations—with bosses and with subordinates
- Salaries—with those to whom we report and with those who report to us

Because negotiating will become an important part of your career, it receives attention in this final section. After some brief background information, the focus is on six guidelines that will steer you toward successful negotiations, both when you are hired and also at other points in your career. The main example used in this section is a salary negotiation for an entry-level position.

How has the art of negotiating changed recently? In the past, the process was often characterized by words like *trickery, intimidation,* and *manipulation.* In this game's lexicon, there were *winners* and *losers* and lots of warlike imagery. Participants were seen as battlefield adversaries who took up extreme positions, defended and attacked each other's flanks, finally agreed reluctantly to some middle ground, and then departed, wounded and usually uncertain of who had won the battle.

Today, the trend is away from this war-zone approach with its "I win, you lose" mentality. As a negotiator, you must enter the process searching for common ground for a very practical reason: Long-term relationships are at stake. In later negotiations, you are much more likely to achieve success if the present negotiation helps both parties. This goal—"We both win"—requires a new set of practices at the negotiation table.

Specifically, six guidelines should drive the negotiation process, all of which embody the viewpoint that successful negotiations involve honest communication wherein both parties benefit. Try to weave these six guidelines into the style of negotiating that you develop.

>> Negotiating Guideline 1: Think Long Term

Enter every negotiation with a long-term strategy for success. You must establish and nurture a continuing relationship with the person on the other side of the table. Later dealings might depend on mutual understandings and goodwill that result from your first meeting. First impressions do count.

How might such long-term thinking apply to actual contract discussions for jobs, especially for your first position after graduating? If you are fortunate enough to be in demand in the job market, you have the leverage to discuss salary expectations and other benefits during an interview. Such discussions often are characterized by you and the employer sharing details about your expectations and the employer's offer. You should enter such sessions with a realistic idea of what you can command in the marketplace. Neither sell yourself short nor harbor inflated ideas of your worth. Your college or university placement office should be able to provide information about salary ranges and benefit options for graduates in your field and organizations in your region.

Of course, the "real world" of the job hunt is such that the supply of new talent may overshadow the demand. You may be so glad to receive a good offer that you hesitate to jeopardize it by attempting to negotiate. Yet, ironically, you can damage your long-term interests in an organization by being overly timid before accepting an offer. Even if there is little or no room for salary negotiation, you should engage in a wide-ranging discussion that allows you to explore options for your contract and learn about features of the position. This dialog helps you learn about the organization. It also gives the employer a healthy respect for your ability to ask serious questions about your career.

Whatever your bargaining position, take advantage of the opportunity to discuss features of your job and the organization. Questions like those that follow may yield important information for you and show your interest in developing a long-term relationship with the employer:

- What philosophy underlies the firm's approach to management?
- What is the general timetable for career advancement?
- Where will your specific job lead?
- What opportunities exist for company-sponsored training?
- How will you be evaluated, and how often?

Employers respect applicants who have done enough homework to ask informed questions about the firm's employment practices. Both parties benefit from a frank and detailed discussion. You get what you need to make an informed decision about the firm, and your potential employer can showcase the organization and observe your ability to ask perceptive questions.

>> Negotiating Guideline 2: Explore Many Options

The negotiation process sometimes begins with only two options—your salary objective and the employer's offer—with seemingly little room for movement. You can escape this *either/or* trap by working to explore many options in the early stages of contract negotiation. This technique opens both parties to a variety of possible solutions and keeps the discussion rolling.

For example, assume that M-Global recently decided to add a new computer systems engineer to the staff at the corporate office in Baltimore. As a college senior about to graduate with a degree in computer science, you applied for the job and had a good first interview. The next week you are called back for a second interview and are offered a job, with a starting salary that (you are told) reflects the standard salary at the firm for new engineers with no experience. However, your research suggests that entry-level jobs in your field should pay a full $3,000 more than the M-Global offer. Although this difference concerns you, you have heard good things about the working environment at M-Global and would like to join the firm.

If you immediately were to state your need for the higher starting salary, the negotiation might be thrown into the either/or trap that leaves little room for agreement. Instead, you should keep the conversation going by putting additional options on the table and asking open-ended questions (i.e., questions that require more than a "yes" or "no" answer). For example, you could temporarily put aside your salary objective and ask how M-Global arrived at the offer figure. While giving the M-Global representative the chance to get facts on the table, this strategy also gives you opportunities to develop and then offer alternatives other than the two salary figures. The discussion might lead to options like these: (1) starting at the lower salary but moving to a higher figure after a successful 90-day trial period; (2) starting at the lower figure but receiving an enhanced stock-option package on being hired; or (3) starting at a slightly higher salary, but giving up the standard moving allowance offered to entry-level employees.

The point is that you must be careful to avoid rigidity. Consider possibilities other than the two ideal goals both parties brought to the negotiation table.

>> Negotiating Guideline 3: Find the Shared Interests

If you succeed in keeping options open during the negotiation process, you will discover points on which you agree. Psychologically, it is to your advantage to draw attention to these points rather than to points of conflict. Finding shared interests helps establish a friendship that, in turn, makes your counterpart more willing to compromise.

Let's go back to the preceding M-Global example. Assume you are continuing to discuss a number of salary options but have reached no agreement. Chances for closure may increase if you temporarily stop discussing salary and instead search for points, however minor, on which you agree. For example, you could ask about job tasks in the position. When you learn that new engineers spend about 25% of their workday writing reports, you comment that your college training included two electives in technical writing, along with a senior-level research report. The M-Global representative praises the extra effort you made to prepare for the communication tasks in a technical profession.

This discussion about writing, although brief, highlights information that may have been missed during M-Global's early reviews of your application. The company's interest in good writing overlaps with the extra effort you gave to this discipline in college. That shared interest may motivate the company to offer a salary figure closer to what you desire. At the very least, you have reinforced the decision M-Global officials made to offer you the job over three other finalists.

>> Negotiating Guideline 4: Listen Carefully

Despite shared interests, negotiations often return to basic differences. An effective technique at this point is to seek information on the rationale behind your counterpart's views. It furthers the negotiation and, in fact, your own case to ask questions and then listen carefully to the answers coming from the other side of the table.

How are we helped by asking questions? Let's return to the M-Global example. When you are confronted with the salary offer, ask how M-Global arrived at that figure. Your question may uncover what is really behind the offer. Did M-Global recently make similar offers to other applicants? Is M-Global aware of national salary surveys that tend to support your request? Asking such questions benefits both you and the entire negotiation process in four ways:

- You give your counterparts the opportunity to explain their views (thus breaking out of the either/or cycle).
- You discover what motivates them (making it more likely that you will reach consensus).
- You expose careless logic and unsupported demands.
- You move closer to objective standards on which to base negotiations.

From your persistent questioning, careful listening, and occasional responses, information may emerge that would otherwise have remained buried. You may discover, for example, that M-Global is basing its salary offer on data pertaining to another part of the country, where both salaries and costs of living are lower. That would give you the opportunity to argue for a higher starting salary on the basis of regional differences in compensation.

>> Negotiation Guideline 5: Be Patient

In the old hard-sell negotiations, participants frequently pushed for quick decisions, often to the regret of at least one of the parties. The better approach is to slow down the process. For example, you might want to delay agreement on a final salary figure until a later meeting, giving both you and your counterpart the chance to digest the conversation and consider options.

The main benefit of slowing down the process is to prevent basing decisions on the emotionalism of the moment. When objectivity takes a backseat to emotions in any negotiation—with an applicant, a client, a spouse, or a vendor—it is always best to put on the brakes, for two reasons:

- Good negotiated settlements should stand the test of time. When one party feels pressured, mistakes are made.
- Well-thought-out decisions are more likely to produce better long-term relationships, a major goal of your negotiations.

>> Negotiating Guideline 6: Do Look Back

Conventional wisdom has it that once you negotiate an agreement, you should not look back to second-guess yourself, because it only makes you less satisfied with what cannot be changed. That kind of thinking assumes that negotiations are spontaneous phenomena that cannot be analyzed, which is not true. If you have conducted your negotiations methodically, you will have much to gain from postmortems—particularly if they are in writing. Keep a negotiation journal to review before every major negotiation starts. Besides reminders, this journal should contain a short summary of previous negotiations. Make these entries immediately after a session ends, being sure to answer the following questions:

- What options were explored before a decision was made?
- What shared interests were discovered?
- Did you emphasize these shared interests?
- What questions did you ask?
- How did you show that you were listening to responses?

So do look back. Analyze every negotiation to discover what went right and what went wrong during the proceedings. Like other communication skills, such as writing and speaking, the ability to negotiate improves with use. With a few basic guidelines in mind and a journal on which to reflect, you will discover the power of friendly persuasion.

>>> Chapter Summary

This chapter surveys the entire process of searching for jobs, from performing your initial research to negotiating a contract. As a first step in the process, use the Internet and other sources of information to learn about occupations and specific employers that interest you. Second, write letters and resumes that get attention and respond to specific needs of employers. You can choose from chronological, functional, or combined resume formats, using the patterns of organization and style that best highlight your background. Third, prepare carefully for your job interview, especially in anticipating the questions that may be asked. Then perform with confidence. Also, do not forget to send a thank-you letter or email soon after the interview. Finally, use the negotiation phase of the job-search process to begin building a long-term relationship with your future employer, rather than focusing only on short-term gains of your first contract.

>>> Learning Portfolio

Communication Challenge "20-Something—Have Degree, Won't Travel"

Zach Lipkowski has been out of Mountain State College for almost a year and is still looking for his first real job. In the meantime, he has become a part-time ski instructor and worked in construction. With the help of a cheap apartment and a loan from his parents, he is managing to cover his bills, but he certainly hasn't used his biology degree from Mountain State. Now he finally has a break, having learned about an opening at M-Global's Denver office. What follows is a description of Zach's background, along with some questions related to his application letter, resume, job interview, and an assignment for a written response to the Challenge.

A Checkered College Career

At Mountain State, Zach majored in biology, a field that had interested him since childhood and that he thought promised a good career. His boyhood hobby of studying poison arrow frogs kindled his interest in the life sciences. It also led to a field trip to the Costa Rican rain forest in high school with a Denver biological group. He even helped a well-known frog expert on a study, providing the Denver researcher with observations about exotic frogs Zach kept as pets at home. So biology was a good fit.

In the first couple years of college, Zach considered many career options. Perhaps he could enter veterinary or medical school, or he might do research for industry or the government in environmental science. Another possibility was graduate school in biology toward a teaching career. By the time he was a junior, however, it was clear his grades probably were not high enough to get into medical or vet school, or a Ph.D. program in biology. He graduated with a 2.5 overall and a 2.9 in his major—respectable, but not stellar. Some low grades in subjects like math, history, and English had hurt his average.

Despite some mediocre grades, he had the chance to work on an interesting senior project in which he studied *microbe digesters*, organisms used to clean up pollution sites by converting toxins into nontoxic substances through the chemical digestive process. These critters are one of the newest environmental remediation techniques. Although only one member of a six-person team, he helped place microbes and then gather data at the site—a paper mill with toxic water. A team member who was a graduate student actually wrote most of the report, but Zach helped draft the summary of fieldwork. His modest part in the project had further inspired his hands-on interest in biology.

Zach Meets the Job Market—Twice

Although Zach did not seem destined to inhabit the ivy-covered walls of graduate school, he thought he would have a good shot at entry-level jobs in environmental science, and he was right. He applied for many environmental science openings and got two good offers, both in the Midwest.

After much thought he turned down both offers, wanting to stay in Colorado because his mother was ill and, frankly, because he had always wanted to work in Colorado for his career. He was confident that an opportunity would open up as the Denver job market improved. When his mother recovered from her illness, Zach started working in August for a small construction firm as a day laborer, stopping for several months to work at a ski resort. He also took off a month for the trip he had always dreamed about—a fishing and canoeing venture to British Columbia.

By May, a year after graduation, he was regretting not having taken one of the job offers a year earlier. Most of his college friends were working in their fields, even if only at modest jobs, yet he was still doing the same kind of work he had done during summers in college. So Zach decided to reapply for entry-level jobs in environmental consulting firms. These jobs would involve the following main tasks: (1) visiting sites that needed environmental remediation, (2) overseeing lab tests, (3) forming conclusions and recommendations based on these tests, (4) writing reports that summarized activities, and (5) writing proposals and giving presentations to seek new work.

An Offer Brings Self-Assessment

This week, Zach's college placement center called him about an opening as an environmental scientist at M-Global's Denver office. He planned to send a job letter and resume and hoped for a chance to interview.

Taking the advice of a placement specialist at his alma mater, Zach evaluated his skills, abilities, and experience before writing his new letter and resume. He jotted down these basic points about his background, some of which were detailed previously:

Positive
- Good work ethic
- Long-time interest in applied biology
- Took four-course sequence in environmental science—two courses short of a minor in the field

- Enjoys outdoors and travel
- Worked on microbe project in college, helping to write report

Neutral or Negative

- Did not take professional position after college
- Had mediocre college grade point average
- Focused more on straight biology than on applied environmental science
- Got "C" in required technical writing course
- Had no outstanding extracurricular activities—study, part-time work in the dining hall, and "winding down" seemed to take most time

Questions and Comments for Discussion

Put yourself in Zach's shoes. As he prepares for the job application process, what advice would you give him on the following issues? Refer to the guidelines in the chapter in formulating your answers.

1. This chapter cites specific goals for cover letters. What should Zach emphasize in his letter to M-Global?
2. Zach plans to submit a one-page resume with his letter. What format should he select? Why?

3. Re-read Zach's list of positive and neutral or negative features. Do you agree with his self-assessment? What features of his background will be of most concern to M-Global? Why? What can Zach do to reduce the impact of negative items in his background during the entire application process?
4. Specifically, how should Zach deal with (a) his mediocre grade point average and (b) the year he has spent doing work unrelated to his intended profession?
5. If his letter and resume interest M-Global, Zach will interview with Ken Pierson, an M-Global project manager who knows very little about biology or microbes. His field is geology. Yet he manages projects that use the talents of a wide range of scientists, engineers, and technicians. What advice would you give Zach to make the best possible impression during the interview?

Write About It

Assume the role of Zach. Write a one-page letter of application addressed to Carl Jensen, Director of Human Resources for the Denver branch of M-Global. Find ways to present negatives (such as Zach's year of experiences since graduation) as strengths.

Collaboration at Work Planning for Success

General Instructions

Each Collaboration at Work exercise applies strategies for working in teams to chapter topics. The exercise assumes you (1) have been divided into teams of about three to six students, (2) will use team time inside or outside of class to complete the case, and (3) will produce an oral or written response. For guidelines about writing in teams, refer to pages 24–28.

Background for Assignment

Some students attend college for its own sake because they love learning; others may like to learn but mainly view college as a stepping-stone to a career. Because most students are in the second group, they give a good deal of thought to what they will do with their working lives. If they don't, they should. Achieving success in a career starts with establishing a careful plan for getting from here to there. It is a process that involves considerably more than gaining a degree.

Team Assignment

For this exercise, you and your team members must brainstorm about a strategy for getting a particular job or entering a particular career desired by one of your team's members. (You can do the exercise for several careers, if you have time.) Following are the three steps:

1. Share information about the ideal jobs of all team members.
2. Choose one position or career that best fits the assignment (i.e., one that lends itself to being achieved in an incremental process that is doable).
3. Generate a list of specific steps for getting the position or entering the career (include deadlines and criteria for success at each stage).

If you have time to conduct some research and have access to the Internet, go to one of the websites listed on page 555 in this chapter for assistance in your research.

Assignments

1. Job Letter and Resume

Find a job advertisement in the newspaper, on the Internet, or at your college placement office. The ad should match either qualifications you have now or those you plan to have after you complete the academic program on which you are now working. Write a job letter and resume that respond to the ad. Submit the letter, resume, and written advertisement to your instructor.

If useful for this assignment and if permitted by your instructor, you may fictionalize part of your resume so that it lists a completed degree program and other experience not yet acquired. This way, the letter and resume reflect the background you would have if you were applying for the job. Choose the resume format that best fits your credentials.

As an alternative, write a letter and resume to apply for an internship in your major field. To find out about internships, contact your department or campus internship director, or ask about internships at your college placement office.

2. Job Interview

Pair up with another classmate for this assignment. First, exchange the letters, resumes, and job ads referred to in assignment 1. Discuss the job ads so that you are familiar with the job being sought by your counterpart, and vice versa. Then perform a role-playing exercise during which you act out the two interviews, one person as applicant and the other as interviewer.

Option: Include a third member in your team. Have this person serve as a recorder, providing an oral critique of each interview at the end of the exercise. Then collaborate among the three of you in producing a written critique of the role-playing exercise. Specifically, explain what the exercise taught you about the main challenges of the job interview.

3. Follow-Up Letter

Write a follow-up letter to the interview that resulted from assignment 2.

4. Follow-Up Letter—M-Global Projects

Last month, you submitted a job letter and resume for a position with the Barlow Group in Dallas, Texas. Now the firm has written to express interest in your application. It wants to know more about some summer employment you mentioned on your resume—the project you worked on was similar to some of Barlow's projects.

Assume the summer work in question was as student-in-training on one of the projects described at the end of chapter 2. Write a letter that briefly describes the project and your participation in it. Use information from the project, along with invented details about the activities you completed as an assistant. Even if your tasks were not especially glorified—manual labor or office support, for example—strive to describe learning experiences that would be meaningful to your reader. You are writing Daniel C. Yates, Barlow Consulting Group, 600 Industrial Way, Dallas, TX 75221.

5. Negotiation for Entry-Level Job

As in assignment 2, pair up with another student. Assume that the letters, resumes, interviews, and follow-up letters from the preceding assignments resulted in a second interview for one of you (i.e., select one of the positions, with one of you acting as applicant and the other as interviewer).

The topic of this second interview is the position being offered to you. After talking with your team member about the context of this simulated interview, conduct a negotiation session wherein the two of you discuss one or more aspects of the position being offered (salary, benefits, travel schedule, employee orientation, training arrangement, career development, etc.).

6. Negotiation with M-Global Client

In this exercise, you and a classmate will simulate a negotiation session between Sharon Gibbon, an M-Global training manager at the Cleveland office, and Bernard Claxton, training director of Cleveland's Mercy Hospital. Study the following details before beginning your 10- or 15-minute discussion.

Option: Collaborate with your teammate in writing an evaluation of this role-playing exercise. Explain the major obstacles encountered by both Gibbon and Claxton, and describe the techniques attempted by both parties to overcome these obstacles.

General Background: Gibbon recently submitted a proposal to Claxton, offering to have M-Global conduct three hazardous-waste seminars for the plant staff at Mercy Hospital. Claxton calls Gibbon to say that he wants to go ahead with the seminars, contingent on some final negotiations between the two. Claxton and Gibbon agree to meet in a few days, presumably to iron out a final agreement. Gibbon wants the contract, and Claxton's staff needs the training; however, the deal won't be sealed until they have their discussion and resolve several issues. Following are their respective points of view.

Gibbon's Viewpoint: Sharon Gibbon offered to have M-Global teach five one-day seminars for a fee of $15,000 ($3,000 per seminar). Each seminar will be team-taught by two of M-Global's certified industrial hygienists, Tom Rusher and Susan Sontack. They are expert trainers with quite a bit of field experience in the identification and safe use of hazardous chemicals and other wastes. The $3,000 course fee is standard for M-Global's hazardous-waste seminars, although the company has on rare occasions given 10% discounts for any of the same seminars after the first one for the same client. Gibbon is interested in picking up Mercy Hospital as a client, but she also recognizes that the two instructors she has committed for the seminars may be needed for jobs the company has not yet scheduled. She is leery of cutting fees for Mercy Hospital when there may be other full-fee work right around the corner.

Claxton's Viewpoint: Claxton knows that the hospital staff must have hazardous-waste training to conform to new county and city regulations, and he heard from other hospitals that M-Global has the best training in the business. Yet he has real problems spending $15,000 on five seminars. Proposals from other firms were in the $9,000 to $12,000 range for the five seminars, and Claxton's training budget is modest. Although the other firms that submitted proposals did not share M-Global's reputation, they too offered team-taught seminars by certified industrial hygienists. Although Claxton prefers to hire M-Global and although he knows that good training is worth the money, he is hoping to get Gibbon to lower M-Global's fee when they meet for their negotiation session. He knows that one benefit he can offer M-Global is continued training contracts from the hospital, because the high employee turnover necessitates frequent training in handling hazardous waste. In addition, there may be other training opportunities for M-Global at the hospital once Claxton completes his upcoming needs assessment of the staff training program.

7. Ethics Assignment

Searching for employment presents job seekers with some ethical challenges. A few "ethically challenged" individuals paint the portrait in their resume of someone who only remotely resembles the real thing. Certainly, lying and deception occur, but most writers simply want to present what they have accomplished and learned in the best possible light. As this chapter suggests, the resume is no time to be overly modest. With the goal of supportable self-promotion

in mind, evaluate the degree to which the following resume entries are accurate representations of the facts that follow them.

1. **Resume Entry:** June–September 2007—Served as apprentice reporter for a Detroit area weekly newspaper.

   Reality: Worked for a little over three months as a fact-checker for a group of reporters. Was let go when the assistant editor decided to offer the apprentice position to another, more-promising individual with more journalistic experience.

2. **Resume Entry:** July 2006—Participated in university-sponsored trip to Germany.

   Reality: Flew to Germany with two fraternity brothers for a two-day fraternity convention in Munich, after which the three of you toured Bavaria for a week in a rental car.

3. **Resume Entry:** Summer 2006, 2007, 2008—Worked for Berea Pharmacy as a stock clerk, salesperson, and accountant.

   Reality: Helped off and on with the family business, Berea Pharmacy, during three summers while in college—placing merchandise on shelves, working the cash register, and tallying sales at the end of the day. Your father had regular help so you were able to spend at least half of each summer camping with friends, playing in a softball league, and retaking a couple of college courses.

8. International Communication Assignment

There are many opportunities to work abroad, whether in internships, through a contracting firm, or through direct hiring. Using a website for a professional organization in your major field or a website such as www.internabroad.com, find an overseas internship or employment opportunity that interests you. Research and write a report on the cultural practices of the country in which the internship or employment is located. Your instructor will indicate whether your report will be oral or written. Information can be acquired from sources such as the following:

- Faculty and students who have visited the country
- Internet sites on other nations and on international communication
- Friends and colleagues familiar with the country you have chosen
- Books and articles on international communication and working overseas

9. A.C.T. N.O.W. Assignment (Applying Communication To Nurture Our World)

The purpose of this assignment is to assist members of the college or university community who want help writing job letters and resumes. It is a team assignment. First, use this chapter or any other source gathered by your team members to establish a short set of written guidelines for job correspondence. (Be sure to acknowledge sources on the document you prepare.) Second, meet with your instructor to review the guidelines document, making sure all members of your team have a good grasp of the material. Third, submit the document for consideration for publication in the campus newspaper, human resources newsletter, or similar campus publication.

201 Edge Drive
Norcross PA 17001
March 14, 2008

Mr James Vernon Personnel Director
M-Global Inc
105 Halsey Street
Baltimore MD 21212

Dear Mr. Vernon:

My academic advisor, Professor Sam Singleton, informed me about an electrical-engineering opening at M-Global, where he worked until last year. I am writing to apply for the job.

I understand that M-Global is making a major effort to build a full-scale equipment development laboratory. That prospect interests me greatly, because of my academic background in electrical engineering. At Northern Tech, I took courses in several subjects that might be useful in the lab's work—for example, microprocessor applications, artificial intelligence, and fiber optics.

Also, related work at Jones Energy & Automation, Inc., has given me experience building and developing new electronics systems. In particular, my work as an assembler taught me the importance of precision and quality control. I'd like the opportunity to apply this knowledge at M-Global.

Personal business will take me to Baltimore April 8–10. Could you meet with me on one of those days to discuss how M-Global might use my skills? Please let me know if an interview would be convenient at that time.

Enclosed is a resume that highlights my credentials. I hope to be talking with you in June.

Sincerely,

Donald Vizano

Donald Vizano

Enclosure: Resume

■ **Model 16–1** ■ Job letter (modified block) and chronological resume

Donald Vizano
201 Edge Drive
Norcross PA 17001
(300) 555-7861
dvizano@nct.edu

OBJECTIVE: A full-time position in electrical engineering,
with emphasis on designing new equipment in
automation and microprocessing

EDUCATION: 2002–2008 Bachelor of Science in Electrical
Engineering (expected June 2008)
Northern College of Technology,
Shipley, PA 3.5 GPA (out of 4.0 scale)

Major Courses:

Fiber Optics   Artificial Machine Intelligence
Robotic Systems   Communication Control Systems
Microprocessor Control   Microcomputer Applications
Microcomputer Systems   Digital Control Systems
                                              Semiconductor Circuits & Devices

Related Courses:

C++   Programming Languages
Business Communication   Engineering Economy
Industrial Psychology   Technical Communication

Other Skills: Fluent in Spanish

ACTIVITIES
AND HONORS: Institute of Electrical and Electronic Engineering (IEEE)
Dean's List, 8 quarters.

EMPLOYMENT:
2004–2008   Electronic Assembler (part-time)
Jones Energy & Automation, Inc.
Banner, PA

2003–2004   Lab Monitor (part-time)
Computer Services
Northern College of Technology
Shipley, PA

REFERENCES: Available upon request

■ **Model 16–1** ■ continued

1523 River Lane
Worthville OH 43804
August 6, 2008

Mr Willard Yancy
Director, Automotive Systems
XYZ Motor Company, Product Development Division
Charlotte NC 28202

Dear Mr. Yancy:

Recently I have been researching the leading national companies in automotive computer systems. Your job ad in the July 6 *National Business Employment Weekly* caught my eye because of XYZ's innovations in computer-controlled safety systems. I would like to apply for the automotive computer engineer job.

Your advertisement notes that experience in computer systems for machinery or robotic systems would be a plus. I have had extensive experience in the military with computer systems, ranging from a digital communications computer to an air traffic control training simulator. In addition, my college experience includes courses in computer engineering that have broadened my experience. I am eager to apply what I have learned to your company.

My mechanical knowledge was gained from growing up on my family's dairy farm. After watching and learning from my father, I learned to repair internal combustion engines, diesel engines, and hydraulic systems. Then for five years I managed the entire dairy operation.

With my training and hands-on experience, I believe I can contribute to your company. Please contact me at 614/555-2731 if you wish to arrange an interview.

Sincerely,

James M. Sistrunk

James M. Sistrunk

Enclosure: Resume

■ **Model 16–2** ■ Job letter (block style) and chronological resume

James M Sistrunk
1523 River Lane
Worthville OH 43804
(614) 555-2731
jmsistrunk@tmail.com

Professional Objective:

To contribute to the research, design, and development of automotive computer control systems

Education:

B.S., Computer Engineering, 2005–present
Columbus College, Columbus, Ohio
Major concentration in Control Systems with minor in Industrial Engineering. Courses included Microcomputer Systems, Digital Control Systems, and several different programming courses.

Computer Repair Technician Certification Training, 2002–2003
U.S. Air Force Technical Training Center, Keesler Air Force Base, Biloxi, MS
General Computer Systems Option with emphasis on mainframe computers. Student leader in charge of processing and orientation for new students from basic training.

Career Development:

Computer Repair Technician, U.S. Air Force, 2001–2003
Secret Clearance

Responsibilities and duties included:
- Repair of computer systems
- Preventative maintenance inspections
- Diagnostics and troubleshooting of equipment

Accomplishments included:
- "Excellent" score during skills evaluation
- Award of an Air Force Specialty Code "5" skill level

Assistant Manager, Spring Farm, Wootan, Ohio, 1996–2001
Responsible for dairy operations on this 500-acre farm. Developed management and technical skills; learned to repair sophisticated farm equipment.

Special Skills:

Adobe Creative Suite
Microsoft Word
Object Oriented Languages
C++ Programming

References:

Available upon request

■ **Model 16–2** ■ continued

456 Cantor Way #245
Gallop Minnesota 55002
September 3, 2008

Ms Judith R Gonzalez
American Hospital Systems
3023 Center Avenue
Randolf Minnesota 55440

Dear Ms. Gonzalez:

My placement center recently informed me about the Management Trainee opening with Mercy Hospital. As a business major with experience working in hospitals, I wish to apply for the position.

Your job advertisement notes that you seek candidates with a broad academic background in business and an interest in hospital management. At Central State College, I've taken extensive coursework in three major areas in business: finance, marketing, and personnel management. This broad-based academic curriculum has provided a solid foundation for a wide variety of management tasks at Mercy Hospital.

My summer and part-time employment also matches the needs of your position. While attending Central State, I've worked part-time and summers as an assistant in the Business Office at Grady Hospital. That experience has acquainted me with the basics of business management within the context of a mid-sized hospital, much like Mercy.

The enclosed resume highlights the skills that match your Management Trainee opening. I would like the opportunity to talk with you in person and can be reached at 612-555-1111 for an interview.

Sincerely,

Denise Ware Sanborn

Denise Ware Sanborn

Denise Ware Sanborn
456 Cantor Way #245
Gallop Minnesota 55002
612-555-1111
sanborndenise@cscm.edu

Objective
Entry-level management position in the health care industry. Seek position that includes exposure to a wide variety of management and business-related tasks.

Education
Bachelor of Arts Degree, June 2008
Central State College
Gallop, Minnesota

Major: Business Administration
Grade Point Average: 3.26 of possible 4.0, with 3.56 in all major courses
All college expenses financed by part-time and summer work at Grady Hospital in St. Paul, Minnesota.

Skills and Experience
Finance
    Helped with research for three fiscal year budgets
    Developed new spreadsheet for monthly budget reports
    Wrote accounts payable correspondence
Marketing
    Solicited copy from managers for new brochure
    Designed and edited new brochure
    Participated in team visits to ten area physicians
Personnel
    Designed new performance appraisal form for secretarial staff
    Interviewed applicants for Maintenance Department jobs
    Coordinated annual training program for nursing staff

Awards
2008 Arden Award for best senior project in the Business Administration Department (paper that examined Total Quality Management)

Dean's list for six semesters

References
Academic and work references available upon request.

■ **Model 16–3** ■ continued

2389 Jenson Court
Gulfton MS 39200
(601) 555-1111
February 18, 2008

Mr Nigel Pierce Personnel Director
Structural Systems Inc
105 Paisley Way
Jackson MS 39236

Dear Mr. Pierce:

I am writing in response to your ad for a technical representative in the July 13 (Sunday) edition of the *Jackson Journal*. I believe my experience in construction and my degree in civil engineering technology make me an excellent candidate for this position.

I am very familiar with your products for the wood construction market. The laminated beams and floor joists your company manufactures were specified by many of the architects I have worked with during my co-op experience at Mississippi College. Work I have done in the residential and small commercial construction industry convinced me of the advantages of your products over nominal lumber.

Enclosed is my resume, which focuses on the skills gained from my co-op work that would transfer to your firm. I look forward to meeting you and discussing my future with your company.

Sincerely,

Todd L. Fisher

Todd L. Fisher

Enclosure: Resume

■ **Model 16–4** ■ Job letter (modified block) and functional resume

Todd L Fisher
2389 Jenson Court
Gulfton MS 39200
(601) 555-1111
tlfish@ail.com

PROFESSIONAL OBJECTIVE

Use my education in civil engineering and my construction experience to assume a technical advisory position.

EDUCATION

Mississippi College
Hart, Mississippi; Bachelor of Science, Civil Engineering Technology
June 2007, GPA: 3.00 (out of 4.00)

PROFESSIONAL EXPERIENCE

Financed education by working as co-op student for two Jackson construction firms for 18 months.

Design Skills

Assisted with the layout and design of wall panels for Ridge Development condominium project.

Created layout and design for complete roof and floor systems for numerous churches and small commercial projects.

Computer Skills

Introduced computerization to the design offices of a major construction company (HP hardware in HPbasic operating system).

Designed trusses on Sun workstations in the UNIX operating system. Operated as the system administrator for the office.

Leadership Skills

Instructed new CAD (computer-assisted design) operators on the operation of design software for panel layout and design.

Designed and implemented management system for tracking jobs in plant.

REFERENCES

References available upon request.

■ **Model 16–4** ■ continued

Karen S Patel
300 Park Drive
Birtingdale NY 20092
(210) 555-2112
KSPatel@tmail.net

OBJECTIVE

Position as in-house technical writer and as trainer in communication skills

EDUCATION

Sumpter College, Marist, Vermont
M.S. in Technical Communication, GPA: 4.0, December 2007

Warren College, Aurora, New York
M.A. in English, Cum Laude, June 2005

University of Bombay, India
B.A. in English, First Class Honors, June 2001

EMPLOYMENT
Editing/Writing

Public Relations Office, Sumpter College 2006–present
Administrative Assistant: Write press releases and conduct interviews. Publish news stories in local newspapers and in *Sumpter Express.* Edit daily campus newsletter.

Hawk Newspapers, Albany, New York, 2003–2004
Warren College Internship: Covered and reported special events; conducted interviews; assisted with proofreading, layout, headline count. Scanned newspapers for current events; conducted research for stories. Published feature stories.

Teaching/Research

Sumpter College, Marist, Vermont, 2006–2007
Teaching Assistant: Tutored English at the Writing Center, answered "Grammar Hotline" phone questions, edited and critiqued student papers, taught English to non-English speakers, and helped students prepare for Regents exams.

Warren College, Aurora, New York, 2004–2005
Teaching Assistant: Taught business writing, supervised peer editing and in-class discussions, held student conferences, and graded students' papers.
Research Assistant: Verified material by checking facts, wrote brief reports related to research, researched information and bibliographies.

COMPUTER SKILLS

WordPerfect, Microsoft Office, Adobe Creative Suite, Adobe Technical Communication Suite

REFERENCES

Available upon request

■ **Model 16–5** ■ Combined resume

SUSAN A MARTIN

SCHOOL ADDRESS
540 Wood Drive
Bama CA 90012
(901) 555-2222

PERMANENT ADDRESS
30 Avon Place
Atlas, CA 90000
(901) 555-6074

EMAIL
smartin@piercecollege.edu

OBJECTIVE
Analyze and solve problems involving natural and pollution
control systems as an Environmental Scientist

EDUCATION
Pierce College, Bama, California
Bachelor of Science, Environmental Science
May 2008, GPA: 3.15 (out of 4.00)

Pleasant Valley College, Barnes, Nevada
Associate in Applied Science, Engineering Science
May 2006, GPA: 3.20 (out of 4.00)

PROFESSIONAL EXPERIENCE
Research Skills:
* Worked as lab assistant in a research project to analyze the
effect of acid rain on frog reproduction in Lake Lane.
* Designed Pierce College computer program to analyze data on
ozone depletion.

Leadership Skills:
* Taught inventory procedures to new employees of Zane's
Office Supply.
* Helped incoming freshmen and transfer students adjust to
Pierce College as dormitory resident assistant.

Organizational Skills:
* Maintained academic department files as student assistant in
Environmental Science Department.
* Organized field trips for Pierce College Mountaineering Club.

HONORS AND ACTIVITIES
Dean's list (five semesters)
President of Cycling Club

EMPLOYMENT HISTORY
Dormitory Resident Assistant, Peirce College, Bama, CA
2007-2008
Trainer, Zane's Office Supply, Bama, CA, 2006-2007

REFERENCES
References and transcripts available upon request

■ **Model 16–6** ■ Combined resume formatted for submission online

Leslie Highland
997 Simmons Drive
Boise Idaho 88822
(208) 555-2233
lnh@btt.net

OBJECTIVE:	A full-time position in architectural design with emphasis on model-making and renderings for future buildings.

EDUCATION:

Boise Architectural College
Boise, Idaho
Bachelor of Science
Architectural Engineering Technology
June 2006

Harvard University
Cambridge, Massachusetts
Certificate in Advance Architectural Delineation
August 2000

ACTIVITIES AND HONORS:

Boise Architectural College
Winner of Senior Design Project
Architectural Engineering Technology

Charter Member of American Society of
Architectural Perspectives

EMPLOYMENT:

2000–2006   **Architectural Designer and Delineator**
Dorsey-Hudson, Architects
Boise, Idaho

1998–2000   **Architectural Designer and Renderer**
Windsor and Associates, Architects
St. Lake, Utah

1995–1998   **Achitectural Renderer and Drafter**
Sanders and Associates, Architects
Provo, Utah

1993–1995   **Architectural Drafter**
Brown Engineering
St. Lake, Utah

REFERENCES: References and portfolio available upon request.

■ **Model 16–7** ■ Resume with graphics—not effective for computer scanning

Chapter | 17 | Style in Technical Writing

>>> Chapter Outline

Overview of Style   589
  Definition of Style   589
  Importance of Tone   589

Writing Clear Sentences   590
  Sentence Terms   591
  Guidelines for Sentence Style   591

Being Concise   592

Being Accurate in Wording   596

Using the Active Voice   598
  What Do Active and Passive Mean?   598
  When Should Active and Passive Voices Be Used?   599

Using Nonsexist Language   600
  Sexism and Language   600
  Techniques for Nonsexist Language   600

Plain English and Simplified English   603
  Plain English   603
  Simplified English   603

Chapter Summary   604

Learning Portfolio   605
  Communication Challenge   605
  Collaboration at Work   606
  Assignments   607

This chapter, as well as the Handbook (Appendix A), focuses on the final stage of the writing process—revising. As you may already have discovered, revision sometimes gets short shrift during the rush to finish documents on time. That's a big mistake. Your writing must be clear, concise, and correct if you expect the reader to pay attention to your message. Toward that end, this chapter offers a few basic guidelines on style. The Handbook contains alphabetized entries on grammar, mechanics, and usage.

After defining style and its importance, this chapter gives suggestions for achieving five main stylistic goals:

- Writing clear sentences
- Being concise
- Being accurate in wording
- Using the active voice
- Using nonsexist language

>>> Overview of Style

This section (1) provides an overview of *style* as it applies to technical writing and (2) defines one particularly important aspect of style—called *tone*—that relates to every guideline in this chapter.

Definition of Style

Just as all writers have distinct personalities, they also display distinct features in their writing. Writing style can be defined as follows:

Style: the features of one's writing that show its individuality, separating it from the writing of others and shaping it to fit the needs of particular situations. Style results from the conscious and subconscious decisions each writer makes in matters like word choice, word order, sentence length, and active and passive voice. These decisions are different from the "right and wrong" matters of grammar and mechanics (see the Handbook). Instead, they are composed of choices writers make in deciding how to transmit ideas.

Style is usually a series of personal decisions you make when you write. As noted in chapter 13, however, much writing is being done these days by teams of writers. Collaborative writing requires individual writers to combine their efforts to produce a consensus style, usually a compromise of stylistic preferences of the individuals involved. Thus personal style becomes absorbed into a jointly produced product. Similarly, many companies tend to develop a company style in documents like reports and proposals.

Importance of Tone

Tone is a major component of style and thus deserves special mention here. Through tone, you express an attitude in your writing—for example, neutral objectivity on the one hand, or unbridled enthusiasm on the other. The attitude evident in your tone exerts great influence over the reader. Indeed, it can determine whether your document achieves its objectives. Much like the broader term *style, tone* refers to the way you say something rather than what you say.

The following adjectives show a few examples of the types of tone or attitude that can be reflected in your writing. Here they are correlated with specific examples of documents:

1. **Casual tone:** Email to three colleagues working with you on a project.
2. **Objective tone:** Formal report to a client in which you present data comparing cost information for replacing the company's computer infrastructure.
3. **Persuasive tone:** Formal proposal to a client in hopes of winning a contract for goods or services.
4. **Enthusiastic tone:** Recommendation letter to a university to accept one of your employees in a master's program.
5. **Serious tone:** Memorandum to employees about the need to reduce the workforce and close an office.
6. **Authoritative tone:** Memo to an employee in which you reprimand him or her for violations of a policy about documenting absences.
7. **Friendly tone:** Letter to long-term clients inviting them to an open house at your new plant location.

Although there are almost as many variations in tone as there are occasions to write documents, one guideline always applies: Be as positive as you can possibly be, considering the context. Negative writing has little place in technical communication. In particular, a condescending or sarcastic tone should be avoided at all costs. It is the kind of writing you will regret. When you stress the positive, you stand the best chance of accomplishing your purpose and gaining the reaction you want from the reader.

Despite the need to make style conform to team or company guidelines, each individual remains the final arbiter of her or his own style in technical writing. Most of us will be our own stylists, even in firms in which in-house editors help clean up writing errors. This chapter helps such writers deal with everyday decisions of sentence arrangement, word choice, and the like. However, although style is a personal statement, you should not presume that anything goes. Certain fundamentals are part of all good technical style in the professional world. Let's take a look at these basics.

>>> Writing Clear Sentences

Each writer has his or her own approach to sentence style, yet everyone has the same tools with which to work: words, phrases, and clauses. This section defines some basic terminology in sentence structure, and then it provides simple stylistic guidelines for writing clear sentences.

Sentence Terms

The most important sentence parts are the subject and verb. The *subject* names the person doing the action or the thing being discussed (e.g., *He* completed the study / The *figure* shows that); the *verb* conveys action or state of being (e.g., She *visited* the site / He *was* the manager).

Whether they are subjects, verbs, or other parts of speech, words are used in two main units: phrases and clauses. A *phrase* lacks a subject or verb or both and it thus must always relate to or modify another part of the sentence (She relaxed *after finishing her presentation. / As project manager,* he had to write the report). A *clause,* however, has both a subject and a verb. Either it stands by itself as a *main clause (He talked to the team)* or it relies on another part of the sentence for its meaning and is thus a *dependent clause (After she left the site, she went home).*

Beyond these basic terms for sentence parts, you also should know the four main types of sentences:

- A *simple sentence* contains one main clause *(He completed his work).*
- A *compound sentence* contains two or more main clauses connected by conjunctions *(He completed his work, but she stayed at the office to begin another job).*
- A *complex sentence* includes one main clause and at least one dependent clause *(After he finished the project, he headed for home).*
- A *compound–complex sentence* contains at least two main clauses and at least one dependent clause *(After they studied the maps, they left the fault line, but they were unable to travel much farther that night).*

Guidelines for Sentence Style

Knowing the basic terms of sentence structure makes it easier to apply stylistic guidelines. Following are a few fundamental guidelines that form the underpinnings for good technical writing. As you review and edit your own writing or that of others, put these principles into practice.

>> Guideline 1: Place the Main Point Near the Beginning

One way to satisfy this criterion for good style is to avoid excessive use of the passive voice (see "Using the Active Voice" on pages 598–599); another way is to avoid lengthy phrases or clauses at the beginnings of sentences. Remember that the reader usually wants the most important information first.

Original: "After reviewing the growth of the Cleveland office, it was decided by the corporate staff that an additional lab should be constructed at the Cleveland location."

Revision: "The corporate staff decided to build a new lab in Cleveland after reviewing the growth of the office there."

>> Guideline 2: Focus on One Main Clause in Each Sentence

When you string together too many clauses with *and* or *but,* you dilute the meaning of your text. However, an occasional compound or compound–complex sentence is acceptable, just for variety.

Original: "The M-Global hiring committee planned to interview Jim Steinway today, but bad weather delayed his plane departure, and the committee had to reschedule the interview for tomorrow."

Revision: "The M-Global hiring committee had to change Jim Steinway's interview from today to tomorrow because bad weather delayed his flight."

>> Guideline 3: Vary Sentence Length, but Seek an Average Length of 15–20 Words

Of course, do not inhibit your writing process by counting words while you write. Instead, analyze one of your previous reports to see how you fare. If your sentences are too long, make an effort to shorten them, such as by making two sentences out of one compound sentence connected by *and* or *but.*

You should also vary the length of sentences. Such variety keeps your reader's attention engaged. Make an effort to place important points in short but emphatic sentences. Reserve longer sentences for supporting main points.

Original: "Our field trip for the project required that we conduct research on Cumberland Island, a national wilderness area off the Georgia Coast, where we observed a number of species that we had not seen on previous field trips. Armadillos were common in the campgrounds, along with raccoons that were so aggressive that they would come out toward the campfire for a handout while we were still eating. We saw the wild horses that are fairly common on the island and were introduced there by explorers centuries ago, as well as a few bobcats that were introduced fairly recently in hopes of checking the expanding population of armadillos."

Revision: "Our field trip required that we complete research on Cumberland Island, a wilderness area off the Georgia Coast. There we observed many species we had not seen on previous field trips. Both armadillos and raccoons were common in the campgrounds. Whereas the armadillos were docile, the raccoons were quite aggressive. They approached the campfire for a handout while we were still eating. We also encountered Cumberland's famous wild horses, introduced centuries ago by explorers. Another interesting sighting was a pair of bobcats. They were brought to the island recently to check the expanding armadillo population."

>>> Being Concise

Some experts believe that careful attention to conciseness could shorten technical documents by 10% to 15%. As a result, reports and proposals would take less time to read and cost less to produce. This section on conciseness offers several techniques for reducing verbiage without changing meaning.

>> Guideline 1: Put Actions in Verbs

Concise writing depends more on verbs than it does on nouns. Sentences that contain abstract nouns that hide actions can be shortened by putting the action in strong verbs instead. By converting abstract nouns to action verbs, you can eliminate wordiness, as the following sentences illustrate:

Wordy: "The *acquisition* of the property was accomplished through long and hard negotiations."

Concise: "The property was *acquired* through long and hard negotiations."

Wordy: "*Confirmation* of the contract occurred yesterday."

Concise: "The contract was *confirmed* yesterday."

Wordy: "*Exploration* of the region had to be effected before the end of the year."

Concise: "The region had to be *explored* before the end of the year."

Wordy: "*Replacement* of the transmission was achieved only three hours before the race."

Concise: "The transmission was *replaced* only three hours before the race."

As the examples show, abstract nouns often end with *-tion* or *-ment* and are often followed by the preposition *of*. These words are not always "bad" words; they cause problems only when they replace action verbs from which they are derived. The following examples show some noun phrases along with the preferred verb substitutes:

assessment of	assess
classification of	classify
computation of	compute
delegation of	delegate
development of	develop
disbursement of	disburse
documentation of	document
elimination of	eliminate
establishment of	establish
negotiation of	negotiate
observation of	observe
requirement of	require
verification of	verify

>> Guideline 2: Shorten Wordy Phrases

Many wordy phrases have become common in business and technical writing. Weighty expressions add unnecessary words and rob prose of clarity. Following are some of the culprits, along with their concise substitutes:

afford an opportunity to	permit
along the lines of	like

an additional	another
at a later date	later
at this point in time	now
by means of	by
come to an end	end
due to the fact that	because
during the course of	during
for the purpose of	for
give consideration to	consider
in advance of	before
in the amount of	of
in the event that	if
in the final analysis	finally
in the proximity of	near
prior to	before
subsequent to	after
with regard to	about

>> Guideline 3: Replace Long Words with Short Ones

In grade school, most students are taught to experiment with long words. Although this effort helps build vocabularies, it also can lead to a lifelong tendency to use long words when short ones will do. Of course, sometimes you want to use longer words just for variety—for example, using an occasional *approximately* for the preferred *about*. As a rule, however, the following long words (in the left column) should routinely be replaced by the short words (in the right column):

advantageous	helpful
alleviate	lessen, lighten
approximately	about
cognizant	aware
commence	start, begin
demonstrate	show
discontinue	end, stop
endeavor	try
finalize	end, complete
implement	carry out
initiate	start, begin
inquire	ask
modification	change

prioritize	rank, rate
procure	buy
terminate	end, fire
transport	move
undertake	try, attempt
utilize	use

>> Guideline 4: Leave Out Clichés

Clichés are worn-out expressions that add words to your writing. Although they once were fresh phrases, they became clichés when they no longer conveyed their original meaning. You can make writing more concise by replacing clichés with a good adjective or two. Following are some clichés to avoid:

as plain as day

ballpark figure

efficient and effective

few and far between

last but not least

leaps and bounds

needless to say

reinvent the wheel

skyrocketing costs

step in the right direction

>> Guideline 5: Make Writing More Direct by Reading It Aloud

Much wordiness results from talking around a topic. Sometimes called *circumlocution,* this stylistic flaw arises from a tendency to write indirectly. It can be avoided by reading passages aloud. Hearing the sound of the words makes problems of wordiness quite apparent. It helps condense all kinds of inflated language, including the wordy expressions mentioned earlier. Remember, however, that direct writing must also retain a tactful and diplomatic tone when it conveys negative or sensitive information.

Indirect:	"We would like to suggest that you consider directing your attention toward completing the project before the commencement of the seasonal monsoon rains in the region of the project area."
Direct:	"We suggest you complete the project before the monsoons begin."
Indirect:	"At the close of the last phase of the project, a bill for your services should be expedited to our central office for payment."
Direct:	"After the project ends, please send your bill immediately to our central office."
Indirect:	"It is possible that the well-water samples collected during our investigation of the well on the site of the subdivision could possibly contain some

chemicals in concentrations higher than is allowable according to the state laws now in effect."

Direct: "Our samples from the subdivision's well might contain chemical concentrations beyond those permitted by the state."

>> Guideline 6: Avoid *There Are, It Is,* and Similar Constructions

There are and *it is* should not be substituted for concrete subjects and action verbs, which are preferable in good writing. Such constructions delay the delivery of information about who or what is doing something. They tend to make your writing lifeless and abstract. Avoid them by creating (1) main subjects that are concrete nouns and (2) main verbs that are action words. Note that the following revised passages give readers a clear idea of who is doing what in the subject and verb positions.

Original: "There are many M-Global projects that could be considered for design awards."

Revision: "Many M-Global projects could be considered for design awards."

Original: "It is clear to the hiring committee that writing skills are an important criterion for every technical position."

Revision: "The hiring committee believes that writing skills are an important criterion for every technical position."

Original: "There were 15 people who attended the meeting at the client's office in Charlotte."

Revision: "Fifteen people attended the meeting at the client's office in Charlotte."

>> Guideline 7: Cut Out Extra Words

This guideline covers all wordiness errors not mentioned earlier. You must keep a vigilant eye for any extra words or redundant phrasing. Sometimes the problem comes in the form of needless connecting words, like *to be* or *that*. Other times it appears as redundant points—that is, those that have been made earlier in a sentence, paragraph, or section and do not need repeating.

Delete extra words when their use (1) does not add a necessary transition between ideas or (2) does not provide new information to the reader. (One important exception is the intentional repetition of main points for emphasis, as in repeating important conclusions in different parts of a report.) The examples in Figure 17–1 display a variety of wordy or redundant writing, with corrections made in longhand.

>>> Being Accurate in Wording

Good technical writing also demands accuracy in phrasing. Technical professionals place their reputations and financial futures on the line with every document that goes out the door. That fact shows the importance of taking your time on editing that deals with the accuracy of phrasing. Accuracy often demands more words, not fewer. The main rule is:

Never sacrifice clarity for conciseness.

■ **Figure 17–1** ■

Editing of wordy or redundant writing

Example 1: Preparing the client's final bill involves ~~the~~ checking ~~of~~ all _project_ invoices ~~for the project.~~

Example 2: The report examined what the M-Global project manager considered ~~to be~~ a technically acceptable risk.

Example 3: During ~~the course of~~ its field work, the M-Global team will ~~be engaged in the process of~~ reviewing all ~~of the~~ notes ~~that have been~~ accumulated in previous studies.

Example 4: ~~Because of his position~~ as head of the _M-Global_ public relations group, ~~at M-Global,~~ he planned ~~such that he would be able~~ to attend the meeting.

Example 5: She believed ~~that the~~ recruiting ~~of~~ more minorities for the technical staff is essential.

Example 6: The department must determine its ~~aims and~~ goals so that they can be included in ~~the~~ _M-Global's 2005_ annual strategic plan ~~produced by M-Global for the year of 2005.~~

Example 7: Most M-Global managers ~~generally~~ agree that all ~~of the~~ company's employees ~~at all the offices~~ deserve ~~at least~~ some ~~degree of~~ training each year ~~that they work for the firm.~~

Careful writing helps to limit liability that your organization may incur. Your goal is very simple: Make sure words convey the meaning you intend—no more, no less. Some basic guidelines to follow include:

>> Guideline 1: Distinguish Facts from Opinions

In practice, this guideline means you must identify opinions and judgments as such by using phrases like _we recommend, we believe, we suggest,_ or _in our opinion._

Example: "In our opinion, spread footings would be an acceptable foundation for the building you plan at the site."

If you want to avoid repetitious use of such phrases, group your opinions into listings or report sections. Thus a single lead-in can show the reader that opinions, not facts, are forthcoming.

Example: "On the basis of our site visit and our experience at similar sites, we believe that (1) _____, (2) _____, and (3) _____."

>> **Guideline 2:** Include Obvious Qualifying Statements When Needed

This guideline does not mean you must be overly defensive in every part of the report; it means that you must be wary of possible misinterpretations.

Example: "Our summary of soil conditions is based only on information obtained during a brief visit to the site. We did not drill any soil borings."

>> **Guideline 3:** Use Absolute Words Carefully

Avoid words that convey an absolute meaning or that convey a stronger meaning than you intend. One notable example is *minimize,* which means to reduce to the lowest possible level or amount. If a report claims that a piece of equipment will *minimize* breakdowns on the assembly line, the passage could be interpreted as an absolute commitment. The reader could consider any breakdown at all to be a violation of the report's implications. If instead the writer had used the verb *limit* or *reduce,* the wording would have been more accurate and less open to misunderstanding.

>>> Using the Active Voice

Striving to use the active voice can greatly improve your technical writing style. This section defines the active and passive voices and then gives examples of each. It also lists some practical guidelines for using both voices.

What Do *Active* and *Passive* Mean?

Active-voice sentences emphasize the person (or thing) performing the action—that is, somebody (or something) does something ("Matt completed the field study yesterday"). Passive-voice sentences emphasize the recipient of the action itself—that is, something is being done to something by somebody ("The field study was completed [by Matt] yesterday"). Following are some other examples of the same thoughts being expressed in first the active and then the passive voice:

■ **Examples:** Active-Voice Sentences:

1. "We *reviewed* aerial photographs in our initial assessment of possible fault activity at the site."

2. "The study *revealed* that three underground storage tanks had leaked unleaded gasoline into the soil."

3. "We *recommend* that you use a minimum concrete thickness of 6 in. for residential subdivision streets."

- **Examples:** Passive-Voice Sentences:

1. "Aerial photographs *were reviewed* [by us] in our initial assessment of possible fault activity at the site."

2. "The fact that three underground storage tanks had been leaking unleaded gasoline into the soil *was revealed* in the study."

3. "*It is recommended* that you use a minimum concrete thickness of 6 in. for residential subdivision streets."

Just reading through these examples gives the sense that passive constructions are wordier than active ones. Also, passive voices tend to leave out the person or thing doing the action. Although occasionally this impersonal approach is appropriate, the reader can become frustrated by writing that fails to say who or what is doing something.

When Should Active and Passive Voices Be Used?

Both the active and passive voices have a place in your writing. Knowing when to use each is the key. Following are a few guidelines that will help:

- *Use the active voice when you want to:*

1. Emphasize who is responsible for an action ("*We recommend* that you consider . . . .")

2. Stress the name of a company, whether yours or the reader's ("*PineBluff Contracting expressed* interest in receiving bids to perform work at . . . ")

3. Rewrite a top-heavy sentence so that the person or thing doing the action is up front ("*Figure 1 shows* the approximate locations of . . . ")

4. Pare down the verbiage in your writing, because the active voice is usually a shorter construction

- *Use the passive voice when you want to:*

1. Emphasize the receiver of the action or the action itself rather than the person performing the action ("*Samples will be sent* directly from the site to our laboratory in Sacramento")

2. Avoid the kind of egocentric tone that results from repetitious use of *I, we,* and the name of your company ("*The project will be directed* by two programmers from our Boston office")

3. Break the monotony of writing that relies too heavily on active-voice sentences

Although the passive voice has its place, it is far too common in business and technical writing. This stylistic error results from the common misperception that passive writing is more objective. In fact, excessive use of the passive voice only makes writing more tedious to read. In modern business and technical writing, strive to use the active voice.

>>> Using Nonsexist Language

Language usually follows changes in culture rather than anticipating such changes. An example is today's shift away from sexist language in business and technical writing—indeed, in all writing and speaking. The change reflects the increasing number of women entering previously male-dominated professions such as engineering, management, medicine, and law. It also reflects the fact that many men have taken previously female-dominated positions as nurses and flight attendants.

This section on style defines sexist and nonsexist language. Then it suggests ways to avoid using gender-offensive language in your writing.

Sexism and Language

Sexist language is the use of wording, especially masculine pronouns like *he* or *him,* to represent positions or individuals who could be either men or women. For many years, it was perfectly appropriate to use *he, his, him*, or other masculine words in sentences such as the following:

- "The operations specialist should check page 5 of his manual before flipping the switch."
- "Every physician was asked to renew his membership in the medical association before next month."
- "Each new student at the military academy was asked to leave most of his personal possessions in the front hallway of the administration building."

The masculine pronoun was understood to represent any person—male or female. Such usage came under attack for several reasons:

1. As previously mentioned, the entry of many more women into male-dominated professions has called attention to the inappropriate generic use of masculine pronouns.

2. Many people believe the use of masculine pronouns in a context that could include both genders constrains women from achieving equal status in the professions and in the culture—that is, the use of masculine pronouns encourages sexism in society as a whole.

Either point supplies a good enough reason to avoid sexist language. Many women in positions of responsibility may read your on-the-job writing. If you fail to rid your writing of sexism, you risk drawing attention toward sexist language and away from your ideas. Common sense argues for following some basic style techniques to avoid this problem.

Techniques for Nonsexist Language

This section offers techniques for shifting from sexist to nonsexist language. When shifting to nonsexist language, many writers have problems with subject–verb agreement. (The *engineer* recorded *their* data.) The strategies that follow help you avoid this problem. Not all these strategies will suit your taste in writing style; use the ones that work for you.

>> Technique 1: Avoid Personal Pronouns Altogether

One easy way to avoid sexist language is to delete or replace unnecessary pronouns:

Example:

Sexist Language: "During *his* first day on the job, any new employee in the toxic-waste laboratory must report to the company doctor for *his* employment physical."

Nonsexist Language: "During *the* first day on the job, each new employee in the toxic-waste laboratory must report to the company doctor for *a* physical."

>> Technique 2: Use Plural Pronouns Instead of Singular

In most contexts you can shift from singular to plural pronouns without altering meaning. The plural usage avoids the problem of using masculine pronouns.

Example:

Sexist Language: "*Each* geologist should submit *his* time sheet by noon on the Thursday before checks are issued."

Nonsexist Language: "*All* geologists should submit *their* time sheets on the Thursday before checks are issued."

Interestingly, you may encounter sexist language that uses generic female pronouns inappropriately. For example, "Each nurse should make every effort to complete *her* rounds each hour." As in the preceding case, a shift to plural pronouns is appropriate: "Nurses should make every effort to complete *their* rounds each hour."

>> Technique 3: Alternate Masculine and Feminine Pronouns

Writers who prefer singular pronouns can avoid sexist use by alternating *he* and *him* with *she* and *her*. When using this technique, writers should avoid the unsettling practice of switching pronoun use within too brief a passage, such as a paragraph or page. Instead, writers may switch every few pages, or every section or chapter.

Although this technique is not yet in common use, its appeal is growing. It gives writers the linguistic flexibility to continue to use masculine and feminine pronouns in a generic fashion. However, one problem is that the alternating use of masculine and feminine pronouns tends to draw attention to itself. Also, the writer must work to balance the use of masculine and feminine pronouns, in a sense to give equal treatment.

>> Technique 4: Use Forms Like *He or She, Hers or His,* and *Him or Her*

This solution requires the writer to include pronouns for both genders.

Example:

Sexist Language: "The president made it clear that each M-Global branch manager will be responsible for the balance sheet of *his* respective office."

Nonsexist Language: "The president made it clear that each M-Global branch manager will be responsible for the balance sheet of *his or her* respective office."

This stylistic correction of sexist language may bother some readers. They believe that the doublet structure of *her or his,* is wordy and awkward. Many readers are bothered even more by the slash formations of *he/she, his/her,* and *her/him.* Avoid using these.

>> Technique 5: Shift to Second-Person Pronouns

Consider shifting to the use of *you* and *your,* words without any sexual bias. This technique is effective only with documents in which it is appropriate to use an instructions-related command tone associated with the use of *you.*

Example:

Sexist Language:	"After selecting *her* insurance option in the benefit plan, each new nurse should submit *her* paperwork to the Human Resources Department."
Nonsexist Language:	"Submit *your* paperwork to the Human Resources Department after selecting *your* insurance option in the benefit plan."

>> Technique 6: Be Especially Careful of Titles and Letter Salutations

Today, most women in business and industry are comfortable being addressed as *Ms.* If you know that the recipient prefers *Miss* or *Mrs.,* use that in you salutation. If a person's gender isn't obvious from the name, call the person's employer and ask how the person prefers to be addressed. (When calling, also check on the correct spelling of the person's name and the person's current job title.) Receptionists and secretaries expect to receive such inquiries.

When you do not know who will read your letter, never use *Dear Sir* or *Gentlemen* as a generic greeting. Such a mistake may offend women reading the letter and may even cost you some business. *Dear Sir or Madam* is also inappropriate. It shows you do not know your audience, and it includes the archaic form *Madam.* Instead, call the organization for the name of a particular person to whom you can direct your letter. If you must write to a group of people, replace the generic greeting with an *Attention* line that denotes the name of the group.

Examples:

Sexist Language:	Dear Miss Finnegan: [to a single woman for whom you can determine no title preference]
Nonsexist Language:	Dear Ms. Finnegan:
Sexist Language:	Dear Sir: *or* Gentlemen:
Nonsexist Language:	Attention: Admissions Committee

No doubt the coming years will bring additional suggestions for solving the problem of sexist language. Whatever the culture finally settles on, it is clear that good technical writing style no longer tolerates the use of such language.

>>> Plain English and Simplified English

When you are writing technical or business documents, you may be asked to use one of two important styles of workplace writing: Plain English or Simplified English. Both of these styles include specific recommendations about sentence structure and word choice, but they are designed for particular audiences and purposes.

Plain English

Plain English is a specific style recommended for United States government documents and for documents such as proposals and reports that are submitted to federal agencies. Although people had been discussing clearer government documents for years, the Plain Language movement gained strong support during the mid-1990s. In 1995, a group of people began creating standards for Plain English in government writing. This group became the Plain Language Action and Information Network (PLAIN).

Plain English guidelines include many of the elements of clear technical communication: audience awareness, good document design, effective use of headings, and clear organization. However, Plain English is most clearly defined by its style recommendations, which include the following:

- Use active voice
- Put actions in strong verbs
- Use *you* to speak directly to the reader
- Use short sentences (no longer than about 20 words)
- Use concrete words
- Use simple and compound sentences with a subject–verb structure
- Make sure that modifiers are clear
- Use parallel structure for parallel ideas
- Avoid wordiness

The Plain Language website at http://www.plainlanguage.gov includes a complete discussion of Plain English with examples and links to other resources.

Simplified English

Simplified English includes many of the same recommendations as Plain English, and it is sometimes confused with Plain English. However, it serves a different purpose and is designed for a different audience. Simplified English, sometimes called *Controlled English* or *Internationalized English,* is designed for the global economy. It is designed for an audience for whom English is a second language, to be easily translatable from English into other languages. A leading organization for the development of Simplified English is the European Association of Aerospace Manufacturers (AECMA), which created the original standard in the 1980s.

Simplified English is designed to be clear and unambiguous, so it recommends specific sentence structures and limited vocabulary. Simplified English includes the following:

- Use only approved words
- Use one word for each meaning (avoid synonyms)
- Use only one meaning for each word (e.g., *close* is used only as a verb)
- Use active voice
- Use strong verbs
- Use articles (*a, an, the*) or demonstrative adjectives (*this, that, these, those*) for clarity
- Avoid strings of more than three nouns
- Use short sentences (less than 20 words)

More information about Simplified English standards is available at http://www.asdste100.org, and an overview of Simplified English is available at http://www.userlab.com/SE.html, which also includes a sample list of approved words at http://www.userlab.com/Downloads/SE.pdf. Because the standards were developed for the aerospace industry, the word lists are specialized for that industry. Other industries are developing their own word lists. A more general word list can be downloaded from the Publications/Documents section of Intecom's website at http://www.intecom.org.

>>> Chapter Summary

Style is an important part of technical writing. During the editing process, writers make the kinds of changes that place their personal stamp on a document. Style can also be shaped (1) by a team, in that writing done collaboratively can acquire features of its diverse contributors; or (2) by an organization, in that an organization may require writers to adopt a particular writing style. Yet the decision-making process of individual writers remains the most important influence on the style of technical documents.

This chapter offers five basic suggestions for achieving good technical writing style. First, sentences should be clear, with main ideas at the beginning and with one main clause in most sentences. Although sentences should average only 15 to 20 words, you should vary sentence patterns in every document. Second, technical writing should be concise. You can achieve this goal by reading prose aloud as you rewrite and edit. Third, wording should be accurate. Fourth, the active voice should be dominant, although the passive voice also has a place in good technical writing. And fifth, the language of technical documents should be free of sexual bias.

Although clarity and conciseness are important to all workplace writing, you may be asked to follow a specific style sheet. Many organizations create their own in-house style guides, and some styles such as Plain English and Simplified English are used for industries or to clarify cross-cultural communication.

>>> Learning Portfolio

Communication Challenge "An Editorial Adjustment"

M-Global, Inc., hired a technical writer/editor at its Cleveland office, the smallest branch in the company. The office finally generates enough reports and proposals to justify the addition, and Evelyn Tobin started the job a month ago. Some of the Cleveland employees who were comfortable with the old system are now having trouble adjusting to having an editor. What follows is some background on the hiring of Evelyn, the changes she is making in office writing, some questions and comments for discussion, and an assignment for a written response to the Challenge.

Winds of Change

For years, the staff at M-Global's Cleveland office handled all of its own writing and editing. Managers, engineers, scientists, accountants, trainers, and others had to draft and edit their own copy. Because they could depend on no one else to help, they gave great attention to the process and prided themselves on the quality of their writing. With the aid of several good secretaries who often corrected grammar while they typed, the documents produced seemed adequate.

The growth of the office, however, increased the number and complexity of the reports and proposals that went out the door. The quality of editing began to decline. Those who observed the trend tied it to the following changes:

- Each writer simply had a higher volume of reports and proposals to complete, to keep the office competitive with similar firms.
- A new mobility in the workforce meant that fewer employees received on-the-job training from old-timers at the office. In fact, more than half the positions requiring a college degree had been filled in the past three years.
- This new workforce came from many different academic backgrounds and from other firms, making it harder than it used to be to impose a set "style" at the office.
- The experienced secretaries, who had been expert editors, retired. Many of them were simply not replaced, because word-processing programs reduced the need for secretarial staff.

The branch manager observed these changes. Perhaps the last straw came when one long-time client returned a report with corrections made in red ink, along with this note: "You guys used to turn out good reports. What's happened?" With that embarrassment, the branch manager quickly hired an in-house technical writer/editor.

New Editor Takes Charge

When Evelyn Tobin started work a month ago, she met with all the staff to discuss her duties. At that meeting, there was general agreement that Evelyn would (1) provide writing advice, (2) perform a style edit for some reports, (3) be the lead writer for key proposals, (4) help with training in the office, and (5) do a quick grammar edit on as many reports as she had time to review.

With a B.S. in Technical Communication and two years of editing experience with a government agency that emphasized Plain English, Evelyn was used to simplifying writing that was confusing, convoluted, or too technical. Although she had not worked with technical firms like M-Global, she assumed all her experience would translate to the new job. As it happened, most of her initial work involved style edits of reports that were to be sent to a mixed readership—some readers had a technical background, but others did not. Following are several changes Evelyn made in the reports, along with the original passages:

1. **Original:** The purpose of the new well is to allow Tank, Inc., to perform monthly water-level monitoring at three locations at the oil field so that the results can be sent to the Water Quality Control Board.

   Evelyn's Revision: The new well will allow Tank, Inc., to monitor water levels at the oil field. Then the data will be sent to the Water Quality Control Board.
2. **Original:** During the drilling of the boring, some soil sampling was performed by our technicians for the purpose of determining the exact location of the water table at the site.

   Evelyn's Revision: While drilling the boring, the technicians sampled soils to locate the water table.
3. **Original:** This letter proposal has been prepared by us for use by whatever attorney you select so that he can present a ballpark figure of costs to the college governing board.

   Evelyn's Revision: We prepared this proposal for whatever attorney you select. Then she can present a cost estimate to the college governing board.
4. **Original:** At this point in time, it is our belief that you should give equal consideration to both alternatives, for both can afford you the opportunity to complete expansion of the office complex prior to summer.

Evelyn's Revision: At this point, we believe you should consider both alternatives. Either one will allow you to expand the office by next summer.

5. **Original:** There are a total of two ways we are recommending that you consider changing the plans in order to minimize the chance for earthquake damage.

   Evelyn's Revision: We recommend two changes to reduce the chance for earthquake damage.

Questions and Comments for Discussion

1. Study the before and after versions carefully.
   - Explain the rationale you think Evelyn would have for each of the changes she made.
   - Given the audience for the documents, was she right to make the changes?
   - Can you see changes in content that the original writer may find unacceptable?
   - Are there any cases where you need more context surrounding the passage to provide adequate answers to the two previous questions? Explain.
2. Would your answers to any of the previous questions change if the only audience for the report had been a team of technical experts?
3. Are there any alternative revisions that you think would be as effective as, or more effective than, the revisions Evelyn made?

4. Suggest what you think would be the best way for Evelyn to convey her revisions to the writer. Would this method change or stay the same as she gains more experience at the office?
5. One employee came to Evelyn for advice on some grammar-checking software. He noted that his software stopped at every passive-voice sentence and suggested an active-voice replacement, and he wondered if he should always make the change. If you were in Evelyn's position, what answer would you give?
6. Discuss the effect that hiring a technical writer/editor might have on an office like M-Global–Cleveland—that is, how might the change affect the corporate culture of a such a company, where the professional staff spends from 25% to 50% of its time writing and editing documents?
7. If you were working at the Cleveland office, how would you feel about having your documents reviewed for style? For grammar?

Write About It

Although M-Global does not often bid on federal projects, it does sometimes write proposals for state or county projects. Assume the role of Evelyn and write a memo that argues for the importance of Plain English and that presents a brief description of Plain English style. Use the resources at http://www.plainlanguage.gov for information about Plain English.

Collaboration at Work *Describing Style*

General Instructions

Each Collaboration at Work exercise applies strategies for working in teams to chapter topics. The exercise assumes you (1) have been divided into teams of about three to six students, (2) will use team time inside or outside of class to complete the case; and (3) will produce an oral or written response. For guidelines about writing in teams, refer to pages 24–28.

Background for Assignment

As this chapter points out, the term *style* refers to the way you choose to express an idea, as opposed to the content of the idea itself. The definition of *style* early in this chapter makes it clear that writers adopt particular styles for differ-

ent contexts. For example, following are three passages that express the same idea in three different ways:

1. The results of the experiment strongly suggest to the team conducting the study that the hypothesis is valid.
2. After evaluating the results of the experiment, we concluded that the hypothesis is valid.
3. We believe the experiment worked.

Team Assignment

Describe how the previous three passages convey information differently to the reader. Can you describe the differences in style? When is one passage more appropriate to use than another?

Assignments

1. Conciseness—Abstract Words

Make the following sentences more concise by replacing abstract nouns with verbs. Other minor changes in wording may be necessary.

a. Verification of the agreement was indicated by the signing of the contract by members of the M-Global corporate staff.

b. The inspectors indicated that observation of the site occurred on July 16, 2005.

c. Negotiation of the final contract was to happen on the day after their arrival.

d. After three hours of discussion, the branch managers agreed that establishment of a new M-Global mission statement should take place in the next fiscal year.

e. Assessment of the firm's progress will happen during the annual meeting of the M-Global Board of Directors.

f. The entire company agreed that classification of employees according to level of education was inappropriate.

g. Documentation of the results of the lab test appeared in the final report.

h. Unlike the previous year, this year the disbursement of stock dividends will occur after the annual meeting.

i. In analyzing the managerial style of the manager, the outside evaluators determined that delegation of authority appeared to be a problem for her.

j. The financial statement showed that computation of the annual revenues had been done properly.

2. Conciseness—Wordy Phrases and Long Words

Condense the following sentences by replacing long phrases and words with shorter substitutes.

a. In the final analysis, we decided to place the new pumping station in the proximity of the old one.

b. Prior to commencing the project, they met to prioritize their objectives.

c. Endeavoring to complete the study on time, Sheila transported the supplies immediately from the field location to the M-Global lab.

d. During the course of his career, he planned to utilize the experience he had gained in the ambulance business.

e. His work with the firm terminated due to the fact that he took a job with another competing firm.

f. In the event that two clients need a crew in Austin next week, we can give consideration to using the same crew for both projects and lowering travel costs for both clients.

g. Jim McDuff was not cognizant of the fact that younger employees felt differently than older employees about the expansion of their office building.

h. To implement the Phoenix asbestos project, we made adjustments in the workload of two engineers so that they could be available to undertake the project in Phoenix.

i. Subsequent to the announcement he made, he held a news conference for approximately one hour of time.

j. At this point in time, she had every hope that her annual bonus would afford her family the opportunity to take an additional family vacation.

3. Conciseness—Clichés and *There Are/It Is* Constructions

Rewrite the following sentences by eliminating clichés and the wordy constructions *there are* and *it is*.

a. They all agreed that the issue had been discussed repeatedly for the past 10 years; thus they did not want to reinvent the wheel during the current study.

b. There are many examples of skyrocketing equipment costs affecting the final budget for a project.

c. It is a fact that most employees at M-Global believe the company has taken a step in the right direction by adding international offices.

d. Needless to say, it is clear that Karen is looking forward to the three-week vacation.

e. She explained to her staff that it was as plain as day that they would have to decrease their labor costs.

f. The prospective client asked for a ballpark figure of the project costs.

g. Last but not least, there was the issue of quality control that he wanted to emphasize in his speech.

h. In these modern times today, there are new approaches that college graduates should take to the job search.

i. Susan ended the meeting by concluding that there were a number of mutually agreeable solutions that could be explored so that the new departments in conflict could peacefully coexist.

j. It is a fact that our boss ended the meeting about a loss of profits by noting that we are all in the same boat.

4. Sentence Clarity

Improve the clarity of the following sentences by changing sentence structures or by splitting long sentences into several shorter ones.

a. Therefore, to collect a sample from above the water table, and thus to follow the directions provided by the client, the initial boring was abandoned and the drill rig was repositioned about two feet away and a new boring was drilled.

b. After capping the soil sample ring with PVC end caps and then notifying all members of the project team, we placed it in a cooler for storage on-site and transportation later to a chemical analytical laboratory.

c. Based on the geotechnical data obtained from the subsurface exploration program, the results of the percolation testing, and the planned plumbing fixtures, the feasibility of installing a leachfield-type on-site sewage-disposal system was evaluated.

d. Percolation test #1 was performed approximately 40 feet east of the existing pump house and percolation test #2 was performed near the base of the slope approximately 65 feet west of the pump house, and then the results were submitted to the builder.

e. We appreciate the opportunity to provide our services on this project and look forward to continuing our relationship with XYZ Trading and Transportation Company when we begin the Zanter Project with your Finance Department next spring.

f. All of the earth materials encountered in our exploration can be used for trench backfill above manhole and pipe bedding, provided they are free of organic material, debris, and other deleterious materials, and they are screened to remove particles greater than six inches in diameter.

g. This study was conducted to identify, to the extent possible, based on available information from the city files and the criteria described in our proposal of June 18, 2008, whether activities near the site may have involved the use, storage, disposal, or release of hazardous or potentially hazardous substances to the environment.

h. The properties consist of approximately 5,000 acres, including those parcels of Heron Ranch owned by American Axis Insurance Company, the unsold Jones Ranch parcels, the village commercial area, the mobile home subdivisions, two condominium complexes, a contractor's storage area, an RV storage area, a sales office, a gatehouse, open space parcels, and the undeveloped areas for future Buildings 1666, 1503, 1990, and 1910.

i. Having already requested permits for the construction of the bathhouse, medical center, maintenance building, boat dock, swimming pool, community building, and an addition to the community building, we still need to apply for the storeroom permit.

j. A report dated May 25, 2007, for the ABC Corporation confirmed that the updated business plan had been completed the previous month, but a new plan had to be submitted by May 25, 2008.

5. Active and Passive Voice Verbs

Make changes in active and passive voice verbs, where appropriate. Refer to the guidelines in the chapter. Be able to supply a rationale for any change you make.

a. It was recommended by the personnel committee that you consider changing the requirements for promotion.

b. No formal report about assets was reported by the corporation before it announced the merger.

c. The graphs showing the differences in depreciation and interest and the net loss on the investment are shown in Appendix A.

d. It has been noted by the Department of Environmental Services that the laundry business was storing toxic chemicals in an unsafe location.

e. The samples from the Scottish Highlands will be sent to M-Global's engineering lab in London.

f. The violation of ethical guidelines was reported by the commissioner to the president of the association.

g. No complete equipment inventory has been made by M-Global's Boston office.

h. It was concluded by the employee committee that M-Global's retirement program needed to be revised.

i. Dirt brought to the site should be evaluated by the engineer on-site before it is placed in the foundation.

j. Due to presence of a good deal of sand at the location, excavations are anticipated by us to be relatively unstable.

6. Sexist Language

Revise the following sentences to eliminate sexist language.

a. The department decided to advertise for a department chairman in three national newspapers.

b. Although each manager was responsible for his own budget, some managers obviously had better accounting skills than others.

c. The company policy manual states that each secretary should submit her time card twice a month.

d. If an hourly worker misses no work for sickness during a calendar year, he will receive a $500 bonus at year's end.

e. Each flight attendant is required to meet special work standards as long as she is employed by an international airline.

f. Typically, a new engineer at M-Global receives his first promotion after about a year.

g. Every worker wonders whether he is saving enough for retirement.

h. If a pilot senses danger, she should abort the takeoff.

i. Upon arriving at the site, an M-Global scientist should make immediate contact with his client representative.

j. [greeting section of a letter] Gentlemen:

7. Advanced Exercise—Conciseness

The following sentences contain more words than necessary. Rewrite each passage more concisely but without changing the meaning. If appropriate, make two sentences out of one.

a. The disbursement of the funds from the estate will occur on the day that the proceedings concerning the estate are finalized in court.

b. During the course of the project that we conducted for Acme Pipe, several members of our project team were in the unfortunate position of having to perform their fieldwork at the same time that torrential rains hit the area, totaling three inches of rain in one afternoon.

c. At a later date we plan to begin the process of prioritizing our responsibilities on the project so that we will have a clear idea of which activities deserve the most attention from the project personnel.

d. Needless to say, we do not plan to add our participation to the project if we conclude that the skyrocketing costs of the project will prohibit our earning what could be considered to be a fair profit from the venture.

e. The government at this point in time plans to discontinue its testing of every item but will undertake to implement testing again in approximately five months.

f. Hazerd, Inc., will endeavor to finalize the modifications of the blueprints for a ballpark figure of about $850.

g. For us to supply the additional supplies that the client wishes to procure from us, the client will have to initiate a change order that permits additional funds to be transferred into the project account.

h. Upon further analysis of the many and varied options that we are cognizant of at this time, it is our opinion that the long-term interests of our firm would be best served by reducing the size of the production staff by 300 workers.

i. Prior to the implementation of the state law with regard to the use of asbestos as a building material, it was common practice to utilize this naturally occurring mineral in all kinds of facilities, some of which became health hazards subsequently.

j. In the event that we are given permission to undertake the research, be sure to make certain to perform an efficient and effective search of available literature in a research facility so that we do not end up, in the final analysis, reinventing the wheel with regard to knowledge of superconductors.

8. Advanced Exercise—General Style Rules

Revise the following sentences by applying all the guidelines mentioned in this chapter. When you change passive verbs to active, it may be necessary to make some assumptions about the agent of the action, because the sentences are taken out of context.

a. Based on our review of the available records, conversations with the various agencies involved, including the Fire Department and the Police Department, and a thorough survey of the site where the spill occurred, it was determined that the site contained chemicals that were hazardous to human health.

b. After seven hours at the negotiation table, the union representatives and management decided that the issues they were discussing could not be resolved that evening, so they met the next day at the hotel complex, at which point they agreed on a new contract that would increase job security and benefits.

c. It is recommended by us that your mainframe computer system be replaced immediately by a newer model.

d. After the study was completed by the research team and the results were published in the company newsletter the following month, the president decided to call a meeting of all senior-level managers to discuss strategies for addressing problems highlighted by the research team.

e. Our project activities can be generally described in this way. The samples were retrieved from the site and then were transported to the testing lab in the containers made especially for this project, and at the lab they were tested to determine their soil properties; the data were analyzed by all the members of the team before findings and conclusions were arrived at.

f. First the old asbestos tile was removed. Then the black adhesive was scraped off. Later the floor was sanded smooth. The wood arrived shortly. Then the floor was installed.

g. The figures on the firm's profit margins in July and August, along with sales commissions for the last six months of the previous year and the top 10 salespersons in the firm, are included in the Appendix.

h. It was suggested by the team that the company needs to invest in modern equipment.

i. It is the opinion of this writer that the company's health plan is adequate.

j. Shortly after the last change in leadership, and during the time that the board of directors was expressing strong views about the direction that the company was taking, it became clear to me and other members of the senior staff that the company was in trouble.

k. Each manager should complete and submit his monthly report by the second Tuesday of every month.

l. After completing our engineering analysis, it is clear that metal fatigue caused the structure to fail.

m. Upon hearing the captain's signal, each flight attendant should complete her checklist of preflight procedures.

n. Our weed-spraying procedure will have minimal impact on shrubbery that surrounds the building site.

o. It was reported today from the corporate headquarters that the health-care plan has been approved by the president.

9. Editing Paper of Classmate

For this assignment, exchange papers with a member of your class. Use either the draft of a current assignment or a paper that was completed earlier in the term. Edit your classmate's work in accordance with this chapter's guidelines on style, and then explain your changes to the writer.

10. Editing Sample Memo

Using the guidelines in this chapter, edit the following memorandum. The assignment can be completed individually or in teams as a team-editing project.

DATE: January 12, 2008
TO: All Employees of Denver Branch
FROM: Leonard Schwartz, Branch Manager
SUBJECT: New Loss-Prevention System

As you may have recently heard, lately we received news from the corporate headquarters of the company that it would be in the best interest of the entire company to pay more attention to matters of preventing accidents and any other safety-related measures that affect the workplace, including both office and field activities related to all types of jobs that we complete. Every single employee in each office at every branch needs to be ever mindful in this regard so that he is most efficient and effective in the daily performance of his everyday tasks that relate to his job responsibilities such that safety is always of paramount concern.

With this goal of safety ever present in our minds, I believe the bottom line of the emphasis on safety could be considered to be the training that each of us receives in his first, initial weeks on the job as well as the training provided on a regular basis throughout each year of our employment with M-Global, so that we are always aware of how to operate in a safe manner. The training vehicle gives the company the mechanism to provide each of you with the means to become aware of the elements of safety that relate to the specific needs and requirements of your own particular job. Therefore, at this point in time I have come to the conclusion in the process of contemplating the relevance of the new corporate emphasis on safety to our particular branch that we need, as a branch, to give much greater scrutiny and analysis to the way we can prevent accidents and emphasize the concern of safety at every stage of our operation for every employee. Toward this end, I have asked the training coordinator, Kendra Jones, to assemble a written training program that will involve every single employee and that can be implemented beginning no later than June of this year. When the plan has been written and approved at the various levels within the office, I will conduct a meeting with every department in order to emphasize the major and minor components of this upcoming safety program.

It is my great pleasure to announce to all of you that effective in the next month (February) I will give a monthly safety award of $100 to the individual branch employee at any level of the branch who comes up with the best, most useful suggestion related to safety in any part of the branch activities. Today I will take the action of placing a suggestion box on the wall of the lunchroom so that all of you will have easy access to a way to get your suggestions for safety into the pipeline and to be considered. As an attachment to the memo you are now reading from me, I have provided you with a copy of the form that you are to use in making any suggestions that are then to be placed in the suggestion box. On the last business day of each month, the box will be emptied of the completed forms for that month, and before the end of the following week a winner will be selected by me for the previous month's suggestion program and an announcement will be placed by me to that effect on the bulletin board in the company workroom.

If you have any questions in regard to the corporate safety program as it affects our branch or about the suggestion program that is being implemented here at the Denver office at M-Global, please do not hesitate to make your comments known either in memorandum form or by way of telephonic response to this memorandum.

? 11. Ethics Assignment

Reread this chapter's "Communication Challenge" concerning Evelyn Tobin, the new editor at M-Global's Cleveland office. Now assume that you, as an electrical engineer at Evelyn's office, have asked for her help in preparing an article for publication in a professional journal. As you hand her the article, you are quick to add you have long-standing problems organizing information and editing well. Two days later the draft appears in your mailbox looking like your first graded paper in English 101 in college. Evelyn has even provided a suggested outline for reorganizing the entire piece. On reading her comments and reviewing the outline, you find that you agree with almost all of her suggestions. You follow her suggestions and proceed to meet with her several times and show her three more drafts, including the final that she edits and proofs.

Feeling that she has done more on your article than she would normally do as part of her job responsibilities, Evelyn diplomatically asks how you plan to acknowledge her work on the final published article. How do you respond to her? Do you list her on the title page as co-author, do you mention her in a footnote as an editor, or do you adopt some other approach? Explain the rationale you give Evelyn after telling her your decision. What are the main ethical considerations in making the decision?

🌐 12. International Communication Assignment

One major problem with international communication occurs when product instructions are written (or translated) by individuals who do not have enough familiarity with the language being used. The problem can be solved by *localization*, or choosing writers or translators who are, in fact, native speakers and writers. For this assignment, locate a set of instructions written in English with stylistic errors that would not have been made by a native speaker/writer. Point out these errors and suggest appropriate revisions.

 13. A.C.T. N.O.W. Assignment (Applying Communication To Nurture Our World)

For this assignment, use the same general context as that described in the A.C.T. N.O.W. exercise in chapter 16 (see p. 576). Using just this textbook, prepare a set of style guidelines to submit for consideration to a campus or community publication. The purpose is to assist individuals who may not have the benefit of a technical writing course or may not have access to current publications in the field. You may use any material from this chapter, as long as you list it as a source somewhere in the document you prepare.

Appendix

>>> **Handbook**

This handbook includes entries on the basics of writing. It contains three main types of information:

1. **Grammar:** the rules by which we edit sentence elements. Examples include rules for the placement of punctuation, the agreement of subjects and verbs, and the placement of modifiers.
2. **Mechanics:** the rules by which we make final proofreading changes. Examples include the rules for abbreviations and the use of numbers. A list of commonly misspelled words is also included.
3. **Usage:** information on the correct use of particular words, especially pairs of words that are often confused. Examples include problem words like *affect/effect, complement/ compliment,* and *who/whom.*

Another editing concern, technical style, is the topic of chapter 17, including guidelines for sentence structure, conciseness, accuracy of wording, active and passive voice, and nonsexist language. Together, chapter 17 and this handbook will help you turn unedited drafts into final revised documents.

This handbook is presented in alphabetized fashion for easy reference during the editing process. Grammar and mechanics entries are in all uppercase; usage entries are in lowercase. Several exercises follow the entries.

A/An

A and *an* are different forms of the same article. *A* occurs before words that start with consonants or consonant sounds. EXAMPLES:

■ a three-pronged plug
■ a once-in-a-lifetime job (*once* begins with the consonant sound of *w*)
■ a historic moment (many speakers and some writers mistakenly use *an* before *historic*)

An occurs before words that begin with vowels or vowel sounds. EXAMPLES:

■ an eager new employee
■ an hour before closing

A lot/Alot

The correct form is the two-word phrase *a lot*. Although acceptable in informal discourse, *a lot* usually should be replaced by more formal diction in technical writing. EXAMPLE: "They retrieved many [not *a lot of*] soil samples from the construction site."

Abbreviations

Technical writing uses many abbreviations. Without this shorthand form, you end up writing much longer reports and proposals without any additional content. Use the following seven basic rules in your use of abbreviations, paying special attention to the first three:

Rule 1: Do Not Use Abbreviations When Confusion May Result

When you want to use a term just once or twice and you are not certain your readers will understand an abbreviation, write out the term rather than abbreviating it. EXAMPLE: "They were required to remove creosote from the site, according to the directive from the Environmental Protection Agency." Even though *EPA* is the accepted abbreviation for this government agency, you should write out the name in full if you are using the term only once to an audience that may not understand it.

Rule 2: Use Parentheses for Clarity

When you use a term more than twice and are not certain that your readers will understand it, write out the term the first time it is used and place the abbreviation in parentheses, and then use the abbreviation in the rest of the document. In long reports or proposals, however, you may need to repeat the full term in key places. EXAMPLE: "According to the directive from the Environmental Protection Agency (EPA), they were required to remove the creosote from the construction site. Furthermore, the directive indicated that the builders could expect to be visited by EPA inspectors every other week."

Rule 3: Include a Glossary When There Are Many Abbreviations

When your document contains many abbreviations that may not be understood by all readers, include a well-marked glossary at the beginning or end of the document. A glossary simply collects all the terms and abbreviations and places them in one location for easy reference.

Rule 4: Use Abbreviations for Units of Measure

Most technical documents use abbreviations for units of measure. Do not include a period unless the abbreviation could be confused with a word. EXAMPLES: mi, ft, oz, gal., in., and lb. Note that units-of-measurement abbreviations have the same form for both singular and plural amounts. EXAMPLES: $\frac{1}{2}$ in., 1 in., 5 in.

Rule 5: Avoid Spacing and Periods

Avoid internal spacing and internal periods in most abbreviations that contain all capital letters. EXAMPLES: ASTM, EPA, ASEE. Exceptions include professional titles and degrees such as P.E., B.S., and B.A.

Rule 6: Be Careful with Company Names

Abbreviate a company or other organizational name only when you are sure that officials from the organization consider the abbreviation appropriate. IBM (for the company) and

UCLA (for the university) are examples of commonly accepted organizational abbreviations. When in doubt, follow rule 2—write the name in full the first time it is used, followed by the abbreviation in parentheses.

Rule 7: Common Abbreviations

The following common abbreviations are appropriate for most writing in your technical or business career. They are placed into three main categories of measurements, locations, and titles.

Measurements.   Use these abbreviations only when you place numbers before the measurement.

ac	alternating current	gal.	gallon
amp	ampere	gpm	gallons per minute
bbl	barrel	hp	horsepower
Btu	British thermal unit	hr	hour
bu	bushel	Hz	hertz
C	Celsius	in.	inch
cal	calorie	j	joule
cc	cubic centimeter	K	Kelvin
circ	circumference	ke	kinetic energy
cm	centimeter	kg	kilogram
cos	cosine	km	kilometer
cot	cotangent	kw	kilowatt
cps	cycles per second	kwh	kilowatt-hour
cu ft	cubic feet	l	liter
db	decibel	lb	pound
dc	direct current	lin	linear
dm	decimeter	lm	lumen
doz *or* dz	dozen	log.	logarithm
F	Fahrenheit	m	meter
f	farad	min	minute
fbm	foot board measure	mm	millimeter
fig.	figure	oz	ounce
fl oz	fluid ounce	ppm	parts per million
FM	frequency modulation	psf	pounds per square foot
fp	foot pound	psi	pounds per square inch
ft	foot (feet)	pt	pint
g	gram	qt	quart

rev	revolution	v	volt
rpm	revolutions per minute	va	volt-ampere
sec	second	w	watt
sq	square	wk	week
sq ft	square foot (feet)	wl	wavelength
T	ton	yd	yard
tan.	tangent	yr	year

Locations.   Use these common abbreviations for addresses (e.g., on envelopes and letters), but write out the words in full in other contexts.

AL	Alabama	MS	Mississippi
AK	Alaska	MO	Missouri
AS	American Samoa	MT	Montana
AZ	Arizona	NE	Nebraska
AR	Arkansas	NV	Nevada
CA	California	NH	New Hampshire
CZ	Canal Zone	NJ	New Jersey
CO	Colorado	NM	New Mexico
CT	Connecticut	NY	New York
DE	Delaware	NC	North Carolina
DC	District of Columbia	ND	North Dakota
FL	Florida	OH	Ohio
GA	Georgia	OK	Oklahoma
GU	Guam	OR	Oregon
HI	Hawaii	PA	Pennsylvania
ID	Idaho	PR	Puerto Rico
IL	Illinois	RI	Rhode Island
IN	Indiana	SC	South Carolina
IA	Iowa	SD	South Dakota
KS	Kansas	TN	Tennessee
KY	Kentucky	TX	Texas
LA	Louisiana	UT	Utah
ME	Maine	VT	Vermont
MD	Maryland	VI	Virgin Islands
MA	Massachusetts	VA	Virginia
MI	Michigan	WA	Washington
MN	Minnesota	WV	West Virginia

WI	Wisconsin	N.W.T.	Northwest Territories
WY	Wyoming	N.S.	Nova Scotia
Alta.	Alberta	Ont.	Ontario
B.C.	British Columbia	P.E.I.	Prince Edward Island
Man.	Manitoba	P.Q.	Quebec
N.B.	New Brunswick	Sask.	Saskatchewan
Nfld.	Newfoundland	Yuk.	Yukon

Titles. Some of the following abbreviations go before the name (e.g., Dr., Ms., Messrs.), whereas others go after the name (e.g., college degrees, Jr., Sr.).

Atty.	Attorney	M.A.	Master of Arts
B.A.	Bachelor of Arts	M.S.	Master of Science
B.S.	Bachelor of Science	M.D.	Doctor of Medicine
D.D.	Doctor of Divinity	Messrs.	Plural of Mr.
Dr.	Doctor (used mainly with medical and dental degrees but also with other doctorates)	Mr.	Mister
		Mrs.	Used to designate married, widowed, or divorced women
Drs.	Plural of Dr.	Ms.	Used increasingly for all women, especially when one is uncertain about a woman's marital status
D.V.M.	Doctor of Veterinary Medicine		
Hon.	Honorable		
Jr.	Junior	Ph.D.	Doctor of Philosophy
LL.D.	Doctor of Laws	Sr.	Senior

Accept/Except

Accept and *except* have different meanings and often are different parts of speech. *Accept* is a verb that means "to receive." *Except* is a preposition or verb and means "to make an exception or special case of." EXAMPLES:

- I *accepted* the service award from my office manager.
- Everyone *except* Jonah attended the marine science lecture.
- The company president *excepted* me from the meeting because I had an important sales call to make the same day.

Advice/Advise/Inform

Advice is a noun that means "suggestions or recommendation." *Advise* is a verb that means "to suggest or recommend." Do not use the verb *advise* as a substitute for *inform*, which means simply "to provide information." EXAMPLES:

- The consultant gave us *advice* on starting a new retirement plan for our employees.

- She *advised* us that a 401(k) plan would be useful for all our employees.
- She *informed* [not *advised*] her clients that they would receive her final report by March 15.

Affect/Effect

Affect and *effect* generate untold grief among many writers. The key to using them correctly is remembering two simple sentences: (1) *affect* with an *a* is a verb meaning "to influence"; (2) *effect* with an *e* is a noun meaning "result." There are some exceptions, however, such as when *effect* can be a verb that means "to bring about," as in, "He effected considerable change when he became a manager." EXAMPLES:

- His progressive leadership greatly *affected* the company's future.
- One *effect* of securing the large government contract was the hiring of several more accountants.
- The president's belief in the future of microcomputers *effected* change in the company's approach to office management. (For a less-wordy alternative, substitute *changed* for *effected change in*.)

Agree to/Agree with

In correct usage, *agree to* means that you have *consented to* an arrangement, an offer, a proposal, and so on. *Agree with* is less constraining and only suggests that you are *in harmony with* a certain statement, idea, person, and the like. EXAMPLES:

- Representatives from M-Global *agreed to* alter the contract to reflect the new scope of work.
- We *agree with* you that more study may be needed before the nuclear power plant is built.

All Right/Alright

All right is an acceptable spelling; *alright* is not. *All right* is an adjective that means "acceptable," an exclamation that means "outstanding," or a phrase that means "correct." EXAMPLES:

- Sharon suggested that the advertising copy was *all right* for now but that she would want changes next month.
- Upon seeing his article in print, Zach exclaimed, *"All right!"*
- The five classmates were *all right* in their response to the trick questions on the quiz.

All Together/Altogether

All together is used when items or people are being considered in a group or are working in concert. *Altogether* is a synonym for "utterly" or "completely." EXAMPLES:

- The three firms were *all together* in their support of the agency's plan.
- There were *altogether* too many pedestrians walking near the dangerous intersection.

Allusion/Illusion/Delusion/Elusion

These similar sounding words have distinct meanings. Following is a summary of the differences:

1. **Allusion:** a noun meaning "reference," as in you are making an allusion to your vacation in a speech. The related verb is *allude*.

2. **Illusion:** a noun meaning "misunderstanding or false perception." It can be physical (as in seeing a mirage) or mental (as in having the false impression that your hair is not thinning when it is).

3. **Delusion:** a noun meaning "a belief based on self-deception." Unlike *illusion*, the word conveys a much stronger sense that someone is out of touch with reality, as in having "delusions of grandeur." The related verb is *delude*.

4. **Elusion:** a noun meaning "the act of escaping or avoiding." The more common form is the verb, *elude*, meaning "to escape or avoid."

Examples:

- His report included an *allusion* to the upcoming visit by the government agency in charge of accreditation.
- She harbored an *illusion* that she was certain to receive the promotion. In fact, her supervisor preferred another department member with more experience.
- He had *delusions* that he soon would become company president, even though he started just last week in the mailroom.
- The main point of the report *eluded* him because there was no executive summary.

Already/All Ready

All ready is a phrase that means "everyone is prepared," whereas *already* is an adverb that means something is finished or completed. EXAMPLES:

- They were *all ready* for the presentation to the client.
- George had *already* arrived at the office before the rest of his proposal team members had even left their homes.

Alternately/Alternatively

Because many readers are aware of the distinction between these two words, any misuse can cause embarrassment or even misunderstanding. Follow these guidelines for correct use.

Alternately.   As a derivative of *alternate, alternately* is best reserved for events or actions that occur "in turns." EXAMPLE: While digging the trench, he used a backhoe and a hand shovel *alternately* throughout the day.

Alternatively.   A derivative of *alternative, alternatively* should be used in contexts where two or more choices are being considered. EXAMPLE: We suggest that you use deep foundations at the site. *Alternatively*, you could consider spread footings that were carefully installed.

Amount/Number

Amount is used in reference to items that *cannot* be counted, whereas *number* is used to indicate items that *can* be counted. EXAMPLES:

- In the last year, we have greatly increased the *amount* of computer paper ordered for the Boston office.
- The last year has seen a huge increase in the *number* [not *amount*] of boxes of computer paper ordered for the Boston office.

And/Or

This awkward expression probably has its origins in legal writing. It means that there are three separate options to be considered: the item before *and/or*, the item after *and/or*, or both items.

Avoid *and/or* because readers may find it confusing, visually awkward, or both. Instead, replace it with the structure used in the previous sentence; that is, write "A, B, or both," *not* "A and/or B." EXAMPLE:

The management trainee was permitted to select two seminars from the areas of computer hardware, communication skills, or both [not *computer hardware and/or communication skills*].

Anticipate/Expect

Anticipate and *expect* are not synonyms. In fact, their meanings are distinctly different. *Anticipate* is used when you mean to suggest or state that steps have been taken beforehand to prepare for a situation. *Expect* only means you consider something likely to occur. EXAMPLES:

- *Anticipating* that the contract will be successfully negotiated, Jones Engineering is hiring three new hydrologists.
- We *expect* [not *anticipate*] that you will encounter semicohesive and cohesive soils in your excavations at the Park Avenue site.

Apt/Liable/Likely

Maintain the distinctions in these three similar words.

1. *Apt* is an adjective that means "appropriate," "suitable," or "has an aptitude for."
2. *Liable* is an adjective that means "legally obligated" or "subject to."
3. *Likely* is either an adjective that means "probable" or "promising" or an adverb that means "probably." As an adverb, it should be preceded by a qualifier such as *quite*.

Examples:

- The successful advertising campaign showed that she could select an *apt* phrase for selling products.
- Jonathan is *apt* at running good meetings. He always hands out an agenda and always ends on time.
- The contract makes clear who is *liable* for any on-site damage.
- Completing the warehouse without an inspection will make the contractor *liable* to lawsuits from the owner.
- A *likely* result of the investigation will be a change in the law. [*likely* as an adjective]
- The investigation will quite *likely* result in a change in the law. [*likely* as an adverb]

Assure/Ensure/Insure

Assure is a verb that can mean "to promise." It is used in reference to people, as in, "We want to *assure* you that our crews will strive to complete the project on time." In fact, *assure* and its derivatives (like *assurance*) should be used with care in technical contexts, because these words can be viewed as a guarantee.

The synonyms *ensure* and *insure* are verbs meaning "to make certain." Like *assure*, they imply a level of certainty that is not always appropriate in engineering or the sciences. When their use is deemed appropriate, the preferred word is *ensure;* reserve *insure* for sentences in which the context is insurance. EXAMPLES:

- Be *assured* that our representatives will be on-site to answer questions that the subcontractor may have.
- To *ensure* that the project stays within schedule, we are building in 10 extra days for bad weather. (An alternative: "So that the project stays within schedule, we are building in 10 extra days for bad weather.")

Augment/Supplement

Augment is a verb that means to increase in size, weight, number, or importance. *Supplement* is either (1) a verb that means "to add to" something to make it complete or

to make up for a deficiency or (2) a noun that means "the thing that has been added."
EXAMPLES:

- The power company supervisor decided to *augment* the line crews in five counties.
- He *supplemented* the audit report by adding the three accounting statements.
- The three accounting *supplements* helped support the conclusions of the audit report.

Awhile/A While

Though similar in meaning, this pair is used differently. *Awhile* means "for a short time." Because "for" is already a part of its definition, it cannot be preceded by the preposition "for." The noun *while*, however, can be preceded by the two words "for a," giving it essentially the same meaning as *awhile*. EXAMPLES:

- Kirk waited *awhile* before trying to restart the generator.
- Kirk *waited for a while* before trying to restart the generator.

Balance/Remainder/Rest

Balance should be used as a synonym for *remainder* only in the context of financial affairs. *Remainder* and *rest* are synonyms to be used in other nonfinancial contexts. EXAMPLES:

- The account had a *balance* of $500, which was enough to avoid a service charge.
- The *remainder* [or *rest,* but not *balance*] of the day will be spent on training in oral presentations for proposals.
- During the *rest* [not *balance*] of the session, we learned about the new office equipment.

Because/Since

Maintain the distinction between these two words. *Because* establishes a cause–effect relationship, whereas *since* is associated with time. EXAMPLES:

- *Because* he left at 3 p.m., he was able to avoid rush hour.
- *Since* last week, her manufacturing team completed 3,000 units.

Between/Among

The distinction between these two words has become somewhat blurred. However, many readers still prefer to see *between* used with reference to only two items, reserving *among* for three or more items. EXAMPLES:

- The agreement was just *between* my supervisor and me. No one else in the group knew about it.

- The proposal was circulated *among* all members of the writing team.
- *Among* Sallie, Todd, and Fran, there was little agreement about the long-term benefits of the project.

Bi-/Semi-/Biannual/Biennial

The prefixes *bi* and *semi* can cause confusion. Generally, *bi* means "every two years, months, weeks, etc.," whereas *semi* means "twice a year, month, week, etc." Yet many readers get confused by the difference, especially when they are confronted with a notable exception such as *biannual* (which means twice a year) and *biennial* (which means every two years).

Your goal, as always, is clarity for the reader. Therefore, it is best to write out meanings in clear prose, rather than relying on prefixes that may not be understood. EXAMPLES:

- We get paid twice a month [preferable to *semimonthly* or *biweekly*].
- The part-time editor submits articles every other month [preferable to *bimonthly*].
- We hold a company social gathering twice a year [preferable to *biannually* or *semi-annually*].
- The auditor inspects our safety files every two years [preferable to *biennially*].

Capital/Capitol

Capital is a noun whose main meanings are (1) a city or town that is a government center, (2) wealth or resources, or (3) net worth of a business or the investment that has been made in the business by owners. *Capital* can also be an adjective meaning (1) "excellent," (2) "primary," or (3) "related to the death penalty." Finally, *capital* can be a noun or an adjective referring to uppercase letters.

Capitol is a noun or an adjective that refers to a building where a legislature meets. With a capital letter, it refers exclusively to the building in Washington, D.C., where the U.S. Congress meets. EXAMPLES:

- The *capital* of Pickens County is Jasper, Georgia.
- Our family *capital* was reduced by the tornado and hurricane.
- She had invested significant *capital* in the carpet factory.
- Their proposal contained some *capital* ideas that would open new opportunities for our firm.
- In some countries, armed robbery is a *capital* offense.
- The students visited the *capitol* building in Atlanta. Next year they will visit the *Capitol* in Washington, D.C., where they will meet several members of Congress.

Capitalization

As a rule, you should capitalize *specific* names of people, places, and things—sometimes called *proper nouns*. For example, capitalize specific streets, towns, trademarks, geologic eras, planets, groups of stars, days of the week, months of the year, names of organizations, holidays, and colleges. However, remember that excessive capitalization—as in titles of positions in a company—is inappropriate in technical writing and can appear somewhat pompous.

The following rules cover some frequent uses of capitals:

1. Major words in titles of books and articles. Capitalize prepositions and articles only when they appear as the first word in titles. EXAMPLES:

   - *For Whom the Bell Tolls*
   - *In Search of Excellence*
   - *The Power of Positive Thinking*

2. Names of places and geographic locations. EXAMPLES:

   - Washington Monument
   - Cleveland Stadium
   - Dallas, Texas
   - Cobb County

3. Names of aircraft and ships. EXAMPLES:

   - *Air Force One*
   - *SS Arizona*
   - *Nina, Pinta,* and *Santa Maria*

4. Names of specific departments and offices within an organization. EXAMPLES:

   - Humanities Department
   - Personnel Department
   - International Division

5. Political, corporate, and other titles that come before names. EXAMPLES:

   - Chancellor Hairston
   - Councilwoman Jones
   - Professor Gainesberg
   - Congressman Buffett

   Note, however, that general practice does not call for capitalizing most titles when they are used by themselves or when they follow a person's name. EXAMPLES:

   - Jane Cannon, a professor in the Business Department.
   - Jim McDuff, president of M-Global, Inc.
   - Chris Presley, secretary of the Oil Rig Division.

Center On/Revolve Around

The key to using these phrases correctly is to think about their literal meaning. For example, you center *on* (not around) a goal, just as you would center on a target with a gun or bow and arrow. Likewise, your hobbies revolve *around* your early interest in water sports, just as the planets revolve around the sun in our solar system. EXAMPLES:

- All her selling points in the proposal *centered on* the need for greater productivity in the factory.
- At the latest annual meeting, some stockholders argued that most of the company's recent projects *revolved around* the CEO's interest in attracting attention from the media.

Cite/Site/Sight

1. *Cite* is a verb meaning "to quote as an example, authority, or proof." It can also mean "to commend" or "to bring before a court of law" (as in receiving a traffic ticket).

2. *Site* usually is a noun that means "a particular location." It can also be a verb that means "to place at a location," as with a new school being sited by the town square, but this usage is not preferred. Instead use a more conventional verb, such as *built*.

3. *Sight* is a noun meaning "the act of seeing" or "something that is seen," or it can be a verb meaning "to see or observe."

Examples:

- We *cited* a famous geologist in our report on the earthquake.
- Rene was *cited* during the ceremony for her exemplary service to the city of Roswell.
- The officer will *cite* the party-goers for disturbing the peace.
- Although five possible dorm *sites* were considered last year, the college administrators decided to build [preferred over *site*] the dorm at a different location.
- The *sight* of the flock of whooping cranes excited the visitors.
- Yesterday we *sighted* five whooping cranes at the marsh.

Complement/Compliment

Both words can be nouns and verbs, and both have adjective forms (*complementary, complimentary*).

Complement.   *Complement* is used as a noun to mean "that which has made something whole or complete," as a verb to mean "to make whole, to make complete," or as an adjective. You may find it easier to remember the word by recalling its mathematical definition: Two *complementary* angles must always equal 90 degrees. EXAMPLES:

- As a noun: The *complement* of five technicians brought our crew strength up to 100 percent.
- As a verb: The firm in Canada served to *complement* ours in that together we won a joint contract.
- As an adjective: Seeing that project manager and her secretary work so well together made clear their *complementary* relationship in getting the office work done.

Compliment.   *Compliment* is used as a noun to mean "an act of praise, flattery, or admiration," as a verb to mean "to praise, to flatter," or as an adjective to mean "related to praise or flattery, or without charge." EXAMPLES:

- As a noun: He appreciated the verbal *compliments*, but he also hoped they would result in a substantial raise.
- As a verb: Howard *complimented* the crew for finishing the job on time and within budget.
- As an adjective: We were fortunate to receive several *complimentary* copies of the new software from the publisher.

Compose/Comprise

These are both acceptable words, with an inverse relationship to each other. *Compose* means "to make up or be included in," whereas *comprise* means "to include or consist of." The easiest way to remember this relationship is to memorize one sentence: "The parts compose the whole, but the whole comprises the parts." One more point to remember: The common phrase *is comprised of* is a substandard, unacceptable replacement for *comprise* or *is composed of.* Careful writers do not use it. EXAMPLES:

- Seven quite discrete layers *compose* the soils that were uncovered at the site.
- The borings revealed a stratigraphy that *comprises* [not *is comprised of*] seven quite discrete layers.

Consul/Council/Counsel

Consul, council, and *counsel* can be distinguished by meaning and, in part, by their use within a sentence.

> *Consul:* A noun meaning an official of a country who is sent to represent that country's interests in a foreign land.
>
> *Council:* A noun meaning an official group or committee.
>
> *Counsel:* A noun meaning an adviser or advice given, or a verb meaning to produce advice.

Examples:

- (Consul) The Brazilian *consul* met with consular officials from three other countries.
- (Council) The Human Resources *Council* of our company recommended a new retirement plan to the company president.
- (Counsel—as noun) After the tragedy, they received legal *counsel* from their family attorney and spiritual *counsel* from their minister.
- (Counsel—as verb) As a communications specialist, Roberta helps *counsel* employees who are involved in various types of disputes.

Continuous/Continual

The technical accuracy of some reports may depend on your understanding of the difference between *continuous* and *continual*. *Continuous* and *continuously* should be used in reference to uninterrupted, unceasing activities. However, *continual* and *continually* should be used with activities that are intermittent, or repeated at intervals. If you think your reader may not understand the difference, you should either (1) use synonyms that are clearer (such as *uninterrupted* for *continuous*, and *intermittent* for *continual*) or (2) define each word at the point you first use it in the document. EXAMPLES:

- We *continually* checked the water pressure for three hours before the equipment arrived, while also using the time to set up the next day's tests.

- Because it rained *continuously* from 10:00 A.M. until noon, we were unable to move our equipment onto the utility easement.

Criterion/Criteria

Coming from the Latin language, *criterion* and *criteria* are the singular and plural forms of a word that means "rationale or reasons for selecting a person, place, thing, or idea." A common error is to use *criteria* as both a singular and plural form, but such misuse disregards a distinction recognized by many readers. Maintain the distinction in your writing. EXAMPLES:

- Among all the qualifications we established for the new position, the most important *criterion* for success is good communication skills.

- She had to satisfy many *criteria* before being accepted into the honorary society of her profession.

Data/Datum

Coming as it does from the Latin, the word *data* is the plural form of *datum*. Although many writers now accept *data* as singular or plural, traditionalists in the technical and scientific community still consider *data* exclusively a plural form. Therefore, you should maintain the plural usage. EXAMPLES:

- These *data* show that there is a strong case for building the dam at the other location.

- This particular *datum* shows that we need to reconsider recommendations put forth in the original report.

  If you consider the traditional singular form of *datum* to be awkward, use substitutes such as, "This item in the data shows . . . " or "One of the data shows that . . . " Singular subjects like *one* or *item* allow you to keep your original meaning without using the word *datum*.

Definite/Definitive

Although similar in meaning, these words have slightly different contexts. *Definite* refers to that which is precise, explicit, or final. *Definitive* has the more restrictive meaning of "authoritative" or "final." EXAMPLES:

- It is now *definite* that he will be assigned to the London office for six months.
- He received the *definitive* study on the effect of the oil spill on the marine ecology.

Discrete/Discreet/Discretion

The adjective *discrete* suggests something that is separate or made up of many separate parts. The adjective *discreet* is associated with actions that require caution, modesty, or reserve. The noun *discretion* refers to the quality of being "discreet," or the freedom a person has to act on her or his own. EXAMPLES:

- The orientation program at M-Global includes a writing seminar, which is a *discrete* training unit offered for one full day.
- The orientation program at M-Global includes five *discrete* units.
- As a counselor in M-Global's Human Resources Office, Sharon was *discreet* in her handling of personal information about employees.
- Every employee in the Human Resources Office was instructed to show *discretion* in handling personal information about employees.
- By starting a flextime program, M-Global, Inc., will give employees a good deal of *discretion* in selecting the time to start and end their workday.

Disinterested/Uninterested

In contemporary business use, *disinterested* and *uninterested* have quite different meanings. Because errors can cause confusion for the reader, make sure not to use the words as synonyms. *Disinterested* means "without prejudice or bias," whereas *uninterested* means "showing no interest." EXAMPLES:

- The agency sought a *disinterested* observer who had no stake in the outcome of the trial.
- They spent several days talking to officials from Iceland, but they still remain *uninterested* in performing work in that country.

Due to/Because of

Besides irritating those who expect proper English, mixing these two phrases can also cause confusion. *Due to* is an adjective phrase meaning "attributable to" and almost always follows a "to be" verb (such as "is," "was," or "were"). It should not be used in place of prepositional phrases such as "because of," "owing to," or "as a result of." EXAMPLES:

- The cracked walls were *due to* the lack of proper foundation fill being used during construction.
- We won the contract *because of* [not *due to*] our thorough understanding of the client's needs.

Each Other/One Another

Each other occurs in contexts that include only two persons, whereas *one another* occurs in contexts that include three or more persons. EXAMPLES:

- Shana and Katie worked closely with *each other* during the project.
- All six members of the team conversed with *one another* regularly through email.

e.g./i.e.

The abbreviation *e.g.* means "for example," whereas *i.e.* means "that is." These two Latin abbreviations are often confused, a fact that should give you pause before using them. Many writers prefer to write them out, rather than risk confusion on the part of the reader. EXAMPLES:

- During the trip, he visited 12 cities where M-Global is considering opening offices—e.g., [or, *for example*] Kansas City, New Orleans, and Seattle.
- A spot along the Zayante Fault was the earthquake's epicenter—*i.e.*, [or *that is*] the focal point for seismic activity.

English as a Second Language (ESL)

Technical writing challenges native English speakers and nonnative speakers alike. The purpose of this section is to present a basic description of three grammatical forms: articles, verbs, and prepositions. These forms may require more intense consideration from international students when they complete technical writing assignments. Each issue is described using the ease-of-operation section from a memo about a fax machine. The passage, descriptions, and charts work together to show how these grammar issues function collectively to create meaning.

Ease of Operation—Article Usage

The AIM 500 is so easy to operate that **a** novice can learn to transmit **a** document to another location in about two minutes. Here's **the** basic procedure:

1. Press **the** button marked TEL on **the** face of **the** fax machine. You then hear **a** dial tone.
2. Press **the** telephone number of **the** person receiving **the** fax on **the** number pad on **the** face of **the** machine.
3. Lay **the** document face down on **the** tray at **the** back of **the** machine.

At this point, just wait for **the** document to be transmitted—about 18 seconds per page to transmit. **The** fax machine will even signal **the** user with **a** beep and **a** message on its LCD display when **the** document has been transmitted. Other more advanced operations are equally simple to use and require little training. Provided with **the** machine are two different charts that illustrate **the** machine's main functions.

The size of **the** AIM 500 makes it easy to set up almost anywhere in **an** office. **The** dimensions are 13 inches in width, 15 inches in length, and 9.5 inches in height. **The** narrow width, in particular, allows **the** machine to fit on most desks, file cabinets, or shelves.

Articles.   Articles are one of the most difficult forms of English grammar for non-native English speakers, mainly because some language systems do not use them. Thus speakers of particular languages may have to work hard to incorporate the English article system into their language proficiency.

The English articles include *a, an,* and *the.*

- *A* and *an* express indefinite meaning when they refer to nouns or pronouns that are not specific. The writer believes the reader does not know the noun or pronoun.

- *The* expresses definite meaning when it refers to a specific noun or pronoun. The writer believes the reader knows the specific noun or pronoun.

ESL writers choose the correct article only when they (1) know the context or meaning, (2) determine whether they share information about the noun with the reader, and (3) consider the type of noun following the article.

The ease-of-operation passage includes 31 articles that represent two types—definite and indefinite. When a writer and a reader share knowledge of a noun, the definite article should be used. On 25 occasions the articles in the passage suggest the writer and reader share some knowledge of a count noun. *Count nouns* are nouns that can be counted (pen, cloud, memo). Examples of non-count nouns are sugar, air, and beef.

For example, the memo writer and the memo recipient share knowledge of the particular model fax machine—the AIM 500. Thus, *the* is definite when it refers to "the fax machine" in the memo. Notice, however, that *document* becomes definite only after the second time it is mentioned ("Lay the document face down ...."). In the first reference to *document, a* document refers to a document about which the writer and reader share no knowledge. The memo writer cannot know which document the reader will fax. Only in the second reference do the writer and reader know the document to be the one the reader will fax.

The indefinite article *a* occurs five times, whereas *an* occurs once. Each occurrence signals a singular count noun. The reader and the writer share no knowledge of the nouns that follow the *a* or *an,* so an indefinite article is appropriate. *A* precedes nouns beginning with consonant sounds. *An* precedes nouns beginning with vowel sounds. Indefinite articles seldom precede non-count nouns unless a non-count functions as a modifier (a beef shortage).

Definite and indefinite articles are used more frequently than other articles; however, other articles do exist. The "generic" article refers to classes or groups of people, objects, and ideas. If the fax machine is thought of in a general sense, the meaning changes. For example, "the fax machine increased office productivity by 33%." *The* now has a generic meaning representing fax machines in general. The same generic meaning can apply to the plural noun, but such generic use requires no article: "Fax machines increased office productivity by 33%." *The* in this instance is a generic article.

Articles from "Ease of Operation" Excerpt

Article	Noun	Type	Comment
The	ATM 500	definite	first mention—shared knowledge
a	novice	indefinite	first mention—no shared knowledge
a	document	indefinite	first mention—no shared knowledge
the	basic procedure	definite	
the	button	definite	
the	face	definite	
the	fax machine	definite	first mention without proper name, with reader/writer shared knowledge
a	dial tone	indefinite	first mention—no shared knowledge
the	telephone number	definite	
the	person	definite	
the	fax	definite	
the	number pad	definite	
the	face	definite	
the	machine	definite	
the	document	definite	
the	tray	definite	
the	back	definite	
the	machine	definite	
the	document	definite	second mention
the	fax machine	definite	
the	user	definite	
a	beep	indefinite	first mention—no shared knowledge
a	message	indefinite	first mention—no shared knowledge
the	document	definite	
the	machine	definite	
the	machine's main	definite	functions
The	size	definite	
the	AIM 500	definite	
an	office	indefinite	first mention—preceding vowel sound—no shared knowledge
The	dimensions	definite	
The	narrow width	definite	
the	machine	definite	

Ease of Operation—Verb Usage

The AIM 500 **is** so easy to operate that a novice **can learn** to transmit a document to another location in about two minutes. Here's the basic procedure:

1. **Press** the button marked TEL on the face of the fax machine. You then **hear** a dial tone.
2. **Press** the telephone number of the person receiving the fax on the number pad on the face of the machine.
3. **Lay** the document face down on the tray at the back of the machine.

At this point, just **wait** for the document to be transmitted—about 18 seconds per page to transmit. The fax machine **will** even **signal** the user with a beep and a message on its LCD display when the document **has been transmitted**. Other more advanced operations **are** equally simple to use and **require** little training. **Provided** with the machine **are** two different charts that **illustrate** the machine's main functions.

The size of the AIM 500 **makes** it easy to set up almost anywhere in an office. The dimensions **are** 13 inches in width, 15 inches in length, and 9.5 inches in height. The narrow width, in particular, **allows** the machine to fit on most desks, file cabinets, or shelves.

Verbs.   Verbs express time in three ways—simple present, simple past, and future. *Wait, waited,* and *will wait* and *lay* ("to put"), *laid,,* and *will lay* are examples of simple present, simple past, and future tense verbs. Verbs in the English language system appear as either regular or irregular forms.

Regular Verbs—Simple Tense   Regular verbs follow a predictable pattern. The form of the simple present tense verbs (*walk*) changes to the simple past tense with the addition of *–ed* (*walked*) and changes to the simple future with the addition of a special auxiliary (helping) verb called a *modal* (*will walk*).

Present	Past	Future
learn	learned	will learn
wait	waited	will wait
press	pressed	will press
signal	signaled	will signal
require	required	will require
provide	provided	will provide
illustrate	illustrated	will illustrate
allow	allowed	will allow

Irregular Verbs—Simple Tense   Irregular verbs do not follow a predictable pattern. Most importantly, the past tense is not created by adding *–ed*. The simple present tense of *lay* ("to put") changes completely in the simple past (*laid*).

Present	Past	Future
is	was	will be
are	were	will be
hear	heard	will hear
do	did	will do
get	got	will get
see	saw	will see
write	wrote	will write
speak	spoke	will speak

Unfortunately, the English verb system is more complicated than that. Verbs express more than time; they can also express *aspect*, or whether an action was completed. The perfect aspect indicates that an action was completed (perfected) and the progressive aspect indicates that an action is incomplete (in progress).

Regular Verbs—Aspect   In regular verbs, the perfect aspect is indicated with the addition of a form of the auxiliary (helping) word *to have* to the simple past tense form. In verb phrases that indicate aspect, tense is always found in the first verb in the verb phrase. For example, "I have walked" is present perfect, and "I had walked" is past perfect. The progressive aspect is indicated with the addition of a form of the auxiliary word *to be* and an *–ing* form of the main verb. The progressive aspect is always regular.

Present Perfect	Past Perfect	Future Perfect
have learned	had learned	will have learned
have waited	had waited	will have waited
have pressed	had pressed	will have pressed
have signaled	had signaled	will have signaled
have required	had required	will have required
have provided	had provided	will have provided
have illustrated	had illustrated	will have illustrated
have allowed	had allowed	will have allowed

Present Progressive	Past Progressive	Future Progressive
is learning	was learning	will be learning
is waiting	was waiting	will be waiting
is pressing	was pressing	will be pressing
is signaling	was signaling	will be signaling
is requiring	was requiring	will be requiring
is providing	was providing	will be providing
is illustrating	was illustrating	will be illustrating
is allowing	was allowing	will be allowing

Irregular Verbs—Aspect   The irregular forms of the perfect aspect can be confusing. The auxiliary verbs are the same as for the regular verb phrases, but the main verb can be inflected in a number of ways. Most dictionaries list this form of the verb after the present and past forms of the verb.

Present Perfect	Past Perfect	Future Perfect
have been	had been	will have been
have been	had been	will have been
have heard	had heard	will have heard
have done	had done	will have done
have gotten	had gotten	will have gotten
have seen	had seen	will have seen
have written	had written	will have written
have spoken	had spoken	will have spoken

Let's examine four specific verb forms in the "Ease of Operation" passage.

1. *Is* represents a being or linking verb in the passage. Being verbs suggest an aspect of an experience or being (existence); for example, "He is still here," and "The fax is broken." Linking verbs connect a subject to a complement (completer); for example, "The fax machine is inexpensive."

2. *Can learn* is the present tense verb *learn* preceded by a modal. Modals assist verbs to convey meaning. *Can* suggests ability or possibility. Other modals and their meanings appear next.

Will	Would	Could	Shall	Should	Might	Must
scientific fact possibility determination	hypothetical	hypothetical	formal will	expectation obligation	possibility	necessity

3. *Here's* shows a linking verb (*is*) connected to its complement (*here*). The sentence in its usual order—subject first followed by the verb—appears as, "The basic procedure is here." Article—adjective—noun—linking verb—complement.

Verbs from "Ease of Operation" Excerpt

Verb	Tense	Number	Other Details
is	present	singular	linking/being (is, was, been)
can learn	present	singular	*can* is a modal auxiliary implying "possibility"
Here's (is)	present	singular	linking/being
Press	present	singular	understood "you" as subject
hear	present	singular	action/transitive

Press	present	singular	understood "you" as subject
Lay	present	singular	irregular (lay, laid, laid) singular— understood "you" as subject
wait	present	singular	understood "you" as subject
will signal	future	singular	action to happen or condition to experience
has been transmitted	present perfect	singular	passive voice—action that began in the past and continues to the present
are	present	plural	linking/being
require	present	plural	action/transitive
Provided are	present perfect	plural	passive voice—action that began in the past and continues to the present
illustrate	present	plural	action/transitive
makes	present	singular	action/transitive
are	present	plural	linking/being
allows	present	singular	action/transitive

4. *Press, Lay,* and *wait* (for) share at least four common traits: present tense, singular number, action to transitive, and understood subject of "you." Although "you" does not appear in the text, the procedure clearly instructs the person operating the fax machine— "you." Action or transitive verbs express movement, activity, and momentum, and may take objects. Objects answer the questions Who? What? To whom? Or for whom? in relation to transitive verbs. For example, "Press the button," "Hear a dial tone," "Press the telephone number," "Lay the document face down." Press what? Hear what? Lay what?

Ease of Operation—Preposition Usage

The AIM 500 is so easy to operate that a novice can learn to transmit a document **to** another location **in** about two minutes. Here's the basic procedure:

1. Press the button marked TEL **on** the face **of** the fax machine. You then hear a dial tone.
2. Press the telephone number **of** the person receiving the fax **on** the number pad **on** the face **of** the machine.
3. Lay the document face down **on** the tray **at** the back **of** the machine.

At this point, just wait **for** the document to be transmitted—**about** 18 seconds **per** page to transmit. The fax machine will even signal the user **with** a beep and a message **on** its LCD display when the document has been transmitted. Other more advanced operations are equally simple to use and require little training. Provided **with** the machine are two different charts that illustrate the machine's main functions.

The size **of** the AIM 500 makes it easy to set up almost anywhere **in** an office. The dimensions are 13 inches **in** width, 15 inches **in** length, and 9.5 inches **in** height. The narrow width, **in** particular, allows the machine to fit **on** most desks, file cabinets, or shelves.

Prepositions.   Prepositions are words that become a part of a phrase composed of the preposition, a noun or pronoun, and any modifiers. Notice the relationships expressed within the prepositional phrases and the ways they affect meaning in the sentences. In the "Ease of Operation" passage, about half the prepositional phrases function as adverbs noting place or time; the other half function as adjectives.

Place or	Location	Time
at	on	before
in	above	after
below	around	since
beneath	out	during
over	underneath	
within	under	
outside	near	
into	inside	

One important exception is a preposition that connects to a verb to make a *prepositional verb—wait for.* Another interesting quality of prepositions is that sometimes more than one can be used to express similar meaning. In the "Ease of Operation" passage, for example, both *on* the tray and *at* the back indicate position. Another way to state the same information is *on* the tray *on* the back.

Prepositions from "Ease of Operation" Excerpt

Preposition	Noun Phrase	Comment
to	another location	direction toward
in	(about) two minutes	approximation of time
on	the face	position
of	the fax machine	originating at or from
of	the person	associated with
on	the number pad	position
on	the face	position
of	the machine	originating at
on	the tray	position
at	the back	position of
of	the machine	originating at
At	this point	on or near the time
for	the document	indication of object of desire
about	18 seconds	adverb = approximation
per	page	for every
with	a beep and a message	accompanying
on	its LCD display	position
with	the machine	accompanying

of	the AIM 500	originating at or from
in	an office	within the area
in	width	with reference to
in	length	with reference to
in	height	with reference to
in	particular	with reference to
on	most desks, file cabinets, or shelves	position

Farther/Further

Although similar in meaning, these two words are used differently. *Farther* refers to actual physical distance, whereas *further* refers to nonphysical distance or can mean "additional." EXAMPLES:

- The overhead projector was moved *farther* from the screen so that the print would be easier to see.
- *Farther* up the old lumber road, they found footprints of an unidentified mammal.
- As he read *further* along in the report, he began to understand the complexity of the project.
- She gave *further* instructions after they arrived at the site.

Fewer/Less

The adjective *fewer* is used before items that can be counted, whereas the adjective *less* is used before mass quantities. When errors occur, they usually result from *less* being used with countable items, as in this *incorrect* sentence: "We can complete the job with less men at the site." EXAMPLES:

- The newly certified industrial hygienist signed with us because the other firm in which he was interested offered *fewer* [not *less*] benefits.
- There was *less* sand in the sample taken from 15 ft than in the one taken from 10 ft.

Flammable/Inflammable/Nonflammable

Given the importance of these words in avoiding injury and death, make sure to use them correctly—especially in instructions. *Flammable* means "capable of burning quickly" and is acceptable usage. *Inflammable* has the same meaning, but it is not acceptable usage for this reason: Some readers confuse it with *nonflammable*. The word *nonflammable*, then, means "not capable of burning" and is accepted usage. EXAMPLES:

- They marked the package *flammable* because its contents could be easily ignited by a spark. (Note that *flammable* is preferred here over its synonym, *inflammable*.)
- The foreman felt comfortable placing the crates near the heating unit, because all the crates' contents were *nonflammable*.

Former/Latter

These two words direct the reader's attention to previous points or items. *Former* refers to that which came first, whereas *latter* refers to that which came last. Note that the words are used together when there are only two items or points—not with three or more. Also, you should know that some readers may prefer you avoid *former* and *latter* altogether, because the construction may force them to look back to previous sentences to understand your meaning. The second example gives an alternative.

- (with former/latter) The airline's machinists and flight attendants went on strike yesterday. The *former* left work in the morning, whereas the *latter* left work in the afternoon.
- (without former/latter) The airline's machinists and flight attendants went on strike yesterday. The machinists left work in the morning, whereas the flight attendants left work in the afternoon.

Fortuitous/Fortunate

The word *fortuitous* is an adjective that refers to an unexpected action, without regard to whether it is desirable. The word *fortunate* is an adjective that indicates an action that is clearly desired. The common usage error with this pair is the wrong assumption that *fortuitous* events must also be *fortunate*. EXAMPLES:

- Seeing M-Global's London manager at the conference was quite *fortuitous*, because I had not been told that he also was attending.
- It was indeed *fortunate* that I encountered the London manager, for it gave us the chance to talk about an upcoming project involving both our offices.

Generally/Typically/Usually

Words like *generally, typically,* and *usually* can be useful qualifiers in your reports. They indicate to the reader that what you have stated is often, but not always, the case. Make certain to place these adverb modifiers as close as possible to the words they modify. In the first example, it would be inaccurate to write *were typically sampled,* because the adverb modifies the entire verb phrase *were sampled.* EXAMPLES:

- Cohesionless soils *typically* were sampled by driving a 2-in.-diameter, split-barrel sampler. (Active-voice alternative: *Typically*, we sampled cohesionless soils by driving a 2-in.-diameter, split-barrel sampler.)
- For projects like the one you propose, the technician *usually* cleans the equipment before returning to the office.
- It is *generally* known that sites for dumping waste should be equipped with appropriate liners.

Good/Well

Although similar in meaning, *good* is used as an adjective and *well* is used as an adverb. A common usage error occurs when writers use the adjective when the adverb is required. EXAMPLES:

- It is *good* practice to submit three-year plans on time.
- He did *well* to complete the three-year plan on time, considering the many reports he had to finish that same week.

Imply/Infer

Remember that the person doing the speaking or writing implies, whereas the person hearing or reading the words infers. In other words, the word *imply* requires an active role; the word *infer* requires a passive role. When you *imply* a point, your words suggest rather than state a point. When you *infer* a point, you form a conclusion or deduce meaning from someone else's words or actions. EXAMPLES:

- The contracts officer *implied* that there would be stiff competition for that $20 million waste-treatment project.
- We *inferred* from her remarks that any firm hoping to secure the work must have completed similar projects recently.

Its/It's

Its and *it's* are often confused. You can avoid error by remembering that *it's* with the apostrophe is used *only* as a contraction for *it is* or *it has*. The other form—*its*—is a possessive pronoun. You can remember this by remembering that other possessive pronouns (mine, his) do not have apostrophes. EXAMPLES:

- Because of the rain, *it's* [or *it is*] going to be difficult to move the equipment to the site.
- *It's* [or *it has*] been a long time since we submitted the proposal.
- The company completed *its* part of the agreement on time.

Lay/Lie

Lay and *lie* are troublesome verbs, and you must know some basic grammar to use them correctly.

1. *Lay* means "to place." It is a transitive verb; thus it takes a direct object to which it conveys action. ("She laid down the printout before starting the meeting.") Its main forms are *lay* (present), *laid* (past), *laid* (past participle), and *laying* (present participle).
2. *Lie* means "to be in a reclining position." It is an intransitive verb; thus it does not take a direct object. ("In some countries, it is acceptable for workers to lie down for a midday nap.") Its main forms are *lie* (present), *lay* (past), *lain* (past participle), and *lying* (present participle).

If you want to use these words with confidence, remember the transitive/intransitive distinction and memorize the principal parts. EXAMPLES:

- (lay) I will *lay* the notebook on the lab desk before noon.
- (lay) I have *laid* the notebook there before.
- (lay) I was *laying* the notebook down when the phone rang.
- (lie) The watchdog *lies* motionless at the warehouse gate.
- (lie) The dog *lay* there yesterday too.
- (lie) The dog has *lain* there for three hours today and no doubt will be *lying* there when I return from lunch.

Lead/Led

Lead is either a noun that names the metallic element or a verb that means "to direct or show the way." *Led* is only a verb form, the past tense of the verb *lead*. EXAMPLES:

- The company bought rights to mine *lead* on the land.
- They chose a new president to *lead* the firm into the twenty-first century.
- They were *led* to believe that salary raises would be high this year.

Like/As

Like and *as* are different parts of speech and thus are used differently in sentences. *Like* is a preposition and therefore is followed by an object—not an entire clause. *As* is a conjunction and thus is followed by a group of words that includes a verb. *As if* and *as though* are related conjunctions. EXAMPLES:

- Gary looks *like* his father.
- Managers *like* John will be promoted quickly.
- If Teresa writes this report *as* she wrote the last one, our clients will be pleased.
- Our proposals are brief, *as* they should be.
- Our branch manager talks *as though* [or *as if*] the merger will take place soon.

Loose/Lose

Loose, which rhymes with "goose," is an adjective that means "unfastened, flexible, or unconfined." *Lose*, which rhymes with "ooze," is a verb that means "to misplace." EXAMPLES:

- The power failure was linked to a *loose* connection at the switchbox.
- Because of poor service, the photocopy machine company may *lose* its contract with M-Global's San Francisco office.

Modifiers: Dangling and Misplaced

This section includes guidelines for avoiding the most common modification errors—dangling modifiers and misplaced modifiers. First, however, we must define the term *modifier*. Words, phrases, and even dependent clauses can serve as modifiers. They serve to qualify, or add meaning to, other elements in the sentence. For our purposes here, the most important point is that modifiers must be connected clearly to what they modify.

Modification errors occur most often with verbal phrases. A phrase is a group of words that lacks either a subject or a predicate. The term *verbal* refers to (1) gerunds (*–ing* form of verbs used as nouns, such as, "He likes skiing"), (2) participles (*–ing* form of verbs used as adjectives, such as, "Skiing down the hill, he lost a glove"), or (3) infinitives (the word *to* plus the verb root, such as, "To attend the opera was his favorite pastime"). Now let's look at the two main modification errors.

Dangling Modifiers.   When a verbal phrase "dangles," the sentence in which it is used contains no specific word for the phrase to modify. As a result, the meaning of the sentence can be confusing to the reader. For example, "In designing the foundation, several alternatives were discussed." It is not at all clear exactly who is doing the "designing." The phrase dangles because it does not modify a specific word. The modifier does not dangle in this version of the sentence: "In designing the foundation, we discussed several alternatives."

Misplaced Modifiers.   When a verbal phrase is misplaced, it may appear to refer to a word that it, in fact, does not modify. EXAMPLE: "Floating peacefully near the oil rig, we saw two humpback whales." Obviously, the whales are doing the floating, and the rig workers are doing the seeing here. Yet because the verbal phrase is placed at the beginning of the sentence, rather than at the end immediately after the word it modifies, the sentence presents some momentary confusion.

Misplaced modifiers can lead to confusion about the agent of action in technical tasks. EXAMPLE: "Before beginning to dig the observation trenches, we recommend that the contractors submit their proposed excavation program for our review." On quick reading, the reader is not certain about who will be "beginning to dig"—the contractors or the "we" in the sentence. The answer is the contractors. Thus a correct placement of the modifier should be, "We recommend the following: Before the contractors begin digging observation trenches, they should submit their proposed excavation for our review."

Solving Modifier Problems.   At best, dangling and misplaced modifiers produce a momentary misreading by the audience. At worst, they can lead to confusion that results in disgruntled readers, lost customers, or liability problems. To prevent modification problems, place all verbal phrases—indeed, all modifiers—as close as possible to the word they modify. If you spot a modification error while you are editing, correct it in one of two ways:

1. Leave the modifier as it is and rework the rest of the sentence. Thus you would change "Using an angle of friction of 20 degrees and a vertical weight of 300 tons, the sliding resistance would be . . . " to the following: "Using an angle of friction of 20 degrees and a vertical weight of 300 tons, we computed a sliding resistance of. . . ."

2. Rephrase the modifier as a complete clause. Thus you would change the previous original sentence to, "If the angle of friction is 20 degrees and the vertical weight is 300 tons, the sliding resistance should be. . . . "

In either case, your goal is to link the modifier clearly and smoothly with the word or phrase it modifies.

Number of/Total of

These two phrases can take singular or plural verbs, depending on the context. Following are two simple rules for correct usage:

1. If the phrase is preceded by *the*, it takes a singular verb because emphasis is placed on the group.

2. If the phrase is preceded by *a*, it takes a plural verb because emphasis is placed on the many individual items.

Examples:

- *The number of* projects going over budget *has* decreased dramatically.
- *The total of* 90 lawyers *believes* the courtroom guidelines should be changed.
- *A number* of projects *have* stayed within budget recently.
- *A total of* 90 lawyers *believe* the courtroom guidelines should be changed.

Numbers

Like rules for abbreviations, those for numbers vary from profession to profession and even from company to company. Most technical writing subscribes to the approach that numbers are best expressed in figures (45) rather than words (forty-five). Note that this style may differ from that used in other types of writing, such as this textbook. Unless the preferences of a particular reader suggest that you do otherwise, follow these common rules for use of numbers in writing your technical documents:

Rule 1: Follow the 10-or-Over Rule

In general, use figures for numbers of 10 or more, words for numbers under 10. EXAMPLES: Three technicians at the site / 15 reports submitted last month / one rig contracted for the job.

Rule 2: Do Not Start Sentences with Figures

Begin sentences with the word form of numbers, not with figures. EXAMPLE: "Forty-five containers were shipped back to the lab."

Rule 3: Use Figures as Modifiers

Whether higher or lower than 10, numbers are usually expressed as figures when used as modifiers with units of measurement, time, and money, especially when these units are abbreviated. EXAMPLES: 4 in., 7 hr, 17 ft, $5 per hr. Exceptions can be made when the unit is not abbreviated. EXAMPLE: five years.

Rule 4: Use Figures in a Group of Mixed Numbers

Use only figures when the numbers grouped together in a passage (usually *one* sentence) are both higher and lower than 10. EXAMPLE: "For that project they assembled 15 samplers, 4 rigs, and 25 containers." In other words, this rule argues for consistency within a writing unit.

Rule 5: Use the Figure Form in Illustration Titles

Use the numeric form when labeling specific tables and figures in your reports. EXAMPLES: Figure 3, Table 14–B.

Rule 6: Be Careful with Fractions

Express fractions as words when they stand alone, but as figures when they are used as a modifier or are joined to whole numbers. EXAMPLE: "We have completed two-thirds of the project using the $2\frac{1}{2}$-in. pipe."

Rule 7: Use Figures and Words with Numbers in Succession

When two numbers appear in succession in the same unit, write the first as a word and the second as a figure. EXAMPLE: "We found fifteen 2-ft pieces of pipe in the machinery."

Rule 8: Only Rarely Use Numbers in Parentheses

Except in legal documents, avoid the practice of placing figures in parentheses after their word equivalents. EXAMPLE: "The second party will send the first party forty-five (45) barrels on or before the first of each month." Note that the parenthetical amount is placed immediately after the figure, not after the unit of measurement.

Rule 9: Use Figures with Dollars

Use figures with all dollar amounts, with the exception of the context noted in Rule 8. Avoid cents columns unless exactness to the penny is necessary.

Rule 10: Use Commas in Four-Digit Figures

To prevent possible misreading, use commas in figures of four digits or more. EXAMPLES: 15,000; 1,247; 6,003.

Rule 11: Use Words for Ordinals

Usually spell out the ordinal form of numbers. EXAMPLE: "The government informed all parties of the *first, second,* and *third* [not *1st, 2nd,* and *3rd*] choices in the design competition." A notable exception is tables and figures, where space limitations could argue for the abbreviated form.

Oral/Verbal

Oral refers to words that are spoken, as in "oral presentation." The term *verbal* refers to spoken or written language. To prevent confusion, avoid the word *verbal* and instead specify your meaning with the words *oral* and *written*. EXAMPLES:

- In its international operations, M-Global, Inc., has learned that some countries still rely on *oral* [not *verbal*] contracts.
- Their *oral* agreement last month was followed by a *written* [not *verbal*] contract this month.

Parts of Speech

The term *parts of speech* refers to the eight main groups of words in English grammar. A word's placement in one of these groups is based on its function within the sentence.

Noun. Words in this group name persons, places, objects, or ideas. The two major categories are (1) proper nouns and (2) common nouns. Proper nouns name specific persons, places, objects, or ideas, and they are capitalized. EXAMPLES: Cleveland; Mississippi River; M-Global, Inc.; Student Government Association; Susan Jones; Existentialism. Common nouns name general groups of persons, places, objects, and ideas, and they are not capitalized. EXAMPLES: trucks, farmers, engineers, assembly lines, philosophy.

Verb. A verb expresses action or state of being. Verbs give movement to sentences and form the core of meaning in your writing. EXAMPLES: explore, grasp, write, develop, is, has.

Pronoun. A pronoun is a substitute for a noun. Some sample pronoun categories include (1) personal pronouns (I, we, you, she, he), (2) relative pronouns (who, whom, that, which), (3) reflexive and intensive pronouns (myself, yourself, itself), (4) demonstrative pronouns (this, that, these, those), and (5) indefinite pronouns (all, any, each, anyone).

Adjective. An adjective modifies a noun. EXAMPLES: horizontal, stationary, green, large, simple.

Adverb. An adverb modifies a verb, an adjective, another adverb, or a whole statement. EXAMPLES: *s*oon, generally, well, very, too, greatly.

Preposition. A preposition shows the relationship between a noun or pronoun (the object of a preposition) and another element of the sentence. Forming a prepositional phrase, the preposition and its object can reveal relationships such as location ("They went *over the hill*"), time ("He left *after the meeting*"), and direction ("She walked *toward the office*").

Conjunction. A conjunction is a connecting word that links words, phrases, or clauses. EXAMPLES: and, but, for, nor, although, after, because, since.

Interjection. As an expression of emotion, an interjection can stand alone ("Look out!") or can be inserted into another sentence.

Passed/Past

Passed is the past tense of the verb *pass*, whereas *past* is an adjective, a preposition, or a noun that means "previous" or "beyond" or "a time before the present." EXAMPLES:

- He *passed* the survey marker on his way to the construction site.
- The *past* president attended last night's meeting. [adjective]
- He worked *past* midnight on the project. [preposition]
- In the distant *past*, the valley was a tribal hunting ground. [noun]

Per

Coming from the Latin, *per* should be reserved for business and technical expressions that involve statistics or measurement—such as *per annum* or *per mile*. It should not be used as a stuffy substitute for "in accordance with." EXAMPLES:

- Her *per diem* travel allowance of $90 covered hotels and motels.
- During the oil crisis years ago, gasoline prices increased by more than 50 cents *per gallon*.
- *As you requested* [not *per your request*], we have enclosed brochures on our products.

Per Cent/Percent/Percentage

Per cent and *percent* have basically the same usage and are used with exact numbers. The one word *percent* is preferred. Even more common in technical writing, however, is the use of the percent sign (%) after numbers. The word *percentage* is only used to express general amounts, not exact numbers. EXAMPLES:

- After completing a marketing survey, M-Global, Inc., discovered that 83 *percent* [or 83%] of its current clients have hired M-Global for previous projects.
- A large *percentage* of the defects can be linked to the loss of two experienced quality-control inspectors.

Practical/Practicable

Although close in meaning, these two words have quite different implications. *Practical* refers to an action that is known to be effective. *Practicable* refers to an action that can be accomplished or put into practice, without regard for its effectiveness or practicality. EXAMPLES:

- His *practical* solution to the underemployment problem led to a 30% increase in employment last year.
- The department head presented a *practicable* response, because it already had been put into practice in another branch.

Principal/Principle

When these two words are misused, the careful reader notices. Keep them straight by remembering this simple distinction: *Principle* is always a noun that means "basic truth, belief, or theorem." EXAMPLE: "He believed in the principle of free speech." *Principal* can be either a noun or an adjective and has three basic uses:

- **As a noun meaning "head official" or "person who plays a major role."** EXAMPLE: We asked that a *principal* in the firm sign the contract.

- **As a noun meaning "the main portion of a financial account upon which interest is paid."** EXAMPLE: If we deposit $5,000 in *principal*, we will earn 9 percent interest.

- **As an adjective meaning "main or primary."** EXAMPLE: We believe that the *principal* reason for contamination at the site is the leaky underground storage tank.

Pronouns: Agreement and Reference

A pronoun is a word that replaces a noun, which is called the *antecedent* of the pronoun. EXAMPLES: this, it, he, she, they. Pronouns, as such, provide you with a useful strategy for varying your style by avoiding repetition of nouns. Following are some rules to prevent pronoun errors:

Rule 1: Make Pronouns Agree with Antecedents

Check every pronoun to make certain it agrees with its antecedent in number—that is, both noun and pronoun must be singular, or both must be plural. Of special concern are the pronouns *it* and *they*. EXAMPLES:

- Change "M-Global, Inc., plans to complete their Argentina project next month" to this sentence: "M-Global, Inc., plans to complete its Argentina project next month."

- Change "The committee released their recommendations to all departments" to this sentence: "The committee released its recommendations to all departments."

Rule 2: Be Clear About the Antecedent of Every Pronoun

There must be no question about what noun a pronoun replaces. Any confusion about the antecedent of a pronoun can change the entire meaning of a sentence. To avoid such reference problems, it may be necessary to rewrite a sentence or even use a noun rather than a pronoun. Do whatever is necessary to prevent misunderstanding by your reader. EXAMPLE: Change "The gas filters for these tanks are so dirty that they should not be used" to this sentence: "These filters are so dirty that they should not be used."

Rule 3: Avoid Using *This* as the Subject Unless a Noun Follows It

A common stylistic error is the vague use of *this,* especially as the subject of a sentence. Sometimes the reference is not clear at all; sometimes the reference may be clear after several readings. In almost all cases, however, the use of *this* as a pronoun reflects poor technical style and tends to make the reader want to ask, "This what?" Instead, make the subject of your sentences concrete, either by adding a noun after the *this* or by recasting the sentence. EXAMPLE: Change "He talked constantly about the project to be completed at the

Olympics. This made his office-mates irritable" to the following: "His constant talk about the Olympics project irritated his office-mates."

Punctuation: General

Commas.   Most writers struggle with commas, so you are not alone. The problem is basically threefold. First, the teaching of punctuation has been approached in different, and sometimes quite contradictory, ways. Second, comma rules themselves are subject to various interpretations. And third, problems with comma placement often mask more fundamental problems with the structure of a sentence itself.

Start by knowing the basic rules of comma use. The rules that follow are fairly simple. If you learn them now, you will save yourself a good deal of time later because you will not be questioning usage constantly. In other words, the main benefit of learning the basics of comma use is increased confidence in your own ability to handle the mechanics of editing. (If you do not understand some of the grammatical terms that follow, such as *compound sentence,* refer to the section on sentence structure.)

Rule 1: Commas in a Series

Use commas to separate words, phrases, and short clauses written in a series of three or more items. EXAMPLE: "The samples contained gray sand, sandy clay, and silty sand." According to current U.S. usage, a comma always comes before the "and" in a series. (In the United Kingdom, the comma is left out.)

Rule 2: Commas in Compound Sentences

Use a comma before the conjunction that joins main clauses in a compound sentence. EXAMPLE: "We completed the drilling at the Smith Industries location, and then we grouted the holes with Sakrete." The comma is needed here because it separates two complete clauses, each with its own subject and verb (*we completed* and *we grouted*). If the second *we* had been deleted, there would be only one clause containing one subject and two verbs ("we completed and grouted"). Thus no comma would be needed. Of course, it may be that a sentence following this comma rule is far too long; do not use the rule to string together intolerably long sentences.

Rule 3: Commas with Nonessential Modifiers

Set off nonessential modifiers with commas at the beginning, middle, or end of sentences. *Nonessential modifiers* are usually phrases that add more information to a sentence, rather than greatly changing its meaning. When you speak, there is often a pause between this kind of modifier and the main part of the sentence, giving you a clue that a comma break is needed. EXAMPLE: "The report, which we submitted three weeks ago, indicated that the company would not be responsible for transporting hazardous wastes." But—"The report that we submitted three weeks ago indicated that the company would not be responsible for transporting hazardous wastes." The first example includes a nonessential modifier, would be spoken with pauses, and therefore uses separating commas. The second example includes an essential modifier, would be spoken without pauses, and therefore includes no separating commas.

Rule 4: Commas with Adjectives in a Series

Use a comma to separate two or more adjectives that modify the same noun at the same level of detail. To help you decide if adjectives modify the same noun equally, use this test: If you can reverse their positions and still retain the same meaning, then the adjectives modify the same word and should be separated by a comma. EXAMPLE: "Jason found the old, rotted gaskets."

Rule 5: Commas with Introductory Elements

Use a comma after introductory phrases or clauses of about five words or more. EXAMPLE: "After completing the topographic survey of the area, the crew returned to headquarters for its weekly project meeting." Commas like the one after *area* help readers separate secondary or modifying points from your main idea, which of course should be in the main clause. Without these commas, there may be difficulty reading such sentences properly.

Rule 6: Commas in Dates, Titles, Etc.

Abide by the conventions of comma usage in punctuating dates, titles, geographic place names, and addresses. EXAMPLES:

- May 3, 2006, is the projected date of completion. (However, note the change in the military form of dates: We will complete the project on 3 May 2006.)
- John F. Dunwoody, Ph.D., has been hired to assist on the project.
- M-Global, Inc., has been selected for the project.
- He listed Dayton, Ohio, as his permanent residence.

Note the need for commas after the year *2006,* the title *Ph.D.,* the designation *Inc.,* and the state name *Ohio.* Also note that if the day had not been in the first example, there would be no comma between the month and year and no comma after the year.

Semicolons.   The semicolon is easy to use if you remember that it, like a period, indicates the end of a complete thought. Its most frequent use is in situations where grammar rules would allow you to use a period but where your stylistic preference is for a less abrupt connector. EXAMPLE: "Five engineers left the convention hotel after dinner; only two returned by midnight."

One of the most common punctuation errors, the comma splice, occurs when a comma is used instead of a semicolon or period in compound sentences connected by words such as *however, therefore, thus,* and *then.* When you see that these connectors separate two main clauses, make sure either to use a semicolon or to start a new sentence. EXAMPLE: "We made it to the project site by the agreed-on time; however, [or " . . . time. However, . . . "] the rain forced us to stay in our trucks for two hours."

As noted in the "Lists" entry, there is another instance in which you might use semicolons. Place them after the items in a list when you are treating the list like a sentence and when any one of the items contains internal commas.

Colons.   As mentioned in the "Lists" entry, you should place a colon immediately after the last word in the lead-in before a formal list of bulleted or numbered items. EXAMPLE: "Our field study involved these three steps:" or "In our field study, we were asked to:" The colon may come after a complete clause, as in the first example, or it may split a grammatical construction, as in the second example. However, it is preferable to use a complete clause before a formal list.

The colon can also be used in sentences in which you want a formal break before a point of clarification or elaboration. EXAMPLE: "They were interested in just one result: quality construction." In addition, use the colon in sentences in which you want a formal break before a series that is not part of a listing. EXAMPLE: "They agreed to perform all on-site work required in these four cities: Houston, Austin, Laredo, and Abilene." However, note that there is no colon before a sentence series without a break in thought. EXAMPLE: "They agreed to perform all the on-site work required in Houston, Austin, Laredo, and Abilene."

Apostrophes.   The apostrophe can be used for contractions, for some plurals, and for *possessives*. Only the last two uses cause confusion. Use an apostrophe to indicate the plural form of a word as a word. EXAMPLE: "That redundant paragraph contained seven *area*'s and three *factor*'s in only five sentences." Although some writers also use apostrophes to form the plurals of numbers and full-cap abbreviations, the current tendency is to include only the *s*. EXAMPLES: 7s, ABCs, PCBs, P.E.s.

As for possessives, you probably already know that the grammar rules seem to vary, depending on the reference book you are reading. Following are some simple guidelines:

Possessive Rule 1

Form the possessive of multisyllabic nouns that end in *s* by adding just an apostrophe, whether the nouns are singular or plural. EXAMPLES: actress' costume, genius' test score, the three technicians' samples, Jesus' parables, the companies' joint project.

Possessive Rule 2

Form the possessive of one-syllable, singular nouns ending in *s* or an *s* sound by adding an apostrophe plus *s*. EXAMPLES: Hoss's horse, Tex's song, the boss's progress report.

Possessive Rule 3

Form the possessive of all plural nouns ending in *s* or an *s* sound by adding just an apostrophe. EXAMPLES: the cars' engines, the ducks' flight path, the trees' roots.

Possessive Rule 4

Form the possessive of all singular and plural nouns not ending in *s* by adding an apostrophe plus *s*. EXAMPLES: the man's hat, the men's team, the company's policy.

Possessive Rule 5

Form the possessive of paired nouns by first determining whether there is joint ownership or individual ownership. For joint ownership, make only the last noun possessive. For individual ownership, make both nouns possessive. EXAMPLE: "Susan and Terry's project was entered in the science fair; but Tom's and Scott's projects were not."

Quotation Marks.   In technical writing, you may want to use this form of punctuation to draw attention to particular words, to indicate passages taken directly from another source, or to enclose the titles of short documents such as reports or book chapters. The rule to remember is this: Periods and commas go inside quotation marks; exclamation marks, question marks, semicolons, and colons go outside quotation marks.

Parentheses.   Use parentheses carefully, because long parenthetical expressions can cause the reader to lose the train of thought. This form of punctuation can be used when you (1) place an abbreviation after a complete term, (2) add a brief explanation within the text, or (3) include reference citations within the document text (as explained in chapter 14). The period goes after the closing parenthesis when the parenthetical information is part of the sentence, as in the previous sentence. (However, it goes inside the closing parenthesis when the parenthetical information forms its own sentence, as in the sentence you are reading.)

Brackets.   Use a pair of brackets for the following purposes: (1) to set off parenthetical material already contained within another parenthetical statement and (2) to draw attention to a comment you are making within a quoted passage. EXAMPLE: "Two M-Global studies have shown that the Colony Dam is up to safety standards. (See Figure 4—3 [Dam Safety Record] for a complete record of our findings.) In addition, the county engineer has a letter on file that will give further assurance to prospective homeowners on the lake. His letter notes that 'After finishing my three-month study [he completed the study in July 2007], I conclude that the Colony Dam meets all safety standards set by the county and state governments.'"

Hyphens.   The hyphen is used to form certain word compounds in English. Although the rules for its use sometimes seem to change from handbook to handbook, those that follow are the most common:

Hyphen Rule 1

Use hyphens with compound numerals. EXAMPLE: twenty-one through ninety-nine.

Hyphen Rule 2

Use hyphens with most compounds that begin with *self.* EXAMPLES: self-defense, self-image, self-pity. Other *self* compounds, like *selfhood* and *selfsame,* are written as unhyphenated words.

Hyphen Rule 3

Use hyphens with group modifiers when they precede the noun but not when they follow the noun. EXAMPLES: A well-organized paper, a paper that was well organized, twentieth-century geotechnical technology, bluish-gray shale, fire-tested material, thin-bedded limestone.

However, remember that when the first word of the modifier is an adverb ending in *—ly,* place no hyphen between the words. EXAMPLES: carefully drawn plate, frightfully ignorant teacher.

Hyphen Rule 4

Place hyphens between prefixes and root words in the following cases: (1) between a prefix and a proper name (ex-Republican, pre-Sputnik); (2) between some prefixes that end with a vowel and root words beginning with a vowel, particularly if the use of a hyphen would prevent an odd spelling (semi-independent, re-enter, re-elect); and (3) between a prefix and a root when the hyphen helps to prevent confusion (re-sent, not resent; re-form, not reform; re-cover, not recover).

Punctuation: Lists

As noted in chapter 4 ("Page Design"), listings draw attention to parallel pieces of information whose importance would be harder to grasp in paragraph format. In other words, use lists as an attention-getting strategy. Following are some general pointers for punctuating lists. (See pages 111–112 in chapter 4 for other rules for lists.)

You have three main options for punctuating a listing. The common denominators for all three are that you (1) always place a colon after the last word of the lead-in and (2) always capitalize the first letter of the first word of each listed item.

Option A: Place no punctuation after listed items.   This style is appropriate when the list includes only short phrases. More and more writers are choosing this option, as opposed to option B. EXAMPLE:

In this study, we will develop recommendations that address these six concerns in your project:

- Site preparation
- Foundation design
- Sanitary-sewer design
- Storm-sewer design
- Geologic surface faulting
- Projections for regional land subsidence.

Option B: Treat the list like a sentence series.   In this case, you place commas or semicolons between items and a period at the end of the series. Whether you choose option A or B largely depends on your own style or that of your employer. EXAMPLE:

In this study, we developed recommendations that dealt with four topics:

- Site preparation,
- Foundation design,
- Sewer construction, and
- Geologic faulting.

Note that this option requires you to place an *and* after the comma that appears before the last item. Another variation of option B occurs when you have internal commas within one or more of the items. In this case, you must change the commas that follow the listed items into semicolons. Yet you still keep the *and* before the last item. EXAMPLE:

Last month we completed environmental assessments at three locations:

- A gas refinery in Dallas, Texas;
- The site of a former chemical plant in Little Rock, Arkansas; and
- A waste pit outside of Baton Rouge, Louisiana.

Option C: Treat each item like a separate sentence.   When items in a list are complete sentences, you may want to punctuate each one like a separate sentence, placing a period at the end of each. You *must* choose this option when one or more of your listed items contain more than one sentence. EXAMPLE:

The main conclusions of our preliminary assessment are summarized here:

- At five of the six borehole locations, petroleum hydrocarbons were detected at concentrations greater than a background concentration of 10 mg/kg.
- No PCB concentrations were detected in the subsurface soils we analyzed. We will continue the testing, as discussed in our proposal.
- Sampling and testing should be restarted three weeks from the date of this report.

Regrettably/Regretfully

Regrettably means "unfortunately," whereas *regretfully* means "with regret." When you are unsure of which word to use, substitute the definitions to determine correct usage. EXAMPLES:

- *Regrettably*, the team members omitted their resumes from the proposal.
- Hank submitted his resume to the investment firm, but, *regrettably*, he forgot to include a cover letter.
- I *regretfully* climbed on the plane to return home from Hawaii.

Respectively

Some good writers may use *respectively* to connect sets of related information. Yet such usage creates extra work for readers by making them reread previous passages. It is best to avoid *respectively* by rewriting the sentence, as shown in the several following options. EXAMPLES:

Original: Appendices A, G, H, and R contain the topographical maps for Sites 6, 7, 8, and 10, respectively.

Revision—Option 1: Appendix A contains the topographical map for Site 6; Appendix G contains the map for Site 7; Appendix H contains the map for Site 8; and Appendix R contains the map for Site 10.

Revision—Option 2: Appendix A contains the topographical map for Site 6; Appendix G for Site 7; Appendix H for Site 8; and Appendix R for Site 10.

Revision—Option 3: Topographical maps are contained in the appendices, as shown in the following list:

Appendix	Site
A	6
G	7
H	8
R	10

Set/Sit

Like *lie* and *lay*, *sit* and *set* are verbs distinguished by form and use. Following are the basic differences:

1. *Set* means "to place in a particular spot" or "to adjust." It is a transitive verb and thus takes a direct object to which it conveys action. Its main parts are *set* (present), *set* (past tense), *set* (past participle), and *setting* (present participle).

2. *Sit* means "to be seated." It is usually an intransitive verb and thus does not take a direct object. Its main parts are *sit* (present), *sat* (past), *sat* (past participle), and *sitting* (present participle). It can be transitive when used casually as a direction to be seated. ("Sit yourself down and take a break.")

Examples:

■ He *set* the computer on the table yesterday.

■ While *setting* the computer on the table, he sprained his back.

■ The technician had *set* the thermostat at 75 degrees.

■ She plans to *sit* exactly where she sat last year.

■ While *sitting* at her desk, she saw the computer.

Sic

Latin for "thus," *sic* is most often used when a quoted passage contains an error or other point that might be questioned by the reader. Inserted within brackets, *sic* shows the reader that the error was included in the original passage and that it was not introduced by you. EXAMPLE: "The customer's letter to our sales department claimed that 'there are too [*sic*] or three main flaws in the product.'"

Spelling

All writers find at least some words difficult to spell, and some writers have major problems with spelling. Automatic spell-checking software helps solve the problem, but you must still remain vigilant during the proofreading stage. One or more misspelled words in an otherwise well-written document may cause readers to question professionalism in other areas.

However, you should keep your own list of words you most frequently have trouble spelling. Like most writers, you probably have a relatively short list of words that give you repeated difficulty.

Stationary/Stationery

Stationary means "fixed" or "unchanging," whereas *stationery* refers to paper and envelopes used in writing or typing letters. EXAMPLES:

- To perform the test correctly, one of the workers had to remain *stationary* while the other one moved around the job site.
- When she began her own business, Julie purchased *stationery* with her new logo on each envelope and piece of paper.

Subject–Verb Agreement

Subject–verb agreement errors are quite common in technical writing. They occur when writers fail to make the subject of a clause agree in number with the verb. EXAMPLE: "The nature of the diverse geologic deposits are explained in the report." (The verb should be *is,* because the singular subject is *nature.*)

Writers who tend to make these errors should devote special attention to them. Specifically, isolate the subjects and verbs of all the clauses in a document and make certain that they agree. Following are seven specific rules for making subjects agree with verbs:

Rule 1: Subjects Connected by *And* Take Plural Verbs

This rule applies to two or more words or phrases that, together, form one subject phrase. EXAMPLE: "The site preparation section and the foundation design portion of the report are to be written by the same person."

Rule 2: Verbs After *Either/Or* or *Neither/Nor* Agree with the Nearest Subject

Subject words connected by *either* and *or* (or *neither* and *nor*) confuse many writers, but the rule is very clear. Your verb choice depends on the subject nearest the verb. EXAMPLE: "He told his group that neither the three reports nor the proposal was to be sent to the client that week."

Rule 3: Verbs Agree with the Subject, Not with the Subjective Complement

Sometimes called a *predicate noun* or *adjective,* a subjective complement renames the subject and occurs after verbs such as *is, was, are,* and *were.* EXAMPLE: "The theme of our proposal is our successful projects in that region of the state." However, the same rule would permit this usage: "Successful projects in that part of the state are the theme we intend to emphasize in the proposal."

Rule 4: Prepositional Phrases Do Not Affect Matters of Agreement

As long as, in addition to, as well as, and *along with* are prepositions, not conjunctions. A verb agrees with its subject, not with the object of a prepositional phrase. EXAMPLE: "The manager of human resources, along with the personnel director, is supposed to meet with the three applicants."

Rule 5: Collective Nouns Usually Take Singular Verbs

Collective nouns have singular form but usually refer to a group of persons or things (e.g., *team, committee, crew*). When a collective noun refers to a group as a whole, use a singular verb. EXAMPLE: "The project crew was ready to complete the assignment." Occasionally, a collective noun refers to the members of the group acting in their separate capacities. In this case, either use a plural verb or, to avoid awkwardness, reword the sentence. EXAMPLE: "The crew were not in agreement about the site locations" or, "Members of the crew were not in agreement about the site locations."

Rule 6: Foreign Plurals Usually Take Plural Verbs

Although usage is gradually changing, most careful writers still use plural verbs with *data, strata, phenomena, media,* and other irregular plurals. EXAMPLE: "The data he asked for in the request for proposal are incorporated into the three tables."

Rule 7: Indefinite Pronouns Like *Each* and *Anyone* Take Singular Verbs

Writers often fail to follow this rule when they make the verb agree with the object of a prepositional phrase instead of with the subject. EXAMPLE: "Each of the committee members are ready to adjourn" (incorrect). "Each of the committee members is ready to adjourn" (correct).

To/Too/Two

To is part of the infinitive verb form or is a preposition in a prepositional phrase. *Too* is an adverb that suggests an excessive amount or that means "also." *Two* is a noun or an adjective that stands for the numeral "2." EXAMPLES:

- He volunteered *to* go [infinitive verb] *to* Alaska [prepositional phrase] *to* work [another infinitive verb form] on the project.
- Stephanie explained that the proposed hazardous-waste dump would pose *too* many risks *to* the water supply. Scott made this point, *too*.

Utilize/Use

Utilize is simply a long form for the preferred verb "use." Although some verbs that end in *–ize* are useful words, most are simply wordy substitutes for shorter forms. As some writing teachers say, "Why use 'utilize' when you can use 'use'?"

Which/That

Which is used to introduce nonrestrictive clauses, which are defined as clauses not essential to meaning (as in this sentence). Note that such clauses require a comma before the *which* and a slight pause in speech. *That* is used to introduce restrictive clauses that are essential to the meaning of the sentence (as in this sentence). Note that such clauses have no comma before the *that* and are read without a pause. *Which* and *that* can produce different meanings, as in the following examples:

■ Our benefits package, *which* is the best in our industry, includes several options for medical care.

■ The benefits package *that* our firm provides includes several options for medical care.

■ My daughter's school, *which* is in Cobb County, has an excellent math program.

■ The school *that* my daughter attends is in Cobb County and has an excellent math program.

Note that the preceding examples with *that* might be considered wordy by some readers. Indeed, such sentences often can be made more concise by deleting the *that* introducing the restrictive clause. However, delete *that* only if you can do so without creating an awkward and choppy sentence.

Who/Whom

Who and *whom* give writers (and speakers) fits, but the importance of their correct use probably has been exaggerated. If you want to be one who uses them properly, remember this basic point: *Who* is a subjective form that can only be used in the subject slot of a clause; *whom* is an objective form that can only be used as a direct object or other nonsubject noun form of a sentence. You can check which word you should use by substituting *he* and *him*. Use *who* when you would use *he* and use *whom* when you would use *him*. EXAMPLES:

■ The man *who* you said called me yesterday is a good customer of the firm. (The clause "who . . . called me yesterday" modifies "man." Within this clause, *who* is the subject of the verb "called." Note that the subject role of *who* is not affected by the two words "you said," which interrupt the clause.)

■ They could not remember the name of the person *whom* they interviewed. (The clause "whom they interviewed" modifies "person." Within this clause, *whom* is the direct object of the verb "interviewed.")

Who's/Whose

Who's is a contraction that replaces *who is*, whereas *whose* is a possessive adjective. EXAMPLES:

■ *Who's* planning to attend the annual meeting?

■ Susan is the manager *who's* responsible for training.

- *Whose* budget includes training?
- Susan is the manager *whose* budget includes training.

Your/You're

Your is an adjective that shows ownership, whereas *you're* is a contraction for *you are*. EXAMPLES:

- *Your* office will be remodeled next week.
- *You're* responsible for giving performance appraisals.

Exercise 1: Grammar and Mechanics

The following passages contain a variety of grammatical and mechanical errors covered in the handbook. The major focus is punctuation. Rewrite each passage.

1. Some concerns regarding plumbing design are mentioned in our report, however, no unusual design problems are expected.

2. An estimate of the total charges for an audit and for three site visits are based on our standard fee schedules.

3. The drill bit was efficient cheap and available.

4. The plan unless we have completely misjudged it, will increase sales markedly.

5. Our proposal contains design information for these two parts of the project; Phase 1 (evaluating the 3 computers) and Phase 2 (installing the computer selected).

6. If conditions require the use of all-terrain equipment to reach the construction locations, this will increase the cost of the project slightly.

7. An asbestos survey was beyond the scope of this project, if you want one, we would be happy to submit a proposal.

8. Jones-Simon Company, the owners of the new building, were informed of the problem with the foundation.

9. Also provided is the number and type of tests to be given at the office.

10. Calculating the standard usages by the current purchase order prices result in a downward adjustment of $.065.

11. Data showing the standard uses of the steel, including allowances for scrap, waste and end pieces of the tube rolls, are included for your convenience at the end of this report in Table 7.

12. This equipment has not been in operation for 3 months, and therefore, its condition could not be determined by a quick visual inspection.

13. Arthur Jones Manager of the Atlanta branch wrote that three proposals had been accepted.

14. The generator that broke yesterday has been shipped to Tampa already by Harry Thompson.

15. The first computer lasted eight years the second two years.

16. He wants one thing out of their work speed.

17. On 25 September 2008 the papers were signed.

18. On March 23 2009 the proposal was accepted.

19. The meeting was held in Columbus the Capital of Ohio.

20. M-Global, Inc. completed its Indonesia project in record time.

21. He decided to write for the brochure then he changed his mind.

22. Interest by the Kettering Hospital staff in the development of a master plan for the new building wings have been expressed.

23. However much he wants to work for Gasion engineering he will turn the job down if he has to move to another state.

24. 35 computer scientists attended the convention, but only eleven of them were from private industry.

25. Working at a high salary gives him some satisfaction still he would like more emotional satisfaction from his job.

26. His handwriting is almost unreadable therefore his secretary asked him to dictate letters.

27. Any major city especially one that is as large as Chicago is bound to have problems with mass transit.

28. He ended his speech by citing the company motto; "Quality first, last, and always".

29. Houston situated on the Gulf of Mexico is an important international port.

30. The word *effect* is in that student's opinion a difficult one to use.

31. All persons who showed up for the retirement party, told stories about their association with Charlie over the years.

32. The data that was included in the study seems inconclusive.

33. My colleague John handled the presentation for me.

34. Before he arrived failure seemed certain.

35. While evaluating the quality of her job performance a study was made of her writing skills by her supervisor.

36. I shall contribute to the fund for I feel that the cause is worthwhile.

37. James visited the site however he found little work finished.

38. There are three stages cutting grinding and polishing.

39. The three stages are cutting grinding and polishing.

40. Writers occasionally create awkward verbs *prioritize* and *terminate* for example.

41. Either the project engineers or the consulting chemist are planning to visit with the client next week.

42. Besides Gerry Dave worked on the Peru project.

43. The corporation made a large unexpected gift to the university.

44. The reason for his early retirement are the financial incentives given by his employer.

45. Profit, safety and innovation are the factors that affect the design of many foundations.

46. No later than May 2012 the building will be finished.

47. Each of the committee members complete a review of the file submitted by the applicant.

48. The team completed their collaborative writing project on schedule.

49. Both the personnel officers and the one member of the quality team is going to attend the conference in Fargo.

50. He presented a well organized presentation but unfortunately the other speakers on the panel were not well-prepared.

Exercise 2: Usage

For each of the following passages, select the correct word or phrase from the choices within the parentheses. Be ready to explain the rationale for your choice.

1. John (implied, inferred) in his report that TransAm Oil should reject the bid.

2. Before leaving on vacation, the company president left instructions for the manner in which responsibilities should be split (among, between) the three vice presidents.

3. Harold became (uninterested, disinterested) in the accounting problem after working on it for 18 straight hours.

4. A large (percent, percentage) of the tellers is dissatisfied with the revised work schedule.

5. The typist responded that he would make (less, fewer) errors if the partner would spell words correctly in the draft.

6. From her reading of the annual report, Ms. Jones (inferred, implied) that the company might expand its operations.

7. The president's decision concerning flextime will be (effected, affected) by the many conversations he is having with employees about scheduling difficulties.

8. His (principal, principle) concern was that the loan's interest and (principle, principal) remain under $500.

9. Throughout the day, his concentration was interrupted (continuously, continually) by phone calls.

10. He jogged (continuously, continually) for 20 minutes.

11. Five thousand books (compose, comprise) his personal library.

12. The clients (who, whom) he considered most important received Christmas gifts from the company.

13. The company decided to expand (its, it's) operations in the hope that (its, it's) the right time to do so.

14. The (nonflammable, flammable, inflammable) liquids were kept in a separate room, because of their danger.

15. They waited for (awhile, a while) before calling the subcontractor.

16. Caution should be taken to (ensure, insure) that the alarm system will not go off accidentally.

17. The new floors (are comprised of, are composed of, comprise) a thick concrete mixture.

18. He (expects, anticipates) that 15 new employees will be hired this year.

19. The main office offered to (augment, supplement) the annual operating budget of the Boston office with an additional $100,000 in funds.

20. It was (all together, altogether) too late to make changes in the proposal.

21. The arbitrator made sure that both parties (agreed to, agreed with) the terms and conditions of the contract before it was submitted to the board.

22. Option 1 calls for complete removal of the asbestos. (Alternately, Alternatively), Option 2 would only require that the asbestos material be thoroughly covered.

23. They had not considered the (amount, number) of cement blocks needed for the new addition.

24. (Due to, Because of) the change in weather, they had to reschedule the trip to the project site.

25. The health inspector found (too, to) many violations in that room (to, too).

26. They claimed that the old equipment (used, utilized) too much fuel.

27. Gone are the days when a major construction job gets started with a handshake and (a verbal, an oral) agreement.

28. The complex project has 18 (discreet, discrete) phases; each part deals with confidential information that must be handled (discretely, discreetly).

29. He was (definitive, definite) about the fact that he would not be able to complete the proposal by next Tuesday.

30. He usually received (complementary, complimentary) samples from his main suppliers.

31. To (lose, loose) a client for whom they had worked so hard was devastating.

32. It was (fortunate, fortuitous) he was there at the exact moment the customer needed to order a year's worth of supplies, for the sales commission was huge.

33. Among all the information on the graph, he located the one (data, datum) that shows the price of tuna on the Seattle market at 5 P.M. on August 7.

34. Each (principle, principal) of the corporation was required to buy stock.

35. He returned to the office to (assure, ensure) that the safe was locked.

Photo Credits

Chapter 1

p. 1 Ryan McVay/Getty Images–Photodisc; p. 7 AJA Productions/Getty Images Inc.–Image Bank; p. 17 Photos.com; p. 22 Getty Images–Stockbyte; p. 25 Robert Daly/Image Bank/Getty Images.

Chapter 2

p. 34 Anthony S. Lojacono/Creative Eye/MIRA.com; p. 38 VCL/Chris Ryan/Taxi/Getty Images; p. 42 © Copyright Jose Luis Pelaez, Inc./CORBIS; p. 45 Photolibrary.com; p. 66 David R. Frazier/Photo Researchers, Inc.; p. 67 Harold Sund/Getty Images, Inc.–Image Bank; p. 70 David A. Ponton/Creative Eye/MIRA.com; p. 72 EyeWire Collection/Getty Images–Photodisc; p. 73 Daisuke Morita/Getty Images, Inc.–Photodisc.

Chapter 3

p. 74 Michael McQueen/Getty Images, Inc.–Image Bank; p. 76 Ryan McVay/Getty Images, Inc.–Photodisc; p. 84 EyeWire Collection/Getty Images–; p. 89 Zefa Collection/Corbis Zefa Collection.

Chapter 4

p. 101 © Bettmann/CORBIS All Rights Reserved; p. 103 Spencer Grant/PhotoEdit Inc.; p. 107 Brady/Pearson Education/PH College; p. 117 Keith Brofsky/Getty Images Inc.–Stone Allstock.

Chapter 5

p. 126 Getty Images–Stockbyte; p. 128 Photos.com; p. 134 Steve Gorton © Dorling Kindersley; p. 140 P Crowther/S Carter/Getty Images Inc.–Stone Allstock; p. 144 Hiep Vu/Masterfile Stock Image Library.

Chapter 6

p. 170 Nick Koudis/Getty Images, Inc.–Photodisc; p. 173 Photodisc/Getty Images; SuperStock, Inc.

Chapter 7

p. 193 Photos.com; p. 195 Andrew Olney/Getty Images/Digital Vision; p. 200 Keith Brofsky/Getty Images, Inc.–Photodisc.

Chapter 8

p. 224 Ken Reid/Getty Images, Inc.–Taxi; p. 226 EyeWire Collection/Getty Images–Photodisc; 227 Photolibrary.com; p. 233 EyeWire Collection/Getty Images–Photodisc; p. 236 John A. Rizzo/Getty Images, Inc.–Photodisc.

Chapter 9

p. 258 EyeWire Collection/Getty Images–Photodisc; p. 261 Larry Dale Gordon/Getty Images Inc.–Image Bank; p. 263 Susan Van Etten/PhotoEdit Inc.; p. 269 EyeWire Collection/Getty Images–Photodisc; p. 278 Tony Camacho/Photo Researchers, Inc.; p. 286 Medford Taylor/National Geographic Image Collection.

Chapter 10

p. 302 Getty Images–Stockbyte; p. 313 Antonio M. Rosario/Getty Images Inc.–Image Bank; p. 317 EyeWire Collection/Getty Images–Photodisc.

Chapter 11

p. 366 Robert Pierce/Stock Connection; p. 368 VCL/ANTONIO MO/Getty Images, Inc.–Taxi; p. 371 Tony Freeman/PhotoEdit Inc.; p. 375 Andy Crawford © Dorling Kindersley; p. 382 Michael Newman/PhotoEdit.

Chapter 12

p. 400 © Dorling Kindersley.

Chapter 13

p. 447 Getty Images–Stockbyte; p. 450 LWA/Stone/Getty Images; p. 454 Ronnie Kaufman/CORBIS–NY; p. 465 RNT Productions/Corbis/Bettmann.

Chapter 14

p. 469 Getty Images–Digital Vision; p. 471 Fisher/Thatcher/Getty Images Inc.–Stone Allstock; p. 478 Britt Erlanson/Image Bank/Getty Images; p. 482 SuperStock, Inc.; p. 490 (top) Ryan McVay/Getty Images, Inc.–Photodisc; (bottom) Michael Matisse/Getty Images, Inc.–Photodisc; p. 491 (top) Craig Brewer/Getty Images, Inc.–Photodisc; (left) Gary Ombler © Dorling Kindersley; (right) Chris Knapton/Photo Researchers, Inc.; (bottom) U.S. Department of Energy; p. 492 (top) Michael Crockett © Dorling Kindersley;

(bottom) Michael Newman/PhotoEdit Inc.; p. 494 (top) James Lemass/Photolibrary.com; (bottom) Gary Ombler © Dorling Kindersley.

Chapter 15

p. 495 Keith Brofsky/Getty Images, Inc.–Photodisc; p. 499 Getty Images–Stockbyte; p. Mauritius, GMBH/Phototake NYC; p. 525 © Daly & Newton/Getty Images; p. 531 Bluestone Productions/Superstock Royalty Free.

Chapter 16

p. 551 Getty Images–Stockbyte; p. 552 James Woodson/Getty Images/Digital Vision; p. 558 David Young–Wolff/PhotoEdit Inc.; p. 562 Color Day Production/Image Bank/Getty Images; p. 567 Getty Images, Inc. –Stockbyte.

Chapter 17

p. 588 Getty Images–Stockbyte; p. 590 Getty Images–Stockbyte; p. 603 Getty Images–Stockbyte.

Index

A/an, 613
Abbreviations, 614–617
ABC format, 80–83, 130–131
  for email, 144
  for equipment evaluation reports, 234
  for explanations, 197–199
  for feasibility studies, 317–318
  for formal reports, 261–262
  for informal reports, 228–229
  for instructions, 201
  for job letters, 558
  for lab reports, 237
  for memos, 139–140
  for negative letters, 135
  for neutral letters, 136
  for positive letters, 134
  for problem analyses informal report, 232
  for progress/periodic reports, 235, 236
  for proposals, formal, 311
  for proposals, informal, 308
  for recommendation reports, 233, 234
  for sales letters, 137–138
Abstract, 80, 81–82
  for feasibility studies, 318
  for informal reports, 229
  for proposals, informal, 308
Abstracts, writing research
  descriptive, 538, 539
  guidelines, 539–541
  informational, 538, 539
  jargon, avoiding, 541
  main points, highlighting, 539
  outlines, sketching, 539–540
  purpose statement, 540
  style, 540
Academic writing
  differences between technical communication and, 3
  features of, 2
Accept/except, 617

Active voice, 598–599
Advice/advise/inform, 617–618
Advisers, 16
AeroSpace and Defence Industries Association, 42
Affect/effect, 618
Agree to/agree with, 618
All right/alright, 618
All together/altogether, 618–619
Allusion/illusion/delusion/elusion, 619
A lot/a lot, 613
Already/all ready, 619
Alternately/alternatively, 619–620
American Psychological Association (APA), 532, 533, 534
Amount/number, 620
Analogies, use of, 179
Analysis, 7
And/or, 620
Anticipate/expect, 620
Appendices
  in formal reports, 266, 271
  in proposals, formal, 316
Apt/liable/likely, 621
Archivist, 450
Argument, 7
Assure/ensure/insure, 621
Attachments
  in formal reports, 266, 271
  in informal reports, 231
  in letters or memos, 129, 132
  in proposals, formal, 316
  in proposals, informal, 310
Audiences
  See also Readers
  captive, 2
  decision-making levels, 16–17
  for web pages/websites, analysis of, 369, 371
Augment/supplement, 621–622
Awhile/a while, 622

Backplanning, 451–452
Balance/remainder/rest, 622

Bar charts
  arrangement of bars, 425
  break lines, 424
  examples of confusing, 435–437
  guidelines for using, 423–426
  number of bars in, 423
  spacing between bars, 424–425
Because/since, 622
Beginnings, 77–79
Between/among, 622–623
Bi-/semi-, 623
Bibliography, 18
Body, 80, 82
Boldface, use of, 112
Brainstorming, 25
Bullets
  in instructions, 202–203
  in lists, 111
  oral presentations and alternatives to, 480
Business climate, 36–37

Capital/capitol, 623
Capitalization, 623–624
Cause and effect, 86
Cautions, in instructions, 203–206
Center on/revolve around, 624–625
Charts, 402
  bar, 423–426, 435–437
  flow, 428–430
  junk, 479
  line, 426–427
  oral presentations and presenting, 478–481
  organization, 430–431
  pie, 420–423, 437–439
  schedule, 452–453
Chicago Manual of Style, 533
Circumlocution, 595–596
Cite/site/sight, 625
Classification, 84–85
Clichés, avoiding, 595
Closing sentence, 88

Collaborative writing
  See also Teamwork
  approaches to, 449
  communication in, 454–455
  defined, 448
  finances, managing, 453–454
  guidelines for, 450
  meetings, running effective, 456–459
  members in, 450–451
  modular, 455–456
  planning, 451
  schedule charts, using, 452–453
  steps, 449–450
  subject matter experts, 459–460
  time and money, budgeting, 451–452
Color, 114–115
  complementary, 411
  cost and time of using, 408–409
  in graphics, 408–413, 480
  guidelines for using, 411–413
  hues, 411
  primary, 411
  secondary, 411
  style sheet, developing a, 409–411
  terms, 411
  tertiary, 411
Common knowledge, 46–47, 527
Communication
  See also Oral communication/
    presentation
  collaborative writing and, 454–455
  global, 42–43
Comparison/contrast, 85
Complement/compliment, 625–626
Compose/comprise, 626
Computer conference, 454
Computers
  page design and use of, 115–117
  team writing and use of,
    28, 454
Conciseness, 592–596
Conclusions, 80, 83
  in feasibility studies, 319
  in formal reports, 270–271
  in informal reports, 230–231
  in proposals, formal, 315
  in proposals, informal, 310
Consul/council/counsel, 626
Content
  adjusting, 23
  chunking, 371–372

development for web pages/websites,
    371–374
Continuous/continual, 627
Controlled English, 42, 603–604
Copy notation, 130
Copyrights, 47
Corporate culture, 36
Correspondence, guidelines for writing,
    128–133
Council of Science Editors (CSE), 532,
    533, 536–537
Cover/title page
  in formal reports, 263–264
  in proposals, formal, 311–312
Criterion/criteria, 627
Culture
  corporate, 36
  global differences, 37–42
  high-context, 39
  low-context, 38
  organizational, 35–36

Dangers, in instructions, 203–206
Data/datum, 627
Decision makers, 16
Decision-making levels, 16–17
Deductive reasoning, 85–86
Definite/definitive, 628
Definitions
  difference between descriptions and,
    171–172
  example of, 177
  expanded, 174, 175–176
  formal, 174–175
  guidelines for writing, 173–176
  informal, 173, 174
  location of, in documents, 176
Descriptions
  accuracy and objectivity of, 178
  analogies, use of, 179
  difference between definitions and,
    171–172
  example of, 180
  graphics, use of, 179
  guidelines for writing, 177–180
  testing effectiveness of, 179–180
  ways of using, 178–179
Descriptive abstracts, 538, 539
Devil's advocate, 450–451
Dialogue approach, 449
Discrete/discreet/discretion, 628

Discussion sections
  in formal reports, 269–270
  for proposals, formal, 315
Disinterested/uninterested, 628
Divide and conquer approach, 449
Division, 85
Documentation styles, 532–537
Drafts, 7
  research, 531
  revising, 23–24
  writing initial, 22–23
Drawings, technical, 402
  detail in, 432
  guidelines for using, 431–434
  labeling, 432–433
  legends, use of, 434
  views, types of, 433–434
Due to/because of, 628–629

Each other/one another, 629
Editing
  for grammar, 24
  informal reports, 231
  for mechanics, 24
  for style, 23–24
e.g./i.e., 629
Electronic databases, 506–508
Email
  ABC format, 144
  addresses, suppressing or
    revealing, 141
  appropriate use and style for, 144–145
  collaborative writing and, 454
  copy notation, 130
  defined, 127
  drafts, 143
  formats, 129–130, 141, 142–143
  guidelines, 140–143
  memos versus, 145–146
  purpose statement, 128
  research and use of, 507
  responding to, 133
Enclosures/attachments, in letters or
    memos, 129, 132
Endings, 77–79
English
  as a Second Language (ESL), 629–637
  plain, 603
  simplified, 42, 603–604
Equal Consideration of Interests (ECI),
    43–44

Equipment evaluation reports
  ABC format, 234
  defined, 234
  example of, 235
Ethics
  commitments, keeping, 44
  do no harm, 44
  guidelines, 43–45
  honesty, 44
  individual responsibility, 44
  writing and legal issues, 45–47
European Association of Aerospace
      Industries, 42
European Association of Aerospace
      Manufacturers (AECMA), 604
Executive summary
  in formal reports, 267–268
  for proposals, formal, 313–314
Experts, as readers, 15
Explanations
  ABC format, 197–199
  detail in, how much, 199–200
  difference between instructions and,
      194, 195
  examples of, 194–196
  flowcharts, use of, 200–201
  guidelines for writing, 197–201
  objective point of view, 199
  readers for, 197

Facsimile reference, 129
Facts versus opinions, 230, 597–598
Feasibility studies
  ABC format, 317–318
  abstract/introduction, 318
  conclusions and recommendations, 319
  content of body, 318
  defined, 303
  example of, 305–306
  features of, 316–317
  format for, 317
  graphics, 319
  meeting with readers, 319
Figures, 402
Findings, defined, 230
Flowcharts
  explanations and use of, 200–201
  guidelines for using, 428–430
  labeling, 430
  legends, providing, 429
  standard symbols, 428

Follow-up letters, 566–567
Fonts
  clarity, 406
  guidelines for using, 405–408
  kerning and leading, 407
  readability, 406
  styles, 406, 407
  tone and, 407–408
  types, 114, 405
Formal proposals. *See* Proposals, formal
Formal reports
  ABC format, 261–262
  appendices, 266, 271
  conclusions and recommendations,
      270–271
  cover/title page, 263–264
  defined, 259
  discussion sections, 269–270
  end materials, 271
  example of, 271, 286–301
  executive summary, 267–268
  illustrations, list of, 266–267
  introductions, 268–269
  letter/memo of transmittal, 264–265
  pagination styles, 262–263
  parts of, 262
  table of contents, 265–266
  when to use, 259–261
Former/latter, 638
Fortuitous/fortunate, 638

Generally/typically/usually, 638
General readers, 15–16
Globalization
  communication, 42–43
  cultural differences, 37–42
Goals, for team writing, 25
Good/well, 639
Grammar, 24, 613–660
Graphics
  bar charts, 423–426, 435–437
  color, 408–413
  data accuracy and validity, 413
  defined, 401
  in descriptions, 179
  drawings, technical, 402, 431–434
  in feasibility studies, 319
  flowcharts, 428–430
  fonts, 405–408
  guidelines for, 413–416
  in instructions, 206

line charts, 426–427
  misuse of, 435–439
  oral presentations and presenting,
      478–481
  organization charts, 430–431
  other terms used, 401–402
  permission to use, 47
  pie charts, 420–423, 437–439
  placement of, 104, 414–415
  reasons for using, 20–22, 82,
      402–405, 413
  tables, 416–420
  text references to, 414
  titles, notes, keys, and source data,
      415–416
Grids, 103–104
Groupthink, 44
Groupware, 454–455

Handbook, 613–660
Hanging indents, 105
Headers and footers, 107, 115
Headings
  decimal, 110
  hierarchy of, 109
  in letters and memos, 129
  outlines for creating, 108
  purpose of, 107
  space/lines above or below, 107
  type size for, 109–110
  wording in, 108
High-context cultures, 39

Illustrations, list of
  in formal reports, 266–267
  for proposals, formal, 313
Illustrations, use of term, 401
Imply/infer, 639
Inductive reasoning, 86
Influencers, 16
Informal proposals. *See* Proposals, informal
Informal reports
  ABC format, 228–229
  abstract/introduction, 229
  appearance of, 228
  attachments, 231
  conclusions, 230–231
  contents of body, 229–230
  defined, 225
  editing, 231
  equipment evaluations, 234–235

Informal reports (*continued*)
facts versus opinions in, 230
guidelines for, 227–231
lab, 236–237
planning, 227
problem analyses, 232–233
progress/periodic, 235–236
recommendation, 233–234
when to use, 225–227
which format to use, 228
Information, 6
See also Organization
beginnings and endings, 77–79
collecting, 17–18
key points, repeating, 79–80
Informational abstracts, 538, 539
Instructions
ABC format, 201
bullets or letters, use of, 202–203
cautions, warning, and dangers, listing, 203–206
difference between instructions and, 194, 195
examples of, 196–197
graphics, use of, 206
grouping tasks, 202
guidelines for writing, 201–207
language for, 202
numbered lists, 201–202
simplicity, need for, 206
steps, breaking up, 202
technical level and, 201
testing usability of, 207
Intecom, 42
Interest grabber, 87
International English, 603–604
Internet. *See* Web, searching the
Interviews
See also Job interviews
conducting, 525–526
preparing, 525
recording results, 526–527
research using, 525–527
Introductions
in feasibility studies, 318
in formal reports, 268–269
in proposals, formal, 314–315
in proposals, informal, 308
Italics, use of, 112
Its/it's, 639

Job interviews
answers to questions likely to be asked, 563–564
assertive approach, 565
dress, 565
eye contact, maintaining, 565
follow-up letters, 566–567
manner, 566
physical preparation, 564
practice, 564
preparation, 562–564
Job letters
ABC format, 55
organization of, 557–558
readers' needs, 556–557
Jobs
application letters, 555–558
interviews, 562–567
negotiation, 567–571
researching occupations and companies, 552–555
resumes, 558–562
Justification, 106

Key points, repeating, 79–80

Lab reports
ABC format, 237
defined, 236
example of, 237
Lay/lie, 639–640
Lead-ins, 87, 112
Lead/led, 640
Legal issues, writing and, 45–47
Letter report
See also Informal reports
defined, 225
Letters
ABC format, 130–131, 134, 135, 136, 137–138
copy notation, 130
defined, 127
enclosures/attachments, 129, 132
follow-up, 566–567
formats, 129
headings, 129
negative, 135–136
neutral, 136–137
positive, 134–135
postscripts, 130
proofreading and editing, 132–133

purpose statement, 128
responding to, 133
sales, 137–138
3Cs strategy, 131
tone of voice, 132
of transmittal, 264–265, 312–313
typists initials, 129
you attitude, 132
Library resources
books, 506
company directories, 511, 513
dictionaries, encyclopedias, and other references, 513
electronic databases, 506–508
newspapers, 511, 512
periodicals, 506–508
Library services
circulation, 506
interlibrary loans, 505
references and information, 505
Like/as, 640
Line charts, guidelines for using, 426–427
Line spacing, 105–106
Listeners, oral communication and knowing, 472
Lists (listings), 83, 88
bullets and numbers, use of, 111
instructions and use of numbered, 201–202
page design and, 111–112
punctuation and capitalization, 112
Loose/lose, 640
Low-context cultures, 38

Main idea, development of, 87
Management style, 36
Managers, as readers, 14–15
Margins, 105
Mechanical errors, 24
Meetings
agendas, use of, 457–458
common problems with, 456–457
guidelines for effective, 457–459
leaders, 458
minutes of, 459
start and end time, 458
summarizing, 459
visuals, use of, 459
Memorandum
ABC format, 130–131, 139–140
copy notation, 130

defined, 127
email versus, 145–146
enclosures/attachments, 129, 132
formats, 129
headings, 129
proofreading and editing, 132–133
purpose statement, 128
responding to, 133
3Cs strategy, 131
tone of voice, 132
of transmittal, 264–265, 312–313
types of, 138–139
typists initials, 129
you attitude, 132
Memo report
See also Informal reports
defined, 225
M-Global Inc. example, 47
corporate and branch offices, 51–57
history of, 48–49
projects, 49–51
writing at, 57–59
Modern Language Association (MLA),
533, 535
Modifiers, dangling and misplaced,
640–642
Modular writing, 88–89, 455–456

Negative letters, 135–136
Negotiation, 567–571
Nervousness, oral presentations and over-
coming, 481–484
Neutral letters, 136–137
Nonsexist language, 600–602
Note cards or paper, oral presentation
outline on, 474
Notes, recording, 18
Number of/total of, 642
Numbers, use of, 88
in lists, 111
Numbers, writing, 642–643

Offshoring, 37
Online library catalogs, 498
advanced search techniques, 500–505
author or title search, 499
Boolean search, 502
citing sources, 503
keyword search, 499–500
positional operators, 502
subject search, 499

truncation, 503
web directories, 503
Operators, as readers, 15, 17
Oral communication/presentation
examples, 470–471, 485
eye contact, maintaining, 477–478,
480–481
filler words, avoiding, 476
gestures and posture, 478
graphics, guidelines for presenting,
478–481
handouts, avoiding, 480
listeners, knowing, 472
nervousness, overcoming, 481–484
note cards or paper, outline on, 474
practice, importance of, 475, 481
preacher's maxim, 472–473
preparing and delivery, guidelines for,
471–478
rhetorical questions, use of, 476–477
room layout, 483
speaking organizations, 484
speaking style, 475
stick to a few main points, 473–474
technology, plan for failure of, 481
Oral/verbal, 644
Organization
ABC format, 80–83
cause and effect, 86
classification, 84–85
comparison/contrast, 85
division, 85
importance of, 75
options for information, 75–76
principles of, 76–80
problem/solution, 86
reasoning, deductive and inductive,
85–86
of sections and paragraphs, tips for, 87–88
sequential, 84
Organizational culture, elements of, 35–36
Organizational history, 36
Organization charts
circular design, 431
connecting boxes, 430
guidelines for using, 430–431
linear design, 430
Outlines
for abstracts, 539–540
basic rules, 20
completing, 18–22

to create headings, 108
research, 531
Outsourcing, 37

Page design
computers, use of, 115–117
defined, 102
elements of, 102–112
fonts and color, 112–115
grids, 103–104
headings, 107–110
italics and boldface, use of, 112
lists, 111–112
white space, 105–107
Pagination styles, for formal reports,
262–263
Paragraphs
indenting, 106
length of, 88, 106, 107
organization of, 87–88
Parts of speech, 644
Passed/past, 645
Passive voice, 598–599
Per, 645
Per cent/percent/percentage, 645
Periodic reports
ABC format, 235, 236
defined, 235
example of, 236
Permissions, written, 47
Pie charts
examples of confusing, 437–439
guidelines for using, 420–423
labeling, 422–423
number of divisions in, 420
orientation of, 421
Plagiarism, 527–528
Plain English, 603
Plain Language Action and Information
Network (PLAIN), 603
Planning, 7
form, 10–11
Planning coordinator, 450
Positive letters, 134–135
Postscripts, 130
Practical/practicable, 645
Principal/principle, 645–646
Problem analyses informal report
ABC format, 232
defined, 232
example of, 232–233

Problem/solution, 86

Progress reports
    ABC format, 235, 236
    defined, 235
    example of, 235–236

Pronouns, agreement and reference,
        646–647

Proofreading and editing,
        132–133

Proposals
    defined, 303
    example of, 303–305

Proposals, formal
    ABC format, 311
    appendices, 316
    conclusion, 315
    cover/title page, 311–312
    discussion sections, 315
    executive summary, 313–314
    illustrations, list of, 313
    introduction, 314–315
    letter/memo of transmittal,
        312–313
    table of contents, 313
    when to use, 306

Proposals, informal
    ABC format, 308
    abstract/introductions, 308
    appearance of, 307–308
    attachments, 310
    conclusion, 310
    content of body, 309
    editing, 310
    formats for, 307
    need, establishing, 309
    planning, 307
    when to use, 306

Punctuation
    general, 647–651
    lists, 651–652

Purpose
    determining, 6–9
    statement, 8–9, 128

Questionnaires
    audience for, 523–524
    preparing, 520–523
    reporting results, 524–525
    research using, 520–525
    Questions, oral presentations and use of
        rhetorical, 476–477

Readers
    analyzing, 9, 12–17
    correspondence and analyzing, 128
    decision-making levels, 16–17
    interest curve, 78
    matrix, 14
    obstacles for, 12–13
    types of, 14–16
    ways to understand, 13
    writing different parts for different, 76–77

Reasoning
    deductive, 85–86
    inductive, 86

Receivers, 16–17

Recommendation reports
    ABC format, 233, 234
    defined, 233
    example of, 233–234

Recommendations, 230
    in feasibility studies, 319
    in formal reports, 270–271

Regrettably/regretfully, 652

Research
    abstracts, 538–541
    bibliography, 18, 529–530
    collecting, 17–18
    documentation styles, 532–537
    drafts, writing, 531
    electronic databases, 506–508
    email, use of, 507
    getting started, 496–498
    interviews, 525–527
    library services and resources, 505–513
    notes, recording, 18, 530–531
    online library catalogs, 498–505
    outlines, 531
    plagiarism, 527–528
    primary, 17
    process, 528–531
    questionnaires, 520–525
    secondary, 17, 18
    sources, acknowledging, 18, 527–528
    sources, citing online, 503
    strategy, devising a, 18
    web, 514–520

Respectively, 652–653

Response, determining, 9

Results statement, 9

Resumes
    activities, recognitions, interests, 561
    education, 559–560

experience, 560–561
    format, 558–559
    objective, 559
    references, 562

Revising, 7
    drafts, 23–24
    tasks, 23

Sales letters, 137–138

Salutation, sexist language in, 602

Schedule charts, using, 452–453

Search engines, 516, 517–518

Sections, 87

Sentences
    active versus passive voice, 598–599
    clichés, avoiding, 595
    conciseness, 592–596
    constructions, those to avoid, 596
    facts versus opinions in, 230, 597–598
    length of, 592
    main clause, 592
    main point, 591
    nonsexist language, 600–602
    parts (terms) of, 591
    qualifying statements, 598
    replacing long words with short ones,
        594–595
    verbs, action, 593
    wording accuracy, 596–598
    wordy phrases, shortening, 593–594
    writing clear, 590–592

Sequence approach, 449

Sequential organization, 84

Set/sit, 653

Settle-Murphy, Nancy, 39

Sexism, how to avoid, 600–602

Sic, 653

Simplified English, 42, 603–604

Singer, Peter, 43

Single sourcing, 88–89

Society for Intercultural Education,
        Training, and Research
        (SIETAR), 39

Sources
    acknowledging, 18, 46–47, 527–528
    citing online, 503

Speaking organizations, 484
    See also Oral communication/presentation

Specialization approach, 449

Speed-read approach, 76–77

Spelling, 653–654

Stationary/stationery, 654

Storyboards, use of, 26, 27

Style and tone
 active versus passive voice, 598–599
 American Psychological Association
  (APA), 532, 533, 534
 circumlocution, 595–596
 conciseness, 592–596
 Council of Science Editors (CSE),
  532, 533, 536–537
 editing for, 23–24
 Modern Language Association (MLA),
  533, 535
 nonsexist language, 600–602
 plain English, 603
 sentences, writing clear, 590–592
 sheets, 115–116
 simplified English, 42, 603–604
 types of, 590
 wording accuracy, 596–598

Subject matter experts, working with,
  459–460

Subject-verb agreement, 654–655

Summaries
 executive, 267–268, 313–314
 writing, 23, 83

Synthesis approach, 449

Table of contents
 in formal reports, 265–266
 for proposals, formal, 313

Tables
 defined, 401
 guidelines for using, 416–420
 informal versus formal, 416
 white space in, 417

Team leader, 450

Teamwork
 cross-functional, 456
 global, 44–45
 meetings, running effective,
  456–459
 subject matter experts, 459–460

Team writing, 24–28
 See also Collaborative writing

Technical communication
 defined, 3
 differences between academic writing
  and, 3
 examples of, 4, 5

features of, 3–6
 flowchart, 7

Technical drawings. *See* Drawings, technical

Templates, 115

Text blocs, arranging, 103–104

3Cs strategy, 131

Title page
 in formal reports, 263–264
 for proposals, formal, 311–312

Titles, sexist language and, 602

To/too/two, 655

Tone. *See* Style and tone

Topic sentence, 87

Trademarks, 47

Transitions, 88

Translations, 42–43
 Transmittal, letter/memo of
 in formal reports, 264–265
 for proposals, formal, 312–313

Tufte, Edward R., 435, 479

Type size, 113

Uniform Resource Locator (URL),
  515, 516–517

U.S. Department of Commerce,
  Commercial Service of, 39

User-centered design, 371

Utilize/use, 655

Visual aids, 401

*Visual Display of Quantitative Information,
 The* (Tufte), 435

Warnings, in instructions, 203–206

Web, searching the
 electronic databases, 506–508
 evaluation skills, 515
 example of, 518–520
 fundamentals, 514–516
 Internet domain extensions, common,
  516
 search engines, 516, 517–518
 subject directories and guides, 518
 Uniform Resource Locator (URL),
  515, 516–517

Web browsers, 515–516

Web directories, 503

Web pages/websites
 accessibility guidelines, 391–393
 audience analysis, 369, 371

content chunking, 371–372
 content development, 371–374
 document conversion issues and
  common file formats, 374
 planning, 368–371
 planning form, 370
 publishing, 393
 purposes of, 369
 role in developing, 367–368
 scripting languages and software
  authoring tools, 373–374
 testing site for user base, 389
 usability checks and system settings,
  389–391
 usability reviews, performing,
  389
 user-centered design, 371

Web pages/websites, design, 381
 conventions and principles,
  382–383
 file formats and graphics, 384–385
 finding a theme and developing graphic
  content, 384
 interface layouts, 385–388

Web pages/websites, structure, 374
 advantages and disadvantages, 378
 breadth, 376
 customized, 375, 376, 378
 depth, 376
 hierarchical, 375–376, 378
 hypertextual, 375, 376, 378
 grouping and arrangement strategies,
  381
 labeling guidelines, 379–381
 linear, 375, 378
 navigation design, 379, 380
 process of developing, 376–379

Which/that, 656

White space, 105–107

Who/whom, 656

Who's/whose, 656–657

Wording accuracy, 596–598

Wordy phrases, shortening,
  593–594

Writer's Handbook, 533

Writing, legal issues and, 45–47

Writing in cross-functional teams, 449

You attitude, 132

Your/you're, 657

Use the following standard abbreviations when making corrections on drafts. Your editing instructions may be (1) for your own use during a later stage of your work, (2) for a keyboard operator, or (3) for a colleague with whom you may be working on a group-writing project.

When possible, place the marks next to the copy to which they refer. When space is a problem, sometimes marks need to go in the left or right page margins.

MARK	MEANING	EXAMPLE
∧	insert	techni̯cal
ℓ	delete	technicƶal
—	delete a word	at his ~~his~~ request
ℓ̂	delete and close space	ground̂water
/ or #	insert space	visit/England or visitEngland
∿	transpose	technïcal
◡	close up space	tech◡nical
≡	use capital	m̲̲̲-Global, Inc.
=	use small caps	p̲.m̲.
/	use lower case	the Ȼommittee
∨	add apostrophe or quotation marks or superscript	M-Globaľs policy
∧	use comma here	M-Global∧Inc.
⊙	use period here	Inc⊙
⑤	use semicolon here	St. Paul, Minnesota⑤
⊙⊙	use colon here	as follows⊙⊙
⊰ ⊱	use parentheses here	⊰See Table 4.⊱
⟨ ⟩	use brackets here	⟨SIC⟩
⩒	use hyphen here	well⩒planned meeting
¶	start new paragraph	. . . today.¶Then he began . . .
No ¶	take out paragraph change	No¶He stated that the firm . . .
(stet)	keep original; disregard editing change	Admiral̲ ~~General~~ Jones
⌐	move right	⌐21 Walnut Street Portland, Ohio (216) 374-0011
⌐	move left	⌐ 21 Walnut Street Portland, Ohio (216) 374-0011
⊔	lower	. . . the defense policy."³
⊓	raise	. . . the defense policy."3
‖	align	‖.34 beams ‖.12 bolts ‖.15 bars ‖.7 hammers
(Sp) or ◯	spell out	(Sp) 3 team members or ③ team members
⁗⁗⁗	remove underline	no s̲i̲g̲n̲i̲f̲i̲c̲a̲n̲t̲ pollution
⌁ or M	add conventional dash	a big change$\frac{1}{M}$and I mean big.
⌁ or N	add small dash	1972$\frac{1}{N}$1982
⟿	run together	. . . gave his firm growth.⟿ Later he phased in . . .

PLANNING FORM

Name: _____ Assignment _____

I. Purpose: Answer each question in one or two sentences.

   A. Why are you writing this document?

   B. What response do you want from readers?

II. Audience

   A. Reader Matrix: Fill in names and positions of people who may read the document

	Decision Makers	Advisers	Receivers
Managers			
Experts			
Operators			
General Readers			

   B. Information on individual readers: Answer these questions about the primary audience for this document. If the primary audience includes more than one reader (or type of reader) and there are significant differences between the readers, answer the questions for each (type of) reader. Attach additional sheets as necessary.

Primary audience:

1. What is this reader's technical or educational background?

2. What main question does this person need answered?

3. What main action do you want this person to take?

4. What features of this person's personality might affect his or her reading?

III. Document

   A. What information do I need to include in the

      1. Abstract?

      2. Body?

      3. Conclusion?

   B. What organizational patterns are appropriate to the subject and purpose?

   C. What style choices will present a professional image for me and the organization I represent?